Neurobiology
Ionic Channels, Neurons, and the Brain

NATO ASI Series

Advanced Science Institutes Series

A series presenting the results of activities sponsored by the NATO Science Committee, which aims at the dissemination of advanced scientific and technological knowledge, with a view to strengthening links between scientific communities.

The series is published by an international board of publishers in conjunction with the NATO Scientific Affairs Division

A	**Life Sciences**	Plenum Publishing Corporation
B	**Physics**	New York and London
C	**Mathematical**	Kluwer Academic Publishers
	and Physical Sciences	Dordrecht, Boston, and London
D	**Behavioral and Social Sciences**	
E	**Applied Sciences**	
F	**Computer and Systems Sciences**	Springer-Verlag
G	**Ecological Sciences**	Berlin, Heidelberg, New York, London,
H	**Cell Biology**	Paris, Tokyo, Hong Kong, and Barcelona
I	**Global Environmental Change**	

PARTNERSHIP SUB-SERIES

1. **Disarmament Technologies**	Kluwer Academic Publishers
2. **Environment**	Springer-Verlag
3. **High Technology**	Kluwer Academic Publishers
4. **Science and Technology Policy**	Kluwer Academic Publishers
5. **Computer Networking**	Kluwer Academic Publishers

The Partnership Sub-Series incorporates activities undertaken in collaboration with NATO's Cooperation Partners, the countries of the CIS and Central and Eastern Europe, in Priority Areas of concern to those countries.

Recent Volumes in this Series:

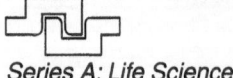

Series A: Life Sciences

Neurobiology
Ionic Channels, Neurons, and the Brain

Edited by

Vincent Torre
Università di Genova
Genova, Italy

and

Franco Conti
Consiglio Nazionale delle Richerche (CNR)
Genova, Italy

Springer Science+Business Media, LLC

Proceedings of a NATO Advanced Study Institute and of
the 23rd course of the International School of Biophysics in Neurobiology,
held May 2 – 12, 1995,
in Erice, Italy

NATO-PCO-DATA BASE

The electronic index to the NATO ASI Series provides full bibliographical references (with keywords and/or abstracts) to about 50,000 contributions from international scientists published in all sections of the NATO ASI Series. Access to the NATO-PCO-DATA BASE is possible in two ways:

—via online FILE 128 (NATO-PCO-DATA BASE) hosted by ESRIN, Via Galileo Galilei, I-00044 Frascati, Italy

—via CD-ROM "NATO Science and Technology Disk" with user-friendly retrieval software in English, French, and German (©WTV GmbH and DATAWARE Technologies, Inc. 1989). The CD-ROM also contains the AGARD Aerospace Database.

The CD-ROM can be ordered through any member of the Board of Publishers or through NATO-PCO, Overijse, Belgium.

Library of Congress Cataloging-in-Publication Data

Neurobiology : ionic channels, neurons, and the brain / edited by
  Vincent Torre and Franco Conti.
      p.   cm. -- (NATO ASI series. Series A, Life sciences ; v.
  289)
    "Proceedings of a NATO Advanced Study Institute and of the 23rd
  course of the International School of Biophysics in Neurobiology,
  held May 2-12, 1995"--T.p. verso.
    "Published in cooperation with NATO Scientific Affairs Division."
    Includes bibliographical references and index.
    ISBN 978-1-4613-7706-1   ISBN 978-1-4615-5899-6 (eBook)
    DOI 10.1007/978-1-4615-5899-6
    1. Neurons--Physiology--Congresses. 2. Ion channels--Congresses.
  3. Neural networks (Neurobiology)--Congresses. 4. Brain-
  -Congresses. I. Torre, Vincent, 1950-   II. Conti, Franco.
  III. North Atlantic Treaty Organization. Scientific
  AffairsDivision. IV. International School of Biophysics in
  Neurobiology (1995 : Erice, Italy) V. Series.
    [DNLM: 1. Neurobiology--congresses. 2. Brain--physiology-
  -congresses. WL 100 N49055 1996]
  QP363.N476 1996
  612.8--dc20
  DNLM/DLC
  for Library of Congress   96-29397
                                                                  CIP

ISBN 978-1-4613-7706-1

© 1996 Springer Science+Business Media New York
Originally published by Plenum Press, New York in 1996
Softcover reprint of the hardcover 1st edition 1996

10 9 8 7 6 5 4 3 2 1

PREFACE

Understanding how the brain works is undoubtedly the greatest challenge for human intelligence and one of the most ambitious goals of contemporary science. We are certainly far from this goal, but significant advancements in several fields of Neuroscience and Neurobiology are being obtained at an increasing pace.

The NATO ASI School in Neurobiology, held in Erice May 2-12, 1995, as the 23rd Course of the International School of Biophysics, provided an update on three basic topics: Biophysics and Molecular Biology of Ion Channels, Sensory Transduction, and Higher Order Functions. Current knowledge on these subjects was covered by formal lectures and critical discussions between lecturers and participants. This book collects original contributions from those scientists who attended the School. Many students presented their results in poster sessions, steering lively informal discussions. A selection of these contributions is also included.

A major portion of the program of the School was devoted to a general overview of current trends of thought and experimental approaches in neurobiology, emphasising the importance of understanding molecular aspects of the elementary events underlying sensory transduction and processing in the nervous system, without indulging however in a pure reductionistic view of such complex phenomena.

Recent studies of molecular biology and the electrophysiology of heterologously expressed ionic channels, have shed new light on the molecular mechanisms underlying ionic permeation of excitable membranes and its regulation by physical and chemical parameters. Voltage-gated channels selectively permeable to Na^+, K^+, and Ca^{2+} have been analysed in great detail and their molecular structure has been revealed. Experiments with ionic channels mutated by genetic engineering techniques have allowed the identification of structural domains and specific amino acids regulating fundamental functions such as selectivity and gating. Surprisingly, the ligand-gated ion channels ultimately responsible for the electrical response of sensory neurons belong to the same family of cation selective voltage-gated proteins, likely sharing with the potassium channel a common genetic ancestor. Similarly, some of the molecular mechanisms underlying secretion and particularly the release of synaptic vesicles have been clarified. Ligand-gated ionic channels activated by neurotransmitters and regulating synaptic transmission belong to a different family of membrane proteins with specific functional and molecular properties.

Our understanding of sensory transduction in photoreceptors, mechanoreceptors, and olfactory neurons is advanced to the point that we may claim to have a good picture of the different mechanisms, both from a functional and a molecular point of view. Photoreceptors and olfactory neurons have developed specialized structures for transducing stimuli. Phototransduction is initiated by the absorption of a photon by a rhodopsin molecule present in the outer segment. Fine dendrites emerging from the cell body of olfactory neurons and usually called cilia are covered by odorant receptors. In both types of neurons, transduction terminates with the modulation of ionic channels gated by second messengers, cyclic GMP in photoreceptors and cyclic AMP in olfactory neurons. The chain of biochemical events in phototransduction and the electrical properties of photoreceptor membrane have been clarified in recent years. Some properties of chemotransduction at the basis of olfaction have been characterized as well. Hair cells also have specially developed structures, called stereocilia, to sense the motion of the liquid in which they are immersed. Mechanotransduction in hair cells is basically different, as it does not involve second messengers. Transduction in these cells is very fast and occurs by a direct link between the stereocilia and the ionic channels underlying the generator current.

The School also discussed general aspects of the information processing in the nervous system and discussed how the limitations of its slow elementary components are overcome by employing a massive degree of parallelism, through the extremely rich set of synaptic interconnections between neurons. The development of models of higher order functions such as memory and learning, has provided a preliminary meaningful understanding of these important aspects of the brain.

The 1995 Course was also an occasion to honour the memory of the late Prof. Antonio Borsellino, who died in November 1992. The founder of the International School of Biophysics at the Ettore Majorana Centre in 1960 and a member of the first council of the International Union of Pure and Applied Biophysics, Antonio Borsellino devoted the second part of his scientific life to the promotion of biophysics as an autonomous branch of life sciences, particularly in Italy, where he is legitimately considered the father of this field. During his earlier activity as a theoretical physicist, his studies contributed to the foundation of quantum electrodynamics, earning him a worldwide reputation and, at the age of 35, the chair of Theoretical Physics at the University of Genova. In the 1950s he felt the challenge of the neurosciences and switched his interests to this field. He would have been particularly pleased by the results discussed in this School and collected in this book. During almost 30 years he has invited the best neuroscientists to lecture in Erice, providing a stimulating and relaxed environment for both young and senior scientists. Many lecturers at the 1995 Course had participated in previous Erice schools organized by Borsellino and recalled that experience with touching remarks. As pupils of Antonio we hope to have been up to the task of organizing for the first time a Biophysics Course on Neuroscience without profiting from his invaluable and unforgettable wisdom.

Vincent Torre
Franco Conti

CONTENTS

Ionic Channels

Sensory Transduction

Neural Networks and Higher Functions

THE ROLE OF THE N-TERMINAL OF THE cGMP-GATED CHANNEL FROM VERTEBRATE RODS

G. Bucossi, M. Nizzari, and V. Torre

Istituto Nazionale per la Fisica della Materia
Genova, Italy

INTRODUCTION

Cyclic nucleotide (CNG) channels (Kaupp et al., 1989) play a fundamental role in phototransduction and chemotransduction (Torre, Ashmore, Lamb and Menini, 1995). Native CNG channels are composed by at least two subunits, usually referred to as alpha and beta subunit (Chen, Peng, Dhallan, Ahamed, Reed and Yau, 1993; Körschen et al., 1995). The alpha subunit is a polypeptide of 690 amino acids, while the full beta subunit is much longer and composed by 1394 amino acids. When heterologously expressed in Xenopus laevis oocytes, the alpha subunit, here referred to as the w.t. channel, forms functional channels, which are activated by cyclic nucleotides. The functional properties of this channel are similar but not identical to those of the native CNG channel. When the alpha and beta subunits are coexpressed in oocytes, ionic channels appear, with properties almost identical to those of the native channel. So far, it has not been possible to have functional channels formed by the beta subunit only. The amino acid sequence of these two polypeptides has a significant degree of homology with those forming voltage gated channels (Jan and Jan, 1990; Heginbotham, Abramson and MacKinnon, 1992; Guy, Durell, Warmke, Drysdale and Ganetzki, 1991; Gouldings et al., 1992; Bonigk et al., 1993; Henn, Baumann and Kaupp, 1995). This structural analogy between voltage gated channels and CNG channels has suggested to analyse the role of glutamate in position 363 of the w.t. channel (Root and MacKinnon, 1993; Eismann, Muller, Heinemann and Kaupp, 1994). Indeed this amino acid controls several features of ionic permeation, such as the sensitivity to external divalent cations, the multi-ion nature of the channel, the single channel conductance and the size of the narrowest section of the pore (Root and MacKinnon, 1993; Eismann et al., 1994; Sesti, Kaupp, Eismann, Nizzari and Torre, 1995; Bucossi et al., 1996).

Native CNG channels, as well as the w.t. channel, do not desensitize or inactivate in the presence of a steady cyclic nucleotide concentration (Karpen, Zimmerman, Stryer and Baylor, 1988), but when glutamate 363 is replaced with alanine, serine or asparagine, the current initially activated in mutant channels E363S, E363A and E363N decreases with time, in a way reminiscent of desensitization of ligand gated channels, or inactivation of voltage gated channels (Bucossi et al., 1996). These results indicate that glutamate 363 also controls the gating of the channel.

In this paper we investigate the role of the N-terminal in the physiology of CNG channels from bovine rods. We will see that when the N-terminal with 154 consecutive amino acids is deleted from the w.t. channel, the truncated channel forms functional channels. The properties of this truncated channel, referred to as T - w.t. channel, are similar but not identical to those of the full w.t. channel. Indeed the single channel conductance and sensitivity to divalent cations in the truncated and in the w.t. channel are almost identical. However the ionic selectivity to Cs$^+$ in the T - w.t. is significantly reduced in comparison with that in the w.t. In addition, in the presence of a steady concentration of cyclic nucleotides the mutant E363A and its truncated form, here referred to as T - E363A, have slightly different properties. These results indicate that the N-terminal plays some role in the way in which the channel operates within the membrane.

METHODS

Mature Xenopus laevis were anaesthetized with 0.2% tricaine methanesulphonate (Sigma) and ovarian lobes were removed surgically. Oocytes from Xenopus laevis were prepared as described in Nizzari, Sesti, Giraudo, Virginio, Cattaneo and Torre, 1993.

Mutagenesis and Oocytes Preparation

Rod mutant T - w.t. is a derivative of p31010 rod wild type which is subcloned into the pT7T3-18U (Pharmacia). In the T - w.t. mutant 5' non translated sequences and 4 to 465 basepairs of cGMP channel codificant region were deleted. As a consequence, in the amino

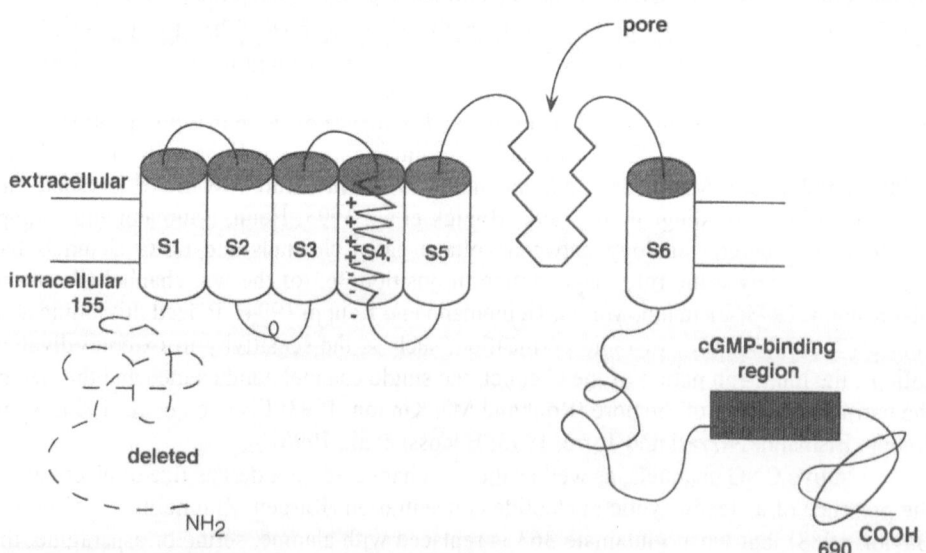

Figure 1. Proposed model of the two-dimensional topology of the CNG channel. The segment indicated by a broken line, corresponding to the amino acids between lysine 2 and valine 155 has been deleted in mutant T - w.t.

acid sequence of mutant T - w.t., the amino acids between lysine 2 and valine 155 were deleted as shown in Fig. 1.

The mRNA was injected into Xenopus laevis oocytes prepared and maintained as described in Nizzari et al., 1993.

Electrical Recordings and Solutions

The pipette solution contained 2 mM EDTA, 10 mM HEPES, 110 mM NaCl buffered to pH 7.6 with TMAOH or with NaOH. A similar solution was used to superfuse the intracellular side of the patch which could contain specified amounts of cGMP. In some control experiments solutions were buffered with TRIS. The technique for measuring macroscopic current was described previously (Sesti et al., 1995). Data were stored on a videotape and analysed off-line.

The single channel conductance was measured from the analysis of amplitude histograms of current fluctuations. When two peaks, corresponding to the closed and open state, could be clearly identified in the amplitude histogram (see for instance Fig. 2B and C), the distance between the two peaks was taken as a measure of the current flowing through a single open channel. Power spectra were computed as described in Sesti et al., 1994.

RESULTS

The mRNA of the w.t. channel and of mutant channels was injected into Xenopus laevis oocytes. The electrical activity of expressed channels was recorded in the inside-out configuration. CNG channels were activated by the addition of cGMP in the bathing medium and the current was recorded in voltage clamp conditions with conventional patch pipettes. The magnitude of the current activated by saturating concentrations of cGMP was significantly lower in membrane patches from oocytes injected with the T - w.t. mRNA than in patches from oocytes injected with the same quantity of w.t. mRNA. Only rarely currents, activated by cGMP, larger than 100 pA (at + 100 mV), could be recorded and it was often possible to have patches containing just one CNG channel.

Single Channel Properties

Fig 2A illustrates current recordings obtained from a membrane patch containing a single T - w.t. channel at membrane voltages between -100 and + 100 mV in the presence of 500μM cGMP. Amplitude histograms of current fluctuations at -100 and + 100 mV are shown in Fig 2B and C respectively.

Visual inspection of current recordings and the analysis of amplitude histograms indicate a single channel conductance of approximately 30 pS. The open state is slightly noisier at negative voltages than at positive voltages. The I-V relation of a single open channel is shown in Fig 2D.

Fig. 3A shows the effect of increasing the concentration of cGMP on channel activity of mutant T - w.t. at -100 mV. The open probability was about .85 in the presence of 500 μM cGMP and half of the maximal activation was observed in the presence of a cGMP concentration between 50 and 100 μM. All these features are also common to the w.t. channel (Nizzari et al., 1993). A quantitative comparison between single channel properties of w.t. and T - w.t. channels is shown in Table 1.

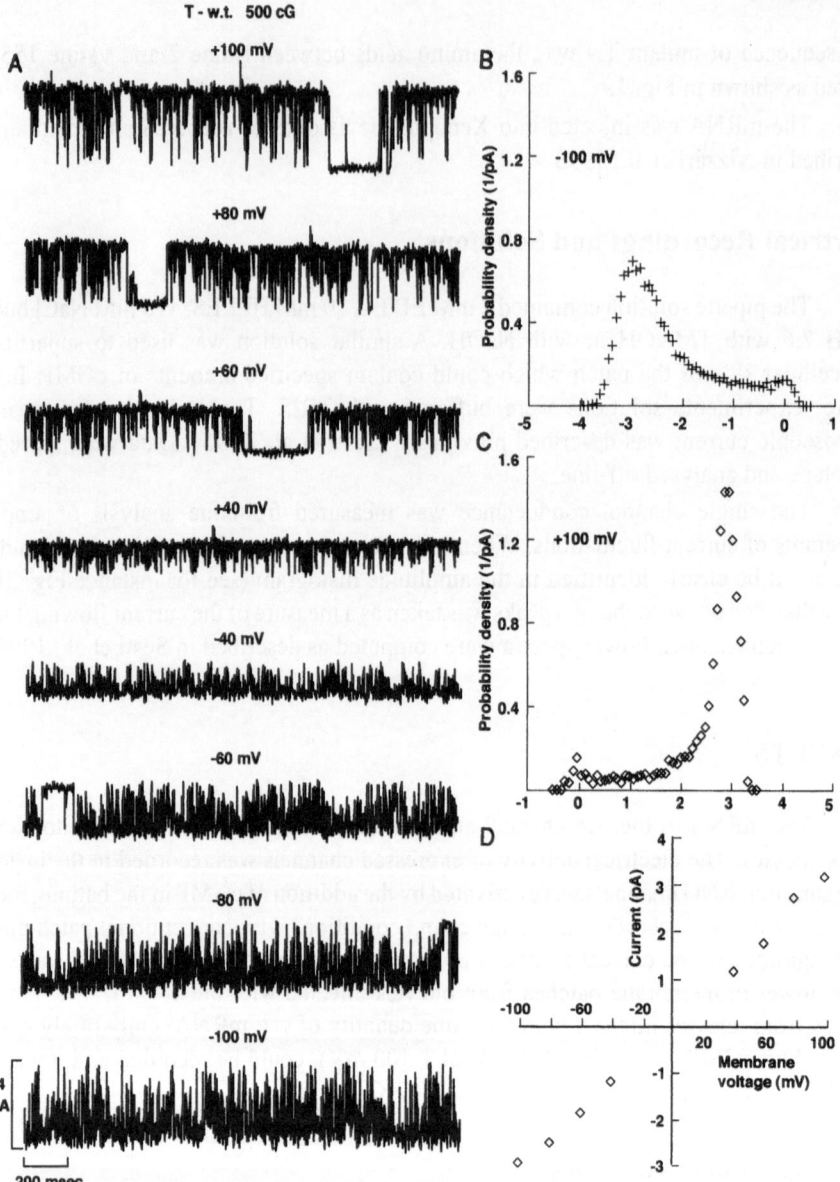

Figure 2. Single channel recordings from mutant T- w.t.. A: current recordings at different membrane voltages from a patch containing a single cGMP gated channel in the presence of 500µM cGMP. B and C amplitude histograms of current fluctuations at + 100 and - 100 mV respectively. D: I-V relation of a single channel from the data shown in A. Data filtered at 2kHz.

Ionic Selectivity

The selectivity sequence based on the reversal potential among alkali monovalent cations of the w.t. channel is $Na^+ > K^+ > Li^+ > Rb^+ > Cs^+$ (Kaupp et al., 1989) and is the same as in the T - w.t. channel. Table 1 reports the reversal potential for biionic solutions, with 110 mM Na^+ in the extracellular medium and an equimolar amount of the tested cation in the intracellular medium. It is evident that the ionic selectivity is almost identical in the two channels with the exception of Cs^+ which is hardly permeable through T - w.t.

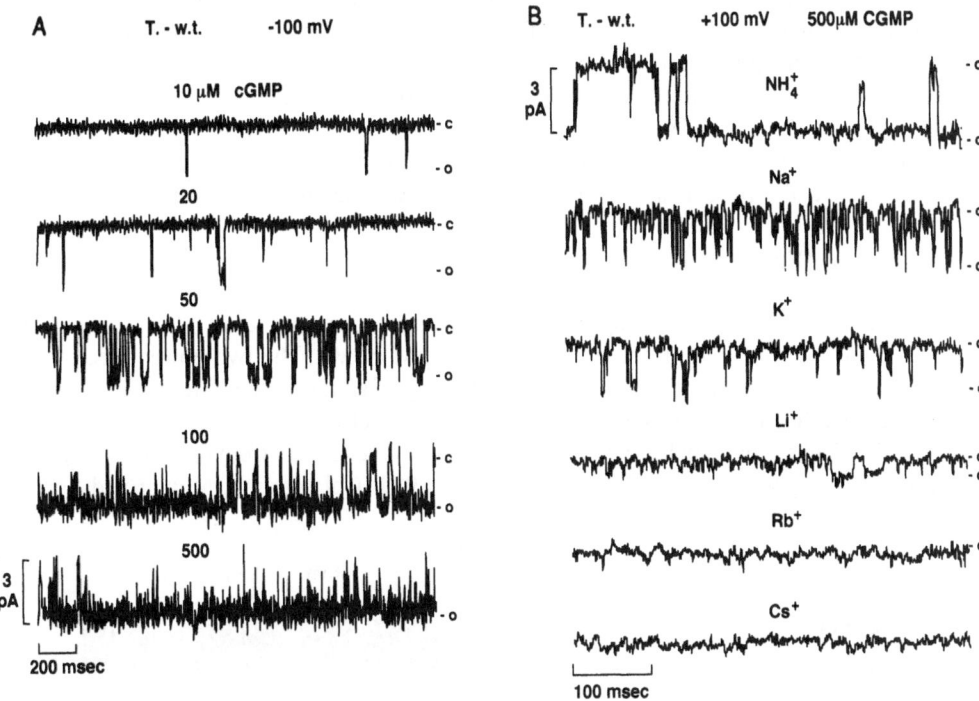

Figure 3. Single channel properties of mutant T - w.t. A: current recodings at -100 mV in the presence of increasing cGMP concentrations. Data filtered at 2kHz. B: current recordings at + 100 mV in the presence of 500μM cGMP. The patch pipette contained 110 mM NaCl and the bathing medium contained an equimolar amount of NH₄Cl, NaCl, KCl, LiCl, RbCl and CsCl. Data filtered at 1 kHz.

Table 1.

	T - w.t.	w.t.
Single channel conductance at - 100 mV	27.8±2.1 pS	30.3±1.2 pS
Single channel conductance at + 100 mV	28.5±1.7 pS	29.7±1.5 pS
Reversal potential of a bionic solution with NH_4^+	-28±4	-32.2±1.7
with K^+	1±1	-2.6±1.2
with Li^+	8±1.5	7.4±0.8
with Rb^+	15.2±3.7	12.3±1.4
with Cs^+	48.8±4.2	25.1±4.5
Single channel conductance at +100 mV		
with NH_4^+	37.3±2	36±3.5
with K^+	26.5±1	27.3±1.8
with Li^+	—	—
with Rb^+	—	—
with Cs^+	n.a.	n.a.
Fraction of current blocked at -40 mV		
by 1 mM Ca^{2+}	45±3%	42±5%
by 1 mM Mg^{2+}	6±3%	8±4%

Fig. 3B illustrates current recordings from a patch containing a single CNG channel, at a holding voltage of +100 mV and in the presence of 500μM cGMP. 110 mM Na$^+$ was present in the patch pipette and equimolar amounts of NH$^+_4$, Na$^+$, K$^+$, Li$^+$, Rb$^+$, and Cs$^+$ were present in the medium bathing the cytoplasmic side of the membrane. The single channel current was 3.4, 2.9, 2.2 and 0.8 pA for NH$^+_4$, Na$^+$, K$^+$ and Li$^+$ respectively. No single channel current was detected in the presence of Rb$^+$ or Cs$^+$. A comparison of single channel currents in the presence of different cations in the w.t. and T - w.t. channels is shown in Table 1.

Gating and Blockage by Divalent Cations

Fig. 4A shows the power spectra of current fluctuations of the w.t. (+) and mutant T - w.t. (◇) at -100 mV and in the presence of 500 μM cGMP. The two power spectra are almost identical, suggesting that the deletion of a large fraction of the N-terminal does not significantly affect the kinetics of the gating of the channel.

Divalent cations, such as Ca^{2+} or Mg^{2+} block CNG channels from rod photoreceptors in a voltage dependent manner (Yau and Baylor, 1989; Colamartino, Menini and Torre,

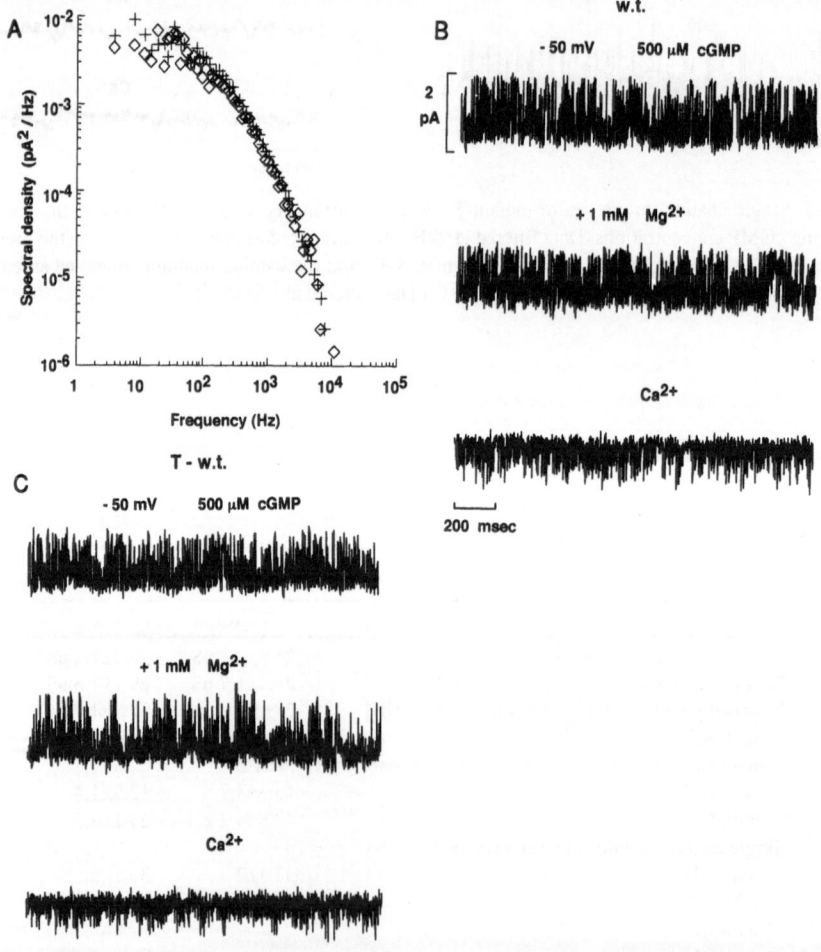

Figure 4. A: Power spectrum of current fluctuations in the w.t. (+) and mutant T - w.t. (◇) channel. Data were filtered at 20 kHz butterworth and were sampled at 60 kHz. Power spectrum computed as the difference of the power spectra in the presence of 500 μM cGMP and in the absence of cGMP. Holding voltage was -100 mV. Single channel recordings from the w.t. (B) and mutant T - w.t. (C) in the absence and in the presence of 1 mM Ca^{2+} or Mg^{2+}. Holding voltage -60 mV in the presence of 100μM cGMP; data filtered at 2kHz.

1991). The blockage efficacy depends on the presence of the divalent cation in the extracellular or in the intracellular medium: the blocking effect is more powerful from the outside (Root and MacKinnon, 1993; Eismann et al., 1994). Fig. 4B and C illustrates single channel recordings from the w.t. and T - w.t. at -50 mV in the absence and in the presence of 1 mM Ca^{2+} and Mg^{2+}. In both channels the single channel current was about 1.4 pA and was not affected by Mg^{2+}. In the presence of 1 mM Ca^{2+} the single channel current was .8 and .85 pA in the w.t and T - w.t. respectively. Also at other membrane voltages between -100 and + 100 mV, the blocking effect of Mg^{2+} and Ca^{2+} in both channels was very similar.

Desensitization in Mutant T - E363A

The current activated at -100 mV by a steady concentration of cGMP in mutant E363A declines within 10 seconds to a steady value which is about 10% of the current initially activated by cGMP. This current decline is not associated to any detectable change in the single channel conductance (Bucossi et al., 1996).

Fig. 5 illustrates current recordings from two different patches at -140 (A) and -160 mV (B). 500µM cGMP was added to the medium bathing the cytoplasmic side of the membrane at the time indicated by the horizontal solid line above traces in A and B. It is evident that the current initially activated by cGMP declines also in mutant T - E363A. Immediately following exposure to cGMP, the activated current was characterized by very rapid transients of some pA, as shown by the traces indicated by the solid circles in C and D. After completion of current decline, different channel openings are observed. Traces indicated by open circles in panels C and D reproduce in more details the section of recordings indicated by the same symbols in A and B. These channel openings have a more squared shape and a conductance of about 10 pS. This change of single channel properties is a distinct feature of desensitization in ligand gated channels and constitutes additional evidence that mutants E363A and T - E363A desensitize.

DISCUSSION

The results reported in this paper indicate that the w.t. and its truncated form have similar electrophysiological properties: the single channel conductance, the blockage of divalent cations and the selectivity sequence in the two channels are identical. It is important to observe, however, that Cs^+ is poorly permeable through T - w.t.: with the patch pipette filled with Na^+ and Cs^+ in the bathing medium, the reversal potential is about 25 mV in the w.t., while it is about 48 mV in the T - w.t.

When glutamate 363 is mutated to an alanine, both mutants E363A and T - E363A exhibit a clear time dependent current decline upon activation with a fixed cGMP concentration. During this current decline the single channel conductance of mutant E363A does not change significantly (Bucossi et al., 1996), but a different behaviour is observed in mutant T - E363A; immediately after the addition of cGMP to the bathing medium, current openings are characterized by very rapid transients with an amplitude ranging between 1 and 4 pA (at -140 or - 160 mV), when the current decline has taken place, square opening can be observed with a single channel conductance of about 10 pS. These results indicate that the current decline observed in mutant T - E363A is associated to a conformational change of the pore region of the channel.

The results presented in this paper indicate that the first 155 amino acids forming the N-terminal have a small influence on the ionic permeation through the cGMP gated channel from bovine rods. This effect could be mediated by a different rearrangement of the two

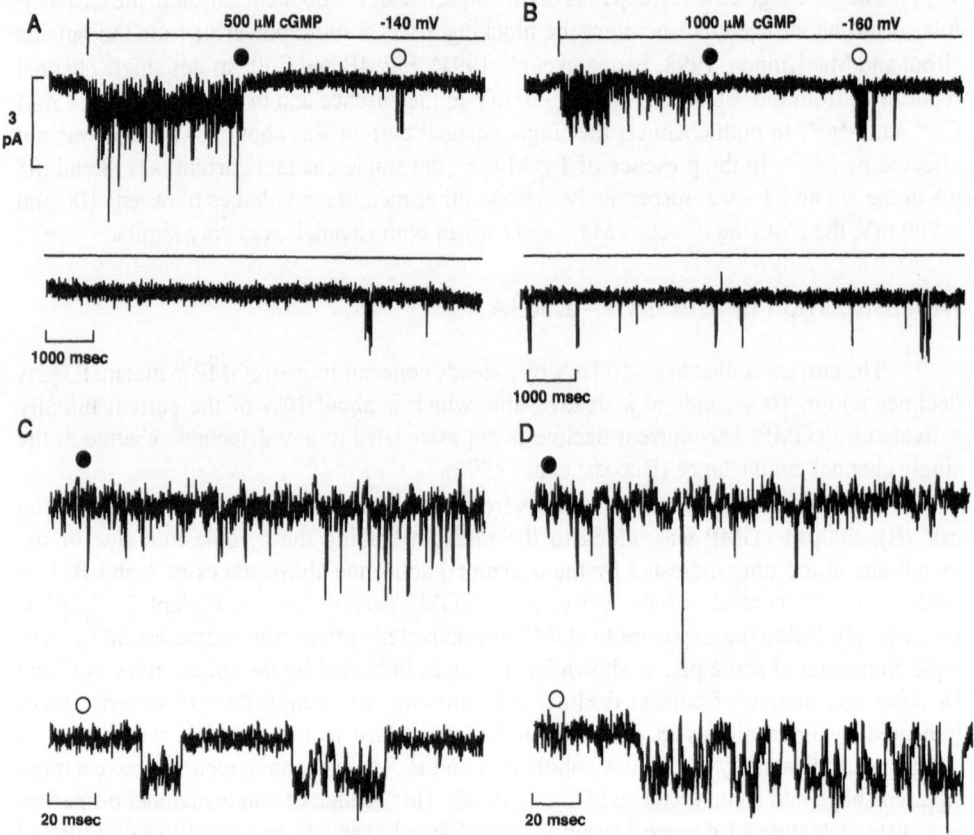

Figure 5. Desensization in mutant T-363A. Current recodings from two patches at -140 (A) and at -160 mV (B). In both panels, the two top traces reproduce current recordings obtained after the addition of 500μM cGMP indicated by the solid lines. The sections indicated by the filled and empty circles in A and B are shown on an enlarged time scale in C and D respectively. Recordings filtered at 500 Hz in A and B and at 4 kHz in C and D.

polypeptides forming the w.t and T - w.t. in the plasma membrane of the oocytes or by a direct influence of the N-terminal on the ionic permeation through the channel.

ACKNOWLEDGMENTS

We are indebted to Dr. A. Menini for her critical reading of the manuscript and to Miss L. Giovanelli who did the art work and checked the English. This research was supported by grants funded by the Human Frontier Science Program and Human Capital and Mobility.

REFERENCES

BÖNIGK, W., ALTENHOFEN, W., MULLER, F., DOSE, A., ILLING, M., MOLDAY, R.S. and KAUPP, U.B. (1993). Rod and cone photoreceptor cells express distinct genes for cGMP-gated channels. Neuron, 10, 865-877.

BUCOSSI, G., EISMANN, E., SESTI, F., NIZZARI, M., SERI, M., KAUPP. U.B. and TORRE, V. (1996). Time dependent current decline in cyclic GMP gated channels caused by point mutations in the pore region. J. Physiol., (493.2, 409-418).

CHEN, T.Y., PENG, Y.W., DHALLAN, R.S.,AHAMED, B.,REED, R.R. and YAU, K-W. (1993). A new subunit of the cyclic nucleotide-gated cation channel in retinal rods. Nature, 362, 764-767.

COLAMARTINO, G., MENINI, A. and TORRE, V.(1991). Blockage and permeation of divalent cations through the cyclic GMP-activated channel from tiger salamander retinal rods. Journal of Physiology, 440, 189-206.

EISMANN, E., MULLER, F., HEINEMANN, S. and KAUPP, B. (1994). A single negative charge within the pore region of a cGMP-gated channel controls rectification, Ca blockage, and ionic selectivity. Proc. Natl.Acad. Sci. USA, 91, 1109-1113.

GOULDING, E.H., NGAI, J., KRAMER, R.H., COLICOS, S., AXEL, R., SIEGELBAUM, S.A. and CHESS, A. (1992). Molecular cloning and single channel properties of the cyclic nucleotide-gated channel from the catfish olfactory neurons. Neuron, 8, 45-58.

GUY, H.R., DURELL, S.R., WARMKE, J., DRYSDALE, R., GANETZKI, B. (1991). Similarities in amino acid sequences of Dorsophila eag and cyclic nucleotide gated channels. Science, 254, 730.

HEGINBOTHAM, L., ABRAMSON, T. and MACKINNON, R. (1992). A functional connection between the pores of distantly related ion channels as revealed by mutant K channels. Science, 258, 11521155.

HENN, D.K., BAUMANN, A. and KAUPP, U.B. (1995) Probing the trans-membrane topology of cyclic nucleotide gated ion channels with a gene fusion approach. Proc. Natl. Acad. Sci. USA, 92, (7425-7429).

JAN, L.Y. and JAN, Y.N. (1990). A superfamily of ion channels, Nature, 345, 672.

KARPEN, J.W., ZIMMERMAN, A.L., STRYER, L. and BAYLOR, D.A. (1988). Gating kinetics of the cyclic GMP-activated channel of retinal rods: flash photolysis and voltage-jump studies. Proc. Natl. Acad. Sci. USA, 85, 1287-1291.

KAUPP, U.B., NIIDOME, T., TANABE, T., TERADA, S., BONIGK, W., STUHMER, W., COOK, N.J., KANGAWA, K., MATSUO, H., HIROSE, T., MIYAIA, T. and NUMA, S. (1989). Primary structure and functional expression from complementary DNA of the rod photoreceptor cyclic GMP-gated channel. Nature, 342,762-766.

KÖRSCHEN, H., ILLING, M., SESTI, F., SEIFERT, R., WILLIAMS, A., GOTZES, S., COLVILLE, S., MILLER, F., DOSE, A., GODDE, M., MOLDAY, L., KAUPP, U.B. and MOLDAY, R.S. (1995). A 240 K protein represents the complete β-subunit of the cyclic nucleotide-gated channel from rod photoreceptor. Neuron, 15, 627-636.

LUHRING, H., HANKE, W., SIMMOTEIT, R. and KAUPP,U.B. (1990). Cation selectivity of the cyclic GMP-gated channel of mammalian rod photoreceptors. In Sensory Transduction, ed. BORSELLINO, A.,CERVETTO, L. TORRE, V., pp. 169-174. Plenum Press, New York.

MENINI, A. (1990). Currents carried by monovalent cations through cyclic GMP-activated channels in excised patches from salamander rods. Journal of Physiology, 424,167-185.

NIZZARI, M., SESTI, F., GIRAUDO, M.T., VIRGINIO, C.,CATTANEO, A and TORRE, V. (1993). Single channel properties of a cloned channel activated by cGMP. Proc. R. Soc.London B, 254, 69-74.

ROOT, M.J. and MACKINNON, R. (1993). Identification of an external divalent binding site in the pore of a cGMP-activated channel. Neuron, 11, 459-466.

SESTI, F., STRAFORINI, M., LAMB, T.D. and TORRE, V.(1994). Properties of single channels activated by cyclic GMPin retinal rods of the tiger salamander. Journal of Physiology, 474, 203-222.

SESTI, F., KAUPP, B.U., EISMANN, E., NIZZARI, M. and TORRE, V. (1995). The multi-ion nature of the cGMP-gated channel from vertebrate rods. Journal of Physiology, (487.1, 17-36).

TORRE, V., ASHMORE, J.F., LAMB, T.D. and MENINI, A. (1995). Transduction and adaptation in sensory receptor cells. J. Neurosci., 15, 7757-7768.

YAU, K.W. and BAYLOR, D.D. (1989). cGMP activated conductance of retinal photoreceptor cells. Annual review of Neuroscience, 12, 289-327.

ZIMMERMAN, A.L. and BAYLOR, D.A. (1992). Cation interactions within the cyclic GMP-activated channel of retinal rods from the tiger salamander. Journal of Physiology, 449, 759-783.

ELECTROPHYSIOLOGICAL STUDIES OF MITOCHONDRIAL CHANNELS

Maria Luisa Campo,[1] Concepción Muro,[1] Henry Tedeschi,[2] and
Kathleen W. Kinnally[3]

[1] Dpto. de Bioquímica y Biología Molecular y Genética
Universidad de Extremadura
Cáceres, Spain
[2] Department of Biological Science
University at Albany, SUNY
Albany, New York
[3] Molecular Medicine, Wadsworth Center
NYS Departments of Health and Biomedical Science
University at Albany, SUNY School of Public Health
Albany, New York

1. INTRODUCTION

Mitochondria are double membrane organelles whose primary function is to fulfill the energy requirements of the cell. This function requires the net transport of large amounts of materials including substrates and adenine nucleotides as well as the import of mitochondrial proteins across the outer and inner membranes. One pathway for the flow of materials is through ion channels. The generally accepted view is that the outer membrane does not provide a permeability barrier for small molecules. This low-selectivity is due, at least in part, to the presence of a large channel called VDAC (voltage dependent anion-selective channel) which is presumed to remain open *in situ* (Colombini *et al.*, 1979, 1989). The inner membrane, on the other hand, is the site of oxidative phosphorylation and is presumed to require a high electrical resistance for efficient energy coupling. It was therefore somewhat surprising when patch-clamp techniques revealed a variety of channels in the inner membrane as shown in Table 1. This apparent paradox is resolved by keeping in mind that, like in the cell membranes of neurons and muscle, the opening of the different mitochondrial channels is subject to regulation by physiological effectors such as voltage, pH, NADH or divalent cations.

This communication will summarize the general characteristics of mitochondrial channels and discuss some of their possible functions.

Table 1. Summary of mitochondrial channel activities

Channel Activity	Size[a] (pS)	Voltage Dependence	Selectivity	Effectors
Outer membrane				
VDAC	650	yes	slight anion	pH, modulator protein
PSC	1250	yes	slight cation	targeting peptides
Inner membrane				
MCC	~1000	yes	slight cation to none	Ca^{2+}, Mg^{2+}, pH, targeting peptides
mCS	107	yes	slight anion	voltage
K^+	9	no	K^+	ATP
ACA	15	no	slight cation	pH, Mg^{2+}
AAA	45	no	slight anion	pH, Mg^{2+}

[a]All measurements made in symmetrical 0.15 M KCl except the K^+ channel was measured in 100 mM KCl in the presence of a gradient (Inoue *et al.*, 1991)

2. OUTER MEMBRANE CHANNELS

The majority of in-depth studies on mitochondrial outer membrane channels have been carried out in reconstituted systems and have lead to the characterization of two channels, VDAC and PSC. While more recent studies have employed patch-clamp techniques on isolated mitochondria, most of the studies of VDAC have been done in planar bilayers while those of PSC have primarily used tip-dip techniques.

2.1. Voltage Dependent Anion-Selective Channel, VDAC

VDAC, or mitochondrial porin, makes up a large percentage of the total outer membrane protein and has a MW of ~30 kDa depending on the species from which it is derived. The high level of expression of VDAC is in keeping with its important function as the primary pathway for solutes across the outer membrane. It is the only mitochondrial channel that has been purified, cloned and sequenced. VDAC is now the object of intense molecular and structural studies (Blachly-Dyson *et al.*, 1989, 1990, 1993, 1994). VDAC is thought to have a beta-barrel structure and therefore is presumed to be similar to bacterial porin (Shao *et al.*, 1994; Mannella *et al.*, 1996). Electron microscopic studies of 2-dimensional crystals indicate a pore opening of ~3 nm (Mannella *et al.*, 1984).

Isoforms of VDAC are prevalent. There are at least four isoforms in human, three in mouse and two in yeast. Correlations between the different isoforms and function are now being made in different cell types. For example, human VDAC1 binds hexokinase while human VDAC2 does not (Blachly-Dyson *et al.*, 1993, 1994). This interaction is expected to facilitate ATP turnover through the energy exchange from ATP to glucose 6-phosphate, a primary energy storage form in some cell-types.

VDAC has a peak conductance of 650 pS and a predominant half-open state of about 300 pS in physiological salt (Colombini, 1979). The open state is slightly anion selective while the half-open state is slightly cation selective. The transition between these two open states is important in terms of VDAC's function as the primary transport pathway across the outer membrane since ATP and ADP are freely permeable only through the fully open state (Benz *et al.*, 1988, 1990; Liu and Colombini, 1992).

VDAC has a rather symmetrical voltage dependence in bilayers closing with increasing potential of either polarity as shown in figure 1. N is a measure of the gating charge to open the channel and varies from 2-4 depending on the species studied . The V_0 is the voltage

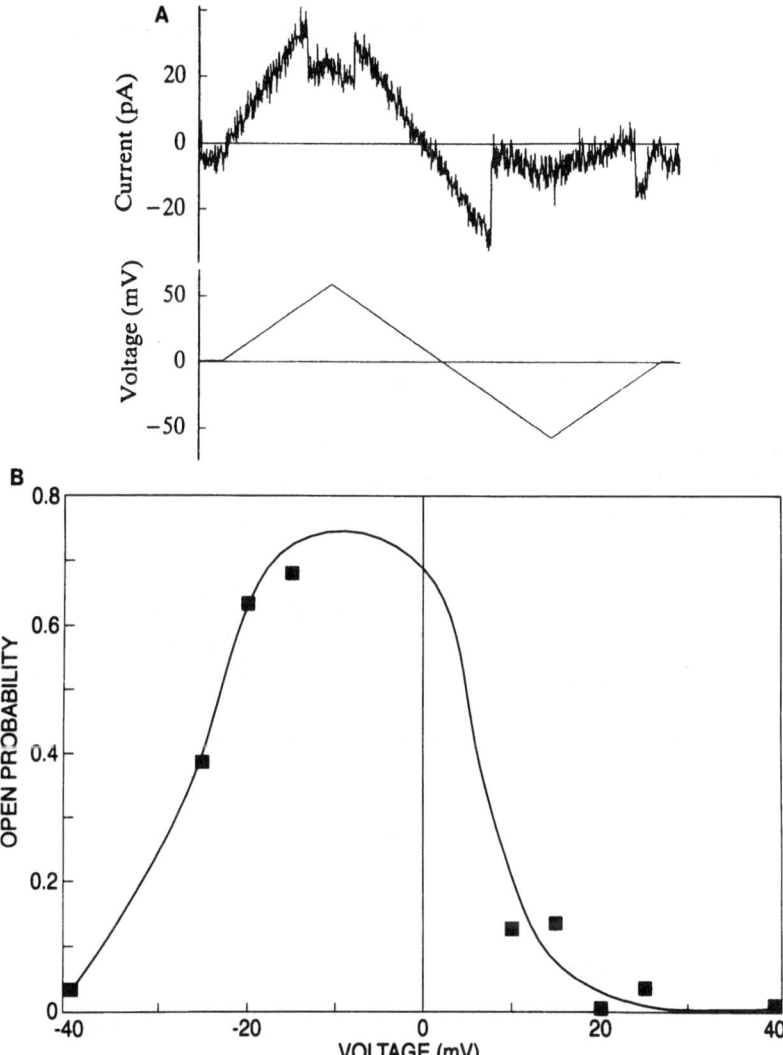

Figure 1. VDAC channel activity and voltage dependence. (A) Current response to a triangle voltage wave (where the ramp is generated by Clampex program and was limited to ±60 mV) shows VDAC closing with higher potentials of either polarity. *Neurospora crassa* outer mitochondrial membranes were solubilized in 2% triton and added to the cis chamber. One apparent channel inserted into the azolectin planar bilayer with symmetrical 150 mM KCl, 5 mM HEPES, 5 mM CaCl$_2$, pH 7.4. Note full closure occurred at negative potentials while half-closure occurred with positive potentials. (B) Current was recorded under voltage-clamp conditions from a patch excised from a liposome reconstituted with purified *Neurospora crassa* VDAC and bathed in symmetrical 150 mM KCl, 5 mM HEPES, pH 7.4. Shown is a plot of the probability that the channel is open at various voltages, calculated as the fraction of total time spent in the fully open (608 pS) state from total amplitude histograms.

at which the open state is occupied half the time and is in the range of ±20-40 mV (Colombini, 1989 and DePinto *et al.*, 1987). These parameters are subject to regulation by physiological effectors, e.g. NADH (Zizi *et al.*, 1994) and a "modulator protein" partially purified from the space between the inner and outer membranes (Liu and Colombini 1992; Lui *et al.*, 1994). Hence the overall outer membrane permeability, thought to be controlled by VDAC, can be modified in response to changing metabolic conditions.

Uncertainties about the high permeability of the outer membrane *in situ* have arisen from recent reports of high resistance seals on isolated mitochondria using patch-clamp techniques (Moran *et al.*, 1992). If VDAC were predominantly open, its high conductance and high density in the outer membrane should preclude the formation of high resistance seals. These observations suggest that, at least in these preparations, most VDAC are completely closed. The previous reports of low resistance seals indicate VDAC can be opened under some experimental conditions (Tedeschi *et al.*, 1987). For a further discussion, see Sorgato and Moran, 1993.

2.2. Peptide Sensitive Channel, PSC

PSC is a slightly cation-selective channel in the mitochondrial outer membrane and has been found in both mammalian and yeast mitochondria. It has been detected in studies of VDAC-deletion mutants and therefore is distinct from VDAC (Thieffry *et al.*, 1990). While most of the studies of PSC employed tip-dip techniques, it has also been recorded from liposomes containing mitochondrial membranes using patch-clamp techniques as well as planar bilayers (Thieffrey *et al.*, 1992, Fevre *et al.*, 1994).

PSC has a peak conductance of over 1 nS in physiological KCl and at least two major subconductance levels; predominant transitions are 500 pS. It is voltage dependent with a higher open probability at cytoplasmic negative potentials (Chich *et al.*, 1991). The gating charge (measure of the effective charge that moves across the membrane to fully open the channel) is about 2 and the V_0 is approximately -20 to 0 mV relative to the cytoplasm depending on the species from which the PSC was derived (Fevre *et al.*, 1994).

A role in protein import for PSC is supported by the transient blockade of this activity by synthetic peptides whose sequences target proteins to mitochondria. The relationship between PSC and the outer membrane protein import apparatus is underway (Henry *et al.*, 1995).

3. INNER MEMBRANE CHANNELS

The electrophysiological studies of the mitochondrial inner membrane are still at an early stage. Several channel activities have been identified by applying patch-clamp techniques to the native inner membrane in a preparation called mitoplasts (mitochondria in which the inner membrane is exposed by either French press or osmotic treatment to break the outer membrane). However, none of the proteins responsible for these activities have been identified and, for the most part, their functions are still subjects of speculation.

3.1. Multiple Conductance Channel, MCC

MCC [also called MMC for *mitochondrial megachannel* by Petronilli *et al.*, (1989)] was originally reported in 1989 (Kinnally *et al.*, 1989) and is a channel activity that has been recorded by patch-clamp techniques in mitoplasts from a variety of organs from several mammals (including human cell lines, Murphy *et al.*, 1995) as well as yeast (Lohret and Kinnally, 1995a, 1995b, Kinnally *et al.*, 1995). It has a peak conductance of ~1 nS and transition sizes in the range of 300 to 600 pS predominate in mammalian and yeast mitoplasts. Mammalian MCC is normally quiescent but can be activated by calcium or high voltage (Kinnally *et al.*, 1991, Zorov *et al.*, 1992).

While the various conductance levels observed might correspond to more than one channel, they have been attributed to a single class. They are activated by similar conditions, e.g. calcium and voltage, and respond similarly to pharmacological agents (Kinnally *et al.*,

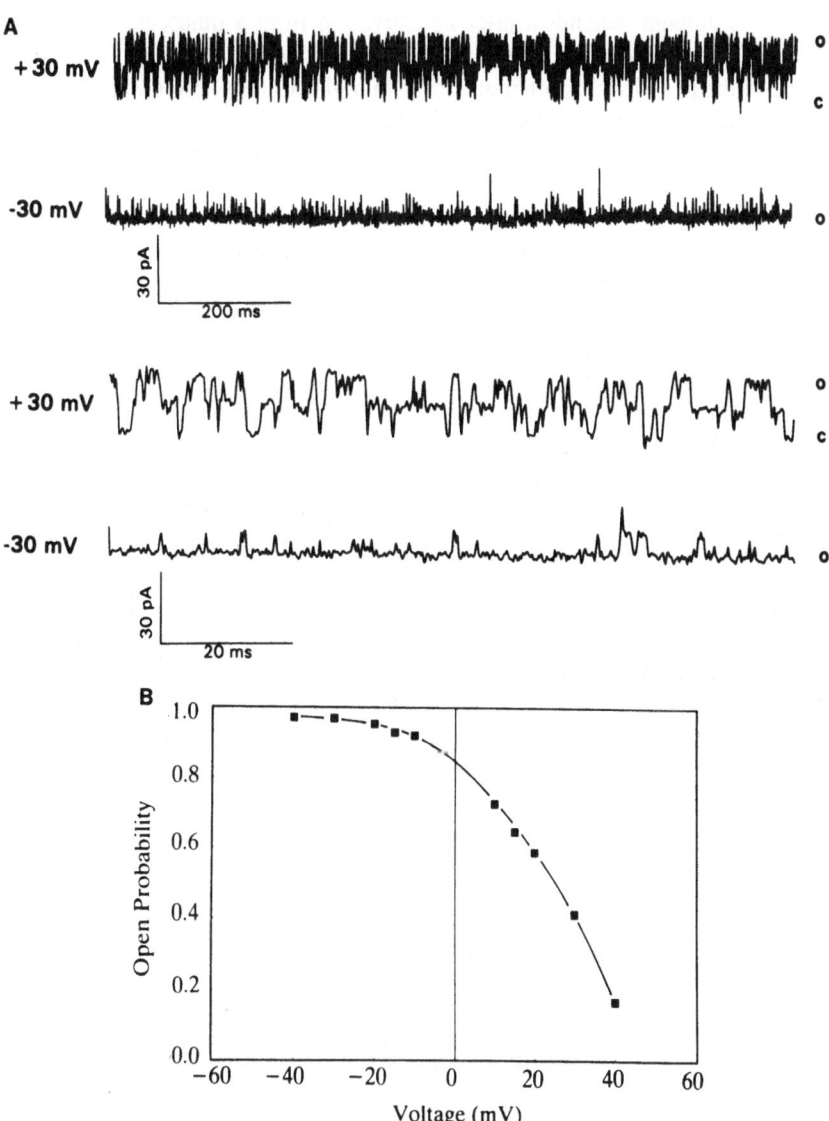

Figure 2. MCC channel activity and voltage dependence. (A) Sample current traces showing MCC channel activity from a murine liver mitoplast. The excised patch was clamped either at +30 or -30 mV. Data was sampled at 5 kHz and bandwidth limited to 2.5 KHz. Mitoplasts were pre-treated with 40 μM FCCP. After excision, the patch was perfused with medium contained 150 mM KCl, 5 mM HEPES, 1 mM EGTA, 40 μM FCCP and 1 mM dithiotreitol. (B) Open probability of the same patch clamped at different voltages. The V_0 was 23 mV and the gating charge was -3.2. C and O refer to close and open states respectively. Other conditions as in fig. 1.

1992, Campo *et al.*, 1995). MCC activity can be reconstituted into liposomes after detergent solubilization of inner membranes while retaining its variety of subconductance levels (Lohret and Kinnally, 1995a). Furthermore, Petronilli *et al.* (1989) showed that conductance levels between 300 and 1300 pS are probably substates of a single channel since several small openings sometimes close in a single large step.

MCC has a slight selectivity for cations over anions. The relative permeability ratio for K^+/Cl^- is 3-6 depending on the species of origin (Lohret and Kinnally, 1995a).

MCC is voltage dependent after its activation from a quiescent closed state (for example by calcium) as shown by the current traces and voltage profile of figure 2. Occupation of the peak level is often greater with high (usually >50 mV) compared to low positive potentials (not shown). There are reports that MCC slowly closes after extended periods at negative potentials and closure is facilitated by higher than physiological salt (Szabó et al., 1993). Closure of MCC with negative potentials is also found after reconstitution of wild-type yeast MCC and is seen in mitoplasts from a yeast strain in which the VDAC gene has been deleted (Lohret and Kinnally, 1995a). Differences in voltage dependence may be related to the regulation of MCC by a variety of physiological effectors some of which might be lost during isolation and/or reconstitution. Furthermore, MCC may reside in different membrane domains whose integrity may be sensitive to experimental conditions, e.g. contact sites (Kinnally et al., 1992).

Voltage dependence can be expressed quantitatively in terms of gating charge and V_0. The gating charge for MCC is -3 to -5 at low positive potentials (Lohret and Kinnally, 1995a). The V_0 varies with the source of mitochondria but is generally in the range of 10-30 mV relative to the mitochondrial matrix. (Lohret and Kinnally, 1995b).

MCC is a very high conductance channel whose uncontrolled opening should decrease coupling of oxidative phosphorylation. Therefore, it is not surprising that MCC's activity is influenced by a variety of physiological effectors. These include divalent cations, pH, ADP and voltage (e.g. see Szabó and Zoratti, 1992). Several pharmacological agents affect MCC including antimycin A, cyclosporin A, and the uncouplers, CCCP and FCCP. In addition, the list of compounds affecting MCC activity include the amphiphilic cations amiodarone, propranolol and quinine, as well as dibucaine (see Campo et al., 1995 for summary).

MCC activity has recently been linked to protein import by the transient blockade of MCC by synthetic peptides whose sequences correspond to targeting signals for mitochondrial protein import (Lohret and Kinnally, 1995b). MCC has several functional similarities to the outer membrane channel PSC (i.e., sizes of conductance transitions and selectivity; for comparison see Kinnally et al., 1995), further implicating a possible role for MCC in protein import. Note that PSC and MCC are thought to be located in different membranes and can be distinguished in protein import mutants (Lohret et al., 1996b).

3.2. Mitochondrial Centum picoSiemen Channel, mCS

The first application of patch-clamp techniques to the native inner mitochondrial membrane resulted in the discovery of a voltage-gated, 100-pS conductance channel (Sorgato et al., 1987) now called mCS. This activity has been recorded in several labs using mitoplasts from a variety of mammalian organs (e.g. mouse liver, heart, and kidney) and more recently from human tissue culture cells (Murphy et al., 1995). However, mCS has not as yet been detected in yeast mitoplasts (Lohret and Kinnally, 1995a). mCS is often recorded from the same patch as MCC suggesting that the two classes of channels coexist in the mammalian mitochondrial inner membrane (Kinnally et al., 1992). While differences have been reported from lab to lab, we have found mCS is normally quiescent but can be activated by manipulations with calcium chelators during mitoplast preparation (Kinnally et al., 1991). However, once activated, mCS is not strongly influenced by calcium in the range of 10^{-9} to 10^{-5} M calcium (Kinnally et al,. 1991).

This channel is voltage dependent, closing with matrix negative potentials as shown in the current traces and voltage profile of Figure 3. Kinetic analysis of its bursting activity indicates that mCS has multiple open and closed states as well as intermediate subconductance states (Klitsch and Siemen, 1991; Campo et al., 1992, Ballarin et al., 1994). The reported peak conductance ranges from about 90-140 pS with an average of about 110 pS.

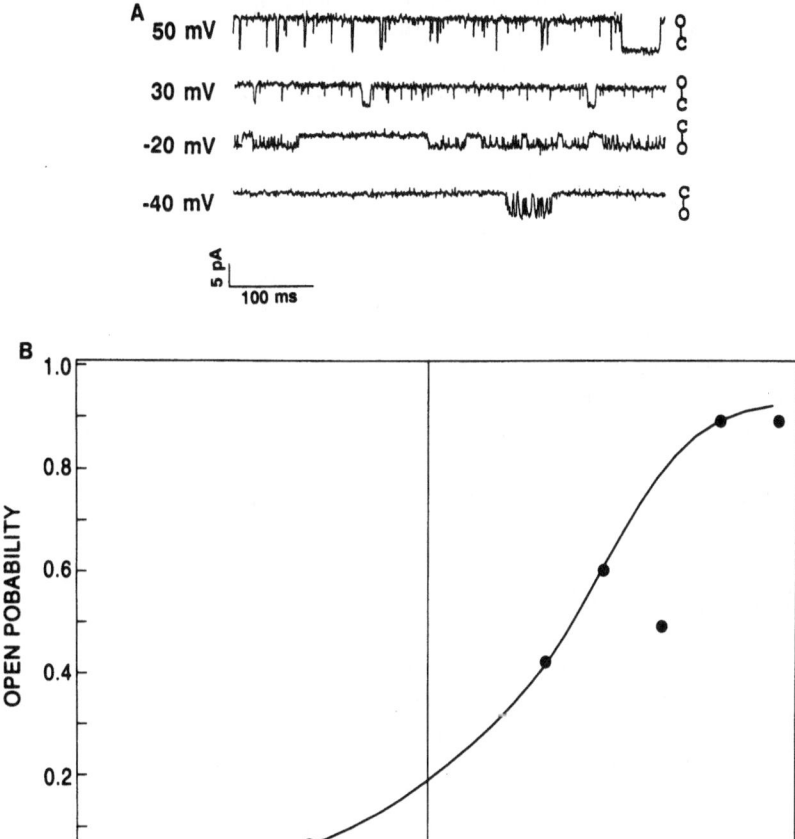

Figure 3. mCS channel activity and voltage dependence. (A) Sample current traces were recorded from an attached patch from a mouse liver mitoplast in 150 mM KCl, 5 mM HEPES, 1 mM EGTA, 2 mM MgCl$_2$, 2.5 μM rotenone, pH 7.4. Data was sampled at 5 KHz and bandwidth limited to 2 KHz. (B) Open probability as a function of voltage was determined from total amplitude histograms of current recorded from a patch excised from a rat heart mitoplast in symmetrical 150 mM KCl, 5 mM HEPES, 1 mM EGTA, 0.75 mM CaCl$_2$, pH 7.4. Other conditions as in fig. 1 and 2.

Substates on the order of 0.25, 0.5, 0.7 and 1.3 times the primary conductance value also have been reported (Kinnally *et al.*, 1993; Klitsch and Siemen, 1991). mCS is slightly anion-selective (P_{Cl-}/P_{K+} = 4.5) in its main 110-pS conductance state and in at least one (50-pS) substate (Kinnally and Tedeschi, 1994).

mCS activity is subject to regulation by physiological effectors. Klitsch and Siemen (1991) have reported that mCS is inhibited by submillimolar levels of di- and trinucleotide phosphates as well as GMP when added to the outside of patched mitoplasts. These results indicate that mCS is not related to the uncoupling protein, thermogenin, which is insensitive to GMP. However, Inoue at al. (1991) report that mCS is not affected by millimolar Mg^{2+} or ATP and micromolar ADP if applied on the matrix side of excised patches (Inoue *et al.*, 1991).

A variety of metals and organic compounds have been tested for their effects on mCS activity. Sorgato *et al.* (1989) have reported that several effectors of other types of channels, e.g. pH, Zn^{2+}, Gd^{3+} and DIDS, have no effect on mCS. However, several pharmacological agents have been found to inhibit mCS activity including amiodarone, propranolol and antimycin A (summarized in Campo et al., 1995).

The function of mCS is a subject of speculation at this time. Some postulated roles for this channel include volume regulation (Klitsch and Siemen, 1991) and thermogenesis.

3.3. ATP-Sensitive K^+ Channel

The primary function of mitochondria is the synthesis of ATP. Therefore, it was exciting to find mitochondria harbor ATP-sensitive channels in their inner membrane. Inoue et al. (1991) originally described the inhibition of a K^+ channel by ATP in fused mitoplasts. This activity has a conductance of about 10 pS in 0.1 M salt and is voltage independent. The localization in the inner membrane is supported by detection of its activity in patches also containing mCS activity. The channel activity is inhibited by 4-aminopyridine and gliben-clamide plus ATP which are known effectors of the plasma membrane ATP-sensitive K^+ channels. A possible role for the ATP-dependent K^+ channel is the regulation or fine-tuning of the mitochondrial membrane potential by changing K^+ permeability in response to the availability of ATP.

3.4. Alkaline pH-Sensitive Channels, ACA and AAA

Two pH-sensitive mitoplast channel activities have been described, both displaying greater open probabilities at alkaline pH and both being activated by depletion of Mg^{2+} (Antonenko et al., 1991b, 1994). One of these channel activities is cation selective (ACA, alkaline-induced cation activity) while the other is slightly anion selective (AAA, alkaline-induced anion activity).

The cation channel ACA has a conductance of about 15 pS in 0.15 M KCl and is relatively voltage insensitive (Antonenko et al., 1992). Its voltage dependence and unit conductance are similar to those of the ATP-sensitive K^+ channel. However, unlike the latter channel, ACA is relatively non-selective for cations and is not affected by 4-aminopyridine and glibenclamide plus ATP. Instead, like mCS and MCC, ACA is inhibited by amiodarone and propranolol. Mg^{2+}-depletion by EDTA is used to induce ACA activity which itself is insensitive to Ca^{2+}. The selectivity and inhibition by Mg^{2+} suggest that ACA may correspond to one of the cation uniporters whose existence was inferred from solution studies and which are implicated in volume homeostasis (e.g. Bernardi et al., 1992).

The anionic channel AAA has an open-state conductance of about 45 pS and, like ACA, is relatively voltage insensitive. AAA has two conductance substates of 1/3 and 2/3 the fully open state and is only slightly selective for different anions (Antonenko et al., 1994). Based on similarities in activation requirements, ACA initially was thought to correspond to IMAC, the inner membrane anion channel inferred from mitochondrial suspension studies. However, there are discrepancies in several functional characteristics of AAA and IMAC, e.g. estimated pore size and degree of anion selectivity (Beavis et al., 1992). AAA may correspond to the low-conductance activity reconstituted from mammalian mitochondrial extracts in bilayers by Hayman et al. (1993) and more recently reported in yeast mitoplasts by Ballarin et al. (1995).

4. DISCUSSION

The application of electrophysiological techniques to mitochondrial membranes has provided a wealth of new information regarding its permeability. The integration of these findings into our understanding of mitochondrial function has led to new ideas about the transport of ions as well the translocation of proteins across these membranes.

4.1. Are There Additional Channels in the Inner and Outer Membranes?

A variety of other transition sizes have been reported in both native and reconstituted electrophysiological studies of mitochondria but are not discussed here. These include the large anion channel recently reported in yeast mitoplasts and a high conductance cation channel from yeast in bilayer studies as well as the low conductance transitions reported from patch-clamping mitochondria (Ballarin *et al.*, 1995; Dihanich *et al.*, 1989; Szabò *et al.*, 1994; Sorgato and Moran, 1995; Moran *et al.*, 1990, 1992). The relationship between these activities and those defined above has not yet been defined.

4.2. Future Studies

It is somewhat surprising that several mitochondrial channels have been described in electrophysiological studies, but the definitive function of these channels remains elusive. While MCC and PSC are probably involved in protein import, the ATP-sensitive K^+ is implicated as an energy-sensor and VDAC functions as the predominant permeability pathway of the outer membrane, speculation is the rule of thumb for the other channels. Furthermore, none of the proteins responsible for these activities (except for VDAC) have been identified despite valiant attempts at biochemical purification. New approaches using molecular techniques appear to hold the key to some of these problems. For example, while the relationship between the permeability transition pore and MCC requires more investigation, the adenine nucleotide translocator is not directly responsible for MCC since deletion of the translocator from yeast mitochondria does not affect MCC activity (Lohret *et al.*, 1996). The application of these techniques to yeast deletion mutants as well as yeast with defined functional flaws will undoubtedly facilitate these studies.

ACKNOWLEDGMENTS

M.L. Campo was supported by Spanish grant DGICYT PB92-0720 and Consejeria de Deporte y Juventud de la Junta de Extremadura and Fondo Social Europeo EIA 94/11. C. Muro was supported by a fellowship from Spanish Ministry of Science and Education. K.W. Kinnally and H. Tedeschi were supported by National Science Foundation grant MCB9117658. We thank Dr. Carmen Mannella for support, discussions, and review of this manuscript.

REFERENCES

Antonenko, Y. N., Kinnally, K. W., Perini, S. & Tedeschi, H. (1991a). Selective effects of inhibitors on mitochondrial inner membrane channels. *FEBS Lett.* 285:89-93.

Antonenko, Y. N., Kinnally, K. W. & Tedeschi, H. (1991b). Identification of anion and cation pathways in the inner mitochondrial membrane by patch clamping of mouse liver mitoplasts. *J. Membr. Biol.* 124:151-158.

Antonenko, Yu., Smith, D., Kinnally, K. W. & Tedeschi, H. (1994). Single channel activity induced in mitoplasts by alkaline pH. *Biochim. Biophys. Acta* 1194:247-254.

Ballarin, C., Sorgato, M. C. & Moran, O. (1994). A minimal kinetic model of the 107 pS channel of the inner membrane of mitochondria. In: *Molecular Biology of Mitochondrial Transport Systems* (Forte, M. & Colombini, M., eds.), Springer Verlag, Berlin, New York, 131-136.

Ballarin, C. & Sorgato, M. C. (1995). An electrophysiological study of yeast mitochondria. *J. Biol. Chem.* 270:19262-19268.

Beavis, A. D. (1992). Properties of the inner membrane anion channel in intact mitochondria. *J. Bioenerg. Biomembr.* 24:77-90.

Benz, R., Kottke, M. & Brdiczka, D. (1990). The cationically selective state of the mitochondrial outer membrane pore: a study with intact mitochondria and reconstituted mitochondrial porin. *Biochim. Biophys. Acta* 1022:311-318.

Benz, R., Wojtczak, L., Bosch, W. & Brdiczka, D. (1988). Inhibition of adenine nucleotide transport through the mitochondrial porin by a synthetic polyanion. *FEBS Lett.* 231:75-80.

Bernardi, P., Zoratti, M. & Azzone, G. F. (1992). Mitochondrial volume homeostasis: regulation of cation transport systems. In: NATO ASI Series, Vol. H 64, *Mechanics of Swelling* (Karalis, T. K., ed.), Springer-Verlag, Berlin Heidelberg, pp. 357-377.

Blachly-Dyson, E., Peng, S. Z., Colombini, M. & Forte, M. (1989). Probing the structure of the mitochondrial channel, VDAC, by site-directed mutagenesis: a progress report. *J. Bioenerg. Biomembr.* 21:471-483.

Blachly-Dyson, E., Peng, S. Z., Colombini, M. & Forte, M. (1990). Selectivity changes in site-directed mutants of the VDAC ion channel: structural implications. *Science* 247:1233-1236.

Blachly-Dyson, E., Zambronicz, E. B., Yu, W. H., Adams, V., McCabe, E. R. B., Adelman, J., Colombini, M. & Forte, M. (1993). Cloning and functional expression in yeast of two human isoforms of the outer mitochondrial membrane channel, the voltage-dependent anion channel. *J. Biol. Chem.* 268:1835-1841.

Blachly-Dyson, E., Baldini, A., Litt, M., McCabe, E. R. B. & Forte, M. (1994). Human genes encoding the voltage-dependent anion channel (VDAC) of the outer mitochondrial memebrane: mapping and identification of two new isoforms. *Genomics* 20:62-67.

Campo, M. L., Tedeschi, H., Muro, C. & Kinnally, K.W. (1995). Mitochondrial membrane channels and their pharmacology. In: *Ion Channel Pharmacology*, (Soria, B. & Ceña, V., ed.), Oxford Univ. Press, in press.

Campo, M. L., Kinnally, K. W., Perini, S. & Tedeschi, H. (1992). The effect of antimycin A on mouse liver inner mitochondrial membrane channel activity. *J. Biol. Chem.* 267:8123-8127.

Chich, J. -F., Goldshmidt, D., Thieffry, M. & Henry, J. -P. (1991). A peptide-sensitive channel of large conductance is localized on mitochondrial outer membrane. *Eur. J. Biochem.* 196:29-35.

Colombini, M. (1979). A candidate for the permeability pathway of the outer mitochondrial membrane. *Nature* 279:643-645.

Colombini, M. (1989). Voltage gating in the mitochondrial channel, VDAC. *J. Membr. Biol.* 111:103-111.

De Pinto, V., Ludwig, O., Krause, J., Benz, R. & Palmieri, F. (1987). Porin pores of mitochondrial outer membranes from high and low eukaryotic cells: biochemical and biophysical characterization. *Biochim. Biophys. Acta* 894:109-119.

Dihanich, M., Schmid, A., Oppliger, W. & Benz, R. (1989). Identification of a new pore in the mitochondrial outer membrane of a porin-deficient yeast mutant. *Eur. J. Biochem.* 181:703-708.

Fèvre, F., Henry, J.-F. & Thieffry, M. (1994). Reversible and irreversible effects of basic peptides on the mitochondrial cationic channel. *Biophys. J.* 66:1887-1894.

Hayman, K. A., Spurway, T. D. & Ashley, R. H. (1993). Single anion channels reconstituted from cardiac mitoplasts. *J. Membr. Biol.* 136:181-190.

Henry, J.-P., Juin, P., Vallette, F. & Thieffry, M. (1995). Characterization and function of the mitochondrial outer membrane peptide sensitive channel. *J. Bioenerg. Biomembr.*, in press.

Inoue, I., Nagase, H., Kishi, K. & Higuti, T. (1991). ATP-sensitive K+ channel in the mitochondrial inner membrane. *Nature* 352:244-247.

Kinnally, K. W., Campo, M. L. & Tedeschi, H. (1989). Mitochondrial channel activity studied by patch-clamping mitoplasts. *J. Bioenerg. Biomembr.* 21:497-506.

Kinnally, K. W., Zorov, D. B., Antonenko, Y. & Perini, S. (1991). Calcium modulation of inner mitochondrial membrane channel activity. *J. Membr. Biol.* 176:1183-1188.

Kinnally, K. W., Antonenko, Y. & Zorov, D. B. (1992). Modulation of inner mitochondrial membrane channel activity. *J. Bioenerg. Biomembr.* 24:99-110.

Kinnally, K. W., Zorov, D. B., Antonenko, Y., Snyder, S., McEnery, M. W. & Tedeschi, H. (1993). Mitochondrial benzodiazepine receptor linked to inner membrane ion channels by nM actions of ligands. *Proc. Natl. Acad. Sci. USA* 90:1374-1378.

Kinnally, K. W. & Tedeschi, H. (1994). Mitochondrial channels: an integrated view. In: *Molecular Biology of Mitochondrial Transport Systems* (Forte, M. & Colombini, M., eds.), Springer Verlag, Berlin, New York, 169-198.

Kinnally, K.W., Lohret, T.A., Campo, M.L. & Mannella, C.A. (1995). Perspectives on the mitochondrial multiple conductance channel. *J. Bioenerg. Biomembr.*, in press.

Klitsch, T. & Siemen, D. (1991). Inner mitochondrial membrane anion channel is present in brown adipocytes but is not identical with the uncoupling protein. *J. Membr. Biol.* 122:69-75.

Liu, M.-Y. & Colombini, M. (1992). A soluble mitochondrial protein increases the voltage dependence of the mitochondrial channel, VDAC. *J. Bioenerg. Biomembr.* 24:41-46.

Liu, M. Y., Togrimson, A. & Colombini, M. (1994). Characterization and partial purification of the VDAC-channel-modulating protein from calf liver mitochondria. *Biochim. Biophys. Acta*, 1185:203-212.

Lohret, T. A. & Kinnally, K. W. (1995). Multiple conductance channel activity of wild-type and VDAC-less yeast mitochondria. *Biophys. J.* 68:2299-2309.

Lohret, T. A. & Kinnally, K. W. Targeting peptides transiently block a mitochondrial channel, *J. Biol. Chem.*, 270:15950-15953.

Lohret, T. A., Murphy, R. C., Drgoñ, T. and Kinnally, K. W. (1996). Evidence that the mitochondrial multiple conductance channel, MCC, is not related to the adenine nucleotide translocator *J. Biol. Chem.*, in press.

Lohret, T. A., Jensen, R. & Kinnally, K. W. (1996). The effect of targeting peptides on a mitochondrial channel, mcc, is altered in the protein import mutant *mas6-1*. *Biophys. J.* 70:A244.

Mannella, C. A., Neuwald, A. F. & Lawrence, C. E. (1996). Detection of likely transmembrane β-strand regions in sequences of mitochondrial pore proteins using the Gibbs sampler. *J. Bioenerg. Biomembr.*, in press.

Mannella, C. A. (1984). Phospholipase-induced crystallization of channels in mitochondrial outer membranes. *Science* 224:165-166.

Moran, O., Sandri, G., Panfili, G., Stuhmer, W. & Sorgato, M. C. (1990). Electrophysiological characterization of contact sites in brain mitochondria. *J. Biol. Chem.* 265:908-913.

Moran, O. & Sorgato, M. C. (1992). High conductance pathways in mitochondrial membranes. *J. Bioenerg. Biomembr.* 24:91-98.

Moran, O., Sciancalepore, M., Sandri, G., Panfilli, E., Bassi, R., Ballarin, C. & Sorgato, M.C. (1992). Ionic permeability of the outer membrane. *Eur. Biophys. J.* 20:311-319.

Murphy, R. C., King, M., Diwan, J.J. & Kinnally, K.W. (1996). Patch-clamp studies of two channels from human mitochondria. *Biophys. J.* 70:A2.

Petronilli, V., Szabò, I. & Zoratti, M. (1989). The inner mitochondrial membrane contains ion-conducting channels similar to those found in bacteria. *FEBS Lett.* 259:137-143.

Shao, L., Van Roey, P., Kinnally, K. W. & Mannella, C. A. (1994). Circular dichroism of isolated mitochondrial channel protein, VDAC: first direct evidence for porin-like secondary structure. *Biophys. J.* 66:A21.

Sorgato, M. C., Keller, B. U. & Stuhmer, W. (1987). Patch clamping of the inner mitochondrial membrane reveals a voltage dependent channel. *Nature* 330:498-500.

Sorgato, M. C. & Moran, O. (1993). Channels in mitochondrial membranes: knowns, unknowns, and prospects for the future. *Crit. Rev. Biochem. Molec. Biol.* 18:127-171.

Szabò, I. & Zoratti, M. (1992). The mitochondrial megachannel is the permeability transition pore. *J. Bioenerg. Biomembr.* 24:111-117.

Szabò, I. & Zoratti, M. (1993). The mitochondrial permeability pore may comprise VDAC molecules. I. The electrophysiological properties of VDAC are compatible with those of the mitochondrial megachannel. *FEBS Lett.* 330:201-205.

Szabò, I., Báthori, G., Wolff, D., Starc, T., Cola, C. & Zoratti, M. (1995). The high-conductance channel of porin-less yeast mitochondria. *Biochim. Biophys. Acta* 1235:115-125.

Thieffry, M., Neyton, J., Pelleschi, M., Fevre, F. & Henry, J.-P. (1992). Properties of the mitochondrial peptide-sensitive cationic chanel studied in planar bilayers and patches of giant lioposomes. *Biophys. J.* 63:333-339.

Tedeschi, H. & Kinnally, K. W. (1994). Mitochondrial channels. In: *Membrane Channels - Molecular and Cellular Physiology* (Peracchia, C., ed.), Academic Press, New York, London, in press.

Tedeschi, H., Mannella, C. A. & Bowman, C. L. (1987). Patch-clamping the outer mitochondrial membrane. *J. Membr. Biol.* 97:21-29.

Zizi, M., Forte, M., Blachly-Dyson, E. & Colombini, M. (1994). NADH regulates the gating of VDAC, the mitochondrial outer membrane channel. *J. Biol. Chem.* 269:1614-1616.

Zorov, D., Kinnally, K. W. & Tedeschi, H. (1992). Voltage activation of heart inner mitochondrial membrane channels. *J. Bioenerg. Biomembr.* 24:119-124.

3

FUNCTIONAL AND STRUCTURAL CONSTITUENTS OF NEURONAL Ca^{2+} CHANNEL MODULATION BY NEUROTRANSMITTERS

E. Carbone,[*] V. Magnelli, V. Carabelli, D. Platano, and G. Aicardi[†]

Dipartimento di Neuroscienze
Corso Raffaello 30
I-10125 Torino, Italy

1. INTRODUCTION

Cytoplasmic Ca^{2+} levels can be effectively regulated by a variety of voltage-dependent Ca^{2+} channels through which Ca^{2+} ions can quickly enter the cell (see Bean, 1989; Tsien, Ellinor & Horne, 1991). As the increase of intracellular Ca^{2+} represents the triggering event of many biological functions, the modulation of Ca^{2+}-entry through voltage-activated Ca^{2+} channels may represent an effective tool to regulate a number of Ca^{2+}-dependent cell activities (see Carbone & Swandulla, 1989). Neurotransmitter and hormone release from pre-synaptic terminals and secretory cells are probably the best examples of cell activities in which Ca^{2+} channel modulation plays a critical role. The released material may either affect its own release by modulating the Ca^{2+} channels controlling the secretion (autocrine regulation) or may interfere with the exocytotic activity of other neurons by inhibiting Ca^{2+} influx at their prejunctional endings (presynaptic inhibition). In both cases Ca^{2+} channel modulation by neurotransmitters and hormones is crucial to the control of secreted products and neuronal functioning. A better knowledge of the molecular mechanisms underlying Ca^{2+} channel modulation by neurotransmitter receptors activation will allow therefore to clarify important issues related to neurosecretion and, in parallel, to improve our understanding of Ca^{2+} channel function.

In this article we will review the most relevant discoveries of the last decade on neuronal Ca^{2+} channel modulation mediated by receptor-activated G proteins and protein

---

[*] Correspondence to: E. Carbone, Dipartimento di Neuroscienze, Corso Raffaello 30, I-10125 Torino, Italy.

[†] *Permanent address*: Dip. di Fisiologia Umana e Generale, P. za S. Donato 2, I-40127 Bologna, Italy. Tel: +39-11-6707786; Fax: 6707708; E-mail: carbone@unito.it

kinase C. Rather than furnishing an exhaustive list of the available literature, we preferred to discuss the molecular basis of these events focusing on the most recent findings derived from the biophysics, pharmacology and molecular biology of Ca^{2+} channel modulation.

2. HISTORICAL BACKGROUND

On a short time scale, the inhibition of neuronal high-threshold (HVA) Ca^{2+} channels induced by neurotransmitters and hormones represents one of the best understood systems of Ca^{2+} channel modulation. The phenomenon was first reported in chick sensory neurons by Dunlap & Fischbach (1978) and subsequently extended to most peripheral and central neurons by an impressive number of reports (see Swandulla, Carbone & Lux, 1991). The main constituents of neurotransmitter action are well established and attributed to a receptor-mediated reaction modulated by either pertussis toxin-sensitive (Dolphin & Scott, 1987; Wanke, Ferroni, Malgaroli, Ambrosini, Pozzan & Meldolesi, 1987) or cholera toxin-sensitive G proteins (Zhu & Ikeda, 1994b). The receptor-activated $G\alpha$ (or $G\beta\gamma$)-subunit of the G protein down-modulates reversibly the HVA Ca^{2+} channels by reducing and delaying their probability of opening. An interesting aspect of this phenomenon is that membrane voltage controls a significant fraction of Ca^{2+} channel inhibition. Strong depolarizations remove quickly the inhibition while resting potentials help to recover the depression (Marchetti, Carbone & Lux, 1986; Bean, 1989: Grassi & Lux, 1989; Elmslie, Zhou & Jones, 1990). A number of recent papers have focused on the basic mechanisms underlying the voltage dependency, voltage independency and selectivity of G protein action (see Hille, 1994). Some points are universally accepted while others are still questioned. Here, we will discuss the converging and contrasting issues focusing on the role that the Ca^{2+} channel modulation by G proteins may play in the control of neurotransmitter and hormone release in synaptic terminals and neurosecretory cells.

3. ACTION OF NEUROTRANSMITTERS ON Ca^{2+} CHANNELS

3.1. Effects on High-Threshold Channels

Neurotransmitter action on HVA Ca^{2+} channels is reversible and mediated by membrane receptors. In human neuroblastoma IMR32 cells (Pollo, Lovallo, Sher & Carbone, 1992) and bullfrog sympathetic neurons (Boland & Bean, 1993) the block of HVA currents by receptor agonists is concentration dependent and develops within few seconds at saturating doses of the agonist. The wash out develops independently of neurotransmitter concentration and is complete between 6 and 10 s. The degree of neurotransmitter inhibition is related to the density of membrane receptors expressed by the cell. For instance, noradrenergic, muscarinic and opioid receptors coexists at different concentrations in IMR32 cells. In these cells, the action of noradrenaline (NA) and oxotremorine (OXO) is usually pronounced (Fig. 1). The inhibition appears stronger at the beginning of the step depolarization and partially relieved toward the end of the pulse. Neurotransmitter action is prevented by receptor antagonists (Fig. 2-right) and, in most cells, the kinetic slowing is limited to non-L-type channels. In differentiated IMR32 cells possessing 20% L-type and 80% N-type channels (Carbone, Sher & Clementi, 1990), the action of OXO and NA is almost absent in cells pre-treated with ω-CTx-GVIA and thus deprived of N-type channels (Fig. 2-left). The same is true for most, but not all, peripheral neurons (Elmslie et al., 1990; Kasai, 1992; Boland & Bean. 1993). L-type channels appear little inhibited in a time-dependent manner by neurotransmitters and the slow-down of channel activation is always additive to the action

Figure 1. Inhibition of HVA currents by adrenergic and muscarinic agonists in human neuroblastoma IMR32 cells. Reversible effects of noradrenaline (NA, 10 μM) and oxotremorine (OXO, 10 μM) on HVA currents at +10 mV in 10 mM Ba^{2+}. The traces were recorded before (C) and after (R) short exposures (1 min) of the cell to the neurotransmitters. Notice the appearance of a slow phase of activation and the absence of inactivation during application of the neurotransmitter. Holding potential -90 mV.

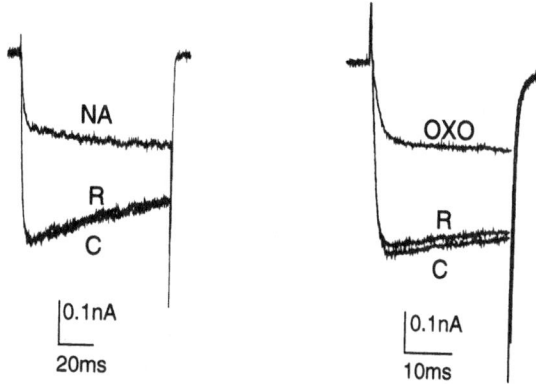

of L-type blockers. This suggests a selective action of neurotransmitters on non-L-type channels. In Fig. 3, oxotremorine action is preserved independently of whether L-type channels are either functioning (left) or blocked by nifedipine (right). The dihydropyridine (DHP) reduces the size of the current without affecting its modified activation-inactivation kinetics by oxotremorine.

The little effects of neurotransmitters on L-type channels observed in some neurons is not, however, a general rule. Inhibitions of L-type channels by noradrenaline, opioids, GABA and ACh have been reported in sympathetic neurons (Bley & Tsien, 1990; Mathie, Bernheim & Hille, 1992), insulin secreting cells (Pollo, Lovallo, Biancardi, Sher, Socci & Carbone, 1993), bovine chromaffin cells (Albillos, Carbone, Gandía, García & Pollo, 1995), hippocampal neurons (Scholz & Miller, 1991) and cerebellar granules (Amico, Marchetti, Nobile & Usai, 1995). In all cases the reversible inhibition of L-type channels caused a current depression with no significant change to the activation-inactivation kinetics.

3.2. Effects on Low-Threshold Channels

At variance with HVA channels, neurotransmitters and G protein activation have variable effects on the low-threshold (LVA, T-type) channels in most neurons. T-type currents may be either unaffected, enhanced or partially inhibited by different levels of G protein activation. An example of a weak action of neurotransmitter on LVA channels is shown in

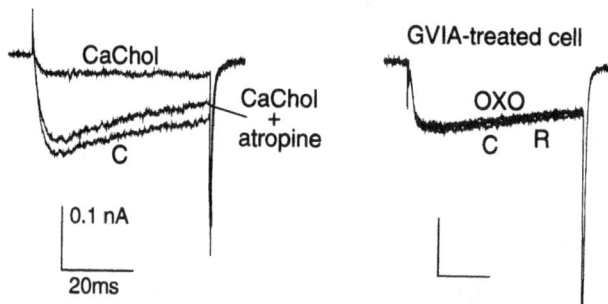

Figure 2. Muscarinic inhibition of HVA currents in IMR32 cells is selective for ω-CTx-GVIA-sensitive (N-type) channels. *Left:* The inhibitory action of carbachol (1 μM) on HVA currents is antagonized by addition of 1 μM atropine to the solution containing the agonist. Test depolarization to +10 mV from -90 mV in 10 mM Ba^{2+}. *Right:* 10 μM oxotremorine has no effect on HVA currents persisting after cell pre-treatment with ω-CTx-GVIA (3.2 μM for 15 min in 2 mM Ca^{2+}).

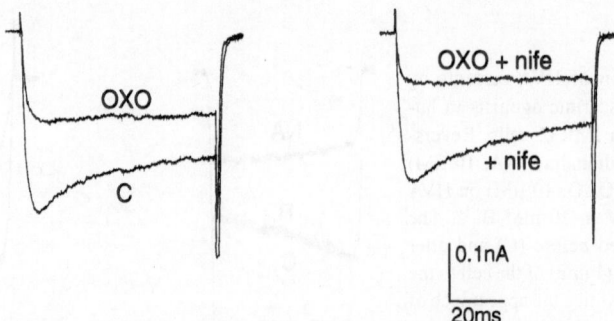

Figure 3. Oxotremorine action is additive to the blocking effects of nifedipine. *Left:* Inhibition of HVA currents by 10 μM oxotremorine (OXO). Test depolarization to +10 mV from -90 mV. *Right:* Addition of nifedipine (*nife*, 10 μM) to the same cell causes a partial depression of the Ba^{2+} current but does not prevent the action of oxotremorine (*OXO + nife*).

the inset of Fig. 4. NA causes a marked depression of the current activating at high voltages (with a maximum around +5 mV) but preserves the early small current originating at low voltages (with a maximum around -25 mV). Small effects on LVA channels have been reported in other neurons using several neurotransmitters (Cox & Dunlap, 1992), but there are clear exceptions to this rule. For instance, dopamine (Marchetti et al., 1986) and opioid agonists (Kasai, 1992; Fisher & Bourque, 1995; Formenti, Arrigoni, Bejan, Avanzini & Mancia, 1995) may cause a significant depression of LVA currents without affecting their time course in peripheral as well as in central neurons.

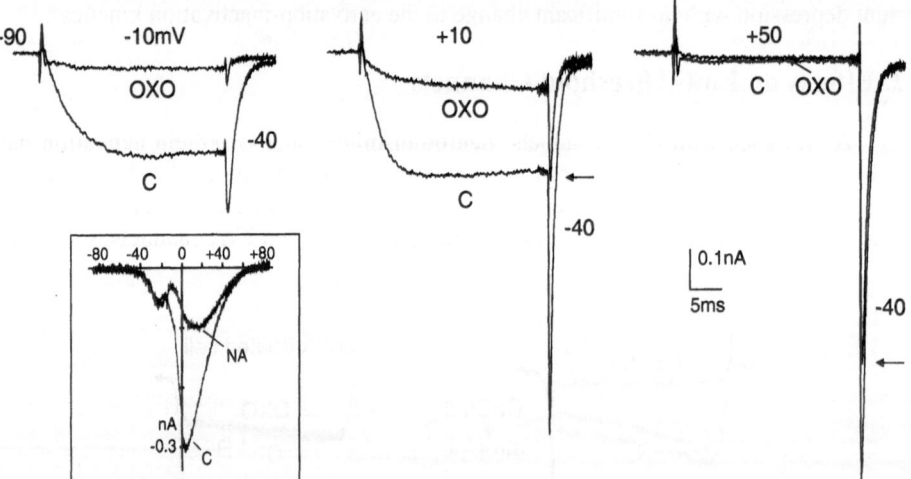

Figure 4. V-dependent inhibition and facilitation of HVA Ba^{2+} currents by oxotremorine and noradrenaline in IMR32 cells. The traces were recorded before (C) and during application of 10 μM OXO in 10 mM Ba^{2+} at the potentials indicated. Step repolarizations to -40 mV; holding potential -90 mV. Tail currents were recorded at a faster sampling rate (20 μs/point) with respect to the current relaxation at test potentials (50 μs/point). The arrows on the middle and right-hand panel indicate the peak amplitude of the "facilitated" tail current with OXO. *Inset:* Selective action of noradrenaline on HVA currents. The traces were recorded in response to ramp commands with a slope of 0.55 mV/ms from -90 mV holding potential in 10 mM Ba^{2+}. The I/V relationships show an early peak at -25 mV (LVA) that is unaffected by 10 μM NA and a second large peak at about +5 mV (HVA) that is strongly depressed by the neurotransmitter. Recordings were corrected for Cd^{2+}-insensitive currents persisting after addition of 200 μM Cd^{2+}.

4. NEUROTRANSMITTER INHIBITION AND VOLTAGE-DEPENDENT FACILITATION

Neurotransmitter inhibition of HVA currents is time- and voltage-dependent. The depression is marked at low voltages and partially relieved at higher depolarizations. In the inset of Fig. 4, NA inhibition in IMR32 cells is more potent at 0 mV (75%) than at +50 mV (28%) and the modified I/V curve is shifted to the right. The time-dependent removal of inhibition with voltage is illustrated by the recordings of the same figure. At -10 mV, depression by OXO is strong (87%) and remains unchanged during the 30 ms step depolarization. At +10 mV, the current reduction is maximal soon after the onset of the pulse but is significantly relieved at the end of the pulse. The modified tail current on return to -40 mV is depressed by only 67% (arrow). At +50 mV, the time course of the inhibited current is hardly separable from the control trace and the inhibition is largely relieved during the 30 ms step depolarization (facilitation). The corresponding tail current is reduced by only 27% (arrow).

Voltage-induced "facilitation" may be significant (>80% at +90 mV) but is usually partial, leaving a residual depression unrecovered by voltage (V-independent). This raises several important questions. Are the V-dependent and V-independent inhibitions controlled by different mechanisms (G proteins or 2nd messengers)? Are they associated to different Ca^{2+} channels? How accurately can the percentage of V-dependent facilitation be measured? Fig. 5 shows a typical voltage protocol used to determine the amount of V-dependent facilitation of Ca^{2+} currents exposed to saturating doses of neurotransmitter (Grassi & Lux, 1989; Elmslie et al., 1990). The current recorded during the test pulse (+10 mV) from the holding potential (trace 1) is compared to the current relaxation at the same potential preceded by a 50 ms step depolarization to +90 mV (trace 2). In control conditions the two current relaxation (1 and 2) are coincident if allowance is made for trace 1 to reach its peak value. With 10 μM OXO the two traces (3 and 4) deviate markedly (Fig. 5B). There is 71% inhibition when comparing the two traces without pre-pulses at the peak (vertical arrows in Fig. 5C), but there is only 16% residual inhibition when comparing the amplitude of trace 2 and 4 at the end of the facilitating pre-pulse (horizontal arrows in Fig. 5D). This is a relatively high degree of facilitation with pre-pulses and, in our view, is the most accurate way to evaluate the percentage of V-dependent modulation. Protocols using a short repolarization (5 to 10 ms) to -90 mV interposed between the pre-conditioning and the test pulse usually underestimate the percentage of facilitation for two reasons. Channel re-inhibition after facilitation develops in a concentration dependent manner with a time constant (τ_{reinh}) of 30 to 50 ms at the holding potential (Lopez & Brown, 1991; Golard & Siegelbaum, 1993; Elmslie & Jones, 1994) and may cause a 28 to 18% channel re-inhibition after 10 ms, respectively. In addition, channel activation at test potentials around 0 mV further delays the estimate of the current facilitation by 3 to 5 ms, which is the time required to reach the peak. Thus, in double pulse protocols with 10 ms repolarization to -90 mV and test pulse to +10 mV, the estimate of channel facilitation at the peak of the facilitated current is delayed by about 13 to 15 ms after the end of the facilitating stimulus. This causes a 40 to 25% amplitude reduction of the facilitated current and a consequent proportional underestimate of the V-dependent inhibition. We think therefore that V-dependent facilitation of Ca^{2+} channels is more accurately estimated at the end of the facilitating pre-pulse. The neurotransmitter-induced depression persisting after pre-pulse represents the best estimate of the V-independent fraction of inhibition not recovered by voltage.

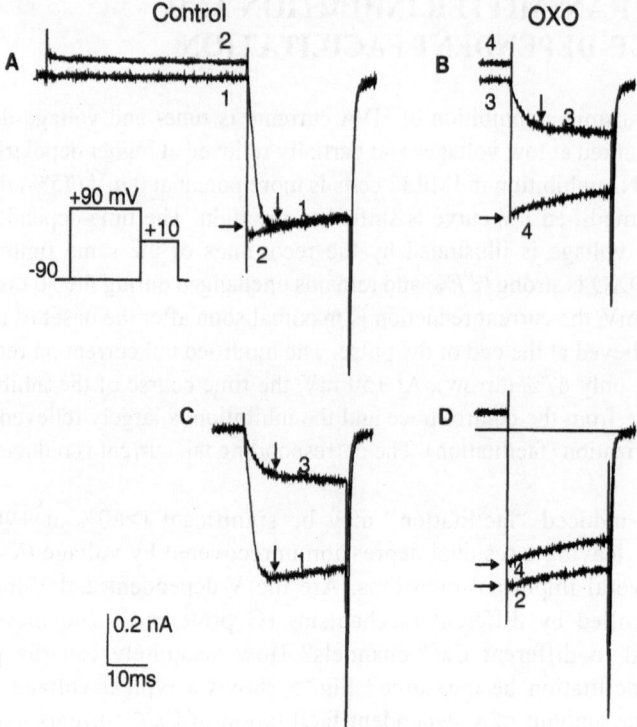

Figure 5. Prepulse-induced facilitation of muscarinic inhibition in IMR32 cells. The two overlapping traces in panel *A* and *B* were recorded without (*trace 1* and *3*) and with (*trace 2* and *4*) a 50 ms conditioning pre-pulse to +90 mV. Test depolarizations to +10 mV; holding potential -90 mV. The double-pulse protocol was delivered before (*A*) and during (*B*) application of 10 μM OXO, with 3 s interval between pulses. The vertical arrows on *trace 1* and *3* indicate the time at which the inhibition is estimated. The horizontal arrows on *trace 2* and *4* indicate the amplitude of the facilitated current relaxation to +10 mV. *Trace 1* and *3* before pre-pulse are shown overlapped in panel C. *Trace 2* and *4* after pre-pulse are shown in panel D. The percentage of voltage-independent inhibition estimated from the ratio of *trace 2* and *4* is 16% while the total inhibition (ratio of trace 1 and 3) is 71%.

5. VOLTAGE-DEPENDENT VERSUS VOLTAGE-INDEPENDENT MODULATION

Variable degrees of V-independent inhibition of Ca^{2+} currents have been reported on most peripheral and central neurons. The steady-state depression insensitive to voltage is commonly associated to N-type channels (Luebke & Dunlap, 1994) but there is also evidence for a similar action on L-type channels in insulin-secreting cells (Pollo et al., 1993), chromaffin cells (Albillos et al., 1995), peripheral and central neurons (Bley & Tsien, 1990; Scholz & Miller, 1991) as well as on T- (Marchetti et al., 1986; Kasai, 1992) and P-type channels (Swartz, 1993). In insulin-secreting and chromaffin cells, for instance, NA and opioids induce both V-independent and V-dependent inhibition that can appear either in isolation or in combination (Pollo et al., 1993; Albillos et al., 1995). V-independent inhibition is largely removed by applications of saturating doses of nifedipine sparing the V-dependent component which is mainly associated to non-L type channels. The same is true for the GABA-induced depression of Ca^{2+} currents in cerebellar granule cells (Amico et al., 1995).

Steady-state inhibition by neurotransmitters has been mostly investigated on N-type currents. It is preserved in the presence of DHP-antagonists and contributes significantly in neurons expressing a small percentage of L-type channels (Cox & Dunlap, 1992). So far, separation of the V-dependent and V-independent depression was mostly based on qualitative grounds. Both pathways are controlled by G proteins, as intracellular GTP-γ-S mimics the two actions and GDP-β-S prevents them. In our experience, however, internal GDP-β-S is extremely effective in preventing the slowing of channel activation but spares some steady-state inhibition by addition of neurotransmitters. This, very likely, is the first direct evidence that the two modulations develop partially through distinct pathways. The second evidence is the lack of correlation between the two effects. Although the presence of channel inactivation does not often allow a clear separation of the slow activation phase from steady-state inhibition, the two modulations may be separately estimated by the conditioning pre-pulse method described above. Other evidences are: 1) The two phenomena have sharply different voltage dependency (Luebke & Dunlap, 1994), 2) Different neurotransmitters cause a prevalence of either the V-dependent or V-independent inhibitions in the same cell (Formenti, Arrigoni & Mancia, 1993), 3) V-dependent and V-independent modulations are regulated by different G protein subunits (Diversé-Pierluissi, Goldsmith & Dunlap, 1995). Kinetic slowing induced by α_2-adrenergic and GABA$_B$ agonists in chick sensory neurons are apparently controlled by G$_o\alpha$ subunits with no involvement of protein kinase C (PKC) (but see below Swartz, 1993; Zhu & Ikeda. 1994a), while steady-state inhibition through α_2-adrenergic receptors is mediated by G$_i\beta\gamma$ subunits and PKC activation. In conclusion, there is evidence for the existence of two modulatory pathway which may sub serve different cellular functions.

The existence of separate V-dependent and V-independent down-modulations of Ca^{2+} currents increases the degree of control of Ca^{2+}-dependent cell functions by endogenous or exogenous membrane receptor agonists. V-dependent inhibitions are expected to be attenuated upon repeated cell stimulation and may be useful in the phasic or use-dependent control of cell function. V-independent depression may be effective in the tonic inhibition of neurotransmitter and hormone secretion, even in the presence of high frequency stimulation or prolonged cell depolarizations.

6. RECONSTITUTED Ca^{2+} CHANNEL MODULATION IN OOCYTES

V-dependent Ca^{2+} channels are oligomeric structures composed of α_1, β, γ and disulfide-linked α_2/δ subunits. The α_1-subunits accounts for most of the channel properties: gatings, ion permeability and toxin sensitivity (see Catterall, 1991). The class A and B α_1-subunit (α_{1A} and α_{1B}) are also shown to be directly modulated by PTX-sensitive G proteins activated by seven-helix transmembrane receptors coexpressed with the α_1-subunits (Bourinet, Soong, Stea, Dubel, Yu & Snutch, 1995). Modulation of α_{1A} and α_{1B} by activated G proteins appears similar to that observed on native Ca^{2+} currents and consists of a marked current inhibition followed by a prolongation of channel activation. In other words, the α_{1A}- and α_{1B}-subunits are the direct target of the activated G proteins and the co-expression of the two subunits with endogenous G proteins allows to mimic the membrane delimited voltage-dependent inhibition of Ca^{2+} channels induced by neurotransmitters in most neurons. Bourinet et al., showed also that transient co-expression of a β-subunit reduces dramatically the V-dependent inhibition and facilitation of α_{1A} and α_{1B} Ba^{2+} currents induced by μ-opioid agonists in Xenopus oocytes. This suggests a regulatory role of β-subunits on Ca^{2+} channel gatings; in line with the idea that β subunits accelerate the activation and inactivation gatings of α_{1C} Ca^{2+} channels (Lacerda et al., 1991).

A regulatory action of endogenous β-subunits on the G protein/Ca^{2+} channel coupling has been also suggested by experiments on Ca^{2+} channel current modulation in sensory neurons using antisense oligonucleotide against β-subunits (Berrow, Campbell, Fitzgerald,

Brickley & Dolphin, 1995). Cell pre-treatment with β-subunits antisense caused a marked Ca^{2+} current decrease in rat DRGs. The reduced currents (N- and L-type), however, were inhibited by $GABA_B$ receptors activation more potently than control currents, suggesting competitive interaction between activated Gα-subunits (or Gβγ) and the Ca^{2+} channel β-subunit for an intracellular site of the channel α_1-subunit. On the light of these results three possible interactions between $G_o\alpha$-, β- and α_1-subunits are postulated: *i*) the β-subunit binds to Gα (or Gβγ), preventing α_1 channel inhibition, *ii*) Gα and β compete for the same site at the α_1-subunit. The presence of functional β-subunits prevents the inhibitory action of Gα, *iii*) Gα and β bind to different sites of the α_1-subunit and the absence of β enhances allosterically the affinity of Gα to the α_1-subunit.

The above findings highlight the regulatory role of β-subunits in the Ca^{2+} channel functioning and open new interesting perspectives for the channel modulation by neurotransmitters. Some of them are important to understand the molecular details of channel gatings. It may be that the β-subunit is the phosphorylation site of intracellular second messengers, like protein kinase C whose activation is shown to prevent the V-dependent inhibition and facilitation of Ca^{2+} channels (see par. 11). It may also be that the degree of β/α_1-subunit coupling, which affects the channel activation-inactivation gatings, depends on the ion flowing through open Ca^{2+} channels (Ca^{2+}, Ba^{2+} or Na^+). This would explain for instance the lack of G proteins effects on Na^+ currents compared to the Ca^{2+} and Ba^{2+} currents which are effectively modulated by neurotransmitters (see par. 9). Future work will certainly clarify some of these issues.

7. MODELS FOR THE V-DEPENDENT INHIBITION AND FACILITATION

Independently of the role of β-subunits, the voltage-dependent down-modulation of native HVA currents by neurotransmitters is usually ascribed to changes in channel gating mediated by activated G-proteins. Internal GTP-γ-S mimics the current depression by external neurotransmitters, while GDP-β-S prevents this action. Identified G proteins (see Dolphin, 1995) are postulated to interact with Ca^{2+} channel gating in a V-dependent manner. Low negative potentials stabilize the binding of the G protein to the channel while strong positive depolarizations favour the normal gating conditions. It is still unclear how the voltage can affect this "protein-protein" interaction. It could be that membrane voltage affects directly the coupling and uncoupling of the two macromolecules by acting on a voltage sensor on the G protein (Swandulla et al., 1991) or that the "voltage-dependency" results from a permanent modification of Ca^{2+} channels, in slow equilibrium with unmodified channels (Kasai,. 1992; Pollo et al., 1992; see scheme 1 and 2 in Fig. 6). In the first cases the modified (M*) channels are facilitated by the voltage-dependent rate constants γ and δ that regulate the modified gating mode. k_1 and k_{-1} are the rate constants regulating the slow equilibrium between normal and modulated channels (see legend Fig. 6). In the second case the facilitation of inhibited channels (H*) to the normal gating mode (C↔O) is further favoured by fast state transitions with voltage-independent rate constants (μ and υ) (Pollo et al., 1992). State transitions are assumed to occur most favourably from open (4υ) rather than from closed (υ) modified channels (Elmslie et al., 1990). "Voltage dependency" may derive also from the reduced affinity of the G protein for open channels and would be, therefore, a consequence of the voltage-sensitivity of channel gatings (Boland & Bean, 1993; Golard & Siegelbaum, 1993). In this case, the inhibited channel is thought to shift from a "reluctant" to a "willing" gating mode during membrane depolarizations (scheme 3). The "willing" mode has increased occupational probability as the G protein dissociates either

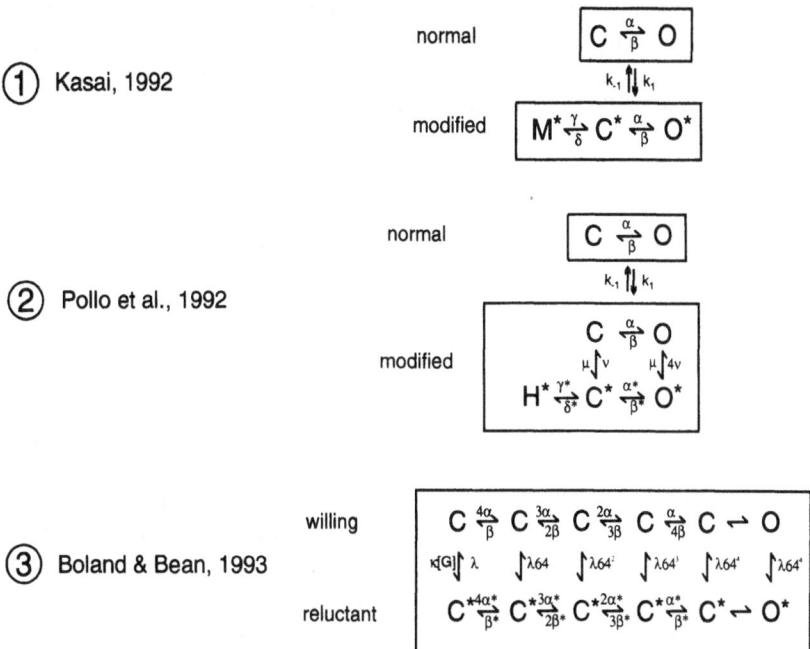

Figure 6. Kinetic models for the Ca^{2+} channel gating modulation by neurotransmitters. In scheme 1, C and O represent the closed and open states of the normal channel which are in equilibrium through the voltage-dependent rate constants α and β (Kasai, 1992). C* and O* are the corresponding closed and open states of the modified channel. M* is the modified (non-conductive) state of the channel and γ and δ are the rate constants producing the slow activation of the channel during depolarization. Scheme 2 is from Pollo et al., 1992. At variance with model 1, the inhibited state of the channel (H*) is allowed to equilibrate with the normal gating mode (C↔O) through the fast transition rate constants μ and υ taken from Elmslie et al., 1990. k$_1$ and k$_{-1}$ are the rate constants regulating the slow equilibrium between normal and modified. channels. They are estimated from the onset and offset of Ca^{2+} channel inhibition during application of saturating doses of neurotransmitters. Scheme 3 is derived from Boland & Bean, 1993. The "willing" and the "reluctant" gating modes are in equilibrium through slow equilibrium rate constants as indicated (see text). The activation kinetics in each mode obey an Hodgkin-Huxley formalism with 4 gating subunits (m^4).

from the closed or from the open state. The off-rate of G protein unbinding (γ) is independent of the activated G protein concentration, [G], but increases by a factor 64 at each sequential activation of one of the 4 channel gating domains, to reach a maximum of 64^4 γ when the channel is fully open (O*). The "reluctant" mode is favoured by G protein binding through the on-rate κ [G] as the channel deactivates on return to resting potentials.

All the above models explain some but not all the features of voltage-dependent inhibition and facilitation. For instance, scheme 1 and 2 account nicely for the time course of facilitation at different neurotransmitter concentrations, for the recovery of channel activation after pre-pulse and for the concentration- and V-independent re-inhibition of open channels. The two models account for the slow equilibrium kinetics (k$_1$, k$_{-1}$) regulating the partial block of N-type channels at saturating doses of neurotransmitters. The two scheme, however, have some shortcomings. The most serious one is that channel activation must always follow channel facilitation through transition M*→O* (scheme 1) or H*→O* (scheme 2). In other words, activation cannot be faster than facilitation while channel activation may develop 5 times faster than facilitation at very positive potentials (+130 to +150 mV) (τ$_{act}$ 1 ms and τ$_{facil}$ 5 ms; Boland and Bean, 1993). This can be partly overcome in scheme 2 by re-designing the on-off rates γ and δ, thus allowing the channels in state H* to reach state O and O* in 1 ms for very positive potentials and then wait for the facilitating transition O*→O to occur with 5 ms time constant (Elmslie et al., 1990). Scheme 3 overcomes this drawback and accounts for most of the kinetic features of the voltage-dependent inhibition and facilitation but fails to mimic the partial block of N-type channels at saturating doses of neurotransmitters and the concentration-independent time constant of slow activation at different doses of neurotransmitters (Kasai, 1992; Pollo et al., 1992; Elmslie & Jones, 1994). While giving a detailed and reasonable account of the G-protein-Ca^{2+} channel interaction the Boland & Bean's model is unable to match the slow equilibrium kinetics between receptor activation and channel inhibition developing within seconds and the fast relaxation between activated G proteins and inhibited Ca^{2+} channels occurring within tens of ms.

New experimental findings have further complicated the overall picture of V-dependent modulation and have made difficult an updated description of the phenomena in terms of kinetic schemes. Thus, rather than proposing new models we find more reasonable to list the relevant points that should be satisfied: 1) V-dependent depression by neurotransmitter is only partial (60 to 75% maximal inhibition) and concentration dependent, 2) Activation of modified channels is slow at around -10 mV and accelerates steeply with voltage (Marchetti et al., 1986). The activation kinetics of modified channels can be very slow at some potential (τ_{act} 40 to 70 ms) and cannot be attributed to a simple voltage shift of the normal channel gating; 3) The amplitude *but not the time constant* of the slow activation is concentration-dependent (Kasai, 1992; Pollo et al., 1992, Elmslie & Jones, 1994). This implies that although the current depression develops through a slow dose-dependent equilibrium reaction, the fast V-dependent removal of inhibition is paradoxically independent of [G]; 4) Facilitation is rate limiting the activation of Ca^{2+} channels at potentials below +60 mV (Pollo et al., 1992). Above +70 mV, activation develops faster ($\tau_{act} \approx 1$ ms) than facilitation ($\tau_{fac} \approx 5$ ms) (Elmslie & Jones, 1990, Boland & Bean, 1993). Thus, modified and unmodified channels first activate and then facilitate, supporting a model in which G proteins dissociate more easily from open than from closed channels (Boland & Bean, 1993). 5) Channel re-inhibition after facilitation is fast (τ_{reinh} 30 to 50 ms) and [G]-independent at potentials around 0 mV where channels are preferentially open (Kasai, 1992; Pollo et al., 1992) but is [G]-dependent at very negative potentials (-80 mV) where channels are mostly closed (Golard & Siegelbaum, 1993; Elmslie & Jones, 1994). This point introduces two complications that none of the proposed models satisfy. Equilibrium of facilitated and re-inhibited channels occurs on a faster time scale (tens of ms) compared to the slow onset of inhibition during application of external neurotransmitters. Onset and offset of inhibition occur on a slower time scale independently of whether the channel is preferentially closed (0.5 s) or open (0.3 s) (see Fig. 4 in Pollo et al., 1992) This suggests that the reactions leading to G proteins activation by receptor occupation is rate limiting the true G protein-Ca^{2+} channel interaction that occurs on a faster time scale when the G-protein is activated. The [G]-dependent re-inhibition at -80 mV implies that the on-rate of G protein binding is driven by mass law and is either favoured by negative voltages or by the closed state of the channel (Elmslie & Jones, 1994). Alternatively, the [G]-independent re-inhibition around 0 mV may be favoured by the less negative potential or by the open state of the channel. The last point sets strict constraints to the model accounting for the "voltage-dependent" modulation of HVA channels. As none of the proposed models satisfies all the above conditions it is reasonable to believe that with proper modifications they could converge to a more general one in which concentration-dependence, voltage-dependence and time course of inhibition, facilitation and re-inhibition are all well fit.

8. Ca^{2+} CHANNEL MODULATION VIEWED THROUGH SINGLE CHANNEL RECORDINGS

The V-dependent and V-independent inhibition by neurotransmitters should also find a correspondence at the single channel level. To date, cell-attached recordings in high Ba^{2+} solutions have shown that neurotransmitters cause a marked reduction of open channel probability interpreted as a shift from a high- (or medium-) to a low-p_o mode mediated by the activated G protein (Delcour & Tsien, 1993). The absence of effects on single channel conductance supports the view that neurotransmitters mainly affect channel gatings rather than channel permeability (Lipscombe, Kongsamut & Tsien, 1989; Shen & Suprenant; 1991; Toselli & Taglietti, 1994). The mode switching model proposed by Tsien's group satisfies most of the observations on the voltage-independent inhibition of macroscopic currents but is unable to explain the fast kinetics of the voltage-dependent facilitation and re-inhibition ($\tau \approx 20$ ms) observed in whole-cell recordings at positive voltages (Fig. 1). If mode switching is conditioned by G proteins binding and unbinding and the average sojourn of the channel in one mode is about 10 s, it seems unlikely that the fast voltage-dependent unbinding of the G protein from the channel in the low-p_o mode causes a shift to the high-p_o mode and produces a current facilitation within 20 ms at positive potentials. Indeed, we were able to prove that, in cell attached patches containing 20 μM NA and 1 μM DPDPE, channel openings are preferentially delayed with a fivefold increase of the latency to first openings with respect to control patches. In addition, test depolarizations to +20 mV preceded by a 50 ms step depolarization to +90 mV allow the recovery of single channel activity by reducing the first latency of openings and increasing p_o (Carbone, Carabelli, Lovallo, Zucker & Magnelli, 1995). The same is true for the noradrenergic modulation of single N-type channels in frog sympathetic neurons (Elmslie & Kelly, 1995). This suggests that also at the single channel level V-dependent facilitation and re-inhibition occur on a fast time scale as predicted by most kinetic models. The absence of serious discrepancies between whole-cell and single channel measurements should therefore stimulate further comparisons of the results by the two methodologies in other neurons, this could possibly overcome some born limitations of cell-attached recordings that do not allow to run test and control experiments on the same channel.

9. CHANNEL GATING VERSUS ION PERMEATION MODULATION

Most groups agree that channel inhibition and facilitation by neurotransmitters is a consequence of channel gating modulation. A recent report, however, proposes that neuro-transmitters inhibition may derive from a partial reduction of channel permeability to divalent cations (Kuo & Bean; 1993). Activated G proteins are thought not only to interfere with the channel subunits responsible for channel gating but also to modify the energy profile controlling ion passage through open channels. Evidence in favour of this idea is based on the following observations: a) G protein inhibition of N-type currents is less effective when the channel carries either inward Na$^+$ or outward Cs$^+$ instead of Ba^{2+}, b) Inhibition is attenuated also when [Ba^{2+}] increases from 3 to 150 mM, c) In the presence of intracellular GTP-γ-S, the single channel conductance estimated by noise analysis on tail currents to -40 mV results 25% smaller if measured before a strong facilitating pre-pulse rather than after. Kuo & Bean propose that the activated G protein at the inner site of the membrane produces a conformational change of the negative protein charges at the outer site of the channel where two high-affinity binding sites for Ba^{2+} (or Ca^{2+}) are located (Kuo & Hess, 1993). The two binding sites are close enough to produce high fluxes of divalent cations by ion-ion repulsion with millimolar $[Ca^{2+}]_o$ or block of Na^+ currents by micromolar $[Ca^{2+}]_o$ (Almers & McCleskey, 1984). In low $[Ba^{2+}]$, the strong binding of divalent cations to the negative sites is partly destabilized by the conformational change induced by the G protein, reducing the effectiveness of ion-ion interactions and consequently the size of Ba^{2+} current. The conformational change induced by the G protein is apparently less critical when either Na^+ or high $[Ba^{2+}]$ are involved. In the first case, Na^+ or Cs^+ ions are expected to be loosely bound to the negative sites of the pore and ion-ion interaction weakly affected by the displaced charges. In the second case, both sites would be occupied by Ba^{2+} ions and the ligand displacement induced by the G protein would result less effective. Ba^{2+} fluxes in 150 mM Ba^{2+} would result weakly depressed by the activated G proteins.

Kuo & Bean's interpretation is very attractive, but requires some consideration. For instance, neurotransmitters do not affect significantly the single channel conductance in 100 mM Ba^{2+} in spite of a marked depression of macroscopic currents (Lipscombe et al., 1989; Shen & Suprenant, 1991; Delcour & Tsien, 1993; Toselli & Taglietti, 1994). The same is true when comparing single channel current amplitude before and after facilitating prepulses in the presence of saturating doses of neurotransmitter (Carbone et al., 1995). In addition, in IMR32 cells we find that Ba^{2+} current depression induced by 20 μM NA in 100 mM Ba^{2+} is only 10 to 15% smaller than that in 5 mM Ba^{2+} while, in agreement with Kuo & Bean. NA inhibition of Na^+ currents through Ca^{2+} channels is far more attenuated (Carabelli, Lovallo & Carbone, unpublished results). Thus, it seems that for G protein-Ca^{2+} channel interaction is more relevant whether the channel conducts Na^+ or Ba^{2+} rather than how many Ba^{2+} ions flow through open channels. How this can occur remains to be understood. There are several alternatives that can be tested before drawing any conclusion. One is that Na^+ and Ba^{2+} permeation through open channels may be regulated by an intramembrane binding site whose energy profile depends more critically on the type of ion flowing (Lux, Carbone & Zucker, 1990) than the energy profile of the two binding sites located at the outer site of the pore (Almers & McCleskey, 1984; Kuo & Hess, 1993). A differential G protein-induced conformational change of the intramembrane binding site for Na^+ and Ba^{2+} would also explain the reduced inhibition of Na^+ currents by neurotransmitter. May be the β/α_1 subunits interaction, crucial for Ca^{2+} channel inhibition, is Ca^{2+}-dependent or regulated by a divalent cation-sensitive second messenger phosphorylation reaction. Alternatively, it might be that the weaker inhibition of Na^+ current by G proteins derives from an increased gating activity of the channel in low Ca^{2+} solutions. If the G protein-Ca^{2+} channel interaction depends on the state of the channel (open or closed), a possible increased flickering of the channel with Na^+ as the main monovalent cation will facilitate the unbinding of the G protein from the more frequently open channel, with consequent relieved inhibition. These and other possibilities need to be tested at the single channel level with solutions of different Na^+ and Ba^{2+} concentrations (10 to 100 mM) that allow accurate estimates of Ca^{2+} channel permeability.

10. Ca^{2+} CHANNEL MODULATION AND CELL DIALYSIS

An issue of interest concerns the role that cell dialysis may play in the maintenance of the V-dependent kinetic slowing of Ca^{2+} channels by neurotransmitters in whole-cell clamped neurons. Somatostatin-induced inhibition of Ca^{2+} currents is reported to be V-dependent with little kinetic slowing when tested on a ciliary ganglion neuron in perforated-patch conditions (Meriney, Gray & Pilar, 1994). Somatostatin action is increased and the kinetic slowing is more prominent when Ca^{2+} currents are recorded from whole-cell clamped neurons. The interpretation of Meriney et al. is that kinetic slowing and voltage dependency are unrelated phenomena and that kinetic slowing, but not the voltage dependency, depends on the wash out of some intracellular factor (a cGMP-dependent protein kinase) that prevents

Ca^{2+} channel inhibition in intact cells. In contrast to this, however, early works on the action of several 2nd messengers have shown that neither cAMP, nor cGMP or IP$_3$ are involved in HVA channel modulation (see Hille, 1994). In addition, Ca^{2+} current recordings in perforated patches show strong voltage dependency and a significant slow down in the presence of external neurotransmitters (Cesare et al., 1994; Luebcke & Dunlap, 1994). Also single channel recordings in cell-attached conditions display clear kinetic slowing of channel activation and a significant voltage-dependent recruitment of inhibited channels by conditioning prepulses (Carbone et al., 1995; Elmslie & Kelly, 1995), suggesting that cell dialysis may not be as critical as indicated. In our view, kinetic slowing depends critically on the time course of channel inactivation, that is faster and more complete in perforated patch than in whole-cell conditions, due to the absence of strong intracellular Ca^{2+} buffers. Fast inactivation may easily distort the slowing down of channel activation that would appear as a steady-state inhibition during neurotransmitter application. Under these conditions it would be wise to normalize the time course of modified currents to the time course of control currents (Kasai, 1992; Elmslie & Jones, 1994). What remains unresolved, in all cases, is the mutual interference of activation and inactivation gating with Ca^{2+} channel modulators (Gα_o subunit versus the α_1/β subunits interactions; see par. 6).

11. PKC ACTION ON Ca^{2+} CHANNEL MODULATION

Given the well established role of G proteins in controlling the voltage-dependent inhibition of high-threshold Ca^{2+} channels by activation of neurotransmitter receptors (see Hille, 1994; Dolphin, 1995), a related question is whether other intracellular messengers are involved in the receptor-Ca^{2+} channel coupling. To date, protein kinase C (PKC) is considered one of the most effective intracellular modulator of neuronal Ca^{2+} channels. The enzyme is activated by increased amounts of diacylglycerol produced by the neurotransmitter-induced hydrolysis of membrane phospholipids (see Nishizuka, 1995). Early works on the PKC action on HVA channels have shown that PKC activation by external application of either diacylglycerol analogues or phorbol esters or by intracellular injection of purified kinases may cause up- or down-regulation of Ca^{2+} currents with little specificity for a Ca^{2+} channel subtype (see Shearman, Sekiguchi, Nishizuka, 1989). Only recently a selective action on different Ca^{2+} channels could be resolved; for instance, the T-type current of rat DRGs is selectively depressed by PKC activation (Schroeder, Fischbach & McCleskey, 1990).

The effects of PKC activation have been studied mostly in the whole-cell recording configuration, i. e., in internally dialyzed cells in which the amount and activity of intracellular messengers may be altered; thus raising some doubts about the physiological relevance of the approach (see Anwyl, 1991). Although the results were not fully consistent with the predictions based on whole-cell experiments, recent studies carried out in cell-attached and perforated patch configuration confirmed that PKC-mediated modulation occurs also in less disturbed conditions (O'Dell & Alger, 1991; see Anwyl, 1991), supporting a physiological rationale to the phenomenon. Single channel recordings in rat hippocampal neurons (O'Dell & Alger, 1991) and frog sympathetic neurons (Yang & Tsien, 1993) revealed that PKC-mediated enhancement of L- and N-type currents is due to an increase of the open channel probability with little or no change to their single channel conductance. In addition, experiments in perforated-patch conditions confirmed that the PKC-mediated Ca^{2+} current increase may be accompanied by acceleration of the current inactivation (Zhu & Ikeda, 1994a).

11.1. Direct Coupling between PKC and Ca²⁺ Channels

Despite most studies have shown the involvement of PKC in neurotransmitter-induced Ca^{2+} channel modulation, only in few cases the PKC-mediated phosphorylation was proven to be directly involved in the Ca^{2+} channels modulation by neurotransmitters. More frequently, PKC exerted only a regulatory action on neurotransmitter effect.

Direct involvement of PKC in receptor-Ca^{2+}channel coupling is based on the observation that: i) PKC activation mimics and prevents the Ca^{2+} channel modulation by neurotransmitter and ii) PKC inhibitors or PKC down-regulation by chronic phorbol ester treatment abolish the effects of neurotransmitter. Examples of this mechanism include the Ca^{2+} channel inhibition by both neuropeptide Y in rat sensory neurons (Ewald, Matthies, Perney, Walker & Miller 1988) and noradrenaline in embryonic chick DRGs (Diversé-Pierluissi & Dunlap, 1993). The latter study also shows that Ca^{2+} channel modulation by neurotransmitter within a single cell may proceeds through different pathways, not necessarily involving PKC activation. In embryonic chick DRGs, for instance, noradrenaline and GABA inhibit N-type Ca^{2+} channels via a PTX-sensitive G protein, but only the noradrenaline effect is directly mediated by PKC.

11.2. PKC Modulation of Ca²⁺ Channel Inhibition by Neurotransmitters

Recent data from different neurons show that neurotransmitter inhibition of Ca^{2+} channels may be partially relieved by PKC activation. Examples include the disruption of: i) glutamate effects in rat hippocampal neurons (Swartz, Merritt, Bean & Lovinger, 1993), ii) adenosine- and GABA$_B$-mediated effect in rat cortical, hippocampal and sensory neurons (Swartz, 1993), iii) noradrenergic-, muscarinic-, somatostatin-, and vasoactive intestinal polipeptide-mediated inhibition in rat sympathetic neurons (Swartz, 1993; Zhu & Ikeda, 1994a; Shapiro, Zhou & Hille. 1995) and, iv) somatostatin inhibition in chick sympathetic neurons (Golard, Role & Siegelbaum, 1993). The mechanism by which PKC interferes with neurotransmitters action is still unknown. In vertebrate central and peripheral neurons, PKC activation prevents the V-dependent inhibition induced by endogenous GTP or by intracellularly applied GTP-γ-S and the current enhancement induced by PKC activation is markedly attenuated by intracellular GDP-β-S (Swartz, 1993; Zhu & Ikeda, 1994a). These findings suggest that PKC-induced disruption of Ca^{2+} channel modulation by neurotransmitters is due to the removal of the G protein-mediated inhibition; but it is still unclear whether the target of the PKC-dependent phosphorylation is the G protein (Katada, Gilman, Watanabe, Bauer & Jacobs, 1985; Pyne, Freissmuth & Palmer, 1992) or the Ca^{2+} channel itself (Ahiljan, Striessing & Catterall, 1991). There is also evidence for a specific action of PKC on the neurotransmitter receptor (Golard et al., 1993).

Despite these uncertanties, however, it remains the physiological significance that these findings anticipate. Removal of Ca^{2+} channel inhibition induced by released neurotransmitter (or hormone) is an effective mechanism by which presynaptic terminals or secretory cells may enhance their efficiency. Whether this occurs through an increased cell activity (V-dependent facilitation) or through the activation of PKC seems irrelevant to the final goal, but underlines the possibility that multiple regulatory mechanisms with different genesis and time courses may variably lead to an increased synaptic efficacy or hormone secretion. Indeed, PKC activation by phorbol esters enhances neurotransmitter release in sympathetic neurons (Wakade, Malhotra, & Wakada, 1985; Majewski, Hoare & Murphy, 1995) and potentiates synaptic transmission in CA$_3$ hippocampal cells (Malenka, Madison & Nicoll, 1986) and neuromuscular junctions (Shapira, Silberberg, Ginsburg & Rahamimott, 1987).

Finally, it should be mentioned that parallel to the fast modulation of Ca^{2+} channel gatings, other forms of Ca^{2+} current facilitation may cause net increases of cytoplasmic Ca^{2+} by recruiting functional Ca^{2+} channels at the plasmalemma. Several neurons and neuroendocrine cells are shown to possess an intracellular pool of N-type Ca^{2+} channels available to recruitment at the cell surface by membrane depolarization and PKC activation (Passafaro, Clementi, Pollo, Carbone & Sher, 1994; Rogers, Passafaro, Richmond, Cooke & Sher, 1995). The same two kinds of stimuli facilitating Ca^{2+} channel gatings within seconds, are thus able to stimulate N-type Ca^{2+} channels translocation to the cell surface on a slower time scale (minutes), further proving the pathways complexity through which Ca^{2+} channel modulation may occur.

12. CONCLUSIONS

Much has been achieved in the past few years in the biophysics and physiology of Ca^{2+} channel modulation. Many groups have contributed to the identification of neuronal Ca^{2+} channel subtypes and to a better understanding of their gating modulation by neurotransmitters and hormones. Some groups have provided molecular descriptions of the different pathways through which Ca^{2+} channel modulation develops. Others have focused on the kinetics and biophysical features of the phenomena and others already succeeded in reconstituting simple pathways of Ca^{2+} channel modulation by cDNA recombinants in oocytes. Much, however, remains to be done. Little is known, for instance, on the role that V-dependent and V-independent Ca^{2+} channel modulation have in the autocrine control of cell exocytosis, or in the regulation of Ca^{2+}-dependent presynaptic inhibition and facilitation. It is predictable that combining the results of different methodologies we shall soon reach a more complete vision of the many processes involved in Ca^{2+} channel modulation and their utility for cell functioning.

ACKNOWLEDGMENTS

We wish to thank our close collaborators: Almudena Albillos, Antonio Garcia, Michele Lovallo, Antonella Pollo and Emanuele Sher for helpful discussions. We are also grateful to Emmanuel Bourinet for sending us a pre-print of his work.

REFERENCES

Ahlijanian, M. K., Striessing, J. & Catterall, W. A. (1991). Phosphorylation of an α_1-like subunit of an ω-conotoxin-sensitive brain calcium channel by cAMP-dependent protein kinase and protein kinase C. *Journal of Biological Chemistry* 266, 20192-20197.

Albillos, A., Carbone, E., Gandía L., García, A. G. & Pollo, A. (1995). Selective voltage-dependent and voltage-independent inhibition of calcium channel subtypes by opioids in bovine chromaffin cells. *Society for Neuroscience Abstract* 21, 1575.

Almers, W. & McCleskey, E. W. (1984). Non-selective conductance in calcium channels of frog muscle: calcium selectivity in a single-file pore. *Journal of Physiology* 353, 585-608.

Amico, C., Marchetti, C., Nobile, M. & Usai, C. (1995). Pharmacological types of calcium channels and their modulation by baclofen in cerebellar granules. *Journal of Neuroscience* (in press).

Anwyl, R.. (1991). Modulation of vertebrate neuronal calcium channels by transmitters. *Brain Research Review* 16, 265-281.

Bean, B. P. (1989). Neurotransmitter inhibition of neuronal calcium currents by changes in channel voltage-dependence. *Nature* 340, 153-156.

Berrow, N. S., Campbell, V., Fitzgerald, E. M., Brickley, K. & Dolphin, A. C. (1995). Antisense depletion of β-subunits modulates the biophysical and pharmacological properties of neuronal calcium. *Journal of Physiology* 482, 481-491.

Bley, K. R. & Tsien, R. W. (1990). Inhibition of Ca^{2+} and K^+ channels in sympathetic neurons by neuropeptides and other ganglionic transmitters. *Neuron* 4, 379-391.

Boland, L. & Bean, B. P. (1993). Modulation of N-type calcium channels in bullfrog sympathetic neurons by luteinizing hormone-releasing hormone: kinetics and voltage-dependence. *Journal of Neuroscience* 13, 516-533.

Bourinet, E., Soong, T. W., Stea, A. & Snutch, T. P. (1996). Determinants of the G-protein-dependent opioid modulation of neuronal calcium channels. *Proceedings of the National Academy of Science (USA)* (in press).

Carbone, E., Carabelli, V., Lovallo, M., Zucker, H. & Magnelli, V. (1995). Voltage-dependent inhibition and facilitation of single N-type channel kinetics by noradrenaline and δ -opioid agonists in IMR32 cells. *Society for Neuroscience Abstract* 21, 514.

Carbone, E., Sher, E. & Clementi, F. (1990). Ca currents in human neuroblastoma IMR32 cells: kinetics, permeability and pharmacology. *Pflügers Archiv* 416, 170-179.

Carbone, E. & Swandulla, D. (1989). Neuronal calcium channels: kinetics, blockade and modulation. *Progress in Biophysics and Molecular Biology* 54, 31-58.

Catterall, W. A. (1993). Structure and function of voltage-gated ion channels *Trends in Neurosciences* 16, 500-506.

Cesare, P., Pollo, A., Magnelli, V., Codignola, A., Sher, E., Clementi, F. & Carbone E. (1994). Opioid-induced inhibition of HVA calcium currents and 5-HT secretion in human small cell lung cancer cells. *Society for Neuroscience Abstracts* 20, 902.

Cox, D. & Dunlap, K.. (1992). Pharmacological discrimination of N-type from L-type calcium current and its selective modulation by transmitters. *Journal of Neuroscience* 12, 906-914.

Delcour, A. H. & Tsien, R. W. (1993). Altered prevalence of gating modes in neurotransmitter inhibition of N-type calcium channels. *Science* 259, 980-984.

Diversé-Pierluissi, M. & Dunlap, K. (1993). Distinct, convergent second messenger pathways modulate neuronal calcium currents. *Neuron* 10, 753-760.

Diversé-Pierluissi, M., Goldsmith, P. K. & Dunlap, K. (1995). Transmitter-mediated inhibition of N-type calcium channels in sensory neurons involves multiple GTP-binding proteins and subunits. *Neuron* 14, 191-200.

Dolphin, A. C. (1995). Voltage-dependent calcium channels and their modulation by neurotransmitters and G proteins. *Experimental Physiology* 80, 1-36.

Dolphin, A. C. & Scott, R. H. (1987). Calcium channel currents and their inhibition by (-)-baclofen in rat sensory neurones. *Journal of Physiology* 386, 1-17.

Dunlap, K. & Fischbach, G. D. (1978). Neurotransmitters decrease the calcium component of sensory neurone action potentials. *Nature* 276, 837-839.

Elmslie, K. S. & Jones, S. W. (1994). Concentration dependence of neurotransmitter effects on calcium current kinetics in frog sympathetic neurones. *Journal of Physiology* 481, 35-46.

Elmslie, K. S. & Kelly, E. (1995). Norepinephrine modulation of single calcium channels in frog sympathetic neurons. *Society for Neuroscience Abstract* 21, 515.

Elmslie, K. S., Zhou, W. & Jones, S. W. (1990). LHRH and GTP-γ-S modify calcium current activation in bullfrog sympathetic neurons. *Neuron* 5, 75-80.

Ewald, D. A., Matthies, H. J. G., Perney, T. M., Walker, M. W. & Miller, R. J. (1988). The effect of down regulation of protein kinase C on the inhibitory modulation of dorsal root ganglion neuron Ca^{2+} currents by neuropeptide Y. *Journal of Neuroscience* 8, 2447-2451.

Fisher, T. E. & Bourque, C. W. (1995). Dynorphin modulates distinct calcium channel subtypes in the somata and axon terminals of a mammalian central neuron. *Society for Neuroscience Abstract* 21, 1574.

Formenti, A., Arrigoni, E., Bejan, A., Avanzini, G. & Mancia, M. (1995). Acetylcholine and enkephalin inhibit high and low voltage activated calcium channels in thalamic relay neurons. *Society for Neuroscience Abstract* 21, 1575.

Formenti, A., Arrigoni, E. & Mancia, M. (1993). Two distinct modulatory effects on calcium channels in adult rat sensory neurons. *Biophysical Journal* 64, 1029-1037.

Golard, A., Role, L. W. & Siegelbaum, S. A. (1993). Protein kinase C blocks somatostatin-induced modulation of calcium current in chick sympathetic neurons. *Journal of Neurophysiology* 70, 1639-1643.

Golard, A. & Siegelbaum, S. A. (1993). Kinetic basis for the voltage-dependent inhibition of N-type calcium current by somatostatin and norepinephrine in chick sympathetic neurons. *Journal of Neuroscience* 13, 3884-3894.

Grassi, F. & Lux, H. D. (1989) Voltage-dependent GABA-induced modulation of calcium currents in chick sensory neurons. *Neuroscience Letters* 105, 113-119.

Hille, B. (1994). Modulation of ion-channel function by G-protein-coupled receptors. *Trends in Neuroscience* 17, 531-532.

Kasai, H.. (1992). Voltage- and time-dependent inhibition of neuronal calcium channels by a GTP-binding protein in a mammalian cell line. *Journal of Physiology* 448, 189-209.

Katada, T., Gilman, A. G., Watanabe, Y., Bauer, S. & Jacobs, K. H. (1985). Protein kinase C phosphorylates the inhibitory guanine-nucleotide-binding regulatory component and apparently suppresses its function in hormonal inhibition of adenylate cyclase. *European Journal of Biochemistry* 151, 431-437.

Kuo, C-C. & Bean, B. P. (1993). G-protein modulation of ion permeation through N-type calcium channels. *Nature* 365, 258-262.

Kuo, C-C. & Hess, P. (1993). Characterization of the high-affinity Ca^{2+} binding sites in the L-type Ca^{2+} channel pore in rat phaeochromocytoma cells. *Journal of Physiology* 466, 657-682.

Lacerda, A. E., Haeyoung, S. K, Ruth, P., Perez-Reyes, E., Flockerzi, V., Hofmann, F., Birnbaumer, L. & Brown, A. M. (1991). Normalization of current kinetics by interaction between the α_1 and β subunits of the skeletal muscle dihydropyridine-sensitive Ca^{2+} channel. *Nature (Lond.)* 352, 527-530.

Lipscombe, D., Kongsamut, S. & Tsien, R. W. (1989). α-adrenergic inhibition of sympathetic neurotransmitter release mediated by modulation of N-type calcium-channel gating. *Nature* 340, 639-642.

López, H. S. & Brown A. M. (1991). Correlation between G protein activation and reblocking kinetics of Ca^{2+} channel current in rat sensory neurons. *Neuron* 7, 1061-1068.

Luebke, J. I. & Dunlap K. (1994). Sensory neuron N-type calcium currents are inhibited by both voltage-dependent and -independent mechanisms. *Pflügers Archiv* 365, 258-262.

Lux, H. D., Carbone E., & Zucker H. (1990). Na$^+$ currents through low-voltage-activated Ca^{2+} channels of chick sensory neurones: block by external Ca^{2+} and Mg^{2+}. *Journal of Physiology* 430, 159-188.

Majewski, H., Hoare, A. & Murphy, T. V. (1995). Endogenous facilitation of noradrenaline release from sympathetic nerves through phospholipase C generation of diacylglycerol and activation of protein kinase C. *Society for Neuroscience Abstract* 21, 2065.

Malenka, R. C., Madison, D. V. & Nicoll, R. A. (1986). Potentiation of synaptic transmission in the hippocampus by phorbol esters. *Nature* 321, 175-177.

Marchetti, C., Carbone, E. & Lux, H. D. (1986). Effects of dopamine and noradrenaline on Ca channels of cultured sensory and sympathetic neurons of chick. *Pflügers Archiv* 406, 104-111.

Mathie, A., Bernheim, L. & Hille, B. (1992). Inhibition of N- and L-type calcium channels by muscarinic receptor activation in rat sympathetic neurons. *Neuron* 8, 907-914.

Meriney, S. D., Gray, D. B. & Pilar, G. R.. (1994). Somatostatin-induced inhibition of neuronal Ca^{2+} current modulated by cGMP-dependent protein kinase. *Nature* 369, 336-339.

Nishizuka, Y. (1995). Protein kinase C and lipid signalling for sustained cellular responses. *FASEB Journal* 9, 484-496.

O'Dell, T. J. & Alger, B. E. (1991). Single calcium channels in rat and guinea-pig hippocampal neurons. *Journal of Physiology* 436, 739-767.

Passafaro, M., Clementi, F., Pollo, A., Carbone, E. & Sher, E. (1994). ω-Conotoxin and Cd^{2+} stimulate the recruitment to the plasmamembrane of an intracellular pool of voltage-operated Ca^{2+} channels *Neuron* 12, 317-326.

Pollo, A., Lovallo, M., Biancardi, E., Sher, E., Socci, C. & Carbone, E. (1993) Sensitivity to dihydropyridines, ω-conotoxin and noradrenaline reveals multiple high-voltage activated Ca^{2+} channels in rat insulinoma and human pancreatic β-cells. *Pflügers Archiv* 423, 462-471.

Pollo, A., Lovallo, M., Sher, E. & Carbone, E. (1992). Voltage-dependent noradrenergic modulation of ω-conotoxin-sensitive Ca^{2+} channels in human neuroblastoma IMR32 cells. *Pflügers Archiv* 422, 75-83.

Pyne, N. J., Freissmuth, M. & Palmer, S. (1992). Phosphorylation of the spliced variant forms of the recombinant stimulatory guanine-nucleotide-binding regulatory protein (G$_S\alpha$) by protein kinase C. *Biochemistry Journal* 285, 333-338.

Rogers, M., Passafaro, M., Richmond, J. E., Cooke, J. M. & Sher, E. (1995). Ca^{2+} and protein kinase C-dependent recruitment of functional Ca^{2+} channels to the surface of RINm5F insulin-secreting cells. *Society for Neuroscience Abstract* 21, 59.

Scholz, K. P., and Miller, R. J. (1991) GABAb receptor-mediated inhibition of Ca^{2+} currents and synaptic transmission in cultured rat hippocampal neurones. *Journal of Physiology,* 444, 669-686.

Schroeder, J. E., Fischbach, P. S. & McCleskey, E. W. (1990). T-type calcium channels: heterogeneous expression in rat sensory neurons and selective modulation by phorbol esters. *Journal of Neuroscience* 10, 947-951.

Shapira, R., Silberberg, S. D., Ginsburg, S. & Rahaminoff, R. (1987). Activation of protein kinase C augments evoked transmitter release. *Nature* 325, 58-60.

Shapiro, M. S., Zhou, J-Y. & Hille, B. (1995). Protein kinases attenuate two membrane-delimited G-protein pathways modulating Ca^{2+} channels, sparing a second-messenger pathway. *Society for Neuroscience Abstract* 21, 516.

Shearman, M. S., Sekiguchi, K & Nishizucka, Y. (1989). Modulation of ion channel activity: a key function of the protein kinase C enzyme family. *Pharmacological Review* 41, 211-237.

Shen, K. Z., & Surprenant, A. (1991). Noradrenaline, somatostatin and opioids inhibit activity of single HVA/N-type calcium channels in excised neuronal membranes. *Pflügers Archiv* 418, 614-616.

Swandulla, D., Carbone, E. & Lux, H. D. (1991). Do calcium channel classifications account for neuronal calcium channel diversity? *Trends in Neurosciences* 14, 46-51.

Swartz, K. J. (1993). Modulation of Ca^{2+} channels by protein kinase C in rat central and peripheral neurons: disruption of G protein-mediated inhibition. *Neuron* 11, 1-20.

Swartz, K. J., Merritt, A., Bean, B. P. & Lovinger, D. M. (1993). Protein kinase C modulates glutamate receptor inhibition of Ca^{2+} channels and synaptic transmission. *Nature* 361, 165-168.

Toselli, M. & Taglietti, V. (1994). Muscarinic inhibition of high-voltage-activated calcium channels in excised membranes of rat hippocampal neurons. *European Biophysical Journal* 22, 391-398.

Tsien, R. W., Ellinor, P. T. & Horne, W. A. (1991). Molecular diversity of voltage-dependent Ca^{2+} channels *Trends in Pharmacological Science* 12, 349-354.

Wakade, A. R., Malhotra, R. K. & Wakada, T. D. (1985). Phorbol ester, an activation of protein kinase C, enhances calcium-dependent release of sympathetic neurotransmitter. *Naunyn-Schmiederberg's Archiv* 331, 122-124.

Wanke, E., Ferroni A., Malgaroli, A., Ambrosini, A., Pozzan, T. & Meldolesi, J. (1987). Activation of muscarinic receptor selectively inhibits a rapidly inactivated Ca^{2+} current in rat sympathetic neurons. *Proceedings of the National Academy of Science (USA)* 84, 4313-4317.

Yang, J. & Tsien, R. W. (1993). Enhancement of N- and L-type calcium channel currents by protein kinase C in frog sympathetic neurons. *Neuron* 10, 127-136.

Zhu, Y. & Ikeda, S. R. (1994a). Modulation of Ca^{2+}-channel currents by protein kinase C in adult rat sympathetic neurons. *Journal of Neurophysiology* 72, 1549-1560.

Zhu, Y. & Ikeda, S. R. (1994b). VIP inhibits N-type Ca^{2+} channels of sympathetic neurons via a pertussis toxin-insensitive but cholera toxin-sensitive pathway. *Neuron* 13, 657-669.

4

FUNCTIONAL, CONFORMATIONAL, AND MOLECULAR MODELLING STUDIES OF VOLTAGE-SENSITIVITY AND SELECTIVITY OF SYNTHETIC PEPTIDES DERIVED FROM ION CHANNELS

Pascal Cosette,[1] Olivier Helluin,[1] Jason Breed,[2] Yehonathan Pouny,[3] Yechiel Shai,[3] Mark S. P. Sansom,[2] and Hervé Duclohier[1]

[1] URA 500 CNRS-Université de Rouen (IFRMP 23)
Boulevard M. de Broglie
76821 Mont-Saint-Aignan, France
[2] Lab. of Molecular Biophysics
University of Oxford, Rex Richards Building
South Parks Road, Oxford, OX1 3QU United Kingdom
[3] Department of Membrane Research and Biophysics
Weizmann Institute of Science, 76100 Rehovot, Israel

1. INTRODUCTION

The demonstration of membrane excitability induced in artificial membranes by alamethicin and basic polypeptides (Mueller & Rudin, 1968) triggered an interest in "simple" model systems consisting of peptides interacting with planar lipid bilayers. This field of research grew as further natural peptides, often endowed with antimicrobial properties, proved to be "channel- or pore-formers" (for review, see e.g. Sansom, 1991). Additionally, ion channels purified from biological tissues were reconstituted into planar lipid bilayers and characterized (Miller, 1986). In parallel synthetic chemical approaches, particularly the "solid-phase technique" (Merrifield, 1963) later supplemented by the "Template-Assembled Synthetic Proteins" (TASP) method (see the review by Tuchscherer & Mutter, 1995), increasingly made available synthetic peptides whose sequences may be modulated as regards their potential channel-forming properties. The main requirements to be met by such peptides, as summarized e.g. by Lear, Wasserman & DeGrado, 1988 and by Spach, Duclohier, Molle & Valleton, 1989, are: a minimal length of 20 residues (if an helix is assumed) to match the thickness of the hydrocarbon core of the bilayer, a sufficient hydrophobicity to interact with the bilayer but also an amphiphilicity so distributed along the sequence as to delineate well defined polar and apolar faces. The last requirement is advantageous both

Neurobiology, edited by Torre and Conti
Plenum Press, New York, 1996

from the point of view of aggregation within the bilayer and for definition of a hydrophilic pathway formed by the juxtaposition of the hydrophilic sectors of the monomers.

Once the amino acid sequences of the voltage-gated ion channels were elucidated together with the associated hydropathy plots and the first topological models, it became evident that some segments might have a potential role in the assembly and function of the channels. Special attention was directed toward the putative membrane-embedded segments of those ion channels. As the hypothesized structure of some of those segments resembles that of natural membrane-interacting polypeptides, the synthetic peptides approach enables to use peptides derived from channels as model systems to investigate pore-forming abilities and other aspects of peptide-membrane interactions. This alternative strategy recognized as such by e.g. Stühmer (1991) was recently reviewed by Montal (1995). However, the relevance of data gained from experiments with synthetic peptides derived from channels to the structure-function of the native protein should be evaluated carefully. It should be taken into account that the functional properties of the segment in the context of the whole protein, which consists of subunits or domains with multiple transmembrane segments, might be different from those of the single transmembrane segments, or might involve more complicated interactions than is evident from experiments with single membrane-embedded segments. On the other hand, recent studies show good correlation between the structure-organization of synthetic peptides and that of the relevant segments within the intact protein (Lovejoy et al. 1992; Barsukov et al. 1992; Gazit & Shai, 1993), so careful interpretation of experiments with synthetic segments may shed light on some aspects which are difficult to tackle by other means with membrane proteins.

The voltage-dependent sodium channel, responsible for the initial inward current during the depolarizing phase of the action potential plays a fundamental role in impulse initiation and propagation in excitable cells. The current chapter present some new data concerning both voltage-sensitivity and ion selectivity of synthetic peptides derived from the sodium channel. The methodology involves the selection of segments with the help of current topological models, their solid-phase synthesis and HPLC purification, followed by measurements of their conformation (CD, circular dichroism in organic solvents as well as in lipid vesicles) and function (conductances induced in planar lipid bilayers). In addition, the ability of the peptides to bind to lipid membranes and to aggregate within them are investigated through fluorescence energy transfer experiments. Finally, molecular modelling and dynamics simulations of the peptides in isolation are employed to reveal conformational preferences of the different parts of the peptides.

Electrophysiological studies of mutated sodium channels expressed in heterologous systems (for a short review, see Stühmer, 1993) have highlighted, within the six putative transmembrane segments of each of the four homologous domain, the essential roles of the S4 segments (helices with basic residues every three positions) as the main voltage-sensors (Stühmer et al. 1989; Auld et al. 1990) and of the P regions (or H5 or SS1-SS2, between the S5 and S6 transmembrane segments) as the selectivity-filter (Terlau et al. 1991; Pusch et al, 1991; Heinemann et al. 1992; Backx et al, 1992; Kontis & Goldin, 1993). However, recent studies show that other parts may be involved, i.e. S2 and S3 in gating and possibly S5 and S6 in ion selectivity (Planells-Cases et al. 1995; Kirsch et al. 1993; Lopez, Jan & Jan, 1994).

To examine the properties of specific segments derived from the sodium channel, peptides resembling the S4-S45 (from domains II and IV) and the four P-regions were synthesized and structurally and functionally characterized. The amino acid sequences, issued from the primary structure of the *Electrophorus electricus* sodium channel (Noda et al. 1984), of these peptides are presented in Table 1.

Table 1. Amino acid sequences of the peptides studied. The positively charged residues in the first family are in bold characters. The presumed β-bends for the second family are underlined and in bold

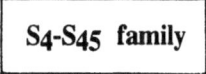

S4-S45$_{IV}$ and fragments :

Ac-TLFRVIR LARIARVL RLIRAAKG IRTLLFALMMS-NH$_2$

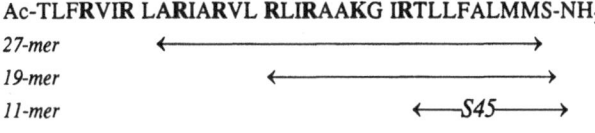

S4-S45$_{II}$:

Ac-SVLRSLRLLRIFKLAKSW**P**TLNILIKI ICNSVGA-NH$_2$

P (H$_5$ or SS$_1$-SS$_2$) family

P$_I$:

Ac-GYTNYDNFAWTFLCLFRLML**QDY**WENLYQMT-NH$_2$

P$_{II}$:

Ac-HMNDFFHSFLIVFRAL**CGEW**IETMWDCMEVG-NH$_2$

P$_{III}$:

Ac-VRWVNLKVNYDNAGMGYLSLLQVST**FKGW**MDIMYA-NH$_2$

P$_{IV}$:

Ac-NFETFGNSMICLFEITT**SAGW**DGLLLPTLNTG-NH$_2$

2. MACROSCOPIC CONDUCTANCE COMPARED FOR S4-S45 AND P PEPTIDES

The functional properties induced by both set of peptides in planar lipid bilayers were first investigated at the macroscopic conductance level so as to screen voltage- and concentration- dependences. In this configuration, virtually solvent-free Montal-Mueller bilayers (Montal & Mueller, 1972) of large area (about 150 μm for the hole diameter), allowing the incorporation of many channels (100-1000) after equilibration with aqueous peptide solutions are submitted to voltage ramps.

2.1. In Symmetrical Ionic Solutions.

Figure 1 illustrates the strikingly different behaviour displayed by two peptides belonging either to the S4-S45 or P families (from the same domain: IV).Within about 20 minutes of equilibration with 100 nM aqueous concentration of S4-S45$_{IV}$ on the *cis*-side, a bilayer formed with neutral lipids develops a typical macroscopic current-voltage (I-V) curve

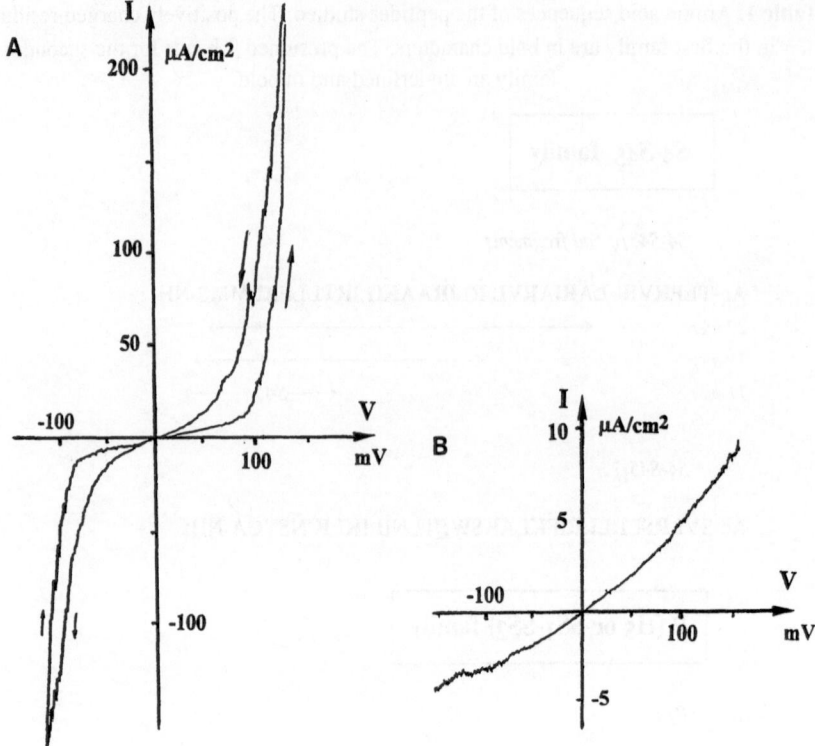

Figure 1. Macroscopic current-voltage curves developed by planar lipid bilayers doped with S4-S45$_{IV}$ (A) and P$_{IV}$ (B). Lipids used for bilayer formation POPC:DOPE (7:3) in (A) and (2:8) in (B). Symmetrical aqueous phases: 500 mM NaCl (in A) or 500 mM KCl (in B) with 10 mM HEPES, pH=7.4. Peptide concentration (*cis*-side): 10^{-7} M in (A) and $2 \cdot 10^{-6}$ M in (B). Voltage ramp sweep: 15 mV/s. (Adapted from Brullemans *et al.* 1994 and Cosette *et al.* 1994).

(Fig. 1A). The current response is symmetrical with respect to zero-voltage and there is a steep exponential branch above a voltage threshold. The voltage-dependence is characterized by Ve, the voltage increment resulting in an e-fold change in conductance. Here, in this example, Ve=19 mV.

This kind of I-V curve is very reminiscent of those displayed by alamethicin and natural or synthetic analogues (Hall *et al.* 1984, Duclohier *et al.* 1992), but a straightforward analysis of the concentration-dependence as performed for the above-mentioned peptides failed to show good correlations. This analysis allows estimation of N, the apparent mean number of peptide monomers per conducting aggregate as N=Va/Ve, where Ve was defined above and Va is the voltage shift of the theshold for an e-fold change in peptide aqueous concentration (Hall *et al.* 1984). Since Ve (in the 10-20 mV) range was found to be better correlated with the background or leak conductance (around 0 mV), the latter reflecting the intrinsic membraneous concentration of the peptide locked into voltage-insensitive conducting aggregates, we used for Va the voltage shift of the threshold for an e-fold change in the leak conductance.

Under these conditions, we found that N was equal to 4 in neutral lipid bilayers. This "apparent mean" number of monomers per conducting aggregate is reduced to 3 in negatively-charged bilayers (POPS-DOPE: 1:1). In the latter system, the current density was also reduced by about one order of magnitude. Presumably, the increased electrostatic binding demonstrated in fluorescence experiments for NBD-S4 with negatively-charged lipid vesicles (Rapaport *et al.* 1992) restricts the voltage-driven transmembrane location of the peptide.

The apparent gating charge q can be derived from Ve through:

$$q = Nzd = (RT/F)/Ve$$

where N is still the mean number of peptide monomer in the conducting aggregate, zd the product of the monomer mobile charge and the fraction of voltage drop the "gate" traverses before achieving activation, R the molar gas constant, T the absolute temperature and F is Faraday's constant. In the best cases (neutral bilayers), q was 3 or thus 0.8 per monomer. This is comparable to alamethicin (Hall *et al.* 1984) on the one hand and to reconstituted BTX-modified (inactivation removed) sodium channels on the other hand (Behrens *et al.* 1989). It is nevertheless smaller than the value of 7 elementary charges derived by Conti and Stühmer (1989) from cloned sodium channels expressed in *Xenopus* oocytes.

As for the macroscopic I-V curve developed by the P peptide (also from domain IV of *Electrophorus electricus*) it was quasi-ohmic with no significant voltage-dependence (Fig. 1B), thus yielding a much lower current density, albeit at a ten-fold higher concentration. The asymmetry or rectification-like property seen in the example of Figure 1B is cancelled when the peptide is added to both *cis*- and *trans*- baths facing the bilayer. Also contrasting with the S4-S45 peptide, the bilayer also had to be richer in PE than in PC headgroups to allow any significant peptide incorporation (as judged from the conductance development). Presumably, the nonlamellar (or some hexagonal phase) tendency of PE may favour the interaction of peptides endowed with a reduced helical content. In any case, the absence of a well marked threshold precluded for peptides P the analysis discussed above for estimating N. However, we found a significant voltage-dependence for macroscopic I-V curves developed by P_{III} (the only one to bear a positive charge in its assumed β-bend and a cluster of positive charges at its N-terminus). This surprising result was confirmed at the single-channel level of investigation (see below)

2.2. Under Ionic Gradients

When the bilayer doped with S4-S45$_{IV}$ was bathed by 500 mM NaCl on the *cis*-side (by convention the positive side) and 450 mM KCl + 50 mM KCl on the *trans*-side, the zero-current or reversal potentials averaged -14 mV, yielding a selectivity or permeability ratio $P_{Na}/P_K = 3$ by application of the Goldman-Hodgkin-Katz equation (Hille, 1984). Similar conditions in the case of P_{IV} failed to produce any substantial shift of the reversal potential, thus implying a lack of ion selectivity for the conducting aggregates formed by this peptide. This was confirmed at the single-channel level (see below), which also demonstrated by contrast limited sodium selectivity for the other three P peptides.

3. SINGLE-CHANNEL CONDUCTANCES

The peptides were investigated for their single-channel conductances with bilayers formed at the tip of patch-clamp pipettes (Hanke *et al.* 1984) and with reduced (as compared to the 'macroscopic' configuration) aqueous peptide concentration. Figure 2A shows an example of single-channel current fluctuations induced by S4-S45$_{IV}$ at two different voltages. The open lifetimes histograms (Fig. 2B) confirm the voltage-dependence seen in the 'macroscopic' configuration. In addition, the probability of the open state Po increased from 0.40 to 0.60 between these two voltages. Single-channel i-V plots point to a single-channel conductance of 8.5 pS in symmetrical 500 mM NaCl and with neutral bilayers. For

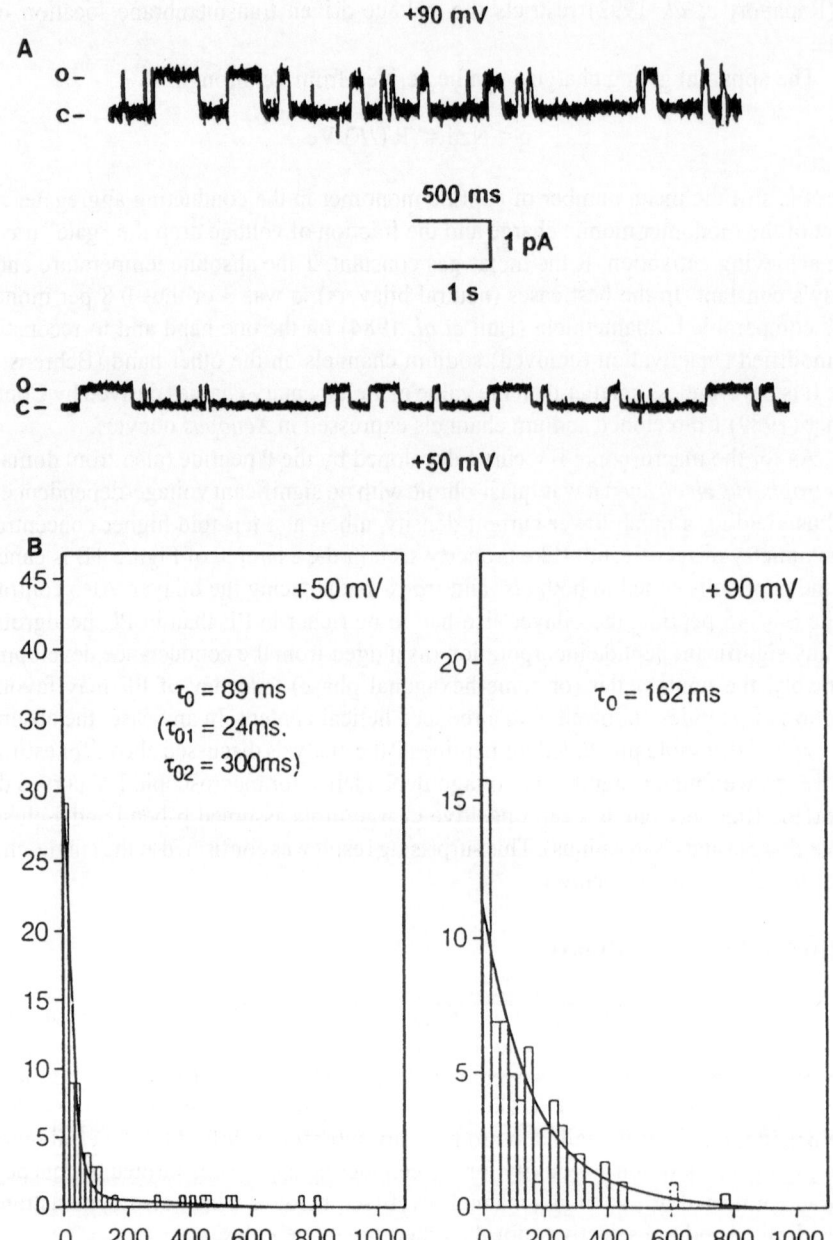

Figure 2. (A): Examples of single-channel traces obtained with S4-S45$_{IV}$ (25 nM in the *cis*-side) at two different voltages. Aqueous phases: 500 mM NaCl both sides. (B): Associated histograms of open channel lifetimes with the mean duration of the open state in the insets. (From Brullemans *et al.* 1994).

negatively-charged bilayers, the slightly lower conductance of 5 pS seems in agreement with the reduced number of monomers per conducting aggregate derived from 'macroscopic' data.

P peptides develop in bilayers conductances of the same order as S4-S45 but lifetime analysis showed a mean open lifetime (τ_o) of about 10 ms except for P$_{II}$, i.e. of much shorter duration (10-fold) than for the other family. Examples of single-channel traces for P$_{III}$ are shown in Figure 3A and the data for several experiments, either in symmetrical 500 mM KCl

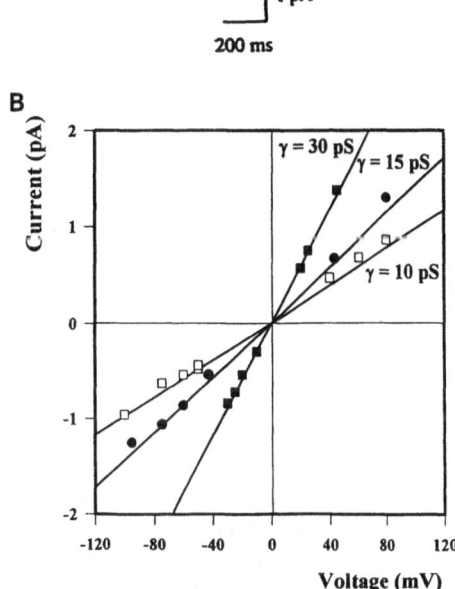

Figure 3. (A): A selection of single-channel traces obtained with P_{III} (25 nM in the *cis*-side) in NaCl or KCl at different applied voltages. (B): P_{III} single-channel current-voltage plot (data pooled from 4 bilayers) with the unit slope conductances indicated on the top right (open squares: in 500 mM KCl, filled circles and squares: two levels in 500 mM NaCl). (Adapted from Duclohier *et al.* 1995).

or 500 mM NaCl are pooled in the i-V plot of Figure 3B. The events collected here represent more than 80% of all observed events, i.e. either integral or non-integral multiples of these conductance levels are very rarely observed. The aggregation state is thus rather well conserved in different experiments. The single-channel conductance of 10 pS observed for P_{III} in symmetrical 500 mM KCl increase to either 15 or 30 pS in symmetrical 500 mM NaCl, the latter open states not being observed simultaneously. The higher level is a "genuine full conductance state" since there is no evidence for intermediate (mid-amplitude) steps to 15 pS (see e.g. upper trace of Fig. 3A). Thus with P_{III}, the sodium selectivity expressed as the permeability ratio (approximatively the conductance ratio) P_{Na}/P_K ranges from 1.5 to 3 (Duclohier *et al.* 1995).

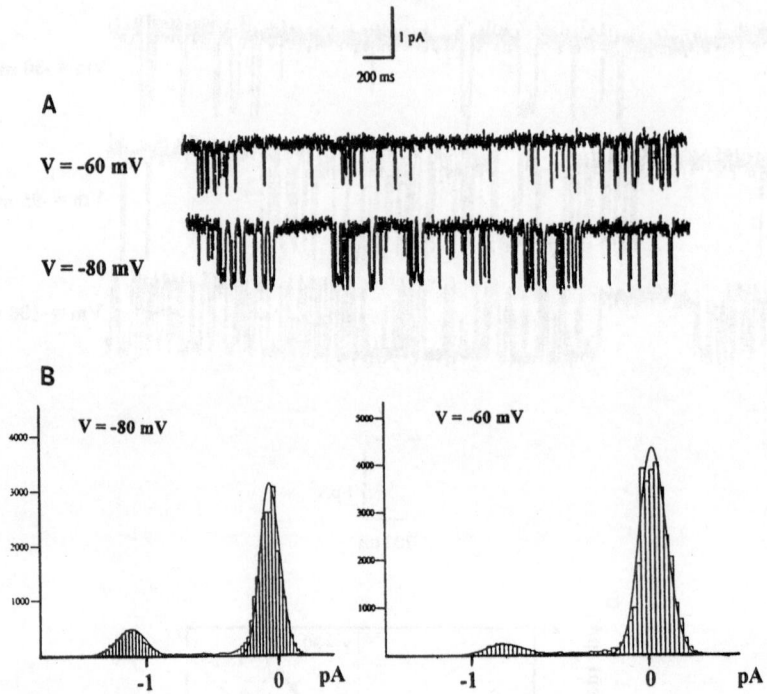

Figure 4. P_{III} exhibits some voltage-dependence. (A): Examples of single-channel traces for two applied voltages obtained in symmetrical 500 mM NaCl, $\gamma = 15$ pS, openings are downward deflections. (B): Associated amplitude histograms showing higher opening probability at the upper voltage.

The finding of some voltage-dependence in the mean duration and probability of the open state for P_{III} (Fig. 4) was quite unexpected but confirmed at the macroscopic conductance level. For instance, the mean open lifetime (τ_o) is 6 and 9 ms and the opening probability P_o is 0.11 and 0.18 for -60 and -80 mV, respectively. By contrast, all the other P peptides did not show any voltage-dependence. Thus for P_{IV}, τ_o is 16 and 17 ms for 50 and 100 mV, respectively. The corresponding values for the opening probability P_o are 0.26 and 0.25.

Single-channel conductances induced with P_I, P_{II} and P_{IV} peptides fall in the same range as discussed above with P_{III} and except for P_{IV}, whose presumed β-turn as predicted by Schetz & Anderson (1993) is electrically neutral (Table 1: bold and underlined section), all the other three Ps exhibit some sodium selectivity, albeit modest. The highest preference for sodium seems correlated with a positive charge on the assumed β-bend (of P_{III}) whilst P_I and P_{II} bear a negative charge in this region. Finally, it is interesting to note that P_{II} and P_{IV}, which are shown by fluorescence energy transfer to form intramembrane heteroaggregates (see below and Pouny & Shai, 1995), display the smallest conductances.

4. STRUCTURAL FEATURES, BINDING TO AND ORGANIZATION WITHIN MEMBRANES

4.1. Secondary Structures Estimated from Circular Dichroism Spectroscopy

Table 2 compares the secondary structure of S4-S45 and P segments both from domain IV in organic solvents and in lipid vesicles. These conformational contents were

Table 2. Comparison of secondary structure contents as estimated from CD in trifluoroethanol (TFE) and in small unilamellar vesicles (SUV)

Peptide	Media	Helix (%)	Sheet (%)	Turn (%)	Random (%)
S4-S45$_{IV}$	TFE	45	15	0	40
	PC-PS (1:1) SUV	55	30	5	10
P$_{IV}$	TFE	35	35	0	30
	PC SUV	0	50	10	40

computed from circular dichroism spectra, using the standards of Chang et al. (1978). Note that the best fit is obtained with one major helix for S4-S45 and several 'delocalized' helix turns for P. The high β-sheet content for P was confirmed by Fourier Tranform Infra-Red spectroscopy: for instance in the case of P$_{II}$, the spectra exhibited a relatively broad major peak at 1636-1639 cm^{-1}, typical of β-conformations, and a minor peak at 1654 cm^{-1} (α-helix). The P$_{II}$ conformational contents were also investigated as a function of the peptide concentration in TFE over two orders of magnitude: the proportions given in Table 2 remained stable from 400 down to 25 μM, below which the helicity slightly overtook the β-sheet content. Thus, it seems that the latter would be favored by interaction-aggregation, as *in situ* in the channel. Upon interaction with lipid vesicles (SUVs), the helicity of peptides from the S4-S45 family tends to increase whilst for the P family the β-conformation is further favored (Table 2). The 10% turn (i.e. 3-4 residues) is plausible for an assumed β-hairpin.

Figure 5 compares the circular dichroism spectra obtained with the four P peptides in two hydrohobic environments: (i) in a mixture of 40% TFE in H$_2$O (Fig. 5A), and (ii) in a membrane mimetic environment of sodium dodecyl sulfate (1% SDS) (Fig. 5B). We found that although the P segments from the different domains differ significantly in their primary sequences, they adopt similar partial α-helical structures (~30%). One exception is P from domain IV,which in this case was significantly elongated on the N-terminal side. The α-helical content of P$_{IV}$ in these measurements is 18% in 40% TFE/water. The discrepancy between CD measurements of P$_{IV}$ in TFE presented in Figure 5 and Table 2 are probably due to the differences in their sequences (compare P$_{IV}$ in Table 2 and below):

KKQGGVDDIFNFETFGNSMICLFEITTSAGWDGLLL

The sequence used here was elongated 10 amino acids toward the N-terminal on the one hand, and 6 amino acids shorter on the C-terminal side on the other hand. Although no definitive conclusion can be inferred and even if some recent models of the pore do support partial helical structures for the P regions (Guy & Durell, in press), it seems reasonable to also assume a significant amount of β-structure.

The slightly higher helical content of S4-S45 of domain IV in polar media, as compared with S4-S45$_{II}$ which contains a proline in its sequence (as do the homologous peptides from domain I and III), is in line with the role of this residue (helix disrupting or kinking, see Fig. 8). However, this helicity reduction is cancelled in membrane-like environment. This may contribute to explain the lack of any detectable effect on the steepness of activation curves upon mutating these Pro residues (from S4) in electrophysiological experiments (Moran *et al.* 1994). Note also that a Pro-Ala substitution in an alamethicin synthetic analogue, an obviously unrelated sequence albeit yielding highly voltage-dependent conductances, did preserve the latter behaviour (Duclohier *et al.* 1992). Finally, the most striking result of a systematic investigation of the conformational contents of S4-S45 and its fragments (see Table 1) as a function of solvent polarity is a sharp α-helix→β-sheet transition (Fig. 6) upon exposure of the S45 moiety (11 residues) from relatively apolar solvents to aqueous environment (Helluin *et al.* 1995). It will be shown below that a Molecular

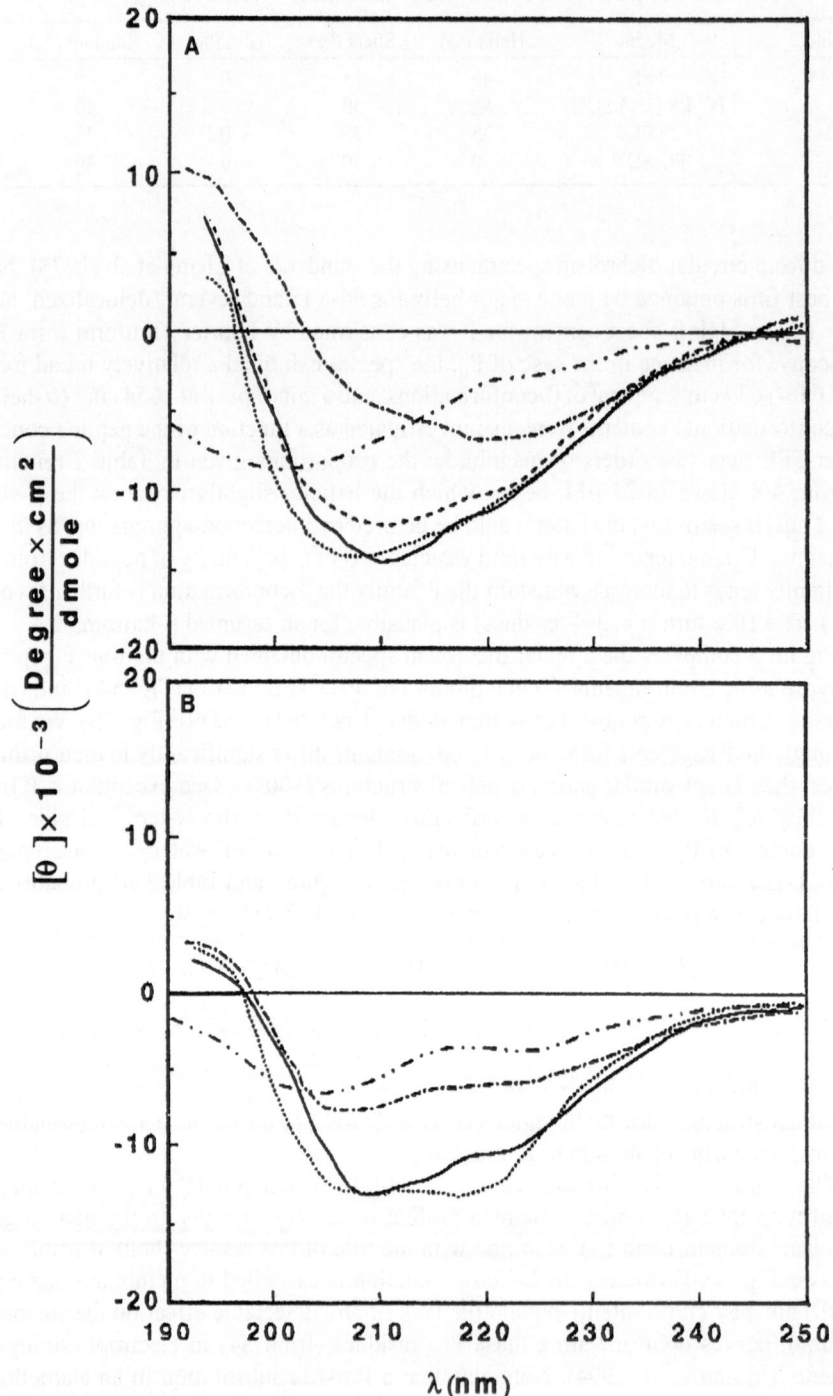

Figure 5. CD spectra of PR segments in : (A) 40% TFE in Water ; 1% SDS in buffer containing 16.7 mM NaCl and 4.3 mM HEPES, pH = 7. Spectra were taken at a peptide concentration of 0.5×10^{-5} to 2.0×10^{-5}M (taken from Pouny and Shai, 1995). Symbols: PR-I ————— ; PR-II ----- ; PR-III — — — ; PR-IV —·—·— ; pre-PR-II —··—··— .

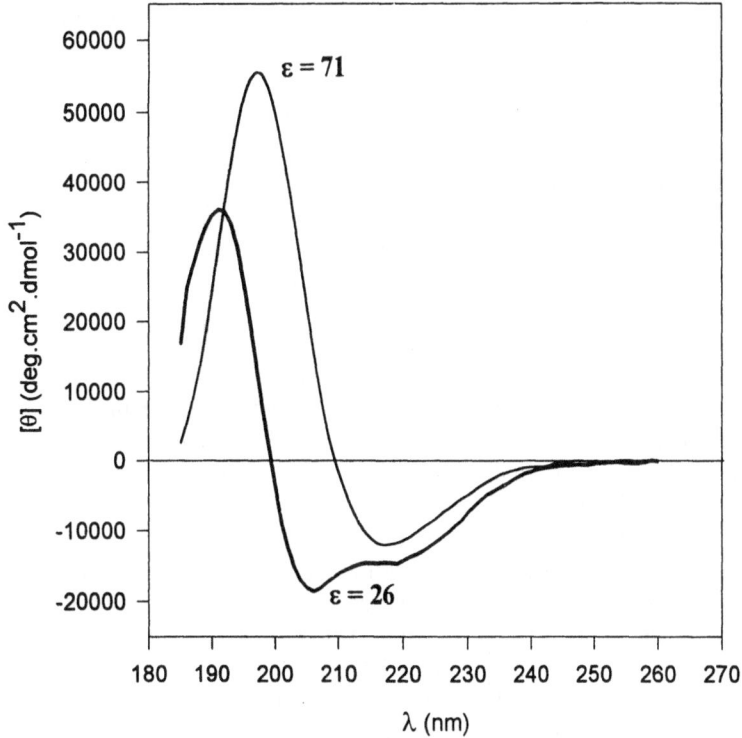

Figure 6. Circular dichroism spectra of the $S45_{11}$ linker (11 residues, see Table 1) as a function of the solvent dielectric constant (ε in Debye units). The latter is modulated through the relative proportions of tri-fluoroethanol (TFE) and H_2O. A 100% TFE solvent yields the spectrum labeled with $\varepsilon = 26$ (essentially identical to the spectrum with 30% TFE) typical for an helical conformation whilst 15% TFE (labeled with $\varepsilon = 71$) yields a β-sheet structure. (Modified from Helluin *et al.* 1995).

Modelling and Dynamics approach confirms at least qualitatively this conformational transition whose potential functional implication will be discussed in the final section.

4.2. Binding of S4 Segments to and Organization within Membranes

A basic property of S4 as a voltage sensor in order to sense variations of the transmembrane electric field should be an ability to move within the lipid bilayer (for review, see e.g. Sigworth, 1993; Franciolini, 1994). Such movements are thought to produce measurable gating currents (Armstrong & Bezanilla, 1973; Keynes & Rojas, 1974) and have been detected with the binding of specific antibodies during depolarization (Sammar, Spira & Meiri, 1992). Among the putative transmembrane helices of the sodium channel, S4 is unique in that every third amino acid is a positively-charged Arg or Lys residue. Not only this segment is highly conserved within the four homologous domains and between sodium channels of different species but also, similar sequences are found in the putative membrane-spanning domains of other voltage-sensing proteins. An NMR study on the S4 segment of the first internal repeat of the rat brain sodium channel indicated that the peptide is predominantly found as an α-helix in organic solvent (Mulvey *et al.* 1989). We have used a spectrophotometric approach to investigate the biophysical properties of a synthetic S4 of the first repeat of the eel sodium channel with the following sequence (Rapaport *et al.* 1992):

RTFRVLRALKTITIFPGLKTIVRA

The peptide was synthesized and labeled selectively at its C- or N-terminal amino acids with either one of the three fluorescent probes: NBD (4-Fluoro-7-nitrobenz-2-oxa-1,3-diazole, to serve as an environmentally-sensitive probe), fluorescein (to serve as an energy donor) and rhodamine (to serve as an energy acceptor).

We found that the S4 peptide incorporates into the lipid bilayer and that both the N- and C-terminals of the segment are located within the acyl-chain region of the membrane, with the N-terminus being situated deeper than the C-terminus. This conclusion was based on the observation that the environment encountered by an NBD group located at either the N- or C-terminal of the peptide was more hydrophobic (emission max. of 522 nm and 528 nm, respectively) than that detected with the probe on the membrane surface (emission max. 533 nm). According to current sodium channel models, although within the membrane, S4 is surrounded by other transmembrane segments, and therefore is not in direct contact with the membrane lipids. However, the fact that S4 is able to interact with the lipid environment makes equally plausible its interaction with the other hydrophobic transmembrane segments of the channel. An interesting finding were the similar partition coefficients obtained with zwitterionic vesicles (PC) and with acidic vesicles (PC/PS): $6.2 \ 10^4$ and $5.1 \ 10^4$, respectively (Fig. 7). Furthermore, both the shape of the binding isotherms (Fig. 7) and resonance energy transfer experiments revealed that S4 can self-assemble within the acidic PC/PS vesicles (Rapaport et al. 1992). This suggests that electrostatic interactions do not play a major role in the binding of the peptide to lipid vesicles and that S4 presents a strong tendency to aggregate when no other transmembrane helices are in its surroundings. This property is different from the one observed with other positivel-charged membranous polypeptides such as the antibacterial peptides dermaseptin (Pouny et al. 1992) and cecropin (Gazit et al. 1994) in which the binding to PC vesicles was about ten fold lower than that obtained with the

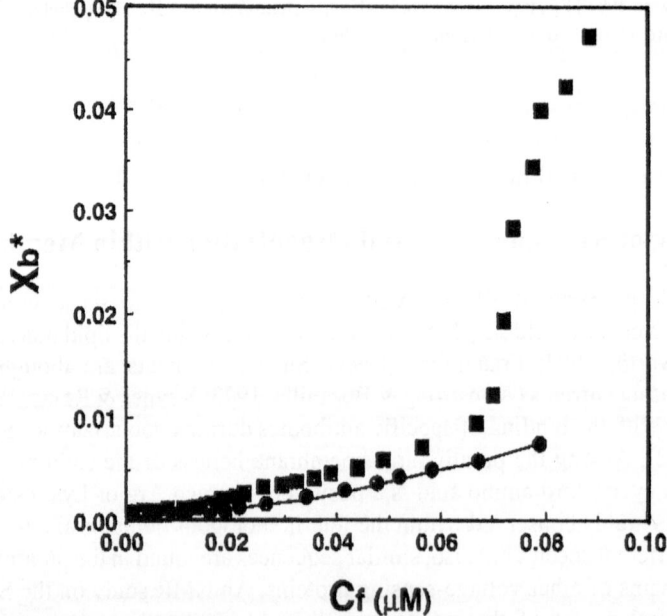

Figure 7. Binding isotherm of S4 by plotting X*b (molar ratio of bound peptide per total lipid) versus Cr (equilibrium concentration of free peptide in the solution) Circles, PC vesicles; squares, PC/PS vesicles (modified from Rapaport et al., 1992).

acidic PC/PS vesicles. The involvement of the homologous S4 of the *Shaker* K$^+$ in the assembly of the potassium channel has been demonstrated by site-directed mutagenesis (Papazian *et al.* 1995; Planells-Cases *et al.* 1995) and by the finding that synthetic S4 of the *Shaker* K$^+$ but not that of the sodium channel co-assembles with the S2 segment of the *Shaker* K$^+$ (Peled *et al.* 1996). Since there is significant homology between the transmembrane segments of both channels, similar interactions might also exist within the sodium channel.

4.3. Binding of P Segments to and Organization within Membranes

As for the P-regions, while four identical segments are assumed to line the pore of potassium channels, the four P-regions of the sodium channel are not identical, although they share significant homology. Furthermore, the P-regions of the *Shaker*-type K$^+$ channel appears to contain ~21 amino acids out of the 40 that build the loop connecting helices 5 and 6. In the sodium channel, the homologous loops are composed of 70-110 amino acids, from which only ~21 are predicted to possess the properties of the P-regions. Studies with synthetic peptides corresponding to the P-region of the *Shaker* K$^+$ channel revealed that they adopt a low level of α-helicity in hydrophobic environment, bind phospholipid membranes, and can self-assemble in their membrane-bound state, but do not coassemble with unrelated membrane-bound peptides (Peled & Shai, 1993). These results support the notion that they are packed in close proximity in the native channel and that they have a possible role in assisting in the correct assembly of the channel. Similar studies were done with the four P-regions of the *Electrophorus electricus* eel sodium channel (Pouny & Shai, 1995). We have synthesized "P-peptides" resembling the consensus sequences of the P-regions of domains I, II, III and the N-terminal form of repeat IV (a.a. 345-370, a.a. 734-759, a.a. 1197-1221 and, a.a.1477-1512, respectively) of the channel. The peptides were also labeled with fluorescent probes and were subjected to the above-mentioned studies, as with S4.

P-peptides were first tested for their ability to interact with phospholipid membranes. We found that although the P segments differ significantly in their primary sequences (Table 1), and also differ from those of the *Shaker* K$^+$ channel, they all bind with the same affinity to phospholipid membranes. This unique property is further emphasized if compared to the results obtained in earlier studies which showed that one amino acid substitution in the homologous P-regions of the *Shaker* K$^+$ channel caused a 5 fold decrease in the partition coefficient of the resultant analogue (Peled & Shai, 1993). Furthermore, slight changes in the amino acid composition of other, biologically active, membrane interacting polypeptides significantly decreased or abolished their binding to membranes (Strahilevitz *et al.* 1994; Gazit *et al.* 1994). This data is in line with the proposal that they have equivalent roles in participating in the formation of the hydrophobic core of the channel. As stated above with the S4 segment, the fact that the P-regions are membrane localized may indicate their potential to interact with the other hydrophobic transmembrane segments of the channel. In addition these results are in line with a "two stage" model for membrane protein folding and oligomerization (Popot *et al.* 1987; Popot & Engelman, 1990). In this "two stage" model, the final structure in membranes results from the packing of smaller elements, each of which reaches thermodynamic equilibrium within the lipid and aqueous phases before packing. This model is supported by a number of studies (see review by Lemmon & Engelman, 1992).

We have further assessed the ability of the P peptides to self-associate or to form heteroaggregates within the membrane milieu. For this purpose two type of experiments were performed: (1) The fluorescence changes of the NBD group upon interaction of a particular NBD-labeled segment with unlabeled segments were measured. Such changes are expected to occur only if the interaction between the segments causes a significant change in the environment of the NBD moiety, either by changing the orientation of the NBD-labeled peptide, or by changing its aggregational state such that quenching or dequenching of the

NBD fluorescence can occur; (ii) Resonance energy transfer experiments (RET) between donor- and acceptor-labelled P peptides. The results of these experiments revealed an interesting observation in that, except the P_I/P_{III} pair, all other combinations form heteroaggregates in their membrane-bound state. Furthermore, as a result of these interactions, the N-termini of P_I and P_{II} moved to more hydrophobic environments. That the coassembly of the P segments in their membrane-bound state is at least partially specific, was demonstrated by the finding that P_I and P_{III} do not coassemble, and neither of the segments coassemble with the unrelated α-helical peptide pardaxin.

5. MOLECULAR MODELLING AND MOLECULAR DYNAMICS

5.1. Aims and Background

The principal aim of these studies is to model possible conformations adopted by sodium channel transmembrane (TM) domain peptides in electrostatic environment which mimic the solvents used in the CD spectroscopic investigations. It is hoped that such simulations will reveal the conformational preferences (e.g. α-helix vs. β-strand) of the different regions of the synthetic peptides. Such information will be of value when attempting the more daunting task of modelling the corresponding TM segments within an intact channel protein. Here, we describe such simulations of S4-S45 peptides from a Na^+ channel.

The first stage of these studies is to generate a completely α-helical model of a S4-S45 peptide, using simulated annealing via restrained molecular dynamics (SA/MD; Kerr et al. 1994). We have previously employed this method to model TM helices from a number of different ion channels (Sankararamakrishnan & Sansom, 1995) and from channel-forming peptides (Kerr & Sansom, 1993). These initial S4-S45 models are then used as starting structures for molecular dynamics (MD) simulations of helix unfolding in order to identify those regions of the peptides with the highest and with the lowest propensity to remain in an α-helical conformation. Finally, we correlate the results of the simulation studies with CD data for the corresponding peptides. All simulations have been carried out using Xplor V3.1 running on DEC Alpha workstations.

5.2. SA/MD Generation of Initial Models

Initial models of $S4-S45_{II}$ and $S4-S45_{IV}$ in all α-helical conformations were generated by SA/MD. For each peptide an ensemble of 100 structures was generated. The Cα template for these simulations was an ideal α-helix extending the full length of the peptide. During the final 5 ps of free MD during SA/MD, no intra-helix distance restraints were applied, thus enabling the structure to drift from an idealized α-helical geometry. Despite this, it is evident that from Figure 8 that both models yielded by SA/MD are almost entirely α-helical. The $S4-S45_{II}$ helix contains a central kink in the vicinity of the proline residue (P19). The mean helix kink angle averaged across the ensemble was 26° (±11°). This should be compared with a kink angle of 34° (±12°) from comparable simulations of the conformation of the Shaker K^+ channel S6 TM helix (Kerr et al. 1995). Interestingly, despite the absence of a central proline residue $S4-S45_{IV}$ exhibited somewhat greater conformational heterogeneity in SA/MD than did $S4-S45_{II}$. However, both models are more than 80% α-helical. In order to explore in more depth possible conformations of these peptides, more prolonged MD simulations were required. Starting structures for such simulations (one structure for each peptide) were obtained by analyzing the SA/MD generated ensembles using PROCHECK (Morris et al. 1992) and in each case the model with the best stereochemistry was used as the initial structure for further MD simulations.

A   B

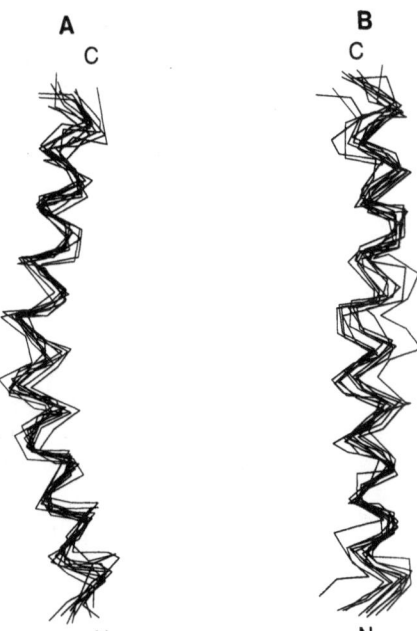

Figure 8. Initial models of (A) S4S45$_{II}$ and (B) S4S45$_{IV}$ as generated by SA/MD. Each diagram shows the superimposed Cα traces of 10 structures selected from the SA/MD-generated ensembles. The N- and C-termini of the helices are labelled. The central kink in the S4S45$_{II}$ helices due to residue P19 is evident.

5.3. MD Simulations and Analysis

The MD simulations were carried out *in vacuo* ie. in the absence of explicit solvent molecules. However, in order to mimic the effects of solvent environment on peptide conformation, simulations were run with different dielectric constants (ε), giving different degrees of screening of electrostatic interactions. Thus, for each peptide four simulations were run: (a) with $ε = 4$ (to mimic an apolar environment, e.g. the hydrocarbon core of a lipid bilayer); (b) with $ε = 25$ (to mimic an environment of intermediate polarity, eg. isopropanol or eg. the interfacial region of a lipid bilayer); (c) $ε = 80$ (to mimic extensive screening of electrostatic interactions by water and by counterions); and (d) $ε = r$, (a widely used approximation of the electrostatic screening experienced by globular proteins in aqueous solution). For each peptide with each dielectric model (a total of 8 simulations) the initial structure was energy minimized, heated from 0 to 300 K for 3 ps (in 50 K, 0.5 ps steps), and equilibrated for 19 ps at 300 K using temperature coupling to a heat bath. The production stage of the MD simulation was run for 300 ps. The timestep in the MD simulations was 0.5 fs, with coordinates saved every 1 ps for subsequent analysis. The *param19* parameter set was employed, with polar hydrogens represented explicitly and apolar hydrogens represented by extended carbon atoms.

All MD simulations started from models of S4-S45$_{II}$ or S4-S45$_{IV}$ which were highly (> 80%) α-helical. CD studies indicate that the helicity of the peptides increases from ca. 40% to ca. 70% as the polarity of the solvent is decreased. Thus, it is of interest to follow the changes in helicity with time for the different dielectric models. Figure 9A shows the percentage helicity trajectories for S4-S45$_{II}$ for three different dielectric models. In all three simulations there is a decrease in the helicity with time as the peptides unfold. For $ε = 80$ unfolding is quite rapid, with an initial drop in helicity to ca. 45% within the first 50 ps. For $ε = 4$ the helicity drops more slowly, but after 300 ps has reached a comparable level. For $ε = r$ the helicity drops much more slowly, reaching ca. 70% after 300 ps. The rapid decline is α-helix content for $ε = 80$ is matched by an increase in β-strand content (Fig. 9B), which reaches a plateau of ca. 11%. For $ε = 4$ the plateau is lower, at ca. 3% β-strand, and for $ε = $

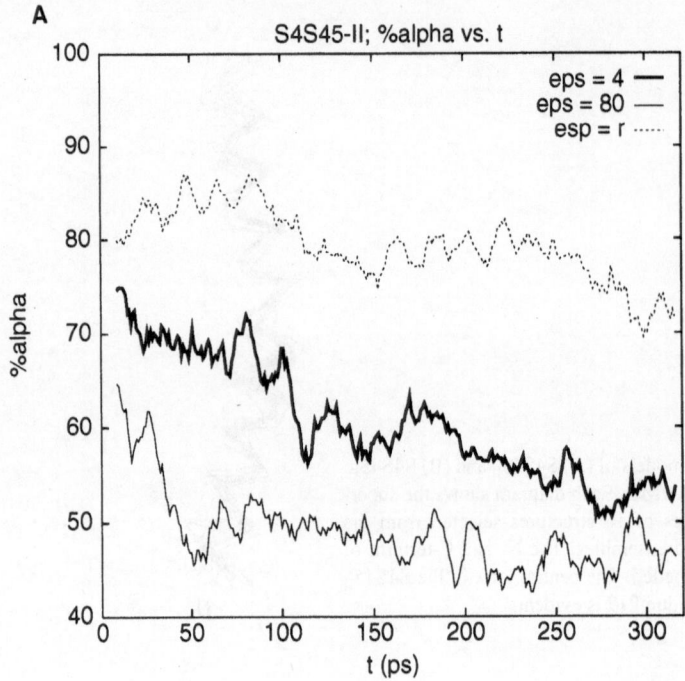

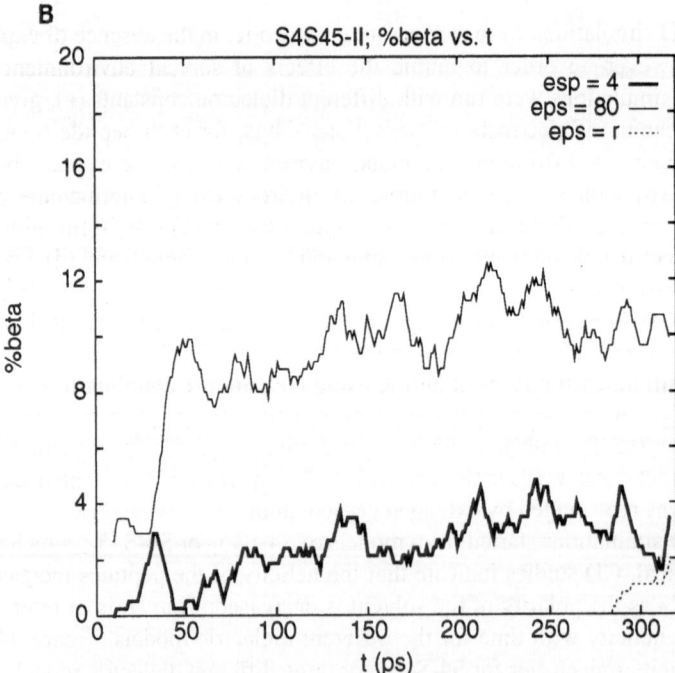

Figure 9. Secondary structure trajectories for the S4S45$_{II}$ MD simulations. (A) shows the percentage α-helix content, averaged over all residues, of S4S45$_{II}$ as a function of time elapsed during the MD simulations. (B) shows the percentage β-strand content for the same simulations. In each graph, the three curves correspond to the ε = 4 (bold line), ε = 80 (thin line) and ε = r (broken line) simulations.

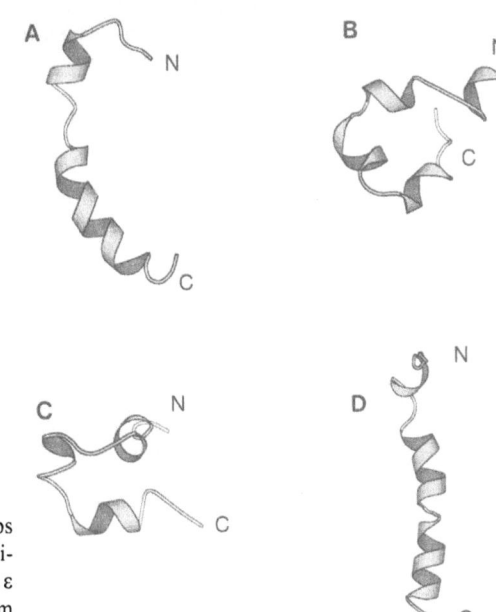

Figure 10. Selected structures from the last 100 ps of the S4S45$_{II}$ MD simulations, indicating the α-helical regions. The diagrams are for the (A) $\varepsilon = 4$; (B) $\varepsilon = 25$; (C) $\varepsilon = 80$; and (D) $\varepsilon = r$ simulations. (Diagram drawn using Molscript; Kraulis, 1991).

r hardly any β-strand is formed. Thus, the different electrostatic models result in different patterns of unfolding of the S4-S45$_{II}$ α-helix. Similar results are seen for S4-S45$_{IV}$ (data not shown), indicating that such unfolding is not simply a consequence of the central proline residue. Interestingly, in the light of the experimental data, the greatest decline in α-helix content is observed for the most polar simulation conditions, i.e. $\varepsilon = 80$.

It is useful to visualize the structures resulting from these simulations. In Figure 10, snapshots of the structures obtained at the end of the four S4-S45$_{II}$ simulations are shown, with α-helices as ribbons and β-strand plus random coil regions as coils. For the least polar simulation ($\varepsilon = 4$, Fig. 10A) S4-S45$_{II}$ is still largely helical, the main loss of helix being at the two termini and in the vicinity of the proline residue. A similar pattern, with a higher overall degree of helicity, is seen for $\varepsilon = r$ (Fig. 10D). In contrast, both of the polar simulations ($\varepsilon = 25$, Fig. 10B; and $\varepsilon = 80$, Fig. 10C) result in considerable loss of α-helix throughout the length of the peptide. Furthermore, for the latter two simulations the peptide collapses to adopt a more compact, less extended structure. Detailed analysis of the pattern of H-bonding in these structures (J. Breed, unpublished results) suggests that extensive H-bonding from cationic sidechains (arginine and lysine) to main-chain carbonyl groups and to hydroxyl sidechains in the S45 region help to stabilize the collapsed structure. Again, similar results (not shown) are obtained with S4-S45$_{IV}$.

Although visualization of structures from MD simulations is informative, it is difficult from such snapshots to assess rigorously the distribution of secondary structure elements along the length of the peptides. To achieve this the structures saved during the final 100 ps of the MD runs were analyzed, residue by residue, in terms of the percentage occupancy of the α-helix and β-strand regions of the Ramachandran plot. The result of this analysis is summarized in Table 3, in which the percentage α-helix and percentage β-strand are listed separately for the S4 (residues 1 to 23) and S45 (residues 24 - 34) regions of each peptide. From these data it is evident that in the least polar simulations ($\varepsilon = 4$ and $\varepsilon = r$) the α content is about the same for the S4 and S45 segments. However, in the more polar simulations ($\varepsilon = 25$ and $\varepsilon = 80$) whereas the percentage α in the S4 segment is approximately

Table 3. Secondary structure contents of S4s and S45s from
molecular modelling

Peptide	Region	ε	%α	%β
S4-S45$_{II}$	S4	4	55 (±7)	2 (±2)
		25	51 (±8)	13 (±3)
		80	51 (±7)	10 (±3)
		r	77 (±6)	0
S4-S45$_{II}$	S45	4	54 (±10)	6 (±3)
		25	50 (±13)	2 (±3)
		80	38 (±13)	10 (±3)
		r	73 (±13)	2 (±3)
S4-S45$_{IV}$	S4	4	54 (±7)	9 (±2)
		25	60 (±8)	11 (±3)
		80	53 (±8)	11 (±3)
		r	79 (±7)	0
S4-S45$_{IV}$	S45	4	60 (±10)	10 (±3)
		25	45 (±10)	22 (±6)
		80	38 (±13)	10 (±6)
		r	70 (±6)	0

The percentage content of α-helix and of β-strand conformations,
averaged over the last 100 ps of the MD simulations, are listed for the
S4 and S45 segments of the peptides.

unchanged from that for ε = 4, the α content of the S45 segment is significantly reduced. This is most evident for ε = 80 for both peptides, where the α content of S45 drops to less than 40% whereas that of S4 remains at above 50%.

Overall, two conclusions emerge from our MD simulations of S4-S45 peptides. The S4 region is relatively stable as an α-helix, regardless of the polarity of the electrostatic environment. In contrast, the S45 region has a tendency to unfold from its initial α-helical conformation if the degree of screening of electrostatic interactions is sufficiently high. This is accompanied by formation of H-bonds from the positively charged sidechains of the α-helical S4 region to the mainchain carbonyls and hydroxyl sidechains of the β-strand/random coil S45 region, resulting in a more compact overall structure. This change in unfolding pattern as ε is increased nicely parallels the CD studies (Helluin *et al.* 1995) which indicate an overall loss of helicity as the polarity of the solvent is increased, and which suggest that S45 may undergo an α to β transition.

DISCUSSION

Initial work with the peptide approach as applied to the sodium channel was concerned with the S3 and S4 transmembrane segments. A first peptide corresponding to S3 from domain I was found to form discrete cation-selective channels but with no evidence for sodium selectivity or voltage dependence (Oiki *et al.* 1988). This study was followed by another one with still a 22-mer peptide but whose sequence corresponded to the S4 segment of domain IV (Tosteson *et al.* 1989). Planar lipid bilayers doped with this peptide showed a voltage-dependent conductance but the individual channel openings had rather large unit conductances, ranging from 70 to 500 pS in 500 mM NaCl, with no selectivity for sodium.

Although some caution is of course needed when extrapolating conclusions drawn from the peptide approach (an especially if restricted to functional studies), we feel that the "integrative and collaborative" study described in this chapter can shed new light to the structure-function of ion channels. Some interesting inferences or propositions can be made. For instance, the polarity-dependent conformational switching of S45 seen in CD experiments and supported by the MD simulations could suggest that the inwardly-directed β-sheet promotion or at least some unfolding previously postulated for S4 during channel activation (Guy & Conti, 1990) would rather involve the S45 moiety. This could also account for the "flexible or spring" region at the junction of S4 and S45 as later modelled (Durell & Guy, 1992). From the selectivity brought about by S45 to S4 aggregates, as shown by Brullemans et al. (1994) and recalled in the first section of this chapter, it is highly plausible that S45 would participate in the inner lining of the pore on the cytoplasmic side. Such involvement of S45 in voltage-gated potassium channels has been independently confirmed in mutagenesis experiments (Slesinger, Jan & Jan, 1993). Thus, S45 may be drawn inside the pore lumen once S4 had been largely exposed to the extracellular side. Upon moving from a membrane to a more aqueous environment (an "hydration step" is also assumed in the gating, see e.g. Rayner et al. 1992), S45 could then undergo some unfolding and possibly adopt a β-sheet structure. We then propose that the pore might be made on the inner side of a β-barrel formed by the four S4 in series on the outer side with the main selectivity filter, presumably made by the anti-parallel arrangement of the short SS1 and SS2 segments (the P-region). Various experimental results on the whole channel reviewed by Guy & Durell (1994) support "the contention of models that the P segments do not span the entire transmembrane region." The overall selectivity thus achieved by these elements in series could then attain values comparable to that of the channel *in situ* (Chandler & Meves, 1965).

Future simulation studies will address such possible conformational changes in more detail via inclusion of explicit solvent molecules in the MD simulations of S4-S45 helix unfolding, and will extend these studies to other TM regions. In the latter context, we have already performed preliminary studies of β-hairpin models of the P region peptide (Cosette *et al.* unpublished results) and have completed a detailed investigation of conformations of the S6 helix of Shaker and the M2 helix of IRK1 (Kerr et al. 1995) K^+ channels.

The fluorescence and resonance energy transfer experiments support the hypothesis that the P regions are packed in close proximity in the tetrameric bundle of the sodium channel. These closely packed P segments may have a role as a structural element that participate in mediating the appropriate association of the hydrophobic core of the sodium channel. Furthermore, the preferential association of P segments might give clue as to their order of organization within the pore, in which the P regions of domain I and III are not adjacent. It is noteworthy that the smallest conductances are displayed by P_{II} and P_{IV} which are shown to be able to interact and form intramembrane heteroaggregates. As the same does not hold for P_I-P_{III}, the presumed "selectivity filter" viewed from above may not have a circular symmetry but may assume a more rectangular geometry, as initially proposed by Hille (1971, 1975). Future conductance experiments with bilayers incorporating various mixtures of the different P peptides will address the question of whether some heteroaggregates could be more sodium selective.

Likewise, eventual molecular recognition between S4-S45 and P peptides from the same or different domains will be investigated both in conductance-selectivity experiments and in resonance energy transfer experiments. The influence of lenghtening both S4-S45 and P peptides, particularly with the adjacent hydrophobic S5 segment and thus the influence of a further anchoring in the membrane, on voltage-sensitivity and ion selectivity will also be studied.

Finally, if the ability of the S4-S45 and P regions to form relatively stable and conducting aggregates within the membrane do not necessarily mean that these aggregates should have the same or similar properties as in the intact channel, they may however serve as a simple model for a synthetic ion channel which is formed by the assembly of several inserted polypeptides (Shai, 1995).

REFERENCES

Armstrong, C. & Bezanilla, F. (1973). Currents related to movement of the gating particles of the sodium channels. *Nature* 242, 459-461.

Auld, V.J., Goldin, A.L., Krafte, D.S., Catterall, W.A., Lester, H.A., Davidson, N. & Dunn, R.J. (1990). A neutral amino acid chnage in segment II S4 dramatically alters the gating properties of the voltage-dependent sodium channel. *Proceedings of the National Academy of Sciences of the USA* 87, 323-327.

Backx, P., Yue, D.T., Lawrence, J.H., Marban, E. & Tomaselli, G.F. (1992). Molecular localization of an ion-binding site within the pore of mammalian sodium channels. *Science* 257, 248-251.

Barsukov, I.L., Nolde, D.E., Lomize, A.L. & Arseniev, A.S. (1992). Three-dimensional structure of proteolytic fragment 163-231 of bacterioopsin determined from nuclear magnetic resonance in solution. *European Journal of Biochemistry* 206, 665-672.

Behrens, M.I., Oberhauser, A., Bezanilla, F. & Latorre, R. (1989). Batrachotoxin-modified sodium channels from squid optic nerve in planar bilayers: ion conduction and gating properties. *Journal of General Physiology* 93, 23-41.

Brullemans, M., Helluin, O., Dugast, J.-Y., Molle, G. & Duclohier, H. (1994). Implication of segment S45 in the permeation pathway of voltage-dependent sodium channels. *European Biophysics Journal* 23, 39-49.

Chang, C.T., Wu, C.S.C. & Yang, J.T. (1978). Circular dichroic analysis of protein conformation: inclusion of β-turns. *Analytical Biochemistry* 91,13-31.

Chandler, WK & Meves, H (1965). Voltage-clamp experiments on internally perfused giant axons. *Journal of Physiology* 180, 788-820.

Conti, F. & Sthumer, W. (1989). Quantal charge redistributions accompanying the structural transitions of sodium channels. *European Biophysics Journal* 16, 73-81.

Cosette, P., Brullemans, M. & Duclohier, H. (1994). Peptide modelling and functional assays of the P-region of voltage-dependent sodium channels. *Biophysical Journal* 66, A281.

Duclohier, H., Molle, G., Dugast, J.-Y. & Spach, G. (1992). Prolines are not essential residues in the "barrel-stave" model for ion channels induced by alamethicin analogues. *Biophysical Journal* 63, 868-873.

Duclohier, H., Cosette, P., Pouny, Y. & Shai, Y. (1995). Sodium selectivity exhibited by synthetic fragments derived from the P-regions of domains I and III of the eel sodium channel. *Biophysical Journal* 68, A265

Durell, S.R. & Guy, H.R. (1992). Atomic scale structure and functional models of voltage-gated potassium channels. *Biophysical Journal* 62, 238-250

Franciolini, F. (1994). The S4 segment and gating of voltage-dependent cationic channels. *Biochimica et Biophysica Acta* 1197, 227-236.

Gazit, E., Lee, W.-J., Brey, P.T. & Shai, Y. (1994). Mode of action of the antibacterial cecropin B2: a spectrofluorimetric study. *Biochemistry* 33, 10681-10692.

Gazit, E. & Shai, Y.(1993). Structural and functional characterization of the a-5 segment of *Bacillus thuringiensis* delta-endotoxin. *Biochemistry* 32, 3429-3436.

Guy, H.R. & Conti, F. (1990). Pursuing the structure and function of voltage-gated channels. *Trends in Neurosciences* 13, 201-206.

Guy, H.R. & Durell, S.R. (1994). Using sequence homology to analyze the structure and function of voltage-gated ion channel proteins. In *Molecular evolution of physiological processes*, Society of General Physiologists Symposium 47, edited by D. M. Fambrough, pp.197-212. Rockefeller University Press.

Guy, H.R. & Durell, S.R. Structural Models of Na⁺, Ca⁺ and K⁺ channels. In *Ion channels and genetic diseases*, Society of General Physiologists Symposium 48, edited by D. Lawson. Rockfeller University Press (in press).

Hall, J.E., Vodyanoy, I., Balasubramanian, T.M. & Marshall, G.R. (1984). Alamethicin: a rich model for channel behavior. *Biophysical Journal* 45, 233-247.

Hanke, W., Methfessel, C., Wilmsen, H.U. & Boheim, G. (1984). Ion channel reconstitution into planar lipid bilayers on glass pipettes. *Bioelectrochemistry and Bioenergetics* 12, 329-339

Heinemann, S.H., Terlau, H., Stühmer, W., Imoto, K. & Numa, S. (1992). Calcium channel characteristics conferred on the sodium channel by single mutations. *Nature* 356, 441-443.

Helluin, O., Breed, J. & Duclohier, H. (1995). Polarity-dependent conformational switching of a peptide mimicking the S4-S5 linker of the voltage-sensitive sodium channel. *Biochimica et Biophysica Acta* (in press).

Hille, B. (1971). The permeability of the sodium channel to organic cations in myelinated nerve. *Journal of General Physiology* 58, 599-619.

Hille, B. (1975). The receptor for tetrodotoxin and saxitoxin. A structural hypothesis. *Biophysical Journal* 15, 615-619.

Hille, B. (1984). Selective permeability: independence. In *Ionic channels of excitable membranes*, pp 226-248. Sinauer Associates Inc., Sunderland, Massachusetts.

Kerr, I.D. & Sansom, M.S.P. (1993). Hydrophilic surface maps of α-helical channel-forming peptides. *European Biophysics Journal* 22, 269-277.

Kerr, I.D., Sankararamakrishnan, R., Smart, O.S. & Sansom, M.S.P. (1994). Parallel helix bundles and ion channels:- molecular modelling via simulated annealing and restrained molecular dynamics. *Biophysical Journal* 67, 1501-1515.

Kerr, I.D., Son, H.S., Sankararamakrishnan, R. & Sansom, M.S.P. (1995). Molecular dynamics simulations of isolated transmembrane helices of potassium channels. *(submitted, 10.9.95)*.

Keynes, R.D. & Rojas, E. (1974). Kinetics and steady-state properties of the charged system controlling sodium conductances in the squid giant axon. *Journal of Physiology* 239, 393-434.

Kirsch, G.E., Shieh, C.C., Drewe, J.A., Vener, D.F. & Brown, A.M. (1993). Segmental exchanges define 4-aminopyridine binding and the inner mouth of K^+ pores. *Neuron* 11, 503-512.

Kontis, K.J. & Goldin (1993). Site-directed mutagenesis of the putative pore region of the rat IIA sodium channel. *Molecular Pharmacology* 43, 635-644.

Kraulis, P.J. (1991). MOLSCRIPT: a program to produce both detailed and schematic plots of protein structures. *Journal of Applied Crystallography* 24, 946-950.

Lear, J.D., Wasserman, Z.R. & DeGrado, W.F. (1988). Synthetic amphiphilic models for protein ion channels. *Science* 240, 1177-1181.

Lemmon, M.A. & Engelman (1992). Helix-helix interactions inside lipid bilayers. *Current Opinion in Structural Biology* 2, 511-518.

Lopez, G.A., Jan, Y.N. & Jan, L.Y. (1994). Evidence that the S6 segment of the Shaker voltage-gated K^+ channel comprises part of the pore. *Nature* 367, 179-182.

Lovejoy, B., Akerfeldt, K.S., DeGrado, W.F. & Eisenberg, D. (1992). Crystallization of proton channel peptides. *Protein Science* 1, 1073-1077.

Merrifield, R.B. (1963). Solid phase peptide synthesis: I. the synthesis of a tetrapeptide. *Journal of American Chemical Society* 89, 2149-2154.

Miller, C. (1986). Ion channel reconstitution: why bother? In *Ionic channels in cells and model systems*, edited by R. Latorre, pp 256-271. Plenum, New York and London.

Montal, M. & Mueller, P. (1972). Formation of bimolecular membranes from lipid monolayers and a study of their electrical properties. *Proceedings of the National Academy of Sciences of the USA* 69, 3561-3566.

Montal, M (1995). Design of molecular function: channels of communication. *Annual Review of Biophysics and Biomolecular Structure* 24, 31-37.

Moran, O., Gheri, A., Zegarra-Moran, O., Imoto, K. & Conti, F. (1994). Proline mutations on the S4 segment of rat brain sodium channel II. *Biochemical and Biophysical Research Communications* 202, 1438-1444.

Morris, A.L., MaCArthur, M.W., Hutchinson, E.G. & Thornton, J.M. (1992). Stereochemical quality of protein structure coordinates. *Proteins: Structure, Function and Genetics* 12, 345-364.

Mueller, P. & Rudin, D.O. (1968). Action potentials induced in bimolecular lipid membranes. *Nature* 217, 713-719.

Mulvey, D., King, J.F., Cooke, R.M., Doak, D.J., Harvey, T.S. & Campbell, I.D. (1989). High resolution ^{1}H NMR study of the solution structure of the S4 segment of the sodium channel protein. *FEBS Letters* 257, 113-117.

Noda, M., Shimizu, S., Tanabe, T., Takai, T., Kayano, T., Ikeda, T., Takahashi, H., Nakayama, H., Kanaoka, Y., Minamino, N., Kangawa, K., Matsuo, H., Raftery, M.A., Hirose, T., Inayama, S., Hayashida, H., Miyati, T.& Numa, S. (1984). Primary styructure of *Electrophorus electricus* sodium channel deduced from cDNA sequence. *Nature* 312, 121-127.

Oiki, S., Danho, W. & Montal, M. (1988). Channel protein engineering: synthetic 22-mer peptide from the primary structure of the voltage-sensitive sodium channel forms ionic channels in lipid bilayers. *Proceedings of the National Academy of Sciences of the USA* 85, 2393-2397.

Papazian, D.M., Shao, X.M., Seoh, S.-A., Mock, A.F., Huang, Y. & Wainstock, D.H. (1995). Electrostatic interactions of S4 voltage-sensor in Shaker K^+ channel. *Neuron* 14, 1293-1301.

Peled, H., Arkin,I.T., Engelman, D.M. & Shai, Y. (1996). Coassembly of synthetic peptides corresponding to transmembrane domains of the Shaker K^+ channel. *40th Annual Meeting of the Biophysical Society, Baltimore.*

Peled, H. & Shai, Y. (1993). Membrane interactions and self-assembly within phospholipid membranes of synthetic segments corresponding to the H-5 region of the Shaker K$^+$ channel. *Biochemistry* 32, 7879-7885.

Planells-Cases, R., Ferrer-Montiel, A.V., Patten, C.D. & Montal, M. (1995). Mutation of conserved negatively-charged residues in the S2 and S3 transmembrane segments of a mammalian K$^+$ channel selectively modulates channel gating. *Proceedings of the National Academy of Sciences of the USA* 92, 9422-9426.

Popot, J.-L. & Engelman, D.M. (1990). Membrane protein folding and oligomerization: the two stage model. *Biochemistry* 29, 4031-4037.

Popot, J.-L., Gerchman, S.E. & Engelman, D.M. (1987). Refolding of bacteriorhodopsin in lipid bilayers. A thermodynamically controlled two-stage process. *Journal of Molecular Biology* 198, 655-676.

Pouny, Y., Rapaport, D., Mor, A., Nicolas, P. & Shai, Y. (1992). Interaction of antimicrobial dermaseptin and its fluorescently labeled analogues with phospholipid membranes. *Biochemistry* 31, 12416-12423.

Pouny, Y. & Shai, Y. (1995). Synthetic peptides corresponding to the four P regions of *Electrophorus electricus* Na$^+$ channel: Interaction with and organization in model phospholipid membranes. *Biochemistry* 34, 7712-7721.

Pusch, M., Noda, M., Stühmer, W. Numa, S. & Conti, F. (1991). Single point mutations of the sodium channel drastically reduce the pore permeability without preventing its gating. *European Biophysics Journal* 20, 127-133.

Rapaport, D., Danin, M., Ghazit, E. & Shai, Y. (1992). Membrane interactions of the sodium channel S4 segment and its fluorescently-labeled analogues. *Biochemistry* 31, 8868-8875.

Rayner, M.D., Starkus, J.G. Ruben P.C. & Alicata, D.A. (1992). Voltage-sensitive and solvent-sensitive processses in ion channel gating: kinetic effect of hyperosmolar media on activation and deactivation of sodium channels. *Biophysical Journal* 61, 96-108.

Sammar, M., Spira, G. & Meiri, H. (1992). Depolarization exposes the voltage sensor of sodium channels to the extracellular regions. *Journal of Membrane Biology* 125, 1-11.

Sankararamakrishnan, R. & Sansom, M.S.P. (1995). Structural features of isolated M2 helices of nicotinic receptors. Simulated annealing via molecular dynamics studies. *Biophysical Chemistry* 55, 215-230.

Sansom, M.S.P. (1991). The biophysics of peptide models of ion channels. *Progress in Biophysics and Molecular Biology* 55, 139-235.

Schetz, J.A. & Anderson, P.A.V. (1993). A reevaluation of the structure in the pore region of voltage-activated cation channels. *Biological Bulletin* 185, 462-466.

Shai, Y. (1995). Molecular recognition between membrane-spanning helices. *Trends in Biochemical Sciences* 20, 460-464.

Sigworth, F. (1993). Voltage gating of ion channels. *Quaterly Review of Biophysics* 27, 1-40.

Slesinger, P.A., Jan, Y.N. & Jan L.Y. (1993). The S4-S5 loop contributes to the ion-selective pore of potassium channels. *Neuron* 11, 739-749.

Spach, G., Duclohier, H., Molle, G. & Valleton, J.-M. (1989). Structure and supramolecular architecture of membrane channel-forming peptides. *Biochimie* 71, 11-21.

Strahilevitz, J., Mor, A., Nicolas, P. & Shai, Y. (1994). Spectrum of antimicrobial activity and assembly of dermaseptin-b and its precursor form in phospholipid membranes. *Biochemistry* 33, 10951-10960.

Stühmer, W. (1991). Structure-function studies of voltage-gated ion channels. *Annual Review of Biophysical Chemistry* 20, 65-78.

Stühmer, W. (1993). Structure and function of sodium channels. *Cellular Physiology Biochemistry* 3, 277-282.

Stühmer, W., Conti, F., Suzuki, H., Wang, X., Noda, M., Yahagi, N., Kuo, H. & Numa. S. (1989). Structural parts involved in activation and inactivation of the sodium channel. *Nature* 339, 597-603.

Terlau, H., Heinemann, S.H., Stühmer, W., Pusch, M., Conti, F., Imoto, K. & Numa, S. (1991). Mapping the site of block by tetrodotoxin and saxitoxin of sodium channel II. *FEBS Letters* 293, 93-96.

Tosteson, M.T., Auld, D.S. & Tosteson, D.C. (1989). Voltage-gated channels formed in lipid bilayers by a positively charged segment of the Na-channel polypeptide. *Proceedings of the National Academy of Sciences of the USA* 86, 707-710.

Tuchscherer, G. & Mutter, M. (1995). Templates in protein de novo design. *Journal of Biotechnology* 41, 197-210.

5

THE HYPERPOLARIZATION-ACTIVATED CURRENT (I_h/I_q) IN RAT HIPPOCAMPAL NEURONS

S. Gasparini,[1] G. Maccaferri,[2] R. D'Ambrosio,[1] and D. DiFrancesco[1]

[1] Università degli Studi di Milano
Dipartimento di Fisiologia e Biochimica Generali, Elettrofisiologia
Via Celoria 26, 20133 Milano, Italy
[2] Unit on Cellular and Synaptic Physiology
Laboratory of Cellular and Molecular Neurophysiology
National Institute of Child Health and Human Development
Bethesda, Maryland 20892-4495

1. INTRODUCTION

The hippocampus is a good model for the study of physiological behaviour of neurons in the Central Nervous System. Its circuitry is well organized and allows a clear identification of the different populations of neurons. In addition, brain slices from the hippocampus were developed to preserve synaptic contacts between neurons [40]. For this reason, in particular in the last twenty years, phenomena of synaptic plasticity (i.e. long term modifications of the efficiency of information processing at synaptic sites) have been extensively studied in this area of the brain [4, 5]. The study of intrinsic ionic conductances of neurones is required to better understand how their behaviour and their firing are controlled. It is especially important to study which currents underlie the membrane potential and hence modulate the excitability of neurons.

In addition to a non-specific leak current, two conductances have been suggested to modulate the membrane potential in CA1 pyramidal cells. A potassium current called $I_{K(M)}$ tends to clamp the membrane potential near E_K, when not blocked by acetylcholine through a muscarinic receptor [8, 25]. A mixed cation current, carried by Na^+ and K^+ and activated by hyperpolarization tends, on the contrary, to depolarize neurons, although whether or not this current is activated at resting potentials is controversial [25, 28, 38]. This last current is characterized by a higher conductance in the inward than in the outward direction and thus produces an inward or "anomalous rectification", because the conductance of the whole membrane increases on hyperpolarization [27]. During a prolonged injection of hyperpolarizing current, a marked depolarizing sag towards the resting membrane potential occurs as a result of its activation. A current with similar features was originally described in cardiac sino-atrial node myocytes, where it was termed "pacemaker" because of its involvement in the generation and control of spontaneous activity [10, 15]. Channels with similar properties have been described in several tissues [1, 30] and were first reported in the nervous system in the hippocampus [25], where the current was termed I_q by the authors. In the thalamus, this current is known as I_h and is functionally important in cellular excitability. I_h is especially important for the control of the firing pattern of thalamic neurons and their modulation by the neurotransmitters serotonin and noradrenaline [31, 37], and by adenosine [34].

The purpose of this review is to describe the characteristics of I_h (or I_q) in hippocampal pyramidal neurons, its physiological role and its modulation by acetylcholine.

2. METHODS

Transverse hippocampal slices were obtained from guinea-pigs and from rats as described in the literature (see ref. 25 and 28) and stored in a holding bath containing artificial cerebrospinal fluid (ACSF) composed of (mM): 120 NaCl, 3.1 KCl, 1 $MgCl_2$, 2 $CaCl_2$, 1.25 KH_2PO_4, 26 $NaHCO_3$, 10 glucose. A single slice was then transferred to the recording chamber were it was held, fully submerged and superfused continuously with ACSF. All solutions were equilibrated with 95% O_2-5% CO_2 to a pH of 7.4.

Under voltage-clamp conditions, in some experiments [28], tetraethylammonium chloride (TEA-Cl 10 mM), 4-amino-pyridine (4-AP 5 mM), tetrodotoxin (TTX 0.5 μM), $BaCl_2$ (1 mM), $NiCl_2$ (0.3 mM) and $CdCl_2$ (0.3 mM) were substituted to equimolar NaCl to block every other current except I_h. In these modified solutions, KH_2PO_4 was eliminated to avoid precipitation and, to compensate for its removal, KCl was raised to 4.35 mM. In other experiments [13, 25], only $BaCl_2$ and TTX were added.

Recording from single CA1 neurones was performed using intracellular [13, 25] or patch clamp techniques [28]: in one case the electrode was filled with 3 M potassium acetate, or caesium acetate in order to reduce potassium conductances (with electrode resistance of 60-100 MΩ). During whole cell recordings the pipette was filled with (mM): 130 K-gluconate, 10 NaCl, 1 $MgCl_2$, 2 adenosine-trisphosphate (ATP), 0.3 guanosine trisphosphate (GTP), 10 N-2-hydroxyethylpiperazine-N'-2-ethanesulfonic acid (HEPES), 0.5 ethylenebis (oxonitrilo) tetra-acetic acid (EGTA) to a final pH of 7.2 (resistance 3-10 MΩ).

UL-FS 49 and Zeneca ZD7288 were kindly provided by Thomae GmbH, Biberach, Germany (Dr. B. Guth) and Zeneca Pharmaceuticals, Macclesfield, UK (Dr. I. Briggs) respectively.

I_h was also recorded from cells in tissue-cultured slices [9] and from dissociated cells in culture [39]. This current has never been studied in acutely dissociated neurons, probably because proteolytic treatments may damage h-channels. Recent evidence has indicated that h-channels can be damaged by external proteolysis in neocortical neurons since the current required several hours after enzymatic dissociation to recover [11].

3. RESULTS

3.1. First Description of a Hyperpolarization-Activated Cation Current in Hippocampal Neurons

A hyperpolarization-activated cationic current in hippocampal neurons was first described by Halliwell & Adams [25], who discovered that, in addition to $I_{K(M)}$, a secondary component was activated during hyperpolarizing steps in voltage-clamp conditions. Small hyperpolarizations from -40 mV gave rise to M-like relaxations, whereas from -65 mV an ohmic behaviour was observed. As hyperpolarizing steps increased, current recorded from -40 mV became smaller due to a decrease in the driving force for potassium, but when the holding potential was held at -65 mV another inwardly relaxing component appeared. They hypothesized that this latter component was due to a time dependent anomalous rectification, responsible for the deviation in the hyperpolarizing potential seen in many neurons, and they called this current I_q. "Q" stands for "queer", which probably echoes the terminology used for the similar hyperpolarization-activated current previously described in cardiac cells, where "f" stands for "funny". They explained their voltage clamp recordings in terms of two separate sets of time- and voltage-dependent channels in the pyramidal cell membrane: the $I_{K(M)}$ current that was recorded between -40 and -70 mV, and I_q, activated at more negative potentials. They found that I_q had a reversal potential more positive than -60 mV and was activated at potentials between -80 and -120 mV. The amplitude and the rate of relaxation increased at more negative voltages.

From a pharmacological point of view, they found this current similar to I_f found in cardiac pacemaker tissue. Indeed I_q was insensitive to tetrodotoxin or Ba^{2+}, or Cd^{2+}, but was blocked by Cs^+ (0.5-3 mM), which was also able to abolish the depolarizing sag in the electrotonic potential [25]. In voltage clamp recordings, when holding potential was close to rest (-70 mV), Cs^+ reduced both the instantaneous and steady-state curves, particularly at negative potential [25].

3.2. Characterization of I_h

3.2.1. I_h is Tonically Active at the Membrane Resting Potential. The reduction of instantaneous component by Cs^+ can have two explanations. Firstly, Cs^+ could block leakage channels as well as I_q. Secondly, there could be q-channels already open at the holding potential. This second hypothesis, suggested by Halliwell and Adams, was confirmed by Maccaferri et al. [28], who found that I_h was elicited by hyperpolarizing steps from a holding potential of -35 mV (fig. 1 a) and was activated below the threshold of -50/-60 mV. Exposure to external Cs^+ (5 mM) caused a complete block of the time-dependent inward current (fig. 1 b). To understand possible effects of Cs^+ on leakage channels two sets of subtracted traces were compared. The Cs^+-dependent component was obtained by subtracting the traces recorded during Cs^+ perfusion from the corresponding control traces (fig. 1 c). To subtract leakage and capacitance, 10 mV hyperpolarizing steps from -35 mV were applied and used, after appropriate scaling, on the whole voltage range investigated (fig. 1 d). Comparing Cs^+- and leak-subtracted traces (fig. 1 c and 1 d), or the I/V plots of the same traces (fig. 1 e), shows that the two subtraction procedures give very similar results. This comparison, therefore, shows that Cs^+ does not affect time-independent components but blocks only the inwardly activating current I_h and thus can be used as a tool to study the functional role of I_h. Moreover Cs^+ was shown to hyperpolarize by about 4 mV the membrane resting potential in unclamped CA1 pyramidal cells [28], also in agreement with the block of an inward component, tonically active at the resting potentials, contributing to maintain the resting potential at more depolarized levels.

3.2.2. Activation Curve of I_h. At first, activation of I_q/I_h at various potential was studied measuring the tail currents and plotting them as a function of the preceding holding potential [25]. The current activated at about -80 mV and saturated at potentials more negative than -120 mV. In subsequent studies a more depolarized activation curve was constructed for the h-current (fig. 2). Since I_h does not show inactivation, the protocol consists of two-steps (fig. 2 a): the current is first activated to a variable degree by variable hyperpolarizing steps (test potentials, from -45 to -135 mV, in 10 mV steps), and is then fully

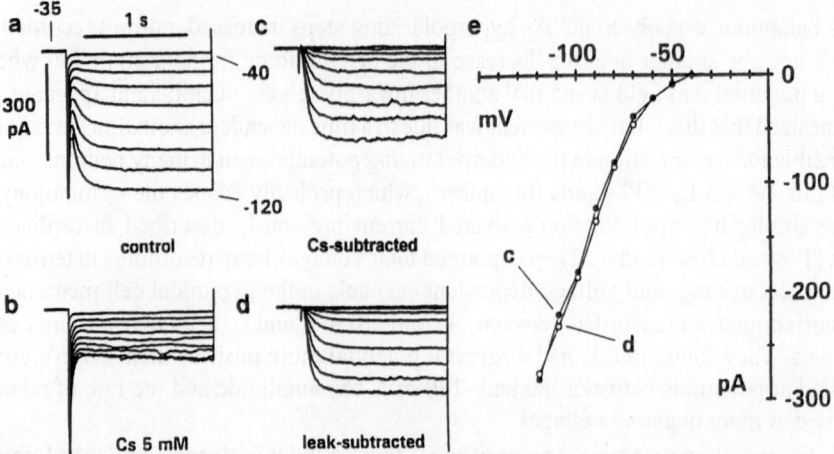

Figure 1. Comparison of I_h traces obtained after Cs^+- and leakage-subtraction in hippocampal CA1 neurons. *a*: Original current traces recorded from a holding potential of -35 mV. *b*: Current traces recorded at the same potentials as in (*a*) in the presence of Cs^+ 5 mM. *c*: Cs^+-sensitive component obtained by subtracting traces in (*b*) from the corresponding traces in (*a*). *d*: Traces obtained from (*a*) after capacitive and linear subtraction. Subtraction was performed using the average of 20 records from -35 to -45 mV, scaled for each potential. *e*: I/V plots obtained from *c* (•) and *d* (O) (from ref. 28, with permission).

activated with a second fixed step to -145 mV). Tails measured at -145 mV are inversely related to the degree of activation of the current at the previous test voltage (fig. 2 b). The threshold for I_h activation was about -50 mV (a value substantially more positive than that previously reported by Halliwell & Adams [25]); saturation occurred at about -140 mV (fig. 2 c) and the midpoint of activation was -98.0 mV [28].

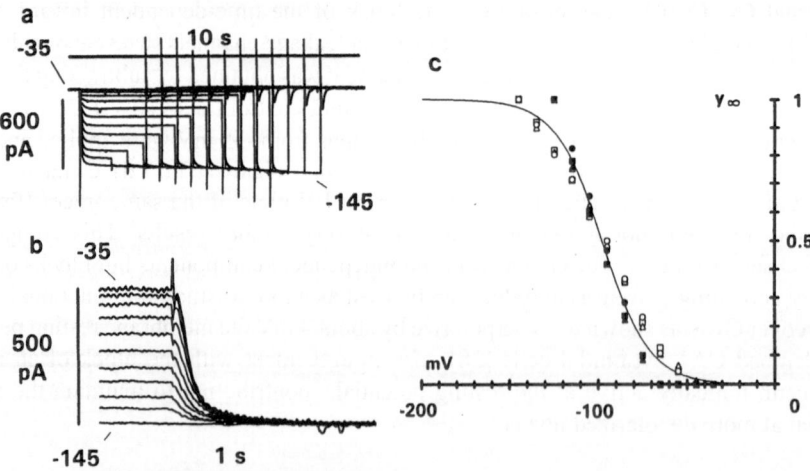

Figure 2. The activation curve of the I_h current as constructed using a method similar to that described by DiFrancesco et al. [18]. *a*: From a holding potential of -35 mV, neurons were hyperpolarized for the time necessary to reach the steady state at each potential and then stepped to -145 mV for 1.75 s to fully activate the current. *b*: Tails recorded at -145 mV are superimposed, after capacity and leakage subtraction. *c*: plot of the I_h activation curve. Open and closed symbols refer to tails measured at -145 mV and -125 mV, respectively. Experimental data were fitted by the Boltzmann equation $y_\infty = 1 / \{\exp [(E - E_h)/ s]\}$ (from ref. 28, with permission).

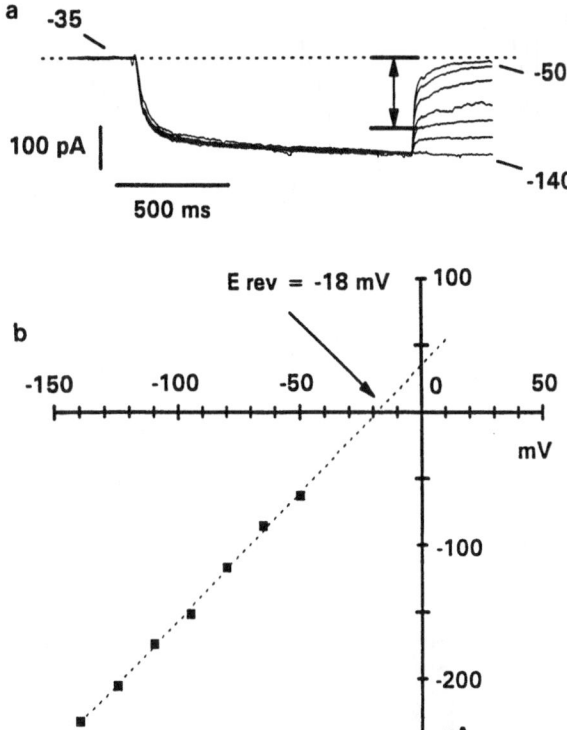

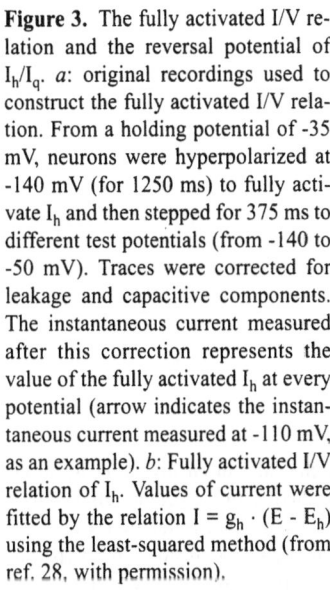

Figure 3. The fully activated I/V relation and the reversal potential of I_h/I_q. *a*: original recordings used to construct the fully activated I/V relation. From a holding potential of -35 mV, neurons were hyperpolarized at -140 mV (for 1250 ms) to fully activate I_h and then stepped for 375 ms to different test potentials (from -140 to -50 mV). Traces were corrected for leakage and capacitive components. The instantaneous current measured after this correction represents the value of the fully activated I_h at every potential (arrow indicates the instantaneous current measured at -110 mV, as an example). *b*: Fully activated I/V relation of I_h. Values of current were fitted by the relation $I = g_h \cdot (E - E_h)$ using the least-squared method (from ref. 28, with permission).

3.2.3. Fully-Activated Curve and Reversal Potential. To determine the slope conductance and the ions responsible for carrying I_h, protocols were performed to establish its fully-activated relation and reversal potential (fig. 3). From a holding potential of -35 mV, I_h was fully activated with a hyperpolarizing step to -140 mV, followed by a step to different test potentials (from -140 to -50 mV) (fig. 3a). The instantaneous current measured at the beginning of the test potential represents the fully activated I_h at every test potential, as it is a function of the fully-activated conductance g_h and the driving force for the current at each potential. The range positive to -50 mV could not be explored due to interference of other components even in the presence of channel blockers. The reversal potential obtained for I_h was approximately -17 mV (with a fully activated conductance of 2.5 nS) (fig. 3b) [28]. H-channels seem therefore to be permeable to Na^+ and K^+ ions as seen also in substitution experiments [28]. Values obtained for the reversal potential, moreover, provide further evidence for the similarity between the hippocampal I_h current and I_f in cardiac tissue.

3.3. Role of I_h on Excitability of CA1 Neurons

3.3.1. Role of I_h on Firing Properties of Hippocampal Neurons. The role of I_h in setting the membrane potential and membrane conductance was shown by using Cs^+ in current clamp conditions. Besides hyperpolarization of neurons, as seen above, Cs^+ induced a decrease of membrane conductance [28]. This result suggests, once again, that Cs^+ blocks an inward current, tonically active at resting potential, and that this component contributes significantly to both the resting voltage level and membrane conductance. Furthermore, a depolarizing pulse able to generate one or more action potentials in control conditions failed to elicit activity in the presence of Cs^+, suggesting that I_h alters excitability. Cs^+, however, did not affect the threshold for the action potential, because if tonic depolarizing current was injected to restore the original resting potential, firing was resumed [28]. This effect was

a

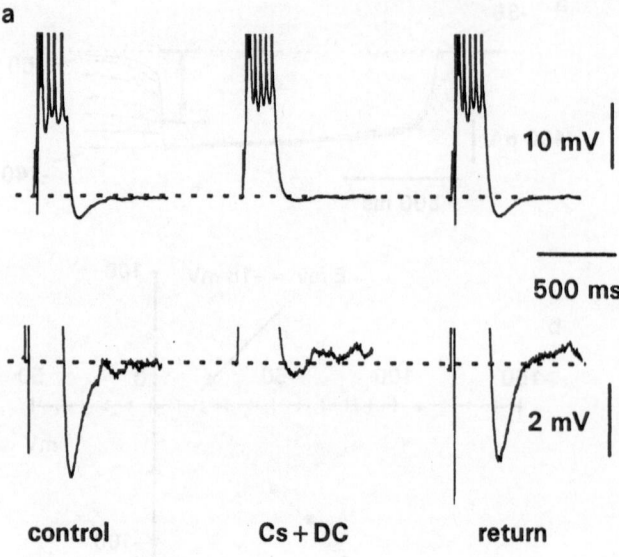

10 mV

500 ms

2 mV

control   Cs + DC   return

b

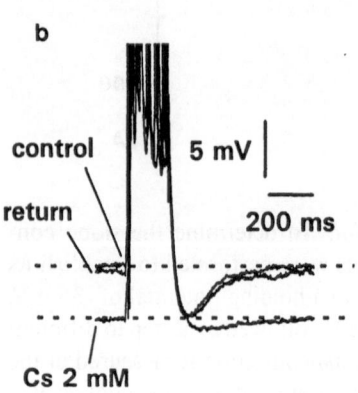

control   5 mV

return   200 ms

Cs 2 mM

Figure 4. Effect of Cs^+ on the Cd^{2+}-insensitive after-hyperpolarization, obtained after a train of action potentials elicited by a 100 pA depolarizing current step of 200-ms duration ($CdCl_2$ 0.3 mM present throughout). *a*:during Cs^+ perfusion, when hyperpolarization induced by the I_h-blocker was compensated by the injection of tonic current (15 pA) to restore the original resting potential, AHP was nearly completely abolished, and then restored after Cs^+-washout. *b*: when hyperpolarization was not compensated, Cs^+ clearly removed the depolarizing phase of AHP, without altering the previous hyperpolarization. In superimposed traces it can be seen that hyperpolarization induced by Cs^+ and AHP have very similar amplitude (from ref. 28, with permission).

therefore due only to the block of I_h and the consequent hyperpolarization induced by Cs^+, such that the membrane potential was further displaced from the threshold for action potentials.

3.3.2. Role of I_h in Ca^{2+}-Independent after-Hyperpolarization. I_h may also be involved in Ca^{2+}-independent after-hyperpolarizations (AHPs) (fig. 4) [28]. In the presence of Cd^{2+}, a depolarizing current step produces a train of action potentials followed by a Ca^{2+}-independent after-hyperpolarization. The AHP was reduced by about 80% during perfusion of Cs^+ and compensation of the hyperpolarization by current injection (fig. 4 a). When hyperpolarization was not compensated, the slow depolarizing phase of AHP was abolished by Cs^+, indicating its dependence on the activation of I_h (fig. 4 b). Moreover, the hyperpolarization of the resting potential induced by perfusion of Cs^+ was comparable to the amplitude of the after-hyperpolarization (fig. 4 b), further evidence that the same current underlies both phenomena.

3.4. Modulation of I_h by Carbachol

Acetylcholine (ACh) in hippocampus is known to have an excitatory action because it reduces three distinct K^+ conductances: 1) $I_{K(M)}$ [25]; 2) a Ca^{2+}-activated K^+ current called

I_{AHP} and responsible for the Ca^{2+}-dependent after-hyperpolarization [29]; 3) a K^+ leak channel ($I_{K,L}$), both time- and voltage-independent [3, 29]. In the presence of Ba^{2+}, known to block both $I_{K(M)}$ and $I_{K,L}$, perfusion of the muscarinic agonist carbachol is still able to generate a depolarization of the membrane potential accompanied by a decrease in input resistance [13]. Carbachol elicited an inward current, that tends to decrease during prolonged perfusion, when the membrane potential is clamped at resting. In addition, an increase of the instantaneous current and a potentiation of I_h was observed when hyperpolarizing voltage steps were applied. Since the response to carbachol tended to decrease, the authors used a ramp protocol that allowed the study of the voltage dependence of the response to carbachol in a short period of time. From this protocol, a current-voltage relation was obtained at different stages of the carbachol response. At first carbachol increased I_h but only at negative potentials, while at more depolarized potentials the two current traces reached the same level. I_q appeared even more potentiated at the peak of the inward current, but, in this case, traces never merged, even at depolarized potentials. Cs^+ was seen to reduce predominantly the negative part of the ramp current traces and to prevent the enhancement of the hyperpolarization-activated inward relaxation by carbachol, again indicating that carbachol potentiates I_q. Thus the current potentiated by carbachol had the same voltage sensitivity as that blocked by Cs^+ and was identified as I_q [13]. In addition, carbachol activated a calcium-dependent non-specific conductance, responsible for the voltage-independent increase of conductance observed in the second phase of the response to carbachol.

In some brain regions I_h was demonstrated to be increased by a cAMP-dependent mechanism through the activation of adenylyl cyclase [6, 37]. In the hippocampus this does not seem to occur since β-adrenergic agonists, known to stimulate production of cAMP, failed to give the same response evoked by carbachol [13]. Instead the effect of carbachol on I_h is antagonized by gallamine [13], a selective antagonist for M_2 receptors [12] and seems to be mediated by the hydrolysis of phosphatidyl-inositol (PI) [13]. It was thus hypothesized that potentiation of I_h by carbachol depends on an elevation of $[Ca^{2+}]_i$, since it requires extracellular Ca^{2+} [13].

3.5. Block of I_h/I_q by Bradycardic Agents

In the last few years, agents known as "specific bradycardic agents" have been developed which decrease heart rate without compromising contractility. Two of them, UL-FS 49 and Zeneca ZD7288 are known to slow the cardiac frequency by blocking I_f current in cardiac cells (sino-atrial node and Purkinje fibres) [7, 16, 48]. These substances have also been shown to block the hyperpolarization-activated current in neurons of the thalamus [35] and substantia nigra [26]. Therefore we studied the effect of UL-FS 49 on I_h/I_q in hippocampal neurons in voltage clamp conditions. From a holding potential of -50 mV cells were hyperpolarized to -100 mV. UL-FS 49 10 μM slowly blocked I_h during activation (fig. 5). Inhibition began about 5 minutes after perfusion of the drug, while the maximum effect was reached after about 20 minutes.

The effect of ZD7288 was studied in current clamp conditions. ZD7288 (10 μM) produced a hyperpolarization of about 6 mV, as shown in the plot of the membrane potential as a function of time in control and during perfusion of the drug (fig. 6). This hyperpolarization is greater than that observed during the perfusion of Cs^+ [28]. This could be due to a voltage-dependence of I_h block by Cs^+, such as it happens in the sino-atrial node where inhibition of I_f is smaller at less hyperpolarized potentials [18]. The sag of the membrane potential during the injection of the hyperpolarizing current and the depolarizing rebound on depolarization can be clearly seen in the control trace (fig. 6, bottom) and are due to slow activation of I_h during hyperpolarization of the membrane potential and its slow deactivation

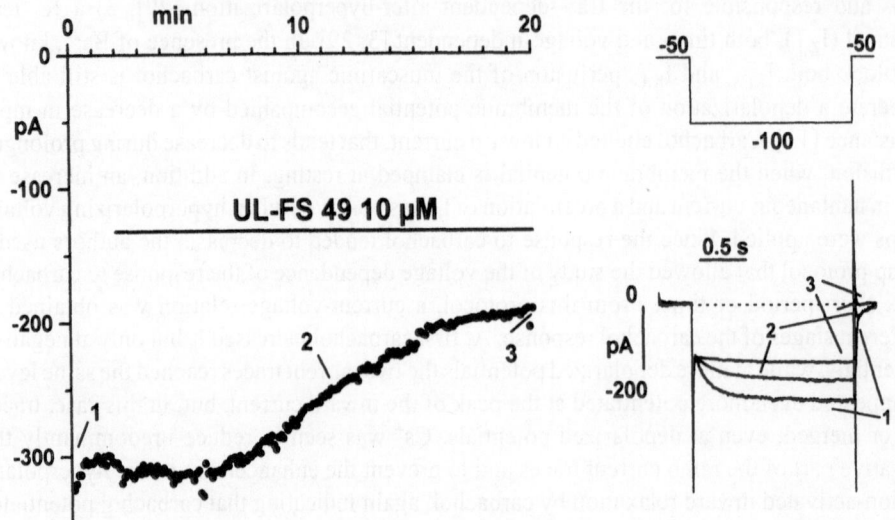

Figure 5. Block of the hyperpolarization-activated current (I$_h$) in a hippocampal CA1 neuron by UL-FS 49 (10 μM). Right: original current traces obtained from the voltage protocol shown in the upper panel, before (1) and during drug perfusion (2, 3). The hyperpolarizing steps were applied every six seconds. Left: time course of I$_h$ amplitude during drug application. The plot was constructed by measuring the current at the end of step to -100 mV, as referred to the holding current. The numbers (1-3) correspond to the current traces shown on the right.

at the resting membrane potential. Those effects disappeared in the presence of ZD7288, indicating a specific block of I$_h$.

4. DISCUSSION

The functional role of the hyperpolarization-activated cation current in the hippocampus depends on the actual threshold for its activation. If activated only below the normal resting potential of hippocampal neurons, as reported by Halliwell & Adams [25], its function would be to simply counterbalance maintained hyperpolarization, such as during GABA$_B$ receptor-mediated IPSPs [32, 42]. During these experiments, however, there may have been a significant contamination by I$_{K(M)}$. On the contrary, if I$_h$ is tonically active at rest, as shown by Maccaferri et al. [28], it would contribute substantially to set the resting membrane potential under normal conditions. Strong evidence supporting the latter view is provided by the about 4 mV hyperpolarization of the resting membrane potential produced by Cs$^+$ (2 mM) and the related reduction of the membrane conductance by about 30%. These two effects, taken together, support the view that Cs$^+$ is able to block an inward component tonically active at resting levels, most likely the hyperpolarization-activated current I$_h$. Its depolarizing action can thus maintain neurons closer to the action potential threshold and allow cell firing even when depolarizing inputs (EPSPs) are small. I$_h$ has indeed been reported to modulate the resting potentials in many neurons [30, 42, 43, 46]. In rat nucleus accumbens neurons, the presence of this current was shown to be correlated to a more depolarized resting potential [47]. In hypoglossal motoneurones, a developmental increase in I$_h$ density could be responsible, at least in part, for the more depolarized resting potential and the higher membrane conductance of neurons from older rats [2]. In the thalamus, the

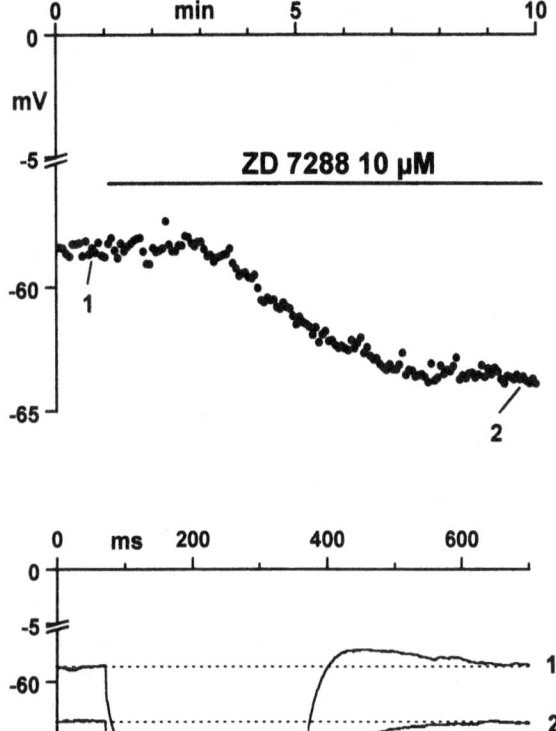

Figure 6. Effect of Zeneca ZD 7288 (10 µM) on membrane potential of hippocampal CA1 neurons. Top: time course of the hyperpolarization of the membrane potential of a CA1 neuron due to the block of I$_h$ current by ZD 7288. Bottom: voltage records before (1) and during (2) drug perfusion. A hyperpolarizing current step (75 pA for 300 ms) was injected every 4 seconds.

shift in the activation curve of I$_h$, due to specific neurotransmitters, was shown to be responsible of changing firing patterns during different activities [31, 37].

I$_h$ is not likely to contribute to the waveforms of individual action potentials, because its activation and deactivation are too slow. On the other hand, its slow deactivation permits a depolarizing rebound of the membrane potential at the end of a hyperpolarizing input and, in some tissues, can bring the neuron to the threshold for an action potential [41]. Finally, in hippocampal pyramidal cells, the slow reactivation of I$_h$, that follows a 200 ms train of action potentials, controls the depolarizing phase of the Ca^{2+}-independent AHP [28].

Pyramidal cells receive cholinergic inputs from fibres originating from neurons of the medial septum [33, 45] suggesting that the action of ACh is important in the modulation of neuronal activity in the hippocampus. Carbachol potentiates I$_h$ in pyramidal cells, although its mechanism is not clear. Colino & Halliwell [13] hypothesized that carbachol shifts the activation curve of the h-current to the positive direction, which could explain the small inward current observed in the initial phase of the carbachol response. Moreover, an increase in the number of channels available for activation was suggested, because of the large increase in the current at -100 mV [13]. At -100 mV, however, I$_h$ is only half-activated [28], and therefore the action of carbachol could be due to a shift of the activation curve.

The fact that the action of ACh on hippocampal I$_h$ is opposite to the one observed in the heart is intriguing [17, 20]. The two actions might be mediated by different effectors. Whereas inhibition of I$_f$ in the heart is mediated by the inhibition of adenylyl cyclase and the consequent decrease in the cAMP [21], in the hippocampus this system does not appear to be involved, since β-adrenergic stimulation does not affect I_h [13]. Very recently, however, it has been reported that norepinephrine and cAMP modulate the hyperpolarization-activated current, independently of protein kinase A [44], a notion fitting very nicely the properties of the cardiac f-current [19].

Potentiation of h-current by carbachol seems to be mediated by the stimulation of PI turnover, and production of inositol (1,4,5)-trisphosphate (IP$_3$), also involved in the inhibition of $I_{K(M)}$ [22]. The consequent increase of $[Ca^{2+}]_i$ could directly activate I_h [13]. An increase of intracellular Ca^{2+}, in fact, has indeed been shown to potentiate I_f in the heart [24], although in this tissue the action of Ca^{2+} is not a direct effect on the channel [49]. In addition, when IP$_3$ was applied intracellularly, it was shown to reduce $I_{K(M)}$, but had no effect on I_h [22].

Another interesting observation is that, in the thalamus, nitric oxide (NO) shifts the activation curve for the hyperpolarization-activated current in the positive direction [36]. NO has been proposed to be a possible retrograde messenger for expression of long-term potentiation [4, 14, 23, 50]. A modulation by NO of hippocampal I_h, if confirmed, could suggest a role for this current in synaptic plasticity.

The presence of a hyperpolarization-activated current, similar to the cardiac I_f, in many areas of the brain may account for certain side-effects associated with the pharmacological treatment of arrhythmias. Specific bradycardic agents, developed to block I_f and known to exert a negative chronotropic effect on cardiac frequency, also block I_h in neurons [26, 35]. Central nervous system side-effects of these substances have to be considered when used therapeutically, but their high specificity makes them invaluable in studies of I_h. Recently QX-314, a quaternary derivative of lidocaine, known as an intracellular blocker of Na^+ channel, was shown to block I_q from inside [38]. Although not as specific as currently used blockers of I_h, this and related substances could be useful to test the effect of blocking I_q in the recorded cell while other neurons are unaffected.

ACKNOWLEDGMENTS

This work was partially supported by C.N.R. (CT 92.02582 to D.D.)

REFERENCES

1. Bader, C. R., Bertrand, D. & Schwartz, E. A. (1982). Voltage-activated and calcium-activated currents studied in solitary rod inner segments from the salamander retina. Journal of Physiology 331: 253-284
2. Bayliss, D. A., Viana, F., Bellingham, M. C. & Berger A. J. (1994). Characteristics and postnatal development of a hyperpolarization-activated current in rat hypoglossal motoneurons in vitro. Journal of Neurophysiology 71: 119-128
3. Benson, D. M., Blitzer, R. D. & Landau, E. M. (1988). An analysis of the depolarization produced in guinea-pig hippocampus by cholinergic receptor stimulation. Journal of Physiology 404: 479-496
4. Bliss, T. V. P. & Collingridge, G. L. (1993). A synaptic model of memory: long-term potentiation in the hippocampus. Nature 361: 31-39
5. Bliss, T. V. P. & Lømo, T. (1973). Long lasting potentiation of synaptic transmission in the dentate area of the anaesthetized rabbit following stimulation of the perforant path. Journal of Physiology 232: 331-356
6. Bobker, D. H. & Williams, J.T. (1989). Serotonin augments the cationic current I_h in central neurons. Neuron 2: 1535-1540
7. BoSmith, R. E., Briggs, I. & Sturgess, N. C. (1993). Inhibitory actions of Zeneca ZD 7288 on whole-cell hyperpolarization activated inward current (I_f) in guinea-pig dissociated sinoatrial node cells. British Journal of Pharmacology 110: 343-349

8. Brown, D. A. & Adams, P.R. (1980). Muscarinic suppression of a novel voltage-sensitive K^+ current in a vertebrate neurone. Nature 183: 673-676

9. Brown, D. A., Gähwiler, B. H., Griffith, W. H. & Halliwell, J. V. (1990). Membrane currents in hippocampal neurons. Progress in Brain Research 83: 141-160

10. Brown, H. & DiFrancesco, D. (1980). Voltage-clamp investigation of membrane currents underlying pacemaker activity in rabbit sino-atrial node. Journal of Physiology 308: 331-351

11. Budde, T., White, J. A. & Kay, A. R. (1994). Hyperpolarization-activated Na^+-K^+ current (I_h) in neocortical neurons is blocked by external proteolysis and internal TEA. Journal of Neurophysiology 72: 2737-2742

12. Burke, R. E. (1986). Gallamine binding to muscarinic M1 and M2 receptors, studied by inhibition of ^3H-pirenzepine and ^3H-quinuclidinylbenzilate binding to rat brain membranes. Molecular Pharmacology 30: 58-68

13. Colino, A. & Halliwell, J. V. (1993). Carbachol potentiates Q current and activates a calcium-dependent non specific conductance in rat hippocampus in vitro. European Journal of Neuroscience 5: 1198-1209

14. Cummings, J. A., Nicola, S. M. & Malenka, R. C. (1994). Induction in the rat hippocampus of long-term potentiation (LTP) and long-term depression (LTD) in the presence of a nitric oxide synthase inhibitor. Neuroscience Letters 176: 110-114

15. DiFrancesco, D. (1991). The contribution of the "pacemaker" current (i_f) to the generation of spontaneous activity in rabbit sino-atrial node. Journal of Physiology 434: 23-40

16. DiFrancesco, D. (1994). Some properties of the UL-FS 49 block of the hyperpolarization-activated current (i_f) in sino-atrial node myocytes. Pflügers Archiv 427: 64-70

17. DiFrancesco, D., Ducouret, P. & Robinson, R. B. (1989). Muscarinic modulation of cardiac rate at low acetylcholine concentrations. Science 243: 669-671

18. DiFrancesco, D., Ferroni, A., Mazzanti, M. & Tromba, C. (1986). Properties of the hyperpolarizing-activated current (I_f) in cells isolated from the rabbit sino-atrial node. Journal of Physiology 377: 61-88

19. DiFrancesco, D. & Tortora, P. (1991). Direct activation of cardiac pacemaker channels by intracellular cyclic AMP. Nature 351: 145-14

20. DiFrancesco, D. & Tromba, C. (1988a) Inhibition of the hyperpolarizing-activated current, i_f, induced by acetylcholine in rabbit sino-atrial node myocytes. Journal of Physiology 405: 477-491

21. DiFrancesco, D. & Tromba, C. (1988b). Muscarinic control of the hyperpolarizing-activated current i_f in rabbit sino-atrial node myocytes. Journal of Physiology 405: 493-510

22. Dutar, P. & Nicoll, R. A. (1988). Classification of muscarinic responses in terms of receptor subtypes and second-messenger systems: electrophysiological studies in vitro. Journal of Neuroscience 8: 4214-4224

23. Garthwaithe, J., Charles, S. L. & Chess-Williams, R. (1988). Endothelium-derived relaxing factor release on activation of NMDA receptors suggests a role as intercellular messenger in the brain. Nature 336: 385-387

24. Hagiwara, N. & Irisawa, H. (1989). Modulation by intracellular Ca^{2+} of the hyperpolarization-activated inward current in rabbit single sino-atrial node cells. Journal of Physiology 409: 121-141

25. Halliwell, J. V. & Adams, P. R. (1982). Voltage-clamp analysis of muscarinic excitation in hippocampal neurons. Brain Research 250: 71-92

26. Harris, N. C., Libri, V. & Constanti, A. (1994). Selective blockade of the hyperpolarization-activated cationic current (I_h) in guinea pig substantia nigra pars compacta neurones by a novel bradycardic agent, Zeneca ZM 227189. Neuroscience Letters 176: 221-225

27. Katz, B. (1949). Les constantes electriques de la membrane du muscle. Archives des Sciences physiologiques 3: 285-299

28. Maccaferri, G., Mangoni, M., Lazzari, A. & DiFrancesco, D. (1993). Properties of the hyperpolarization-activated current in rat hippocampal CA1 pyramidal cells. Journal of Neurophysiology 69: 2129-2136

29. Madison, D. V., Lancaster, B. & Nicoll, R.A. (1987). Voltage clamp analysis of cholinergic action in hippocampus. Journal of Neuroscience 7: 733-741

30. Mayer, M. L. & Westbrook, G. L. (1983). A voltage-clamp analysis of inward (anomalous) rectification in mouse spinal sensory ganglion neurons. Journal of Physiology 340: 19-45

31. McCormick, D. A. & Pape, H.-C. (1990). Noradrenergic and serotoninergic modulation of a hyperpolarization-activated cation current in thalamic relay neurons. Journal of Physiology 431: 319-342

32. Newberry, N. R. & Nicoll, R. A. (1985). Comparison of the action of baclofen with γ-aminobutyric acid on rat hippocampal cells in vitro. Journal of Physiology 360: 161-185

33. Nicoll, R. A. (1985). The septo-hippocampal projection: a model cholinergic pathway. Trends in Neuroscience 8: 533-536

34. Pape, H.-C. (1992). Adenosine promotes burst activity in guinea-pig geniculocortical neurones through two different ionic mechanisms. Journal of Physiology 447: 729-753

35. Pape, H.-C. (1994). Specific bradycardic agents block the hyperpolarization-activated cation current in central neurones. Neuroscience 59: 363-373
36. Pape, H.-C. & Mager, R. (1992). Nitric oxide controls oscillatory activity in thalamocortical neurons. Neuron 9: 441-448
37. Pape, H.-C. & McCormick, D. A. (1989). Noradrenaline and serotonin selectively modulate thalamic burst firing by enhancing a hyperpolarization-activated current. Nature 340: 715-718
38. Perkins, K. L. & Wong, R. K. S. (1995). Intracellular QX-314 blocks the hyperpolarization-activated inward current I_q in hippocampal CA1 pyramidal cells. Journal of Neurophysiology 73: 911-915
39. Segal, M. & Barker, J. L. (1984). Rat hippocampal neurons in culture: potassium conductances. Journal of Neurophysiology 51: 1409-1433
40. Skrede, K. K. & Westgaard, R. H. (1971). The transverse hippocampal slice: a well-defined cortical structure mantained in vitro. Brain Research 35: 589-593
41. Solomon, J. S. & Nerbonne, J. M. (1993a). Hyperpolarization-activated currents in isolated superior colliculus-projecting neurons from rat visual cortex. Journal of Physiology 462: 393-420
42. Solomon, J. S. & Nerbonne, J. M. (1993b). Two kinetically distinct components of hyperpolarization-activated current in rat superior colliculus-projecting neurons. Journal of Physiology 469: 291-313
43. Spain, W. J., Schwindt, P. C. & Crill, W. E. (1987). Anomalous rectification in neurons from rat sensorimotor cortex in vitro. Journal of Neurophysiology 57: 1555-1576
44. Storm, J. F. & Pedarzani, P. (1995). Norepinephrine and cyclic AMP modulates the Q/h-current independently of protein kinase A in hippocampal neurons. Society for Neuroscience Abstracts 21: 209.8
45. Storm-Mathisen, J. (1977). Localization of transmitter candidates in the brain: the hippocampal formation as a model. Progress in Neurobiology 8: 119-181
46. Takahashi, T. (1990). Inward rectification in neonatal rat spinal motoneurones. Journal of Physiology 423: 47-62
47. Uchimura, N., Cherubini, E. & North, R. A. (1990). Cation current activated by hyperpolarization in a subset of rat nucleus accumbens neurons. Journal of Neurophysiology 64: 1847-1850
48. Van Bogaert, P., Goethals, M. & Simoens, C. (1990). Use- and frequency-dependent blockade by UL-FS 49 of the i_f pacemaker current in sheep Purkinje fibres. European Jouranl of Pharmacology 187: 241-256
49. Zaza, A., Maccaferri, G., Mangoni, M. & DiFrancesco, D. (1991). Intracellular calcium does not direcly modulate cardiac pacemaker (i_f) channels. Pflügers Archiv 419: 662-664
50. Zhuo, M., Small, S. A., Kandel, E. R. & Hawkins, R. D. (1993). Nitric oxide and carbon monoxide produce activity-dependent long-term synaptic enhancement in the hippocampus. Science 260: 1946-1950

6

PROPERTIES OF NATIVE AND CLONED CYCLIC NUCLEOTIDE GATED CHANNELS FROM BOVINE

Paola Gavazzo, Cristiana Picco, Letizia Maxia, and Anna Menini

Istituto di Cibernetica e Biofisica
Consiglio Nazionale delle Ricerche and Istituto Nazionale Fisica della
    Materia
Via De Marini 6, 16149, Genova, Italy

1. INTRODUCTION

Signal transduction in vertebrate photoreceptors and olfactory receptor neurons (see for review Torre et al., 1995) involves the activation of non-selective cationic channels directly and cooperatively gated by cyclic nucleotides (CNG channels).

Rod photoreceptors respond to light through the activation of a chain of enzymatic events that stimulates the hydrolysis of cGMP, therefore inducing a decrease in the intracellular concentration of this cyclic nucleotide and hence the closure of a fraction of normally opened CNG channels. Similarly the interaction of odorant molecules with olfactory receptors neurons activates a pathway which produces an increase in cAMP and results in the opening of CNG channels closed at rest .

Ion channels directly activated by cyclic nucleotides have been first described in rod outer segments from frogs (Fesenko et al., 1985), soon after in cones (Haynes & Yau, 1985), olfactory receptor neurons (Nakamura & Gold, 1987) and in other neuronal and non neuronal cells (Dryer & Anderson, 1991; Biel et al., 1993; Biel et al., 1994; Weyand et al., 1994). The purification from bovine retina of a 63 KDa protein behaving as a CNG channel when reconstituted into lipid bilayers (Cook et al., 1987), gave enough sequence information to clone the corresponding cDNA (Kaupp et al., 1989). The heterologous expression of the synthesized mRNA in oocytes resulted in functional channels activated by cyclic nucleotides. Soon after, based on sequence homology, a CNG channel from bovine olfactory receptor neurons has been cloned (Ludwig et al., 1990). CNG channels from retinal rods and olfactory receptor neurons have also been cloned from other species including human, rat and catfish (Dhallan et al., 1990; Dhallan et al., 1992; Goulding et al., 1992).

The primary structures of the rod and olfactory cloned channels show a high degree of homology (60-70%): both are constituted by six hydrophobic membrane spanning segments and contain a cyclic nucleotide-binding domain, a voltage sensor motif and a pore region (see for review Kaupp, 1995).

Neurobiology, edited by Torre and Conti
Plenum Press, New York, 1996

The electrophysiological properties of the cloned channels differ in several aspects from those of the corresponding native channels. Recently a second subunit has been cloned for both rod (Chen et al., 1993; Chen et al., 1994; Koerschen et al.,1995) and olfactory CNG channel (Bradley et al., 1994; Liman & Buck, 1994). The coexpression of the two subunits gives rise to channels that more closely resemble the native channels.

Most of the electrophysiological experiments on native CNG channels have been performed on amphibian species (see for review Menini, 1995; Zimmerman, 1995), but unfortunately cloning and expression of CNG channels from amphibians have not been done yet. The results of experiments from many laboratories clearly show that some electrophysiological properties of both native and cloned channels vary among species. Therefore it is sometimes difficult to distinguish whether some properties of native CNG channels from amphibians differ from those of cloned channels from bovine because of inter species differences or because of the absence of channel subunits or other components. To directly compare electrophysiological properties of native and cloned CNG channels from the same species we performed experiments with the patch clamp technique on both native and cloned rod and olfactory CNG channels from bovine.

2. MATERIALS AND METHODS

Patch-clamp experiments were performed on native and cloned CNG channels from bovine. With native channel we refer to CNG channels studied in their native membrane environments in excised patches from retinal rods and olfactory receptors neurons. With cloned channels we refer to the expression in *Xenopus laevis* oocytes of the mRNA of the α subunit or subunit 1 of rod (Kaupp et al., 1989) or olfactory (Ludwig et al., 1990) CNG channels.

2.1. Isolation of Rod Photoreceptors

Photoreceptor cells have been isolated following the procedure of Schnetkamp & Daemen, (1982). Freshly removed bovine eyes were hemisected under regular light; retinas were removed and gently shaken in 35% sucrose (w/w) in Mammalian Ringer solution (157 mM NaCl, 5 mM KCl, 0.5 mM $MgCl_2$, 0.5 mM $CaCl_2$, 8 mM NaH_2PO_4, 7 mM Na_2HPO_4, pH 6.7). The rod photoreceptors were then isolated by centrifugation; after a first extraction performed at 3000g for 10 min at 4°C, the supernatant containing the photoreceptors was passed through a nylon cloth (50 μm mesh), furtherly diluted in Ringer and centrifuged again at 1500g. The final pellet containing mostly isolated rod and cone photoreceptors was resuspended again in Ringer. An aliquot (0.1 ml) was then transferred to the recording chamber for patch clamp experiments.

2.2. Isolation of Olfactory Receptor Neurons

The ethmoid turbinates located in the bovine just before the cribri form plate present a yellow-brown mucosa where the olfactory receptor neurons can be found. Turbinates were removed from the animal 20-30 minutes after its death and rinsed several times at room temperature in D-Mem (140 mM NaCl, 5.6 mM KCl, 2 mM $MgCl_2$, 2 mM $CaCl_2$, 9.4 mM Glucose, 5 mM Na-Hepes, pH 7.4). Olfactory receptor cells were obtained by a dissociation procedure modified from Maue & Dionne (1987) The mucosal tissue was gently separated with forceps from the supporting cartilage and minced into 2-3 mm pieces. The minced tissue was then incubated in D-Mem containing 0.025% trypsin for 30 40 min; the enzymatic treatment was carried out at 37°C in divalent free media and terminated with 10% of Foetal

Calf Serum. Minced tissue was triturated several times with a fire-polished Pasteur pipette. Several different non-neuronal cell types were produced by the dissociation procedure such as supporting and respiratory cells. Among them olfactory receptor neurons were easily recognized by their morphology having an ovoid soma, a dendrite, a ciliary knob and immobile cilia. The cells released from trituration were diluted with D-Mem and an aliquot (0.1 ml) was loaded in the recording chamber for the patch clamp experiments.

2.3. In Vitro Transcription and Functional Expression

The cDNA specific for the bovine photoreceptor and olfactory channel has been kindly provided to us by Dr. E.Eismann from the laboratory of Prof. U.B. Kaupp.

mRNA were synthesized in vitro with T7 (rod) or T3 (olfactory) RNA phage polymerase using the respective linearized plasmids as a template and injected in *Xenopus* oocytes at a concentration of 0.5 $\mu g/\mu l$. About 50 ng of mRNA were injected in each oocyte. In order to allow channel expression, oocytes were incubated at 19°C for 1-3 days in Barth's solution (88 mM NaCl, 1 mM KCl, 0.82 mM $MgSO_4$, 0.33 mM $Ca(NO_3)_2$, 0.41 mM $CaCl_2$, 2.4 mM $NaHCO_3$, 5 mM Tris-HCl, pH 7.4) additioned with Gentamycin 100 $\mu g/ml$.

2.4. Electrophysiological Recording

Membrane patches were excised in the inside out configuration from the outer segment of retinal rods, or from the soma or dendrite or knob of olfactory receptor neurons, or from oocytes injected with mRNA as described in the previous section. Currents were activated by perfusing the intracellular side of the membrane with solutions containing various concentrations of cGMP or cAMP, using the perfusion system described in Menini and Nunn (1990). The patch pipette solution and the perfusion solutions contained 110 mM NaCl, no added divalent cations, 10 mM Hepes-TMAOH, 1mM EDTA-TMAOH, pH 7.6.

Currents were measured with an Axopatch 1D patch clamp amplifier (Axon Instruments USA), filtered at 1 KHz using a 8-pole low pass Bessel filter, sampled at 2-2.5 KHz and stored on a IBM PC 486 computer. Data analysis was performed using pClamp software (Axon Instruments, Foster City, CA, USA).

3. RESULTS

3.1. Bovine Rod CNG Channels

Currents activated by various concentrations of cGMP or cAMP were measured in excised inside-out membrane patches from the outer segment of bovine rods or from oocytes previously injected with mRNA of the α subunit of bovine rod CNG channel.

Normalized macroscopic currents measured at + 60 mV were plotted in Figure 1 as a function of cyclic nucleotide concentrations for native (A) or cloned (B) bovine rod channels. The relation between activated current and cyclic nucleotide concentration was well described by the Hill equation:

$$I = I_{max} \, c^n/(c^n + K_{1/2}^{\,n}) \tag{1}$$

where I is the activated current, I_{max} the maximal current, c the cyclic nucleotide concentration, $K_{1/2}$ the cyclic nucleotide concentration activating half of the maximal current and n the Hill coefficient.

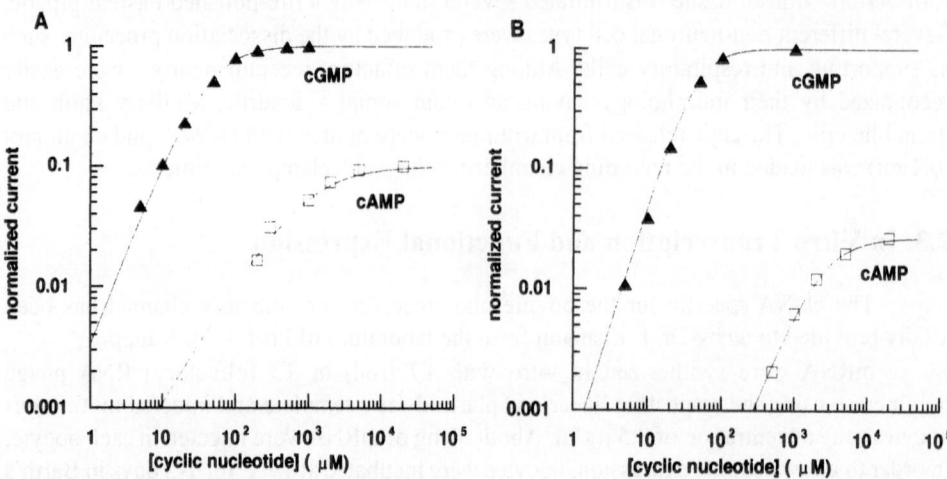

Figure 1. Dose-response curves of the response to increasing concentrations of cGMP and cAMP at +60 mV from the same patch for native bovine rod channels (A) and from two different patches for cloned channels (B) were shown as normalized currents. The continuous lines were the best-fit of the Hill equation (Eq. 1) to the data with the following values: (A) for the native channel $K_{1/2}$(cGMP)=45 μM, n(cGMP)=1.5, I_{max}(cGMP)=582 pA, $K_{1/2}$(cAMP)=908 μM, n(cAMP)=1, I_{max}(cAMP)=58 pA and (B) for the cloned channel $K_{1/2}$(cGMP)=50 μM, n(cGMP)=2, I_{max}(cGMP)=1760 pA for one patch; $K_{1/2}$(cAMP)=2400 μM, n(cAMP)=1.5, I_{max}(cAMP)=153 pA and I_{max}(cGMP)= 5733 pA for the other patch.

A comparison between dose-response curves for native (Fig. 1A) and cloned (Fig. 1B) bovine rod channels reveals some important differences. At +60 mV I_{max} activated by cAMP was 11 ± 6.5 % (N=4) of I_{max} activated by cGMP for the native bovine channels. For the cloned bovine channels I_{max} activated by cAMP was only 2.5 % (N=1) of that activated by cGMP, in agreement with previous measurements (Gordon & Zagotta, 1995). $K_{1/2}$ in the native bovine channels had average values at + 60 mV of 43 ± 2.5 μM (N=3) for cGMP and of 856 ± 72 μM (N=2) for cAMP. $K_{1/2}$ in the cloned channels had higher values: 75 ± 16 μM (N=16) for cGMP and 2380 μM (N=1) for cAMP.

These results indicate that the native bovine rod channels are more sensitive to cAMP than the cloned channels. A comparison between these data and those obtained from the salamander (Menini et al., 1993; see also Fig. 3 from Menini, 1995) shows that there are some differences between species. $K_{1/2}$ for cAMP had similar values (about 850μM) in the bovine and in the salamander, whereas the fraction of maximal current activated by cAMP was lower in the bovine (about 10%) than in the salamander (about 40%).

It is possible to record the activity of single-channels in a membrane patch containing many channels by reducing the concentration of cyclic nucleotide. Figure 2A illustrates recordings of currents activated by 5 μM cGMP or by 500 μM cAMP in an excised patch from a bovine rod at + 60 mV. In the absence of cyclic nucleotides the current trace was quiet while in the presence of cGMP or cAMP brief current transients, typical of single-channel activity, could be observed. Previous studies (Torre et al., 1992; Sesti et al., 1994) on native CNG channels from amphibians have shown that these channels have a rapid flickering that does not allow to determine with certainty the single-channel conductance and makes difficult the kinetics analysis.

Figure 2B shows recordings of single-channels activated by 50 μM cGMP or by 5 mM cAMP from an excised patch containing bovine cloned channels at + 80 mV. By comparing native and cloned channels activated by cGMP it is evident that the flickering characterizing the native channel was not present in the expressed cloned α subunit, which showed well-resolved transitions between the closed and the open state (see also Nizzari et al., 1993).

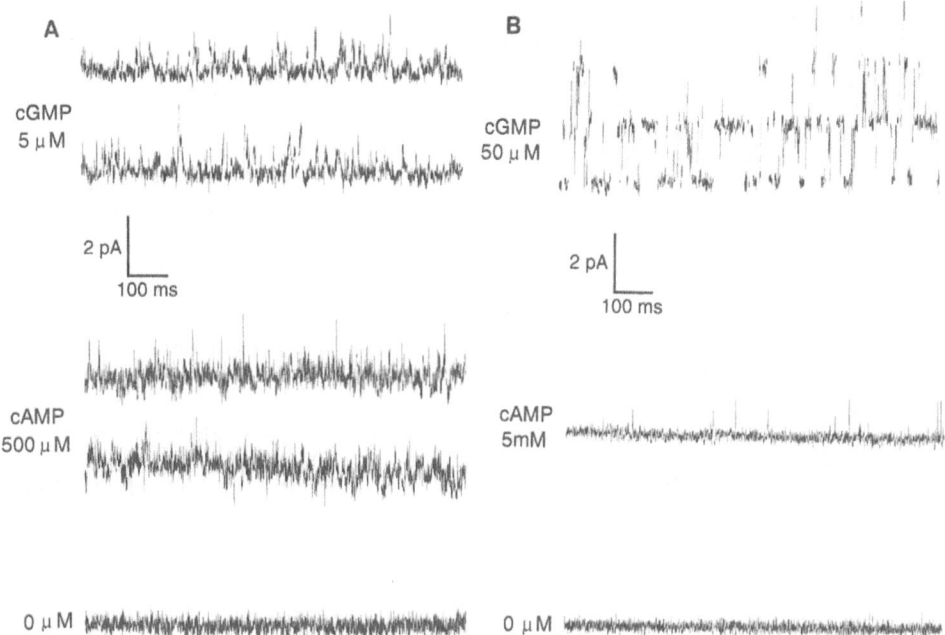

Figure 2. Single channel activity recorded in a membrane patch excised from a bovine rod (A) or from an oocyte (B) injected with the mRNA of the α subunit of the bovine rod channel. (A) Native bovine CNG rod channels were measured in the same patch at +60 mV in the presence of 5 μM cGMP or 500 μM cAMP as indicated in the figure, filtered at 1 kHz and sampled at 2.5 kHz. (B) Cloned bovine CNG rod channels were activated in the same patch at +80 mV by 50 μM cGMP or 5 mM cAMP as indicated in the figure, filtered at 1 kHz and sampled at 2 kHz. In the absence of cyclic nucleotides there was no channel activity as shown in the two traces at the bottom of (A) and (B).

Since the recordings have been performed on the same species for both cloned and native channel, it can be certainly concluded that the different behavior is due to a different channel structure.

Recently a second subunit has been cloned (Chen et al., 1993; Chen et al., 1994; Korschen. et al., 1995) and the coexpression of the two subunits seems to partially reintroduce the flickering behavior characteristic of the native channel. Nevertheless even channels composed of these two subunits do not flicker as much as the native channels do, suggesting that a different stoichiometry could exist *in vivo* or other subunits have not been discovered yet; alternatively post translational modifications could affect the channel in the native environment and not in the heterologous expression system.

3.2. Bovine Olfactory CNG Channels

Currents activated by various concentrations of cGMP or cAMP were measured in excised inside-out membrane patches from the soma or dendrite or knob of bovine olfactory receptor neurons or from oocytes previously injected with mRNA of the α subunit of bovine olfactory CNG channels.

Both in native and cloned (Fig. 3) bovine olfactory CNG channels cAMP and cGMP activated the same I_{max} . From some preliminary experiments $K_{1/2}$ at + 60 mV in the native

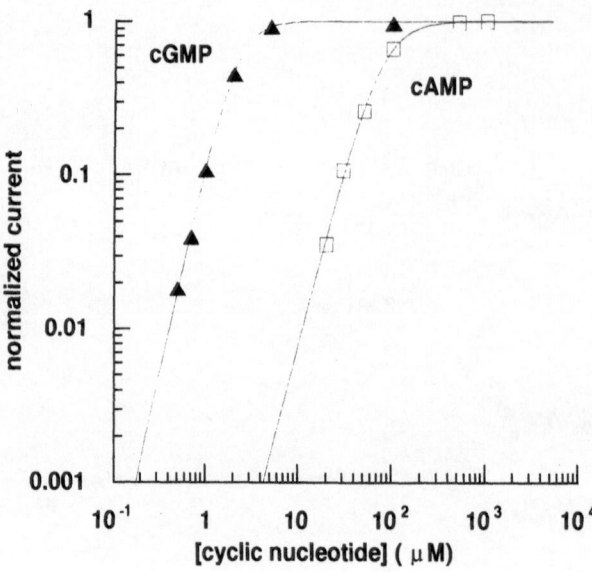

Figure 3. Currents activated by increasing concentrations of cGMP and cAMP were measured on the same patch at +60 mV for cloned bovine CNG channels. Normalized currents as a function of cyclic nucleotide concentrations were well fitted (continuous lines) by the Hill equation (Eq.1) with the following values: $K_{1/2}(cGMP) = 2$ μM, $n(cGMP) = 3$, $I_{max}(cGMP) = 245$ pA and $K_{1/2}(cAMP) = 76$ μM, $n(cAMP) = 2.4$, $I_{max}(cAMP) = 250$ pA.

channels has been estimated to be lower for cGMP (about 1μM) than for cAMP (about 10 μM) (data not shown).

Normalized dose-response curves measured at + 60 mV for the activation of cloned bovine olfactory channels from the same patch are shown in Figure 3. The relation between activated current and cyclic nucleotide concentration was fitted with the Hill equation (Eq.1). $K_{1/2}$ for cGMP had an average value of 1.9 ± 0.3 μM (N=4), while $K_{1/2}$ for cAMP was 60 ± 14 μM (N=3).

Therefore, although cAMP represents the physiological ligand, the olfactory channel is more sensitive to cGMP than to cAMP, resembling in this way the behavior of the photoreceptor CNG channel.

Single-channel activity was recorded and illustrated in Figure 4. Native bovine olfactory channels were recorded at + 60 mV and single-channels were activated by the same low concentration, 0.1 μM, of cGMP or cAMP (Fig. 4A). Cloned olfactory channels were recorded at +80 mV and single-channels were activated by quite different concentrations, 30 μM cAMP or 0.5 μM cGMP.

By comparing figure 2 and figure 4 it is evident that native olfactory channels behave differently from native rod channels: both in the native and in the cloned olfactory channels it is possible to clearly distinguish square opening events (Fig.4). Moreover, unlikely the rod channel, the kinetics of olfactory channels is quite similar for cGMP and cAMP activated currents. From our experiments the conductance of the bovine native channel was about 30 pS at + 60 mV, while for the cloned channel it was about 40 pS at + 80 mV.

Single-channel conductance of the CNG channel from olfactory neurons have been reported to widely vary across species: 12-19 pS for rat and frog (Frings et al., 1992), 30 pS for toad (Kurahashi & Kaneko, 1993), 45 pS for salamander (Zufall et al., 1991) and 55 pS for catfish (Goulding et al., 1992).

4. DISCUSSION

The purpose of the present work was to analyze the behavior of the bovine rod and olfactory CNG channels, both in their native membrane environment and in heterologous

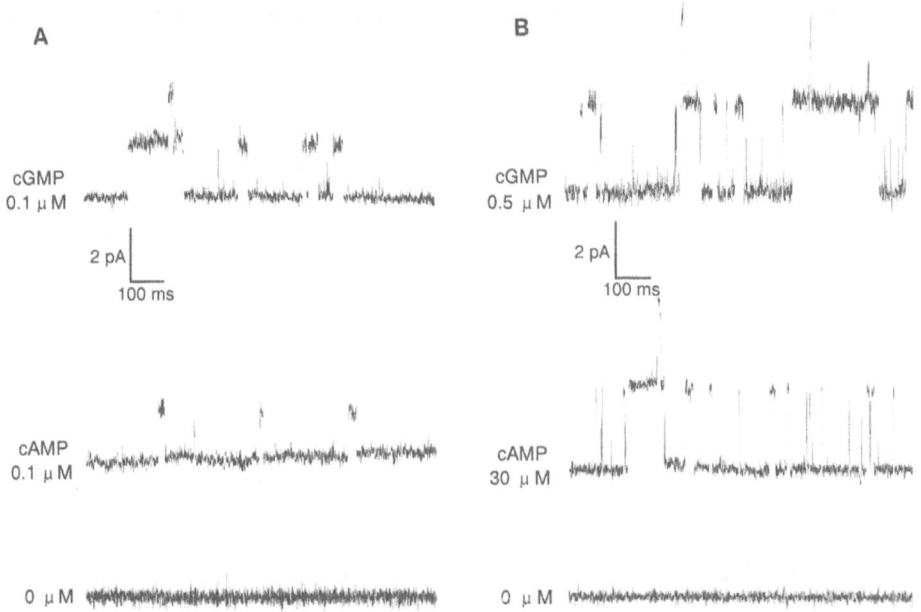

Figure 4. Single channel activity recorded in a membrane patch excised from a bovine olfactory receptor neuron (A) or from an oocyte (B) injected with the mRNA of the α subunit of the bovine olfactory channel. (A) Native bovine CNG olfactory channels were measured in the same patch at +60 mV in the presence of 0.1 μM cGMP or 0.1 μM cAMP as indicated in the figure, filtered at 1 kHz and sampled at 2.5 kHz. (B) Cloned bovine CNG olfactory channels were activated in the same patch at +80 mV by 0.5 μM cGMP or by 30 μM cAMP, filtered at 1 kHz and sampled at 2 kHz. Lowest traces were obtained in absence of cyclic nucleotides.

expression system. The possibility of making a direct comparison of some properties of native and cloned rod and olfactory CNG channels deriving from the same species, confers a particular relevance to the obtained results.

Both native bovine rod and olfactory CNG channels were found to be more sensitive to cAMP than the corresponding cloned channels composed of α subunit (Figs. 1 and 3) , suggesting that cAMP activation could be modulated in the native environments by the expression of other subunits or by the presence of other components.

Activity of native single-channels in bovine rods was found to be very different from that in olfactory neurons. Native rod channels showed very brief opening events and a rapid flickering of the current that make difficult the analysis of the properties of the channel (Fig. 2A), whereas in native olfactory channels it was possible to distinguish well resolved openings, more square and stable (Fig. 3A). The differences in kinetics between rod and olfactory channels were much less pronounced between the respective cloned channels: a comparison of single-channels from rod and olfactory α subunits showed that both had well resolved transitions between the open and the closed state when channels were activated by cGMP (Figs. 2B and 3B).

Therefore at the level of single-channels differences in kinetics between native and cloned channels are more pronounced in rod than in olfactory channels. The complete structures of native channels are likely to be profoundly different in the retinal rods and olfactory neurons, as it is partially suggested by the dissimilarities in the sequences for the recently cloned additional subunits.

Unfortunately the additional subunit for the olfactory channel from bovine has not been cloned yet, and only some properties of the bovine rod channels composed by the two subunits have been characterized. Further studies of the cloned channels together with a continuous comparison with the properties of native channels will be very important for the understanding of the structure of CNG channels in their physiological environment.

ACKNOWLEDGMENTS

We thank E.Eismann and U.B. Kaupp for the cDNA of bovine rod and olfactory CNG channels. We also thank G. Gaggero, D. Magliozzi and P.Guastavino for technical assistance. Special thanks to Mr. M. Canepa who kindly provided us with fresh turbinates and eyes from bovines. This work was supported by HFSP.

5. REFERENCES

BIEL, M., ALTENHOFEN, W., HULLING, R., LUDWIG J., FREICHEL, M., FLOCKERZI, V., DASCAL, N., KAUPP U.B., HOFMANN, F. (1993). Primary structure and functional expression of a cyclic nucleotide-gated channel from rabbit aorta. *FEBS Letters* 329, 134-138.

BIEL, M., ZONG, X., DISTLER M., BOSSE, E., KLUGBAUER, N., MURAKAMI, M., FLOCKERZI, V., HOFMANN, F. (1994). Another member of the cyclic nucleotide-gated channel family, expressed in testis, kidney, and heart. *Proceedings of the National Academy of Sciences of the USA* 91, 3505-3509.

BRADLEY, J., DAVIDSON N., LESTER, H.A., ZINN, K. (1994). Heteromeric olfactory cyclic nucleotide-gated channel confers high sensitivity to cAMP. *Proceedings of the National Academy of Sciences of the USA* 91, 8890-8894.

CHEN, T.Y., PENG, Y.W., DHALLAN, R.S., AHAMED, B., REED, R.R. & YAU, K.W. (1993). A new subunit of the cyclic nucleotide-gated cation channel in retinal rods. *Nature* 362, 764-767.

CHEN, T.Y., ILLING, M., MOLDAY, L.L., HSU, Y.T., YAU, K.W. & MOLDAY, R.S. (1994). Subunit 2 (or β) of retinal rod cGMP-gated cation channel is a component of the 240-kDa channel-associated protein and mediates Ca^{2+}-calmodulin modulation. *Proceedings of the National Academy of Sciences of the USA* 91, 11757-11761.

COOK, N.J., HANKE, W. & KAUPP, U.B. (1987). Identification, purification, and functional reconstitution of the cyclic GMP-dependent channel from rod photoreceptors. *Proceedings of the National Academy of Sciences of the USA* 84, 585-589.

DHALLAN, R.S., YAU, K.W., SHRADER, K.A. & REED, R.R. (1990). Primary structure and functional expression of a cyclic nucleotide-activated channel from olfactory neurons. *Nature* 347, 184-187.

DHALLAN, R.S., MACKE, J.P., EDDY, R.L., SHOWS, T.B., REED, R.R., YAU, K.W. & NATHANS, J. (1992). Human rod photoreceptor cGMP-gated channel: Amino acid sequence, gene structure, and functional expression. *Journal of Neuroscience* 12, 3248-3256.

DRYER, S.E. & HENDERSON, D. (1991). A cyclic GMP-activated channel in dissociated cells of the chick pineal gland. *Nature* 353, 756-758.

FESENKO, E.E., KOLESNIKOV, S.S. & LUYUBARSKY, A.L. (1985). Induction by cyclic GMP of cationic conductance in plasma membrane of retinal rod outer segments. *Nature* 313, 310-313.

FRINGS, S., LYNCH, J.W. & LINDEMANN, B. (1992). Properties of cyclic nucleotide-gated channels mediating olfactory transduction. *Journal of General Physiology* 100, 1-11.

GORDON, S.E. & ZAGOTTA, W.N. (1995). A histidine residue associated with the gate of the cyclic nucleotide-activated channels in rod photoreceptors. *Neuron* 14, 177-183.

GOULDING, E.H., NGAI, J., KRAMER, R.H., COLICOS, S., AXEL, R., SIEGELBAUM, S.A. & CHESS, A. (1992). Molecular cloning and single-channel properties of the cyclic nucleotide-gated channel from catfish olfactory neurons. *Neuron* 8, 45-48.

HAYNES, L.W. & YAU, K.W. (1985). Cyclic GMP-sensitive conductance in outer segment membrane of catfish cones. *Nature* 317, 61-64.

KAUPP, U. B. (1995). Family of cyclic nucleotide gated ion channels. *Current Opinions in Neurobiology* 5, 434-442.

KAUPP, U.B., NIIDOME, T., TANABE, T., TERADA, S., BÖNIGK, W., STÜHMER, W., COOK, N.J., KANGAWA, K., MATSUO, H., HIROSE, T. ET AL. (1989). Primary structure and functional expression from complementary DNA of the rod photoreceptor cyclic GMP-gated channel. *Nature* 342, 762-766.

KÖRSCHEN, H. G., ILLING, M., SEIFERT, R., SESTI, F., WILLIAMS, A., GOTZES, S., COLVILLE, C., MÜLLER, F., DOSÉ, A., GODDE, M., MOLDAY, L., KAUPP, U.B. & MOLDAY, R.S. (1995) A 240 kDa protein represents the complete b subunit of the cyclic nucleotide-gated channel from rod photoreceptor. *Neuron* 15, 627-636.

KURAHASHI, T. & KANEKO, A. (1993). Gating properties of the cAMP-gated channel in toad olfactory receptor cells. *Journal of Physiology* 466, 287-302.

LIMAN, E.R. & BUCK, L.B. (1994). A second subunit of the olfactory cyclic nucleotide-gated channel confers high sensitivity to cAMP. *Neuron* 13, 611-621.

LUDWIG, J., MARGALIT, T., EISMANN, E., LANCET, D. & KAUPP, U.B. (1990). Primary structure of cAMP-gated channel from bovine olfactory epithelium. *FEBS Letters* 270, 24-29.

MAUE, R.A. & DIONNE, V.E. (1987). Preparation of isolated mouse olfactory receptor neurons. *Pflügers Archives* 409, 244-250.

MENINI, A. & NUNN, B.J. (1990). The effect of pH on the cyclic GMP-activated conductance in retinal rods. In A. Borsellino, L. Cervetto and V. Torre (Editors), Sensory Transduction, Plenum Press, New York, 175-181

MENINI, A. (1995). Cyclic nucleotide-gated channels in visual and olfactory transduction. *Biophysical Chemistry* 55, 185-196.

MENINI, A., SANFILIPPO, C. & PICCO, C. (1993). Properties of activation by cAMP and cGMP of the channel in retinal rod outer segments. XXXII Congress of the International Union of Physiological Sciences. Glasgow 1-7 Agosto 1993. 278.32/P

NAKAMURA, T. & GOLD, G.H. (1987). A cyclic nucleotide-gated conductance in olfactory receptor cilia. *Nature* 325, 442-444.

NIZZARI, M., SESTI, F., GIRAUDO, M.T., VIRGINIO, C., CATTANEO, A. & TORRE, V. (1993). Single-channel properties of cloned cGMP-activated channels from retinal rods. *Proceedings of the Royal Society London* 254, 69-74.

SCHNETKAMP, P.P. & DAEMEN, F.J. (1982). Isolation and characterization of osmotically sealed bovine rod outer segments. *Methods in Enzymology* 81, 110-116.

SESTI, F., STRAFORINI, M., LAMB, T.D., TORRE, V. (1994). Gating, selectivity and blockage of single channels activated by cyclic GMP in retinal rods of the tiger salamander. *Journal of Physiology.* 474, 203-222

TORRE, V., STRAFORINI, M., SESTI, F. & LAMB, T.D. (1992). Different channel-gating properties of two classes of cyclic GMP-activated channel in vertebrate photoreceptors. *Proceedings of the Royal. Society London* 250, 209-215.

WEYAND, I., GODDE, M., FRINGS, S., WEINER, J., MULLER, F., ALTENHOFEN, W., HATT, H., & KAUPP, U.B. (1994). Cloning and functional expression of a cyclic-nucleotide-gated channel from mammalian sperm. *Nature* 368, 859-863.

ZIMMERMAN, A. (1995). Cyclic nucleotide gated channnels. *Current Opinions in Neurobiology* 5, 296-303.

ZUFALL, F., FIRESTEIN, S. & SHEPHERD, G.M. (1991). Analysis of single cyclic nucleotide gated channels in olfactory receptor cells. *Journal of Neuroscience* 11, 3573-3580.

7

NMDA-INDUCED MOTOR ACTIVATION IN RODENTS

Non-Aminergic Mechanisms and Long-Lasting Effect

Lydia Giménez-Llort, Sergi Ferré, and Emili Martínez

Department of Neurochemistry
C.S.I.C.
08034 Barcelona, Spain

1. INTRODUCTION

The N-methyl-D-aspartate (NMDA) subtype of the glutamate receptor, mediates slow, Ca^{2+}-linked synaptic excitatory neurotransmission[25,28] and is involved in several physiological[15,24] and pathological situations in the brain[6,21,36]. There is also growing evidence for the involvement of the glutamatergic system in mammal's locomotion, and thus, much experimental data support NMDA receptor plays an important role in motor activity. Most of the behavioural studies are related to the effects of NMDA receptor antagonists which produce an increase of motor activity in rodents after their systemic or intraestriatal administration[7,9,18,29,33]. Antagonistic dopamine/glutamate interactions[14] at presynaptic and postsynaptic levels have been suggested to be important mechanisms for this induced motor activation since it can be inhibited by drugs which produce dopamine depletion or blockade of dopamine receptors[7,9,33]. However, dopamine-independent mechanism could also be involved in this increase of activity, as NMDA receptor antagonist can reverse hypokinesia in monoamine-depleted animals[3,5,17].

Fewer data of literature are related to the effects of systemically administered agonist, N-methyl-D-aspartic acid, which are associated with convulsant activity (invariably associated to excitotoxicity[19,37]) or not (could reflect non-excitotoxic effects[19,21]). Systemic administration of NMDA has been shown to decrease motor activity in rats at low doses[31,35] and to increase it at high doses.[22] A decrease in motor activity after the striatal administration of NMDA has also been reported[30].

We present some results showing that systemically administered subconvulsant doses of NMDA induced: 1: Motor activation in non-reserpinized and reserpinized mice (from Ferré et al., 1994) 2. Long-lasting increase in exploratory activity in rat (from Giménez-Llort et al., 1995).

Neurobiology, edited by Torre and Conti
Plenum Press, New York, 1996

2. METHODS

2.1. Animals

Male mice of the OF1 strain (24-32 g) and male Wistar rats [WI(IOPS AF/Han)] (250-300g) were used. The animals were randomly assigned to the different groups and maintained under laboratory conditions (food and water *ad lib*, $22 \pm 2°C$ and 12L:12D cycles beginning at 0700 h).

2.2. Motor and Exploratory Activity Recording

Four open-field cages ($35.5 \times 35.5 \times 35.5$ cm) were recorded simultaneously in a soundproof, temperature-controlled ($22 \pm 2°C$) experimental room. To record under light or dark conditions, the experimental room was uniformly illuminated with two incandescent lamps (100 W; located 1 m above the floor) or with red lamps, respectively. Motor activity was recorded with a video-computerized system (Videotrack 512, View Point, Lyon) immediately after the animals were placed in the recording cages without any acclimatization period.

The motor activity of groups of three mice (n=1) was analyzed by using a substraction image analysis. The system was set to measure any kind of motor activity (locomotion, rearing, intense grooming, jumps) and to avoid monitoring of very small movements (breathing, non-intense grooming, tremor). The counts were recorded as amount of time (in s 5 min^{-1}) engaged in motor activity during the two hours test.

The exploratory activity of rats when placed in the test cage was recorded. Three different kinds of exploratory movements were defined. Fast (FM) and slow (SM) ambulatory movements consisted of horizontal displacements with a speed greater than 25 cm.s^{-1} and between 12 cm.s^{-1} and 25 cm.s^{-1}, respectively. The percentage of the total 5 minutes session time engaged by the rat in each of these movements was determined. Vertical displacements were measured as total number of rearings (R) per session. The test was done twice a day for 6 days, during both the light and the dark periods of the light-dark cycle. In the first session the animals were administered with NMDA (100 mg.kg^{-1}) and left in the cages for 2h to observe the appearance of convulsions. Convulsant rats were excluded from the study.

2.3. Drugs

Reserpine (Sigma, St.Louis, MO) and D,L-α-methyl-*p*-tyrosine methyl ester hydro-chloride (RBI, Natick, MA) were dissolved in a drop of glacial acetic acid which was made up to volume with 5.5% glucose. *N*-methyl-D-aspartic acid (Sigma) and dizocilpine maleate ((+)-MK-801 hydrogen maleate, RBI) were dissolved in 5.5% glucose. *N*-methyl-D-aspartic acid solutions were adjusted to pH 7.4 with NaOH. The volume of injection was always 10 ml kg^{-1}.

Reserpine (5 or 10 mg.kg^{-1}) and D,L-α-methyl-*p*-tyrosine (200 mg kg^{-1}) were administered s.c. 20 h and 1 h prior to motor activity recording, respectively. NMDA was administered i.p. just before motor activity recording. MK-801 was administered either 1 h prior to (when NMDA was also administered to the same animals) or just before motor activity recording.

2.4. Statistical Analysis

In studies with mice, all values recorded per 5 min were transformed (square root of (counts+0.5))(Andén and Grabowska-Andén, 1988) and analyzed by the "summary measures" method (Mattews et al., 1990), by using the mean of all the transformed data per 3 mice (n=1) as the summary statistic and by using either Student's non-paired t-test or one-way analysis of variance (ANOVA) with post hoc Newman-Keuls comparisons to analyze differences between groups. The level of statistical significance was set at $P<0.05$. $n=4-6$ in all experiments.

To study the differences between the exploratory activity of saline group and NMDA-treated rats, the "summary measures" method was used. The first day's values, the mean of the second and third days' values, and the mean of the fourth to the sixth days' values were used as the summary statistics, and they were subsequently analyzed by ANOVA with post hoc Newman-Keuls comparisons. To determine the differences in the three dependent variables FM, SM and R. a two-factor ANOVA with a within factor (a light-dark factor, with two levels) and a between factor (a treatment factor, with two levels) was applied, separately, for the first day values, the mean of the second and third days values and the mean of the fourth to the sixth days values.

3. RESULTS

3.1. Effect of NMDA and MK-801 on the Motor Activity in Non-Reserpinized Mice

The administration of low doses of NMDA (12.5 and 25 mg.kg^{-1}) induced a significant decrease in motor activity during the first hour of observation compared to controls. With the highest doses (50 and 100 mg.kg^{-1}) this decrease was followed by a significant increase in motor activity during the second hour of observation (Fig.1). The administration of 0.2 and 0.5 mg.kg^{-1} MK-801 induced a significant increase in motor activity during the second hour of observation compared to controls while the lower dose, 0.1 mg.kg^{-1}, did not produce any change. MK-801 2 mg.kg^{-1} induced a significant decrease in motor activity during the first hour of observation compared to controls (Fig.1) and was associated with abnormal breathing (dyspnea) and uncoordinated motor activity (ataxia). No preconvulsant ('wild running') or convulsant activity or death were observed with any dose of NMDA or MK-801.

3.2. Effect of NMDA and MK-801 on the Motor Activity in Long-Term Reserpinized Mice

After 20 h of reserpine administration, mice lost about 20% of their initial body weight and were more vulnerable to the toxic effects of both NMDA and MK-801. Thus, highest doses of both NMDA (100 mg.kg^{-1}) and MK-801 (2 mg.kg^{-1}) induced death, usually preceded by preconvulsant ('wild running') and convulsant activities, in about 25% of these animals.

Significant motor activation were induced by both NMDA (75 and 100 mg/kg) and MK-801 (2 mg.kg^{-1}) (Fig. 2) even when pretreated with a higher dose of reserpine (10 mg.kg^{-1}) plus the dopamine synthesis inhibitor, α-methyl-p-tyrosine (200 mg.kg^{-1}), NMDA 100 mg.kg^{-1} and MK-801 2 mg.kg^{-1} (Fig.3). As shown in figure 4, previous administration of MK-801 0.5 mg.kg^{-1} significantly antagonized the motor activation induced by NMDA 100 mg.kg^{-1}.

3.3. Exploratory Activity After Treatment with Excitatory Amino Acid Receptor Agonists

Only one rat treated with NMDA showed convulsion and was excluded from the study. Data corresponding to FM, SM and R for six days during light and dark periods

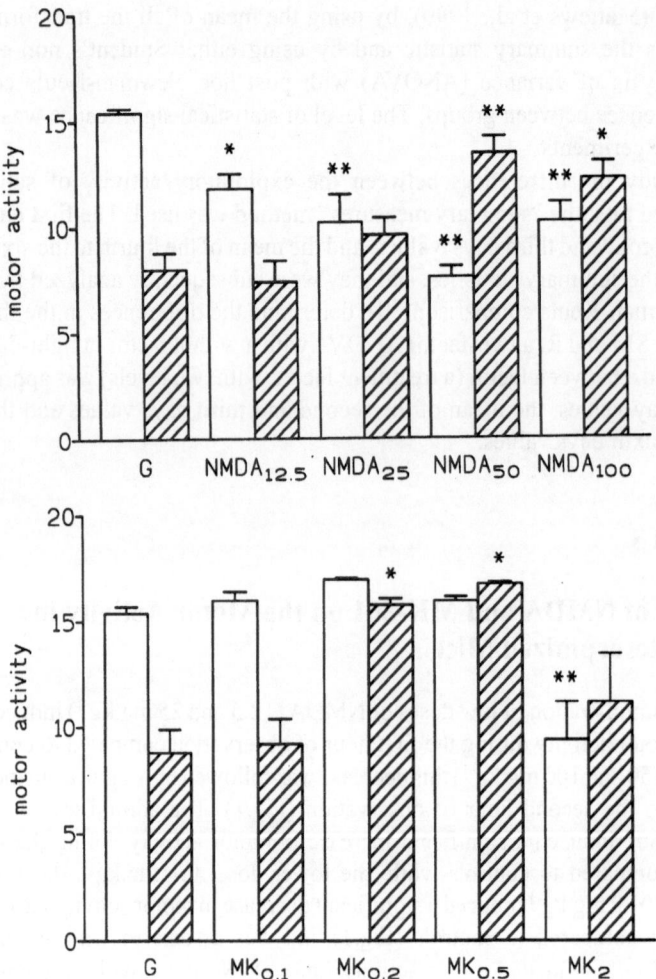

Figure 1. Effect of NMDA and MK-801 on the motor activity of non-reserpinized mice. Means ± S.E.M. of all 5-min transformed data per 3 mice (n=1) from the first 1-h period (open bars) and the second 1-h period of observation (striped bars) from non-reserpinized mice (n=4-6/group). G: glucose; $NMDA_{12.5}$, $NMDA_{25}$, $NMDA_{50}$ and $NMDA_{100}$: NMDA 12.5, 25, 50 and 100 mg.kg^{-1}, respectively; $MK_{0.1}$, $MK_{0.2}$, $MK_{0.5}$ and MK_2: MK-801 0.1, 0.2, 0.5 and 2 mg.kg^{-1}, respectively; *,** significantly different compared to G (ANOVA, P<0.05 or P<0.01, respectively).

of the light-dark cycle are represented in figure 5. In all cases a light-dark factor significance was found (p<0.01 in all cases) due to higher FM, SM and R values during the dark than during the light period of the light-dark cycle. Horizontal and vertical components of motor activity showed a significant positive correlation during both periods of the light-dark cycle in both experimental groups (Pearson's coefficient > 0.5 and $p <$ 0.05 in all cases). A significant decrease induced by NMDA compared to saline was obtained for all the components of exploratory activity (FM, SM and R) in the first day (one-way ANOVA with post-hoc Newman-Keuls comparisons: p<0.01 in all cases). For the second and third days FM values, a significant increase in the NMDA-treated group was obtained (one-way ANOVA with post-hoc Newman-Keuls comparisons: p<0.01). The treatment effect was not found significant for the fourth to sixth days FM, SM or R values.

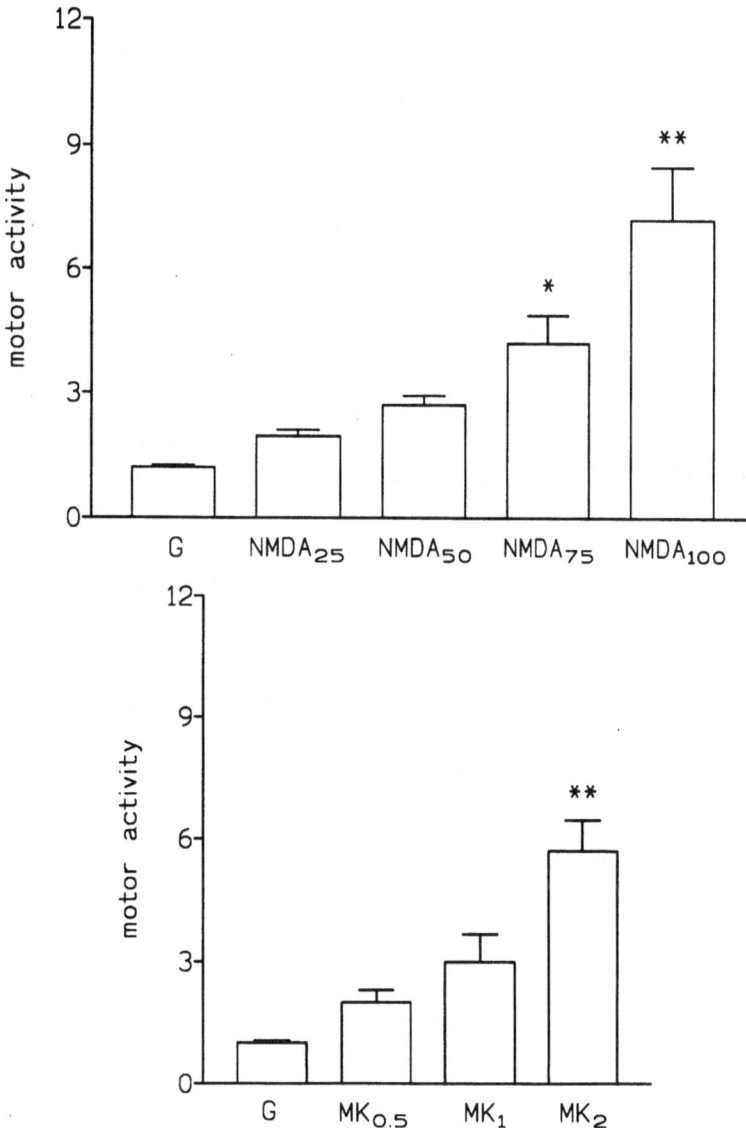

Figure 2. Effect of NMDA and MK-801 on the motor activity of reserpinized mice. Means ± S.E.M. of all 5-min transformed data per 3 mice (n=1) from long-term reserpinized mice (n=4-6/group). G: glucose; $NMDA_{25}$, $NMDA_{50}$, $NMDA_{75}$ and $NMDA_{100}$: NMDA 25, 50, 75 and 100 mg.kg^{-1}, respectively; MK 0.5, MK 1 and MK 2: MK-801 0.5, 1 and 2 mg.kg^{-1}, respectively; *,** significantly different compared to G (ANOVA, $P<0.05$ or $P<0.01$, respectively).

4. DISCUSSION

In the first part of the present study we found that low doses of the non-competitive NMDA receptor antagonist, MK-801, induced-motor activation in non-reserpinized mice. This effect is mediated at a dopamine presynaptic level, as pretreatment with reserpine abolishes this motor activation and the interactions between MK-801 and dopamine receptor agonists in reserpinized mice seem to be antagonistic. Induced-motor activation in long-term reserpinized mice (Fig.2) could only be obtained with high, clearly toxic doses of MK-801

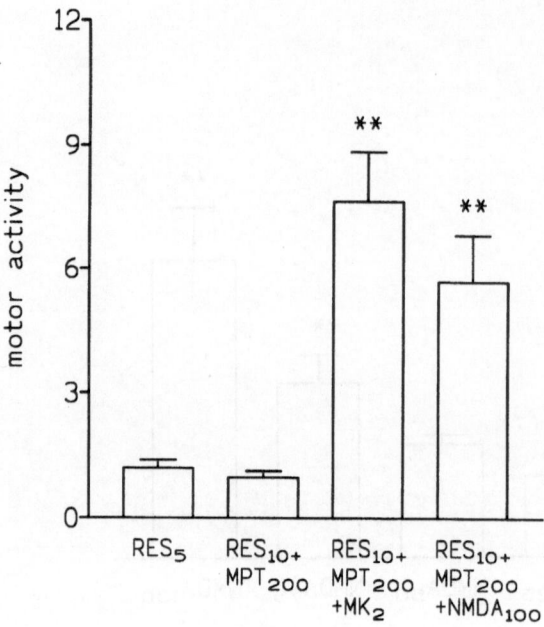

Figure 3. Motor activation induced by NMDA and MK-801 in highly monoamine-depleted mice. Means ± S.E.M. of all 5-min transformed data per 3 mice (*n*=1) from highly monoamine-depleted mice (*n*=4-6/group). RES_5: reserpine 5 mg.kg^{-1}; RES_{10} + MPT_{200}: reserpine 10 mg.kg^{-1} puls α-methyl-*p*-tyrosine 200 mg.kg^{-1}; MK_2: MK-801 2 mg.kg^{-1}; $NMDA_{100}$: NMDA 100 mg.kg^{-1}; ** significantly different compared to RES_5 and to RES_{10} + MPT_{200} (ANOVA, $P<0.01$ in all cases).

(greater than 1 mg.kg-1) which produced motor depression in non-reserpinized mice (Fig.1). This effect of MK-801 in long-term reserpinized mice is dopamine-independent as it was not modified by the treatment with a higher dose of reserpine plus the dopamine synthesis inhibitor, α-methyl-*p*-tyrosine (Fig.4), and could be due to any of the already proposed non-NMDA-linked actions of this compound[27].

In agreement with literature[22,31,35], in non-reserpinized animals, the systemic administration of low doses of NMDA induced motor depression while with high doses this initial depressant effect was followed by a dose-dependent increase in motor activity (Fig.1). In long-term reserpinized mice, NMDA produced motor activation at doses similar to those which induced motor activation in non-reserpinized mice (75 and 100 mg.kg^{-1}), suggesting

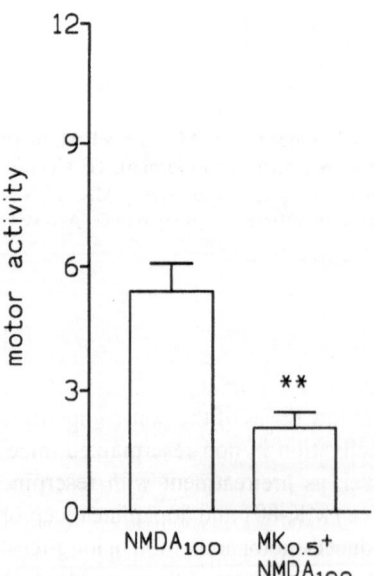

Figure 4. Counteraction by MK-801 of the NMDA induced motor activation in reserpinized mice. Means ± S.E.M. of all 5-min transformed data per 3 mice (n=1) from long-term reserpinized mice (n=4-6/group). $NMDA_{100}$: NMDA 100 mg.kg^{-1}; $MK_{0.5}$: MK-801 0.5 mg.kg^{-1};** significantly different compared to $NMDA_{100}$ (Student's non-paired t-test, P<0.01).

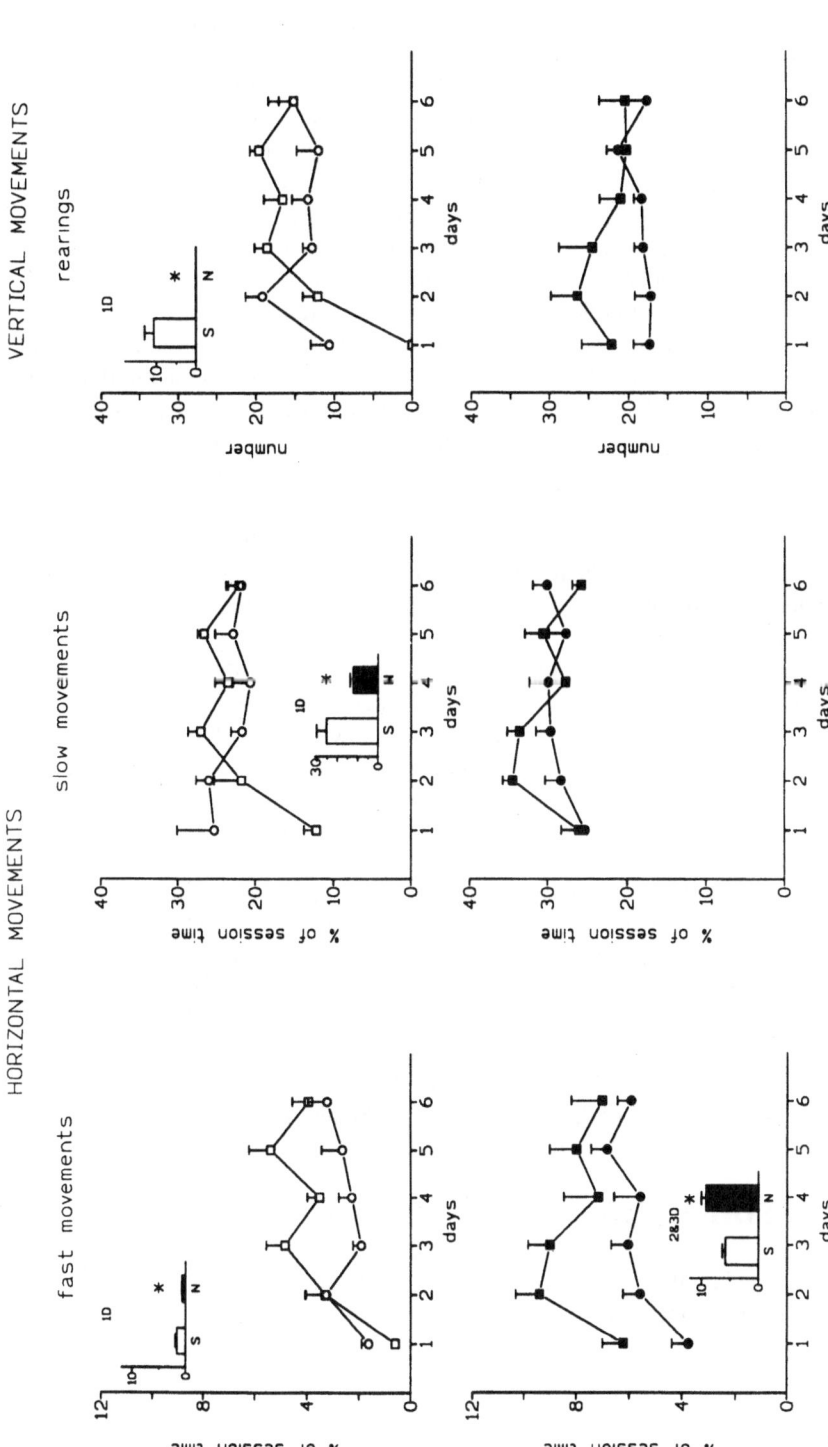

Figure 5. Repeated testing of exploratory activity in NMDA-treated rats. Means ± SEM of fast movements (left graphs), slow movements (central graphs) and rearings (right graphs), during the light period (open symbols) and the dark period (closed symbols) in rats administered saline (circles; *n*=6) or NMDA 100 mg/kg i.p. (squares; *n*=7). Insets: Means ± SEM of fast movements in rats administered saline (white bars) or NMDA (black bars), during the first day (1D) and during the second-third days period (2&3D). 1: significantly different compared to the saline group (ANOVA with post-hoc Newman-Keuls comparisons: *P*<0.01).

that NMDA-induced motor activation could be also mediated at a dopaminergic postsynaptic level. Results of antagonistic effect of a low dose of MK-801 (0.5 mg/kg; fig.4) on NMDA-induced motor activation in long-term reserpinized mice but not modification by the pretreatment with a higher dose of reserpine plus the dopamine synthesis inhibitor, α-methyl-p-tyrosine (Fig.3) also support this hypothesis. The most probable location for the postsynaptic effects of NMDA is at striatal level, because stimulation of NMDA receptors in entopeduncular and nigral neurons would lead to motor depression[14].

Our results of initial NMDA induced motor-depression in mice agree with the significant decrease of exploratory activity in rats shown immediately after the administration of NMDA. The acute decrease in the rat was significant for all the components of exploratory activity (FM, SM and R). Some experimental evidences[13,16,35] suggest it could be due to the ability of NMDA to release adenosine which inhibits motor activity.

Some of the results in mice showed a biphasic effect of NMDA in motor activity, depression followed by activation. In the study of exploratory activity in the rat, the initial motor depression was followed during the next two days (2&3D) by an increased exploratory activity, only significant for fast movements during the dark period[26]. This result could be related to the role of NMDA receptor on long-lasting synaptic effects and neuronal plasticity[8,14,24].

In summary, these results demonstrated that systemically administered subconvulsant doses of NMDA induce motor depression followed by motor activation in mice. NMDA induced motor activation in long-term reserpinized mice at doses which are similar to those causing motor activation in non-reserpinized mice (75 and 100 mg.kg^{-1}), while MK-801 induced motor activation at a dose which was associated with motor depression in non-reserpinized mice (2 mg.kg^{-1}). The NMDA-induced motor activation in long-term reserpinized mice was counteracted by the previous administration of a low dose of MK-801 (0.5 mg.kg^{-1}) and was still present when a stronger dopamine-depleting pretreatment was used. These results suggest that, instead of NMDA receptor antagonists, as it is being recently claimed in the literature, NMDA receptor agonists could counteract motor deficits in dopamine-dependent motor disorders.

A more specific and long-lasting effect of NMDA on motor activity was found when studying exploratory activity in non-convulsant rat during both the light and the dark periods of the light-dark cycle. A single administration of NMDA (100 mg.kg^{-1}, i.p.) produced an acute short-lasting depressant effect on all components of exploratory activity (fast movements, slow movements and rearings), followed during the next two days by a long-lasting increase in exploratory activity, only significant for fast movements in the dark period. This increase in exploratory activity, probably not linked to their excitotoxic action, could represent a behavioural correlate of the NMDA-induced changes in synaptic plasticity[8,14,24].

5. ACKNOWLEDGMENTS

Work supported by the grant from the Spanish Government (FIS, 93/0350). Support from the CIRIT (Generalitat de Catalunya) is also acknowledged (Grup de Recerca de Qualitat).

6. REFERENCES

1. Albin, R.L., R.L. Makowiec, Z.R. Hollingsworth, L.S.Dure, IV, J.V. Penney and A.B. Young, 1992, Excitatory amino acid binding sites in the basal ganglia of the rat: a quantitative autoradiographic study. Neuroscience 46, 35.

2. Andén, N.-E. and M. Grabowska-Andén, 1988, Stimulation of D_1 dopamine receptors reveals direct effects of preferential dopamine autoreceptor agonist B-HT 920 on postsynaptic dopamine receptors, Acta Physiol. Scand. 134, 285.

3. Carlsson, M. and A. Carlsson, 1989, The NMDA antagonist MK-801 causes marked locomotor stimulation in monoamine-depleted mice, J. Neural Transm. 75, 221.

4. Carlsson, M. and A. Carlsson, 1990, Interactions between glutamatergic and monoaminergic systems within the basal ganglia - implications for schizophrenia and Parkinson's disease, Trends Neurosci. 13, 272.

5. Carlsson, M., A. Svensson and A. Carlsson, 1992, Interactions between excitatory amino acids, catecholamines and acetylcholine in the basal ganglia, in: Excitatory Amino Acids, ed. R.P. Simon (Thieme, New York) 189.

6. Choi, D.W., 1988, Glutamate neurotoxicity and diseases of the nervous system, Neuron 1, 623.

7. Clineschmidt, B.V., G.E. Martin, P.R. Bunting and N.L. Papp, 1982, Central sympathomimetic activity of (+)-S-methyl-10,11-dihydro-SH-dibenzo[a,d]cyclohepten-5,10-imine (MK-801), a substance with potent anticonvulsant, central sympathomimetic, and apparent anxiolytic properties, Drug. Dev. Res. 2, 135.

8. Collingridge, G.L. and Singer, W, 1990, Excitatory amino acid receptors and synaptic plasticity. Trends Pharmacol. Sci. 11, 290.

9. Criswell, H.E., K.B. Johnson, R.A. Mueller and G.R. Breese, 1993, Evidence for involvement of brain dopamine and other mechanisms in the behavioural action of the N-methyl-D-aspartic acid antagonist MK-801 in control and 6-hydroxydopamine-lesioned rats, J. Pharmacol. Exp. Ther. 265, 1001.

10. Ferré, S., K. Fuxe, G. Von Euler, B. Johansson and B.B. Fredholm, 1992, Adenosine-dopamine interactions in the brain. Neuroscience 51(3), 501.

11. Geyer, M.A., 1990, Approaches to the characterization of drug effects on locomotor activity in rodents. In: Modern Methods in Pharmacology, Vol. 6, Testing and evaluation of drugs of abuse. New York: Wiley-Liss, Inc., 81.

12. Gerfen, C.R., 1992, The neostriatal mosaic: multiple levels of compartmental organization, Trends Neurosci. 15, 133.

13. Giménez-Llort, L., E. Martínez and S. Ferré, 1995, Dopamine-independent and adenosine-dependent mechanisms involved in the effects of N-methyl-D-aspartate on motor activity in mice, Eur. J. Pharmacol. 275, 171.

14. Greenamyre , J.T., 1993, Glutamate-dopamine interactions in the basal ganglia: relationship to Parkinson's disease, J. Neural Transm. (Gen. Sect.) 91, 255.

15. Harris, E.W., A.H. Ganon and C.W. Cotman, 1984, Long-term potentiation in the hippocampus involves activation of N-methyl-D-aspartate receptors, Brain Res. 323, 132.

16. Hoehn, K. and T.D. White, 1990, N-methyl-D-aspartate, kainate and quisqualate release endogenous adenosine from rat cortical slices. Neuroscience 39, 441.

17. Klockgether, T. and L. Turski, 1990, NMDA antagonists potentiate antiparkinsonian actions of L-dopa in monoamine-depleted rats, Ann. Neurol. 28: 539.

18. Liljequist, S., K. Ossowska, M. Grabowska-Andén and N.-E. Andén, 1991, Effect of the NMDA receptor antagonist, MK-801, on locomotor activity and on the metabolism of dopamine in various brain areas of mice, Eur. J. Pharmacol., 195, 55.

19. Mares, P., L. Velísek, 1992, N-methyl-D-aspartate (NMDA)-induced seizures in developing rats. Dev. Brain Res. 65, 185.

20. Matthews, J.N.S., D.G. Altman, M.J. Campbell and P. Royston, 1990, Analysis of serial measurements in medical research. Brit. Med. J. 230, 730.

21. Meldrum, B. and J. Garthwaite, 1990, Excitatory amino acid neurotoxicity and neurodegenerative disease. Trends Pharmacol. Sci. 11, 379.

22. Metha, A.K. and M.K. Ticku, 1990, Role of N-methyl-D-aspartate (NMDA) receptors in experimental catalepsy in rats, Life Sci. 46, 37.

23. Moore, N.A., A. Blackman, S. Awere and J.L.D. Leander, 1993, NMDA receptor antagonists inhibit catalepsy induced by either D_1 or D_2 receptor antagonists, Eur. J. Pharmacol. 237, 1.

24. Morris, R.G.M., E. Anderson, G.S. Lynch and M. Baudry, 1989, Sélective impairment of learning and blockade of long-term potentiation in vivo by an N-methyl-D-aspartate receptor antagonist AP5, Nature, 319, 774.

25. Nakanishi, S, 1992, Molecular diversity of glutamate receptors and implications for brain function, Science. 258, 597.

26. Norton, S., B. Culver and P. Mullenix, 1975, Development of nocturnal behaviour in albino rats, Behav. Biol. 15, 317.

27. O'Neill, S.K., and G.T. Bolger, 1989, Phencyclidine and MK-801: a behavioural and neurochemical comparison of their interactions with dihydropyridine calcium antagonists, Brain Res. Bull. 22, 611.
28. Seeburg, P.H., 1993, The molecular biology of mammalian glutamate receptor channels, Trends Pharmacol Sci. 14, 297.
29. Schmidt, W.J., 1986, Intrastial injection of DL-2-amino-5-phospho-novaleric acid (AP-S) induces sniffing stereotopy that is antagonized by haloperidol and clozapine, Physcopharmacology, 90, 123.
30. Schmidt, W.J. and Bury, 1988, Behavioural effects of N-methyl-D-aspartate in the anterodorsal striatum of the rat, Life Sci. 43: 545.
31. Schmidt, W.J., M. Bubser and W. Hauber, 1992, Behavioural pharmacology of glutamate in the basal ganglia, J. Neural Transm. (Suppl.) 38, 65.
32. Smith, A.D. and J.P. Bolam, 1990, The neural network of the basal ganglia as revealed by the study of synaptic connections of identified neurons, Trends Neurosci. 13, 259.
33. Svensson, A., E. Pileblad and M. Carlsson, 1991, A comparison between the non-competitive NMDA antagonist dizocilpine (MK-801) and the competitive NMDA antagonist D-CPPene with regard to dopamine turnover and locomotor-stimulatory properties in mice, J. Neural Transm. (GEn. Sect.), 85, 117.
34. Vetulani, J., D. Marona-Lewicka, J. Michaluk, L. Antkiewicz-Michaluk and P. Popik, 1987, Stability and variability of locomotor responses of laboratory rodents. II. Native exploratory and basal locomotor activity of wistar rats, Pol. J. Pharmacol. Pharm. 39, 283.
35. Von Lubitz, D.K.J.E., I.A. Paul, M. Carter and K.A. Jacobson, 1993, Effects of N^6-cyclopentyl adenosine and 8-cyclopenthyl-1,3-dipropylxanthine on N-methyl-D-aspartate induced seizures in mice. Eur. J. Pharmacol. 249, 265.
36. Westbrook, G.L., 1993, Glutamate receptors and excitotoxicity. In: Molecular and Cellular, Vol. 3, Approaches to the Treatment of Neurological Disease, edited by S.G. Waxman. New York: Raven Press, 35.
37. Winn, P., T.W. Stone, M. Latimer, M.H. Hastings, A.J.M. Clark, 1991, A comparison of excitotoxic lesions of the basal forebrain by kainate, quinolate, ibotenate, N-methyl-D-aspartate or quisqualate, and the effects on toxicity of 2-amino-5-phosphonovaleric acid and kynurenic acid in the rat, Brit. J. Pharmacol. 102, 904.
38. Yoshida, Y., T. Ono, A. Kizu, R. Fukushima and T. Miyagishi, 1991, Striatal N-methyl-D-aspartate receptors in haloperidol-induced catalepsy, Eur. J. Pharmacol. 203, 173.

8

TARGET STRIATAL CELLS REGULATE DEVELOPMENT OF MIDBRAIN DOPAMINERGIC NEURONES

Carla Perrone-Capano,[1,2] Giuseppina Amadoro,[1] Angela Tino,[1]
Roberto Pernas-Alonso,[1] Bruno Esposito,[1] and Umberto di Porzio[1]

[1] International Institute of Genetics and Biophysics, CNR
Via Marconi 10, 80125 Naples Italy
[2] Department of General and Environmental Physiology
University of Naples
Via Mezzocannone 8, 80134 Naples, Italy

1. INTRODUCTION

The majority of dopamine (DA)-containing neurones in the mammalian central nervous system (CNS) are located in the midbrain, where they occupy three regions: the substantia nigra, the ventral tegmental area and the retrorubral field (Bjorklund and Lindvall, 1984). Midbrain dopaminergic neurones project to the neostriatum, limbic system and cortex and receive information from multiple structures in the diencephalon and telencephalon. Dopaminergic projections to the striatum form the nigrostriatal system (Figure 1), which plays a central role in the control of movement and several complex behaviours; degeneration of this pathway in men leads to Parkinson's disease, which is characterised by tremor, muscular rigidity, difficulty to initiate movement and loss of postural reflex (Jenner et al., 1992). Dopaminergic projections to the limbic system form the mesolimbic pathway, which sub serves functional roles in emotional balance and reward; over activity of this pathway is associated with schizophrenia and hallucinations. Finally, dopaminergic projections to the neocortex and perifrontal cortex form the mesocortical dopaminergic pathway, which has a role in motivation, attention, planning and social behaviour. Drugs that affect DA neurotransmission are used to treat Parkinson's disease and other movement's disorders, as well as psychotic syndromes including schizophrenia (Ritz et al., 1987; Van Tol et al., 1991; Malmberg et al., 1993; Seeman et al., 1993).

Despite the physiological and clinical importance of the dopaminergic neurones, the mechanisms that direct their development are still largely unknown.

In this chapter we will shortly recall general concepts of brain development and then summarise recent progress in the understanding of the signals by which DA neurones are specified and acquire some of their peculiar characteristics during embryonic development.

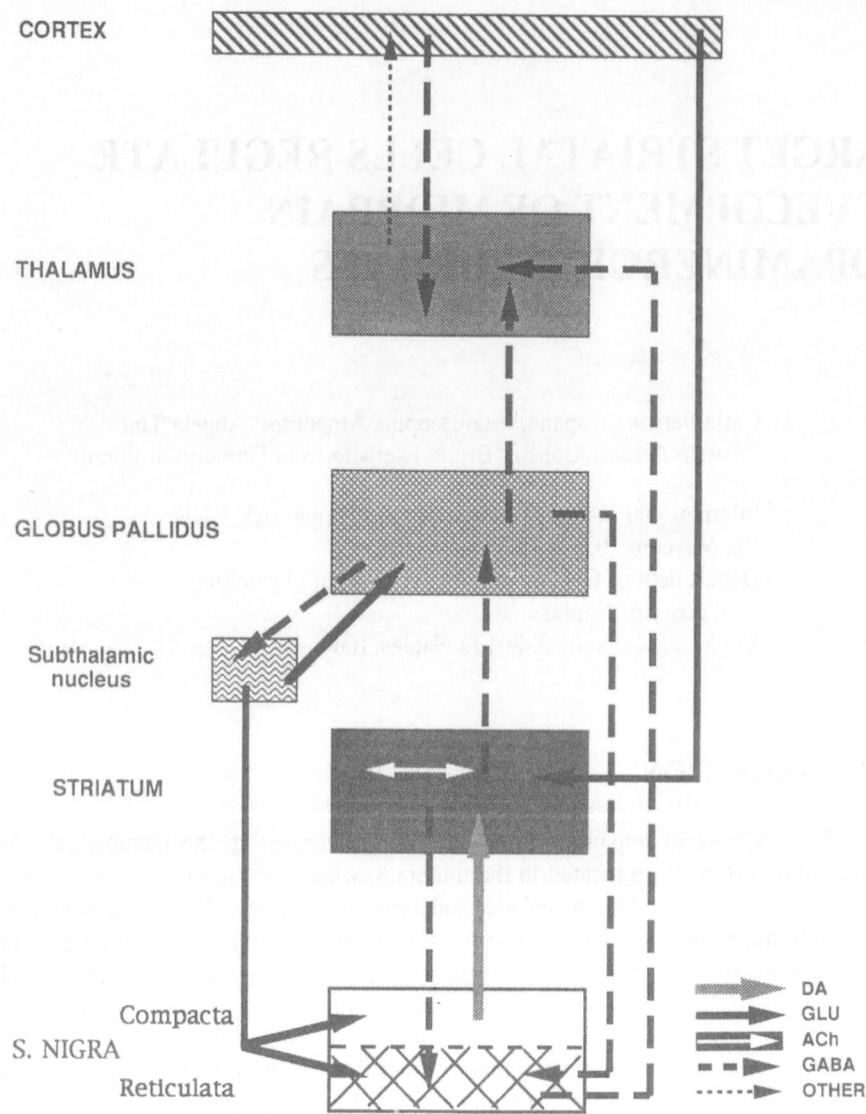

Figure 1. Schematic representation of the mammalian nigro-striatal connections. The major "classical neuro-transmitter" pathways are shown (DA, dopamine; GLU, glutamic acid; Ach, acetylcholine; GABA, γ-aminobutyric acid).

2. BRAIN DEVELOPMENT

The nervous system develops in a series of ordered steps that are precisely timed, with a temporal sequence that is characteristic of each neural structure. The high degree of cellular diversity and nervous functions derives from complex interactions among various cell types. As a result, each neurone connects only with certain target cells and not with others. Development of the nervous system involves not only genetic information but also epigenetic information arising either from within the embryo or from the external environment. The concerted action of various epigenetic factors seems to be critical for enabling a neurone to differentiate appropriately by sequential activation and modulation of specific portions of the genetic program within the developing neuronal precursors (neuroblasts).

Neurones and glial cells of the CNS derive from a specialised region of the ectoderm, the neural plate, which lies along the dorsal midline of the embryo. Early in development the neural plate becomes depressed at the centre (neural groove) and closes to form the neural tube. The lateral margins of the neural plate send into the mesoderm an important contingent of cells known as neural crest cells from which the peripheral nervous system, the adrenal chromoaffin cells and osteoblasts arise (Le-Douarin and Ziller, 1993). The cavity of the neural tube gives rise to the ventricular system of the CNS; the epithelial cells lining the walls of the neural tube (neuroepithelium) proliferate, generating the precursors of glial cells and neurones of the CNS and then migrate to reach the final anatomical destination.

The differentiation of the neural tube into the various regions of the CNS occurs simultaneously in three different ways. On the gross anatomical level, the neural tube constricts to form the chambers of the brain and the spinal cord. At the tissue level, the cell populations within the wall of the neural tube rearrange themselves to form the different functional regions of the CNS. Finally, on the cellular level, neuroepithelial cells themselves differentiate into the numerous types of neurones and glial cells. The neuroblasts do not proliferate uniformly along the length of the neural tube; the cells of the caudal part of the neural tube proliferate to form the spinal cord and the rostral neural tube forms initially three vesicles called the forebrain, the midbain or mesencephalon, and the hindbrain. Later in the embryogenesis the forebrain becomes subdivided into the anterior telencephalon and the more caudal diencephalon; the telencephalon will form the cerebral cortex and the basal ganglia, and the diencephalon will form the thalamus and hypothalamus. The midbrain remains undivided and its lumen becomes the cerebral aqueduct. The hindbrain becomes subdivided into an anterior metencephalon and a more posterior myelencephalon; the metencephalon gives rise to the cerebellum and pons, and the myelencephalon becomes the medulla oblungata. Genes of the homeobox family are responsible for the initial patterning of the CNS (Rubenstein et al., 1994) and are highly conserved throughout evolution.

Further specification of neuronal precursors into the various neuronal cell types is largely determined by environmental cues. Differentiation involves proliferation and generation of specific classes of neurones, migration of cells to characteristic positions, maturation of cells and development of specific interconnections. In order to be effective, each step requires specific signals which act on the cell at a particular stage of development. Within the vertebrate nervous system, individual classes of neurones are found at stereotyped positions, defined by their coordinates along the anterior-posterior and dorso-ventral axes (Jessel and Dodd, 1992). There is evidence that the nothocord (a group of axial mesodermal cells located underneath the neural tube) and the floor plate (a specialised group of neuroepithelial cells at the ventral midline of the neural tube) influence the development of multiple classes of neurones at particular positions in the neural tube (Yamada et al., 1993).

We will focus on the recently identified signals which seem to play a key role in the induction, survival and maturation of midbrain DA neurones.

3. DIFFERENTIATION OF DA NEURONES

3.1 Ontogeny of Midbrain DA Neurones

The ontogeny of midbrain DA neurones has been investigated by immunohistochemistry using antibodies against DA or tyrosine hydroxylase (TH), the rate-limiting enzyme in the catecholamine biosynthetic pathway (Figure 2). In the last years, the presence of TH in

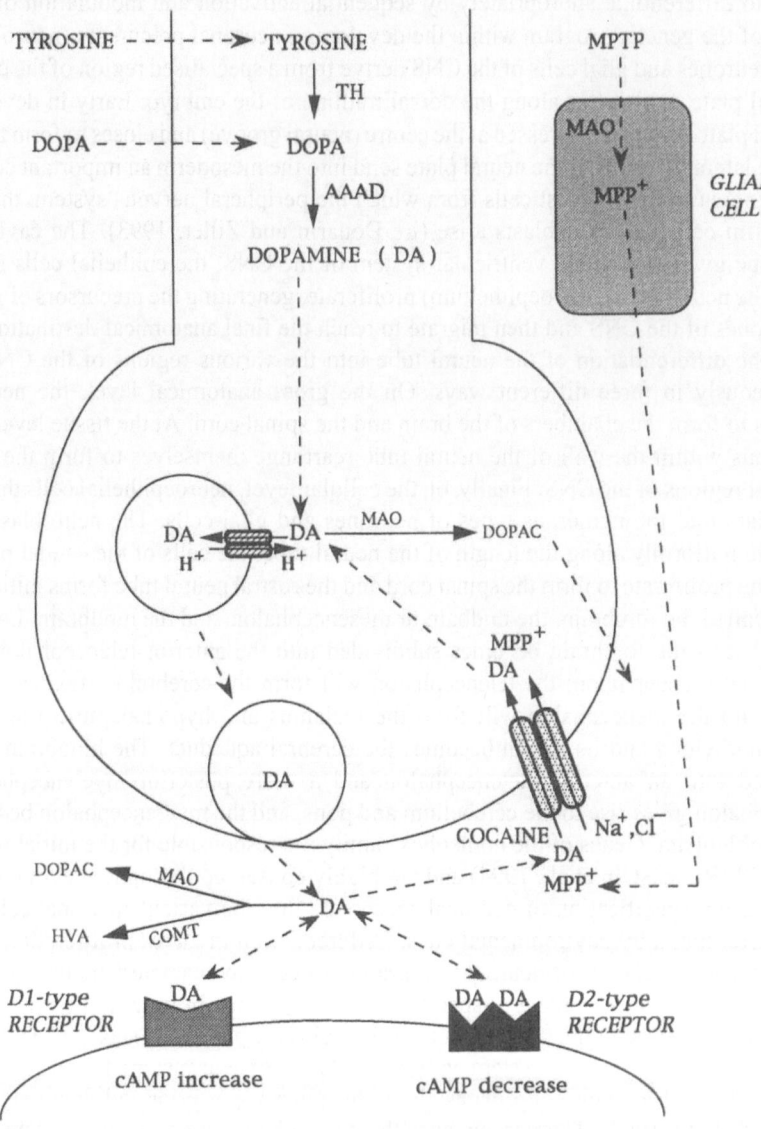

Figure 2. Pathway of dopamine synthesis, storage, release and uptake. Tyrosine is converted to DOPA by tyrosine hydroxylase (TH) and DOPA is converted to dopamine (DA) by aromatic L- amino acid decarboxylase (AAAD). DA is transported into vesicles by the synaptic vesicle monoamine transporter (striped ovals). When vesicles fuse with the presynaptic plasma membrane, DA is released into the synaptic cleft and interacts with postsynaptic D1-type or D2- type receptors (whose activation modulates cAMP level). DA is degraded to 3,4 dihydroxylphenylacetic acid (DOPAC) by monoamine oxidase (MAO) and to homovanillic acid (HVA) by catechol-O-methyltransferase (COMT) or is transported back into the presynaptic terminal by the dopamine transporter (stippled ovals). DA in the cytoplasm can be repackaged into vesicles for release or metabolised. The dopamine transporter is blocked by cocaine and can transport the dopaminergic neurotoxin 1-methyl-4-phenylpyridinium (MPP^+), which derives from 1-methyl-4-phenyl-1,2,3,6-tetrahydropyridine (MPTP), a compound unable to cross the blood-brain barrier, by the action of the glial MAO.

the embryonic brain has been detected also at the mRNA level by the use of sensitive molecular biology techniques. In the CNS, TH^+ neuroblasts will give rise to the dopaminergic neurones of midbrain and the noradrenergic neurones of the brain stem. DA midbrain neurones are generated early during development near the midbrain-hindbrain junction (Voorn et al., 1988) and then migrate extensively in a rostral-ventral direction to their final positions in the ventral midbrain (the substantia nigra, ventral tegmental area and retrorubral field).

In the rodent embryos TH is detectable in developing neuroblasts during mid-gestation. Studies by our group (di Porzio et al., 1990) show that in the mouse midbrain, TH^+ cells and fibers are first apparent by immunocytochemistry around embryonic day (E)8.5-E9. At these stages, TH is detectable only in a few cells which appear to be migrating, thus suggesting that neurotransmitter differentiation may take place prior to the final localisation of the neurones, and independently from specific regional contact. Consistent with our findings, TH mRNA has recently been detected in the whole mouse embryo as early as E8.5 by reverse transcriptase-PCR (Zhou et al., 1995). By E10 the number of TH^+ neurones in the midbrain is increased and by E10.5-E11 their distribution is reminiscent of the A9 (substantia nigra) and A10 (ventral tegmental area) cell groups of the mature mice, although only at E13 they appear as two separate TH^+ areas (di Porzio et al., 1990).

In the rat, TH^+ positive cells (Voorn et al., 1988; Fiszman et al., 1991), as well as TH mRNA (Perrone-Capano et al., 1994) can be identified as early as E12. It is possible that TH expression emerges in the rat at even earlier embryonic stages, but identification and dissection of the midbrain before E12 are technically difficult.

The early appearance of TH^+ cells during rodent embryogenesis suggests a possible morphogenetic or otherwise regulatory function of catecholamine neurotransmitters during mammalian brain development. Catecholamines seem in fact to play a crucial role during ontogeny since targeted disruption of TH (Zhou et al., 1995) and dopamine β-hydroxilase (the enzyme synthesising noradrenaline and adrenaline; Thomas et al., 1995) genes in mice results in mid-gestational lethality (90% and 95%, respectively). A role for catecholamines as morphogenetic factors is in line with a large body of evidence indicating that neurotransmitters can have a trophic effect during embryogenesis.

3.2 Role of the Floor Plate in the Induction of DA Neurones

Recent results (Hynes et al., 1995b) provide compelling evidence showing that DA neurones are born and differentiate by contact with the midbrain floor plate (the already mentioned neuroepithelial cells at the ventral midline of the neural tube). In fact, in E14 rat embryos, TH^+ neurones are found in close proximity to the floor plate cells and E11 explants comprising the midline of the midbrain develop floor plate markers after 1 day in culture and numerous TH^+ neurones after 3 days. Moreover, an exogenous floor plate, but not other embryonic or adult tissues, can induce development of DA neurones in E9 midbrain explants (from which the presumptive endogenous floor plate is removed). This induction is achieved only by direct contact. Further evidence for a physiological role of the floor plate in the specification of DA neurones is provided by an elegant experiment in which transgenic mice developing a second floor plate in vivo, also generate in its proximity an ectopic cluster of DA neurones (Hynes et al., 1995b).

The nature of the inductive signal expressed by floor plate cells able to induce the DA phenotype has been identified in the amino-terminal product of *Sonic Hedgehog* gene (SHH-N) (Hynes et al., 1995a). SHH-N derives from the autoproteolytic cleavage of the Sonic hedgehog (SHH) precursor protein, a signalling molecule implicated in patterning and growth of the vertebrate neural tube (Ericson et al., 1995), which also mediates the induction of motoneurone phenotype by floor plate cells (Perrimon, 1995; Roelink et al., 1995).

Conditioned medium from cells ingegnerized to produce and release SHH-N, as well as purified recombinant SHH-N, is able to induce development of DA neurones in rat midbrain explant cultures, mimicking the effect of floor plate cells in a dose-dependent manner. Moreover, manipulations to increase the activity of cyclic AMP-dependent protein kinase A, which is known to antagonise hedgehog signalling, can block DA neurone induction by floor plate cells (Hynes et al., 1995a). SHH-N becomes predominantly associated with cell surfaces when it is derived from the full length SHH precursor by autoproteolytic cleavage, and DA neurones differentiate in the immediate vicinity of SHH-expressing cells in vivo. These findings strongly suggest that induction of DA neurones in vivo occurs only when they are exposed to sufficient concentration of SHH-N, in close proximity to or in contact with floor plate cells. Although it is now clear that SHH-N is the floor plate-derived inducer of DA neurones, the mechanisms by which it reaches this effect remain to be elucidated. There is however a growing body of evidence suggesting that SHH-N can function as a morphogen to induce by either direct contact or diffusible signals multiple classes of neurones (DA neurones, motoneurones, floor plate cells) in a concentration dependent manner, probably restricted by the neural progenitor position along the anterior-posterior axis (Ericson et al., 1995).

4. TROPHIC FACTORS AND DA NEURONES

The rapidly increasing knowledge about the role of neurotrophic factors for neuronal development, survival, nerve fiber formation and maintenance of neurones in the adult organism has led to a search for factors that might exert such trophic influences on the midbrain DA neurones. As for other neuronal phenotypes, it is plausible that DA neurones, once differentiated, require for their survival and maturation other information and factors in order to refine their phenotype. Most studies searching for "dopaminotrophic" factors have been conducted in vitro, using cultures of foetal DA neurones and measuring features such as TH activity or immunoreactivity, dopamine levels or high affinity DA uptake as markers for DA neurone survival and differentiation.

During the last few years, several polypeptides that promote the survival and/or the maturation of embryonic DA neurones in vitro have been identified. These include basic fibroblast growth factor, epidermal growth factor (Knusel et al., 1990; Casper et al., 1991), and two members of the neurotrophin family: brain-derived neurotrophic factor (BDNF) and neurotrophin 4/5 (NT-4/5) (Hyman et al., 1994). However, transgenic mice in which the high affinity receptor for BDNF and NT-4/5 has been ablated do not show any decrease in the number of midbrain dopaminergic neurones (Klein et al., 1993), and basic fibroblast growth factor and epidermal growth factor were shown to promote the survival of DA neurones indirectly, through astrocytes (Knusel et al., 1990; Casper et al., 1991; Engele and Bohn, 1991). Moreover, none of the previously mentioned factors appears to act specifically and selectively on DA neurones. To date, experimental evidence indicates as promising candidate survival factors for DA neurones three members of the transforming growth factor protein superfamily : glial cell line-derived neurotrophic factor (GDNF; Lin et al., 1993), transform-ing growth factor (TGF) β2 and TGF β3 (Poulsen et al.,1994). GDNF promotes the survival and morphological differentiation of DA neurones in rat embryo midbrain cultures and increases their high affinity DA uptake (2.5-3 fold increase per TH neurone, as compared to control cultures); these effects are specific since GDNF does not increase total neurone or astrocyte number nor does it affect GABAergic and serotoninergic neurones present in the same culture (Lin et al., 1993). TGF β2 and TGF β3 enhance DA neurite outgrowth in vitro and display similar potencies to GDNF on the survival of rat embryonic DA neurones in culture (3 to 5 fold increase in the number of TH⁺ an DA⁺ positive cells as compared with control cultures; Poulsen et al., 1994). Transcripts for TGF β2 and GDNF are found locally in the E15.5 rat ventral midbrain in close proximity to TH⁺ neurones. At birth, the expression of TGF β2 and GDNF in the ventral midbrain is turned off, whereas TGF β2, TGF β3 and GDNF are each up-regulated in different innervation sites for DA neurones (frontal and enthorinal cortex, olfactory bulb and frontal cortex, striatum, respectively). In the adult brain, GDNF mRNA is undetectable, and only weak expression of TGF β2 and TGF β3 can be found in some cortical areas (Poulsen et al., 1994). These findings suggest TGF β2 and GDNF could locally support the survival of embryonic DA neurones while they project to their targets, whilst TGF β2, TGF β3 and GDNF could act at later developmental ages as target-derived factors for distinct subpopulations of DA nerve fibers once they have reached their innervation sites. However the observation that only a subpopulation of cultured DA neurones is rescued in the presence of saturating concentrations of GDNF or TGFs and their action on the same population of embryonic DA cells (Poulsen et al., 1994) indicate that additional physiological "dopaminotrophic" factors may exist.

Despite the lack of GDNF in the adult brain, exogenous administration of this factor has been shown to have a strong protective and reparative effect on midbrain dopamine circuits in vivo, opening interesting therapeutic perspectives for the treatment of Parkinson's disease. Tomac and co-workers (1995) provided biochemical, histochemical and behavioural evidence showing that the adult mouse nigrostriatal system can be protected from the toxic effects of the dopaminergic neurotoxin 1-methyl-4-phenyl-1,2,3,6-tetrahydropyridine (MPTP) in vivo by intracerebral GDNF administration. This neurotrophin also exerts a reparative and regenerative effect when given after MPTP. Moreover, repeated injections of GDNF adjacent to the rat substantia nigra can largely prevent the loss of DA neurones due to retrograde degeneration after transection of their axons (Beck et al., 1995).

5. MATURATION OF VENTRAL MIDBRAIN DA NEURONES

Once neurones acquire their DA phenotype, other signals involved in the maturation of DA properties during embryogenesis are activated. In order to elucidate these mechanisms, we have investigated in the embryonic rat mesencephalon and striatum in vivo the ontogeny of several parameters associated with DA neurone phenotypic expression (Fiszman et al., 1991). In cells acutely dissociated from the embryonic ventral mesencephalon (MES), measurable DA and TH immunostainings are present as early as E 12.5. The number of TH⁺ neurones increases by 15 fold between E13 and E14 and by 2 fold between E14 and E18. This pattern could reflect, in part, a maturation of DA neurones with an increased concentration of TH/cell and, in part, proliferation of DA cells, which undergo their final mitosis at E13-14 (Kono et al., 1991). In MES, DA concentration increases sharply at E16 and reaches a plateau before birth that is ten-fold lower than adult values. In the striatum, DA is first detected at E16, suggesting that DA fibers reach the striatum at this embryonic age. DA catabolites can be detected in MES and striatum only three-four days after DA appearance, suggesting that the lack of DA turnover in early embryogenesis may reflect an absence of DA release from dendrites and axons. In contrast to the early appearance of endogenous MES DA levels, specific high-affinity DA uptake in dissociated MES cells is found only at E16, and increases sharply between E16 and E18, reaching a plateau before birth (Figure 3). Thus, the onset of DA uptake and its subsequent increase, as well as the increase in MES DA synthesis, are concomitant with the arrival of the first DA fibers in the striatum and are only in part correlated with an increase in TH⁺ cell number. These results suggest that the maturation of DA MES presynaptic neurones can be related, at least in part, to interactions with the developing target cells in vivo.

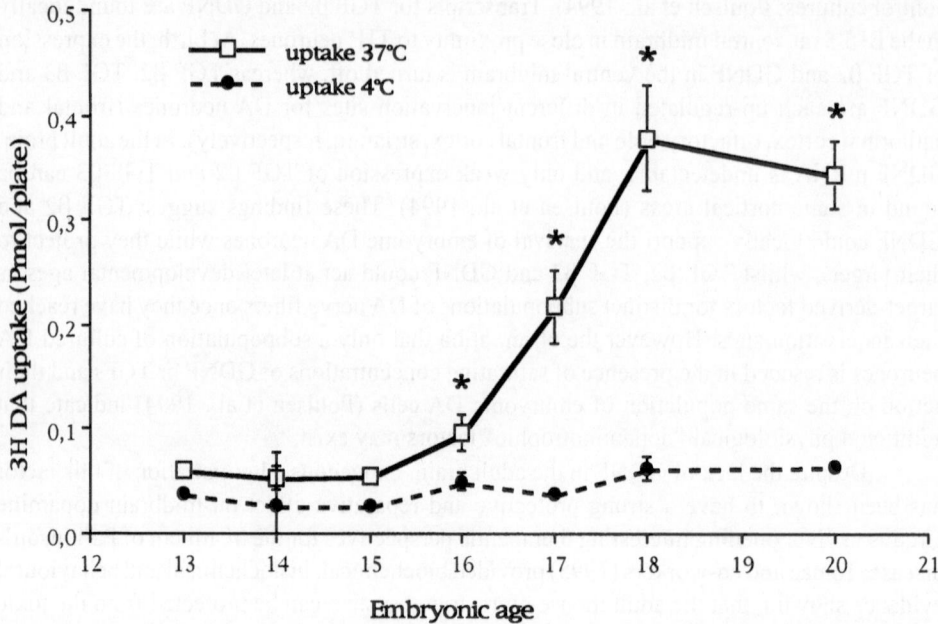

Figure 3. ^{3}H-DA uptake during embryonic development of the ventral mesencephalon. ^{3}H-DA high affinity uptake was measured as previously described (Fiszman et al., 1991) using 50 nM labelled dopamine at 37 °C or at 4 °C (blank) in cells acutely dissociated from rat mesencephalon. Each bar represents the mean ± SEM of separate experiments done in triplicate. Asterisks represent p< 0.05 between uptake values at 37 °C and at 4 °C (ANOVA, Scheff F test).

High-affinity uptake of the released DA into the presynaptic neurone is achieved throughout the activity of the plasma membrane dopamine transporter (DAT), which has been cloned in rat (Giros et al., 1991; Kilty et al., 1991), bovine (Usdin et al., 1991) and man (Giros et al., 1992). DAT is a plasma membrane glycoprotein, member of a multigene family encoding Na[+]-dependent neurotransmitter transporters with 12 putative transmembrane domains (Amara and Arriza, 1993). DAT, in addition to its physiological function, is the site of action of amphetamine and cocaine (Ritz et al., 1987) and it is responsible for the selective accumulation in DA neurones of the dopaminergic neurotoxin 1-methyl-4-phenylpyridinium (MPP[+]; Liu et al., 1992; Schinelli et al.,1988). In the brain, the distribution of DAT mRNA corresponds quite closely to that of dopaminergic neurones (Lorang et al., 1994).

We have recently examined DAT gene expression during prenatal and postnatal development of rat MES and compared it with that of two other genes that play a key role in mesencephalic DA neurotransmission: TH and synaptic vesicle monoamine transporter (SVAT, also known as VMAT2; Perrone-Capano et al., 1994). The latter catalyses transport and storage of neurotransmitters into dense core vesicles in CNS monoaminergic neurones using the electrochemical gradient generated by a vesicular H[+]-ATPase (Liu et al., 1992; Schuldiner, 1994). For comparative purposes we have also analysed the developmental pattern of the neuronal GABA transporter (GAT, subtype 1) gene expression. To perform these experiments, we have used a highly sensitive reverse transcription-PCR (RT-PCR) assay which allows to detect relatively small changes in the levels of the gene transcripts, thus enabling us to study genes with low level of expression. Moreover, this technique allows to compare the expression of various genes and overcome the inherent variations present within any individual PCR by normalising to hypoxantine-phosphoribosyl-transferase (HPRT) mRNA level, used as an internal standard. We found that during development of rat ventral mesencephalon DAT gene transcripts are detected only at around E15, even if the number of amplification cycles in the PCR reaction is increased (Figure 4). On the contrary, the mRNAs for all the other genes examined (HPRT, TH, SVAT and GAT) are already present at E12 (Figure 4). As expected, DAT gene expression is specifically restricted to the ventral MES and could never be detected in other brain areas (striatum or parietal cortex) or tissues (liver) examined, nor in the C6 glioma cell line or in the phaechromocytoma PC12 cells of

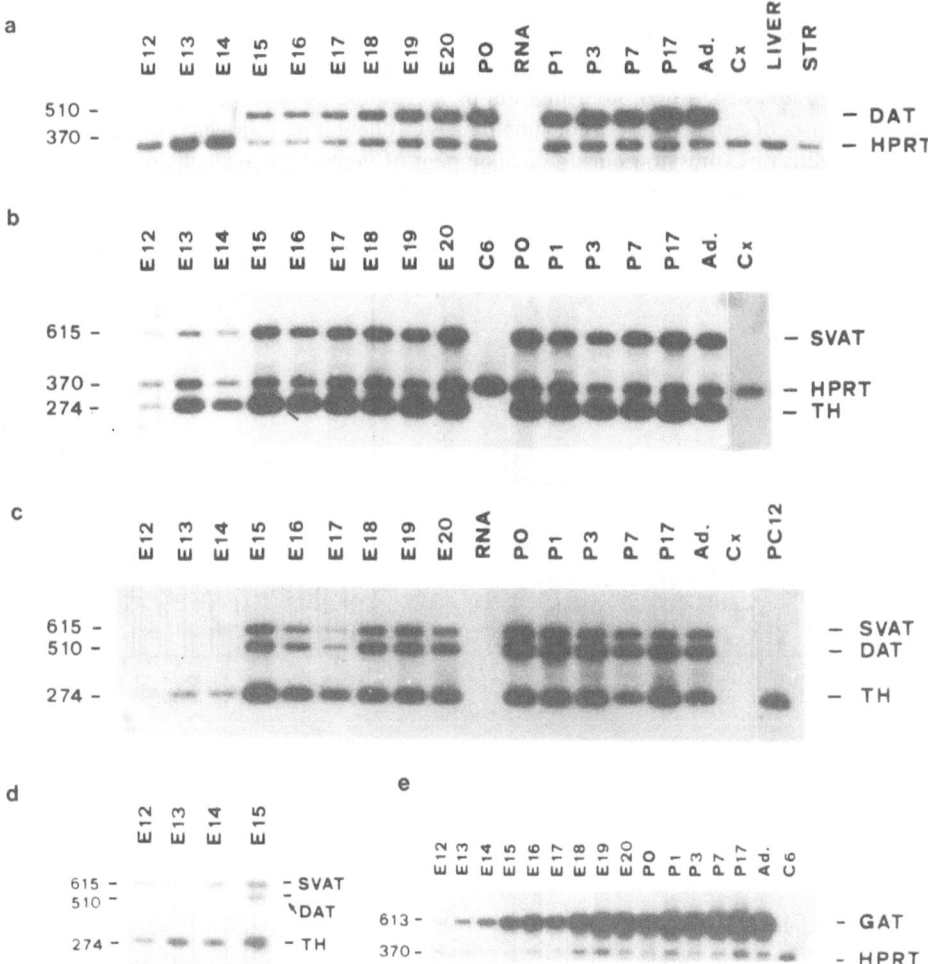

Figure 4. RT-PCR analysis of dopamine transporter (DAT), GABA transporter (GAT), hypoxantine-phosphoribosyl-transferase (HPRT) , tyrosine hydroxylase (TH) and synaptic vesicle monoamine transporter (SVAT) gene expression during development of rat mesencephalon. Specific oligonucleotide sets used to co-amplify cDNAs are: a, DAT primers and HPRT primers; b, SVAT primers, HPRT primers and TH primers; c, SVAT primers, DAT primers and TH primers; d, as in c, but increasing the number of PCR cycles to 29, instead of 24 cycles used in all amplifications; e, GAT primers and HPRT primers. Numbers indicate the embryonic (E) and postnatal (P) age. In panel a, samples E12, E13 and E14 were overexposed in order to highlight the HPRT fragment. Other cDNA samples are: Ad., adult ventral mesencephalon; Cx, adult parietal cortex; LIVER, adult liver; C6, C6 glioma cells; STR, E16 striatum; PC12, PC12 cells. RNA is a non-reverse-transcribed RNA from E20 mesencephalon. Size (in bp) and identity of the PCR products are indicated on the left and on the right, respectively. [From Perrone-Capano et al., 1994; reprinted with permission of Neuroreport].

adrenal origin (Figure 4). DAT mRNA level remains rather constant until birth and shows a sharp increase in the first postnatal days reaching, after two weeks, values comparable with those found in the adult MES (Figure 5a). Regulation of DAT transcription could be a mechanism used to influence DA neurotransmission also during adult life and in pathological conditions. Indeed, DAT gene expression is down-regulated in ageing human substantia nigra (Bannon et al., 1992) and in rat nigral neurones following repeated cocaine admini-stration (Xia et al., 1992; Cerruti et al.,1994). Both GAT and SVAT transcripts show a progressive, modest increase throughout embryonic and early postnatal development reach-ing the highest value in the adult (Figure 5b and 5c). Consistent with our previous TH immunostaining data, TH mRNA is already present at E12 and increases between E12 and E15, when it reaches a level comparable with that found in the postnatal period (Figure 5d). Thus, maturation of DA neurotransmission in developing DA neurones is not synchronous since the onset of genes required for functional DA phenotype follows a complex develop-mental pattern of expression during development of ventral mesencephalon. The delayed onset of DAT gene expression compared to that of the other transcripts examined is

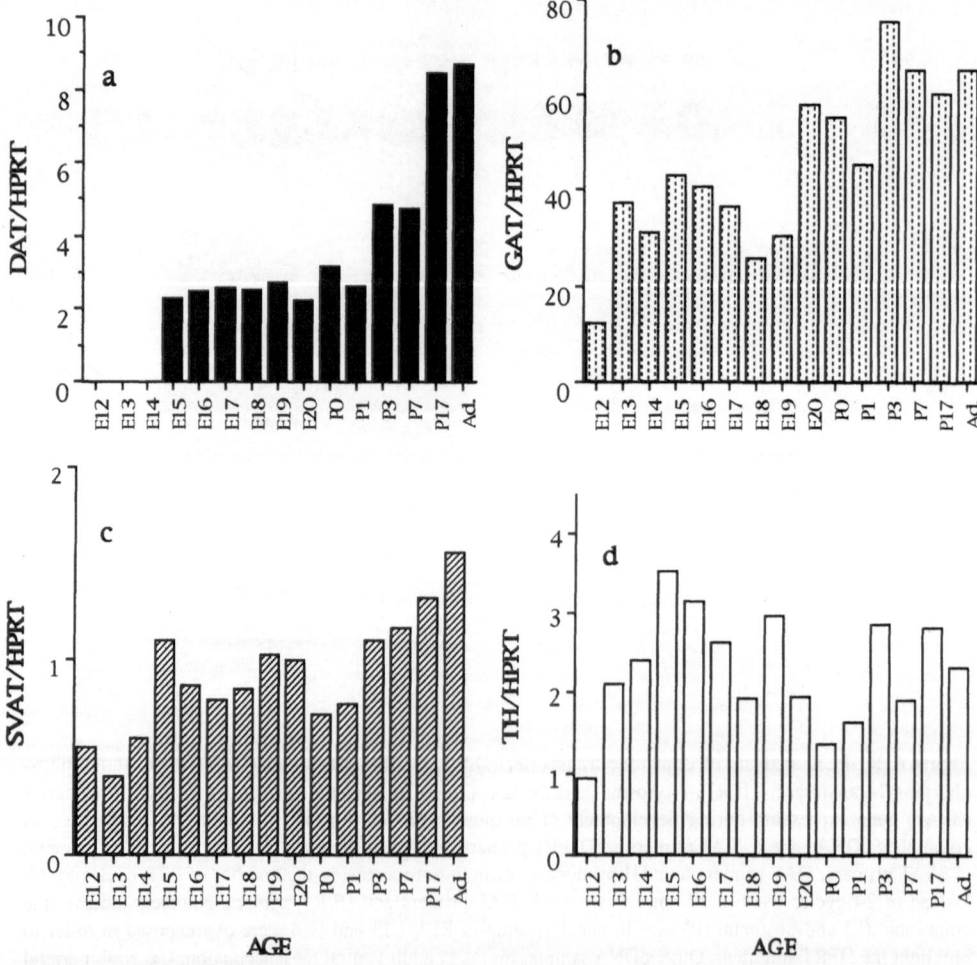

Figure 5. Relative quantitation of DAT, GAT, TH and SVAT mRNAs in rat ventral mesencephalon during development. Values on the left represent the ratios between the yield of any given gene analysed and the internal control HPRT (Perrone-Capano et al., 1994). When PCR reactions were repeated three times (as for DAT) the standard error never exceeded 10% of the mean values.

concomitant with the arrival of the first dopaminergic fibers to the target striatum, suggesting that striatal cells can regulate at transcriptional level key steps in the development of DA functions. Indeed in vitro studies indicate that maturation of DA phenotype can be influenced by environmental cues, since the addition of target striatal cells, but not of non target cells, increases synthesis and uptake of DA (Prochiantz et al., 1979; di Porzio et al., 1980) and DAT transcripts (Perrone-Capano et al., 1994 and unpublished observations) in MES neurones. These findings are in line with the current view of neural development, indicating that the expression of a programme intrinsic to each cell type and the response to extrinsic influences represented by diffusible factors and by interactions with other cells, determine how neurones develop their characteristics and the specificity of their connections.

6. DISCUSSION

One of the most fascinating questions in developmental neurobiology is how neurones are specified, acquire their peculiar characteristics and find their correct connections. In the last years, significant progress has been made in elucidating some of the mechanisms underlying the determination and the maturation of the DA phenotype. It is now clear that midbrain neural plate progenitors give rise to DA neurones by direct contact with SHH-N, the amino-terminal autoproteolytic product of the SHH precursor protein associated with cell surfaces of the floor plate cells. Moreover, trophic molecules such as GDNF, TGF β2 and TGF β3 seem to be physiological and specific factors for the survival, development and maturation of DA neurones, since in the developing CNS they are expressed sequentially as local and target-derived molecules.

The role of target tissue during the ontogeny of the ventral mesencephalon is becoming elucidated at molecular level and it is well established that post-synaptic striatal target cells stimulate the maturation of the DA phenotype, modulating DA synthesis, high-affinity uptake and DAT gene expression.

Increasing knowledge of the DA phenotype will help in elucidating the basic mechanisms of their development, survival, maturation and function and will open interesting therapeutic perspectives for the treatment of diseases related to dysfunction of DA neurotransmission.

ACKNOWLEDGMENTS

The authors thank Dr. F. Morelli for critical reading of the manuscript, T. Vespa and R. Vito for technical assistance. This work is supported by grants from the PP FF Consiglio Nazionale delle Ricerche Ingegneria Genetica and Invecchiamento, Unione Italiana Lotta alla Distrofia Muscolare-Telethon Italy, AIRC and British Council.

REFERENCES

Amara SG, Arriza JL (1993) Neurotransmitter transporters: three distinct gene families. Curr Opin Neurobiol 3: 337-344.

Bannon MJ, Poosch MS, Xia Y, Goebel DJ, Cassin B, Kapatos G (1992) Dopamine transporter mRNA content in human substantia nigra decreases precipitously with age. Proc Natl Acad Sci USA 89: 7095-7099.

Beck KD, Valverde J, Alexi T, Poulsen K, Moffat B, Vandlen RA, Rosenthal A, Hefti F (1995) Mesencephalic dopaminergic neurons protected by GDNF from axotomy-induced degeneration in the adult brain. Nature 373: 339-341.

Bjorklund A, Lindvall O (1984) Dopamine containing systems in the CNS. In: Bjorklund A, Hokfelt T (eds) Handbook of Chemical Neuroanatomy. Elsevier, Amsterdam, pp. 55-122.

Casper D, Mytilineou C, Blum M (1991) EGF enhances the survival of dopamine neurons in rat embryonic mesencephalon primary cell culture. J Neurosc Res 30: 372-381.

Cerruti C, Pilotte NS, Uhl G, Kuhar MJ (1994) Reduction in dopamine transporter mRNA after cessation of repeated cocaine administration. Mol Brain Res 22:132-138.

di Porzio U, Daguet MC, Glowinski J, Prochiantz A (1980) Effect of striatal cells on in vitro maturation of mesencephalic dopaminergic neurons grown in serum-free conditions. Nature 288: 370-373.

di Porzio U, Zuddas A, Cosenza-Murphy DB, Barker JL (1990) Early appearance of tyrosine hydroxylase immunoreactive neurons in the mesencephalon of mouse embryos.Intern J Devl Neurosci 8:523-532.

Engele J, Bohn MC (1991) The neurotrophic effects of fibroblast growth factors on dopaminergic neurons in vitro are mediated by mesencephalic glia. J Neurosci 11: 3070-3078.

Ericson J, Muhr J, Plackzek M, Lints T, Jessel TM, Edlund T (1995) Sonic hedgehog induces the differentiation of vental forebrain neurons: a common signal for ventral patterning within the neural tube. Cell 81: 747-756.

Fiszman ML, Zuddas A, Masana MI, Barker JL, di Porzio U (1991) Dopamine synthesis precedes dopamine uptake in embryonic rat mesencephalic neurons. J Neurochem 56: 392-399.

Giros B, El Mestikawy S, Bertrand L, Caron LG (1991) Cloning and functional characterization of a cocaine-sensitive dopamine transporter. FEBS Lett 295: 149-154.

Giros B, El Mestikawy S, Godinot N, Zheng K, Han H, Yang FT, Caron MG (1992) Cloning, pharmacological characterization, and chromosome assignment of the human dopamine transporter. Mol Pharmacol 42: 383-390.

Hyman C, Juhasz M, Jackson C, Wright P, Ip NY, Lindsay R (1994) Overlapping and distinct actions of the neurotrophins BDNF, NT-3, and NT-4/5 on cultured dopaminergic and GABAergic neurons of the ventral mesencephalon. J Neurosci 14: 335-347.

Hynes M, Porter JA, Chiang C, Chang D, Tessier-Lavigne M, Beachy PA, Rosenthal A (1995a) Induction of midbrain dopaminergic neurons by sonic hedgehog. Neuron 15: 35-44.

Hynes M, Poulsen K, Tessier-Lavigne M, Rosenthal A (1995b) Control of neuronal diversity by the floor plate: contact-mediated induction of midbrain dopaminergic neurons. Cell 80: 95-101.

Jenner P, Schapira AH, Marsden CD (1992) New insights into the cause of Parkinson's disease.Neurology 42: 2241-2250.

Jessel TM, Dodd J (1992) Floor-plate derived signals and the control of neural cell pattern in vertebrates. Harvey Lect 86: 87-128.

Kilty JE, LorAng D, Amara SG (1991) Cloning and expression of a cocaine-sensitive rat dopamine transporter. Science 254: 578-579.

Klein R, Smeyne RJ, Wurst W Long LK, Auerbach BA, Joyner AL, Barbacid M (1993) Targetet disruption of the trkB neurotrophin receptor gene results in nervous system lesions and neonatal death. Cell 75: 113-122.

Knusel B, Michel PP, Schwaber JS, Hefti F (1990) Selective and non selective stimulation of central cholinergic and dopaminergic development in vitro by nerve growth factor, basic fibroblast growth factor, epidermal growth factor, insulin and the insulin-like growth factors I and II. J Neurosci 10: 558-570.

Kono T, Takada M, Wu JY, Kitai ST (1991) Double immunohystochemical detection of transmitter phenotype of proliferating cells using bromodeoxyuridine. Neurosci Lett 132: 113-116.

Le-Douarin NM, Ziller C (1993) Plasticity in neural crest differentiation. Curr Opin Cell Biol 5: 1036-1043.

Lin LH, Doherti DH, Lile JD, Becktesh S, Collins F (1993) GDNF: a glial cell line-derived neurotrophic factor for midbrain dopaminergic neurons. Science 260: 1130-1132.

Liu Y, Peter D, Roghani A, Schuldiner S, Privé GG, Eisenberg D, Brecha N, Edwards RH (1992) A cDNA that suppresses MPP$^+$ toxicity encodes a vesicular amine transporter. Cell 70: 539-551.

Lorang D, Amara SG, Simerly RB (1994) Cell-type-specific expression of catecholamine transporters in the rat brain.J Neurosci 14: 4903-4914.

Malmberg A, Jackson DM, Erickson A, Mohell N (1993) Unique binding characteristics of antipsychotic agents interacting with human dopamine D2A, D2B, and D3 receptors.Mol Pharmacol 4 : 749-754.

Perrimon N (1995) Hedgehog and beyond. Cell 80: 517-520.

Perrone-Capano C, Tino A, di Porzio U (1994) Target cells modulate dopamine transporter gene expression during brain development. Neuroreport 5: 1145-1148.

Poulsen KT, Armanini MP, Klein RD, Hynes MA, Phillips HS, Rosenthal A (1994) TGF β2 and TGF β3 are potent survival factors for midbrain dopaminergic neurons. Neuron 13: 1245-1252.

Prochiantz A, di Porzio U, Kato A, Berger B, Glowinski J (1979) In vitro maturation of mesencephalic dopaminergic neurons from mouse embryos is enhanced in presence of their striatal target cells. Proc Natl Acad Sci USA 76: 5387-5391 .

Ritz MC, Lamb RJ, Goldberg SR, Kuhar MJ (1987) Cocaine receptors on dopamine transporters are related to self-administration of cocaine. Science 237: 1219-1223.

Roelink H, Porter JA, Chiang C, Tanabe Y, Chang DT, Beachy PA, Jessel TM (1995) Floor plate and motor neuron induction by different concentrations of the amino-terminal cleavage product of Sonic hedgehog autoproteolysis. Cell 81: 445-455.

Rubenstein JLR, Martinez S, Schimamura K, Puelles L (1994) The embryonic vertebrate forebrain: the prosomeric model. Science 266:578-580.

Schinelli S, Zuddas A, Kopin I, Barker JL, di Porzio U (1988) MPTP metabolism and MPP$^+$ uptake in vitro in mesencephalic dopaminergic neurons from the mouse embryo. J Neurochem 50: 1900-1907.

Schuldiner S (1994) A molecular glimpse of vesicular monoamine transporters. J Neurochem 62: 2067-2078.

Seeman P, Guan HC, van Tol HH (1993) Dopamine D4 receptors elevated in schizophrenia. Nature 365: 441-445.

Thomas SA, Matsumoto AM, Palmiter RD (1995) Noradrenaline is essential for mouse fetal development. Nature 374: 643-646.

Tomac A, Lindqvist E, Lin LFH, Ogren SO, Young D, Hoffer BJ, Olson L (1995) Protection and repair of the nigrostriatal dopaminergic system by GDNF in vivo. Nature 373: 335-339.

Usdin TB, Mezey E, Chen C, Brownstein MJ, Hoffman BJ (1991) Cloning of the cocaine-sensitive bovine dopamine transporter. Proc Natl Acad Sci USA 88: 11168-11171.

Van Tol HH, Bunzow JR, Guan HG, Sunahara RK, Seeman P, Niznick HB, Civelli O (1991) Cloning of the gene for a human dopamine D4 receptor with high affinity for the antipsychotic clozapine.Nature 350: 610-614.

Voorn P, Kalsbeek A, Jorritsma-Byham B, Groenewegen HJ (1988) The pre- and postnatal development of the dopaminergic cell groups in the ventral mesencephalon and the dopaminergic innervation of the striatum of the rat. Neuroscience 25: 857-887.

Xia Y, Goebel DJ, Kapatos G, Bannon MJ (1992) Quantitation of rat dopamine transporter mRNA: effects of cocaine treatment and withdrawal. J. Neurochem 59: 1179-1182.

Yamada T, Pfaff SL, Edlund T, Jessel TM (1993) Control of cell pattern in the neural tube: motor neuron induction by diffusible factors from nothocord and floor plate. Cell 73: 673-686.

Zhou Q-Y, Quaife CJ, Palmiter RD (1995) Targeted disruption of the tyrosine hydroxylase gene reveals that catecholamines are required for mouse fetal development. Nature 374:640-643.

9

COMPARISON OF "NEAR MEMBRANE" AND BULK CYTOPLASMIC CALCIUM CONCENTRATION IN SINGLE CARDIAC VENTRICULAR MYOCYTES DURING SPONTANEOUS CALCIUM WAVES

A. W. Trafford, M. E. Díaz,[*] S. C. O'Neill, and D. A. Eisner

Department of Veterinary Preclinical Science
The University of Liverpool
P.O. Box 147, Liverpool, L69 3BX, United Kingdom

ABSTRACT

In this paper we discuss differences in the time course of changes of intracellular calcium concentration ($[Ca^{2+}]_i$) occurring in the bulk cytoplasm and adjacent to the surface membrane (subsarcolemmal or 'fuzzy' space) during spontaneous oscillatory release of Ca^{2+} from the sarcoplasmic reticulum (SR). Sarcolemmal Na-Ca exchange current and $[Ca^{2+}]_i$ were measured in single voltage clamped rat ventricular myocytes.

Spontaneous Ca^{2+} release from the SR resulted in a transient inward current which developed and decayed more quickly than the corresponding changes in $[Ca^{2+}]_i$ measured using the Ca^{2+} sensitive fluorescent indicators Indo-1, Fluo-3 and Calcium Green-1. The discrepancy in the time course of changes in current and $[Ca^{2+}]_i$ results in a hysteresis between $[Ca^{2+}]_i$ and current.

A similar hysteresis was observed if $[Ca^{2+}]_i$ was raised with caffeine. The hysteresis between current and $[Ca^{2+}]_i$ was removed by low pass filtering the current record with a time constant of 132 ms.

Digital video imaging was performed to allow simultaneous measurement of $[Ca^{2+}]_i$ at all points of the cell during spontaneous release of Ca^{2+} from the SR. The hysteresis between current and $[Ca^{2+}]_i$ remained even after the spatial and temporal properties of the Ca^{2+} wave and any non linear relationship between current and fluorescence and $[Ca^{2+}]_i$ were accounted for.

Using a model in which there is a barrier to diffusion of Ca^{2+} between the subsarcolemmal and bulk compartments the hysteresis between current and $[Ca^{2+}]_i$ can be ac-

---

[*] Tel: + 44-151-794-4244; Fax: + 44-151-794-4243; email: m.diaz@liv.ac.uk

Neurobiology, edited by Torre and Conti
Plenum Press, New York, 1996

counted for. The calculated subsarcolemmal Ca^{2+} concentration rises before, and to a higher level than the measured bulk cytoplasmic Ca^{2+} concentration. The delay introduced by this diffusion barrier is equivalent to a time constant of 133 ms.

The subsarcolemmal space introduced in this paper may be equivalent to the 'fuzzy space' previously suggested to be important in controlling SR Ca^{2+} release.

1. INTRODUCTION

1.1. Calcium Movements during Cardiac Excitation-Contraction Coupling

Contraction in cardiac muscle is regulated by the movement of Ca^{2+} ions in and out of the SR. In cardiac preparations where the sarcolemmal involvement in excitation contraction coupling has been removed by chemical skinning, Ca release from the SR has been shown to be regulated by the changes of free Ca concentration in the solution bathing the SR (Fabiato & Fabiato, 1975; Fabiato, 1983). In intact myocytes, with a functional sarcolemma, the accepted source of Ca, which gates the release of Ca from the SR, is the transmembrane flux of Ca carried by I_{Ca} during depolarization (Fabiato, 1985; Nabauer *et al.* 1989). However, Leblanc and Hume (1990) showed that, even in the absence of Ca entry through the L-type Ca channel, membrane depolarization could elicit release of Ca from the ryanodine sensitive internal store (the SR) in a manner dependent upon: (i) Na entry through a TTX sensitive channel, and, (ii) extracellular Ca. These findings are indicative of a Ca induced trigger mechanism for Ca release from the SR that involves Ca entry into the cell by reverse mode Na-Ca exchange during depolarization (Lederer *et al.* 1992). Under these conditions the rise of $[Ca^{2+}]_i$ necessary for triggering Ca release from the SR occurs as a result of Na entry through TTX sensitive Na channels activating the Na-Ca exchanger to transport Ca into the cell.

The Na-Ca exchanger uses the electrochemical gradient for Na to pump Ca out of the cell with a stoichiometry of 3Na : 1Ca. Therefore, for each Ca^{2+} extruded from the cell there is a net gain of one positive charge generating an inward current (Miura & Kimura, 1989; Kimura *et al.* 1986). At membrane potentials (E_m) positive to the reversal potential for the Na-Ca exchanger the pump operates in the opposite direction bringing Ca into the cell generating an outward current. However, during a normal action potential with a normal $[Na^+]_i$ the exchanger does not bring adequate Ca into the cell by itself to activate SR Ca release (Cannell *et al.* 1987). Of the factors determining the direction of operation of the Na-Ca exchanger then the most likely variable, other than membrane potential, which would change in favour of Ca entry during EC coupling is $[Na^+]_i$.

By considering a typical cardiac myocyte of volume 20 pl having a peak I_{Na} of 50 nA and time constant of inactivation of 1 ms (Brown *et al.* 1981) then the Na entering the cell on the I_{Na} would produce an increase of $[Na^+]_i$ of approximately 25 µM throughout the whole cytoplasm (Lederer *et al.* 1992). Such a small change of $[Na^+]_i$ above the normal cytoplasmic concentration of ~ 10 mM would be insufficient to produce a significant increase of Ca entry by the Na-Ca exchanger (Miura & Kimura, 1989). However, if the Na entry into the cell occurs into a restricted space adjacent to the sarcolemma as opposed to diffusing freely throughout the entire cytoplasm then the rise of Na concentration detected by the Na-Ca exchanger could be much greater and therefore favour reverse mode operation. For example, if Na entry were restricted to within 100 Å (equivalent to 0.3% of cell volume) of the surface membrane, then the rise of subsarcolemmal Na concentration ($[Na^+]_m$) would be

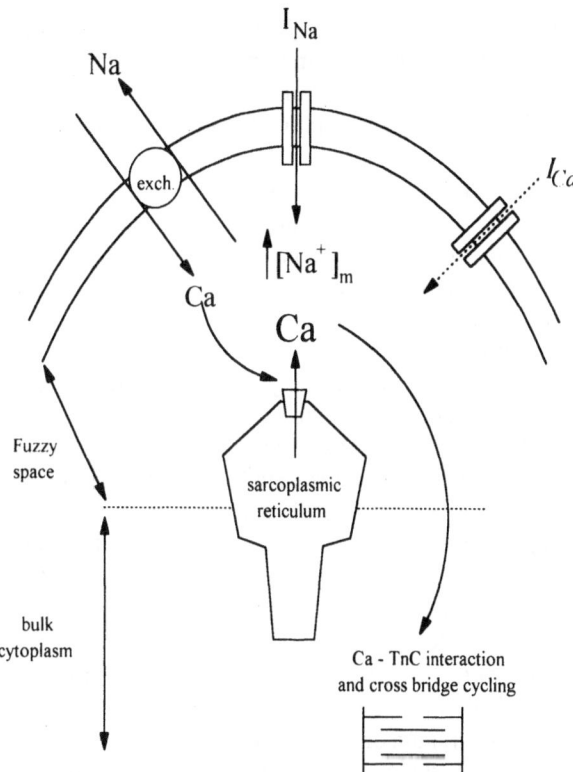

Figure 1. Sodium current induced SR calcium release. Na entering the cell on the Na current produces a localised increase in Na concentration within the fuzzy space adjacent to the surface membrane. This localised increase in Na concentration activates the Na-Ca exchanger and produces reverse mode Ca entry on the exchanger. Ca entering the cell in this manner (in the absence of L-type Ca channel activity) triggers SR Ca release. The Ca released from the SR diffuses into the bulk cytoplasmic compartment and produces myofilament activation and contraction.

approximately 8 mM. This doubling of $[Na^+]_m$ would then be sufficient to promote adequate Ca entry on the Na-Ca exchanger to trigger SR Ca release.

1.2. Near Membrane Concentration Gradients

The results of Leblanc and Hume (1990) therefore suggest that there may be a space adjacent to the surface membrane, termed the subsarcolemmal or fuzzy space (Lederer *et al.* 1992), which is functionally significant during cardiac excitation contraction coupling. The fuzzy space is characterised by restricted diffusion to Na. As a result, the $[Na^+]_m$ in this space sensed by the Na-Ca exchanger is greater, at least in the short term following membrane depolarization, than that in the bulk cytoplasm ($[Na^+]_i$). Figure 1 shows a schematic representation of the proposed involvement of I_{Na} in triggering Ca release from the SR. The Na channel, Na-Ca exchanger, L-type Ca channel and the RyR are all likely to have preferential access to this space given the complex interactions between transsarcolemmal ion movements and the feedback mechanisms of the SR (Stern, 1992; Stern & Lakatta, 1992).

Ionic concentration gradients between the surface membrane and the bulk cytoplasm have also been inferred from a series of experimental results where a discrepancy between changes of Ca concentration in the bulk cytoplasm ($[Ca^{2+}]_b$) and Ca activated currents has been noticed in coronary artery smooth muscle cells (Stehno-Bittel & Sturek, 1992; Ganitkevich & Isenberg, 1996), cardiac atrial myocytes (Lipp *et al.* 1990) and pancreatic acinar cells (Osipchuk *et al.* 1990). Such concentration gradients have also been measured directly using X-ray microprobe analysis in cardiac muscle (Wendt-Gallitelli *et al.* 1993; Wendt-Gallitelli & Isenberg, 1991; Isenberg & Wendt-Gallitelli, 1990) and portal vein smooth muscle (Bond *et al.* 1984).

The existence of concentration gradients between the surface membrane and the main cytoplasmic compartment is likely to have important consequences for cardiac EC coupling in general and in particular the activation and inactivation of membrane associated channels, pumps and exchangers. The aims of this paper are, firstly, to compare the time course of changes of Ca concentration in the bulk cytoplasmic compartment with the changes of Ca concentration estimated to occur near to the sarcolemma ($[Ca^{2+}]_m$) in single rat ventricular myocytes, and, secondly, to account for any differences in the measured changes in Ca concentration which may arise between the two compartments.

In order to infer the changes of Ca concentration that are occurring adjacent to the surface membrane Stehno-Bittel & Sturek (1992) and Osipchuk *et al* (1990) measured the current generated by Ca activated K and Cl channels respectively. Elevation of the cellular Ca load in cardiac muscle is associated with the appearance of spontaneous (not associated with an action potential) inward currents and contractions (Kass *et al.* 1978; Kass *et al.* 1978; Eisner & Lederer, 1979; Mechmann & Pott, 1986; Fedida *et al.* 1987). These groups attributed the inward current to the activity of the Na-Ca exchanger arising as a consequence of spontaneous release of Ca from the SR. Berlin *et al* (1989) later demonstrated that the inward current was indeed attributable to spontaneous release of Ca from the SR which commenced in one discrete region of the cell and then propagated through the cell as a wave of CICR. In order to measure the timecourse of changes of $[Ca^{2+}]_m$ we have used the current generated by the Na-Ca exchanger during spontaneous release of Ca from the SR.

Spontaneous release of Ca from the SR is believed to be a phenomenon related to Ca overload in cardiac myocytes (Stern *et al.* 1988). This type of oscillatory release of Ca is seen in approximately fifty percent of single rat ventricular myocytes voltage clamped at a resting membrane potential of -80 mV. In those cells that do not exhibit oscillatory behaviour it is possible to produce release of Ca from the SR into the subsarcolemmal space by application of caffeine to the superfusate (Callewaert *et al.* 1989; Ashley *et al.* 1977). The measurement of Na-Ca exchange current during spontaneous or caffeine induced release of Ca from the SR avoids confusion from voltage and time dependent currents that would be activated if Ca were released in response to a depolarizing voltage clamp pulse.

We have measured changes in the inward current generated by the Na-Ca exchanger during spontaneous oscillatory or caffeine induced release of Ca from the SR. The experiments were performed under voltage clamp conditions using either the perforated patch or whole cell techniques (for details of experimental solutions see Varro *et al* (1993) and Trafford *et al* (1995). Na-Ca exchange current was measured either during spontaneous or caffeine induced release at a resting membrane potential of -80 (perforated patch) or - 40 mV (whole cell technique).

Changes in $[Ca^{2+}]_b$ were measured with either Indo-1 using a photomultiplier based system or by digital video imaging using Fluo-3 or Calcium Green-1. Using Indo-1 the ratio of emission $R_{400/500}$ was used as an indication of $[Ca^{2+}]_i$ when fluorescence was excited at 340 nm, the Indo-1 records were low pass filtered at 100 Hz and stored on a VCR based system (Medical Systems PCM-8) (O'Neill *et al.* 1990). For digital video imaging the fluorescence records were stored on video tape at video frame rate (25 Hz) and analyzed using a frame grabber and coprocessor (Data Translation, DT2681 & DT2858) (Trafford *et al.* 1993). In order to correct for non-uniformities in cell thickness, dye loading and illumination intensity the images were normalized by dividing by a resting image obtained before Ca release (Cheng *et al.* 1993). For digital video imaging at a single wavelength using

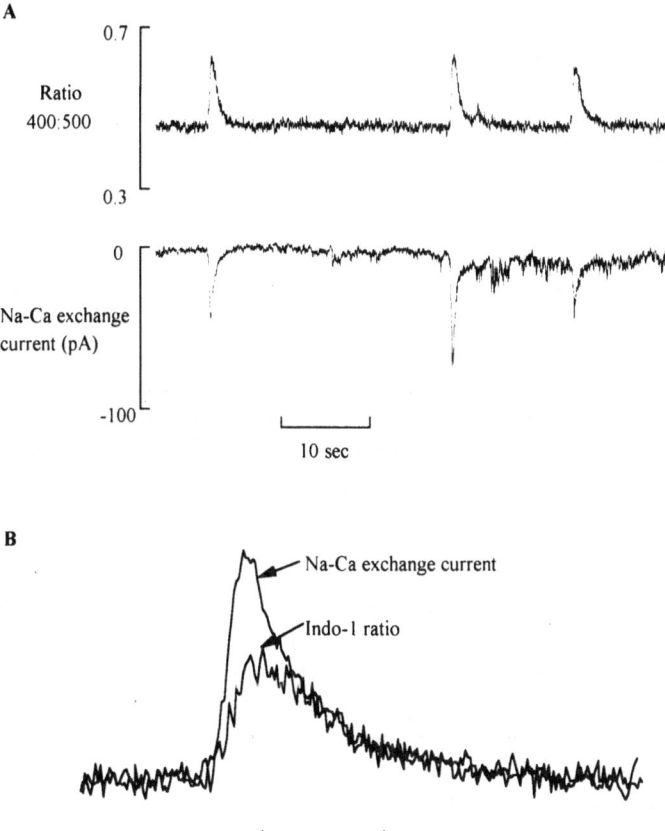

Figure 2. The relationship between current and Ca oscillations. A) Simultaneous measurements of Indo-1 ratio and Na-Ca exchange current for three spontaneous oscillatory releases of Ca from the SR. B) Expanded record of current and ratio for a single oscillation. The current record has been inverted and scaled with respect to the ratio so that the final decaying phases of the two records are superimposed. The cell was voltage clamped at a resting membrane potential of -80 mV.

Fluo-3 or Calcium Green-1 movement artifacts were prevented by addition of 2,3-Butanedione Monoxime (10 - 20 mM) to the superfusate.

2. RESULTS

The timecourses of I_{Na-Ca} and Indo-1 fluorescence for three spontaneous oscillatory releases of Ca from a single voltage clamped myocyte are shown in Figure 2. Each of the oscillatory increases of Ca commences in one region of the cell and propagates throughout the cell as a wave of CICR (Takamatsu & Wier, 1990). Some of the Ca released from the SR during the wave of CICR is extruded from the cell by the electrogenic Na-Ca exchanger generating an inward current as shown in the lower panel of Figure 2A. The Indo-1 ratio and I_{Na-Ca} for the first oscillation are shown on an expanded timescale in Figure 2B. In this panel, the I_{Na-Ca} record has been inverted and superimposed on the Indo-1 ratio record. The current record has been scaled with respect to the Indo-1 record such that the final decaying phases of current and ratio are superimposed. In Figure 2B it is clear that there is a discrepancy between the timecourse of the I_{Na-Ca} and $[Ca^{2+}]_b$ as measured with Indo-1. The figure shows that, when the current and ratio records are normalised in this way, I_{Na-Ca} rises and initially decays faster than the accompanying change of $[Ca^{2+}]_b$ detected by the fluorescent indicator Indo-1. The temporal discrepancy shown here between the timecourses of I_{Na-Ca} and $[Ca^{2+}]_b$ suggests that at the onset of spontaneous oscillatory release the Ca concentration detected by the Na-Ca exchanger in the subsarcolemmal compartment may be different to that detected by the fluorescent indicator in the cytoplasmic (or bulk) compartment. The direction of the temporal discrepancy also suggests that the changes of Ca concentration in the subsarcolemmal compartment are occurring earlier than those in the bulk cytoplasmic compartment.

The temporal discrepancy between current and calcium is shown more clearly in Figure 3 where I_{Na-Ca} is plotted against Indo-1 ratio for a single spontaneous oscillation. The figure shows that for a given $[Ca^{2+}]_b$, in terms of ratio units, the inward current generated by the Na-Ca exchanger is greater when Ca is rising in the bulk compartment than when it is falling (arrows indicate the direction of change of $[Ca^{2+}]_b$). This results in a hysteresis between I_{Na-Ca} and $[Ca^{2+}]_i$. This type of hysteresis was observed in 15 out of 21 cells (71 %) where spontaneous oscillations of current and Ca were examined in this way.

For the experiments shown in Figures 2 and 3 $[Ca^{2+}]_b$ was measured photometrically using the calcium sensitive fluorescent indicator Indo-1. Using this approach it is not possible to measure the spatial distribution of $[Ca^{2+}]_b$ within the cell because Ca measurements represent the average changes of fluorescence within the field of view (Wier *et al.* 1987). Therefore, if the hysteresis between $[Ca^{2+}]_b$ and I_{Na-Ca} arises due to any non uniform distribution of $[Ca^{2+}]_b$ throughout the cell (Takamatsu & Wier, 1990) this method of measuring $[Ca^{2+}]_b$ would not be able to account for the spatial distribution of $[Ca^{2+}]_b$ within the cell. In order to account for any effect of the non uniform distribution of $[Ca^{2+}]_b$ seen

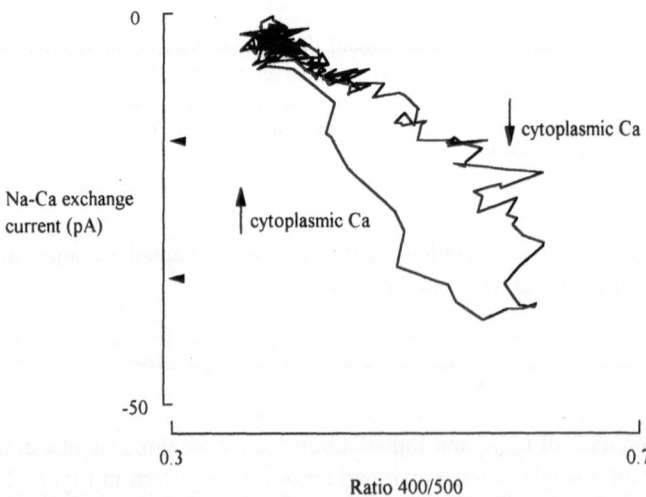

Figure 3. Hysteresis plot during a spontaneous oscillation. The relationship between Na-Ca exchange current and Indo-1 ratio during a spontaneous propagating wave of CICR in a voltage clamped myocyte (holding potential-80 mV). Na-Ca exchange current is plotted against Indo-1 ratio. During the downward limb of the hysteresis loop the calcium concentration in the bulk cytoplasm detected by the fluorescent indicator is increasing and results in a larger inward current than in the upper limb of the loop when the calcium concentration in the bulk compartment is falling.

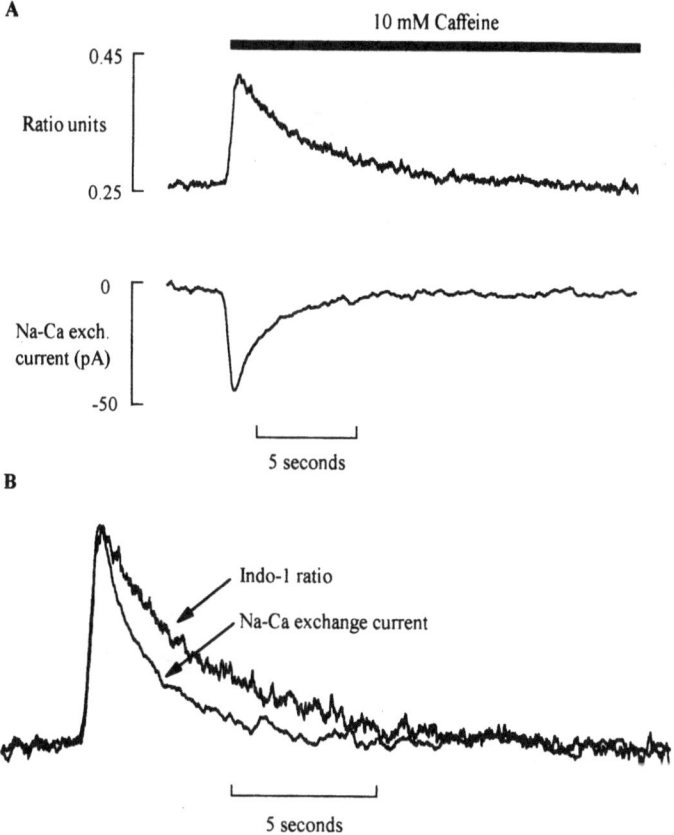

Figure 4. The effect of Caffeine on Indo-1 ratio and Na-Ca exchange current. **A)** Simultaneous measurements of Indo-1 ratio and Na-Ca exchange current, Caffeine was applied to the cell for the time indicated by the solid bar. **B)** Na-Ca exchange current and Indo-1 ratio shown on an expanded time scale, the current record has been inverted and scaled with respect to the ratio record such that the initial rising phases of current and ratio are superimposed. Changes in current and ratio were recorded at a membrane holding potential of -40 mV.

during spontaneous oscillatory release it is possible to use two separate approaches. The first is to produce a synchronous rise of $[Ca^{2+}]_b$ throughout the cytoplasm, and, the second, would be to use imaging techniques to measure the distribution of $[Ca^{2+}]_b$ within the bulk compartment.

It is possible to produce a spatially synchronous rise of $[Ca^{2+}]_b$ in response to a depolarizing voltage clamp command. However, with this method the early changes in I_{Na-Ca} would be obscured by the much larger Na, Ca and K currents and therefore it would not be possible to accurately infer the changes in subsarcolemmal Ca concentration. A second method is to produce release of Ca from the SR by rapidly applying caffeine to the cell. The application of caffeine was observed to produce a synchronous rise of $[Ca^{2+}]_b$, at least within the 40 ms sampling interval of the video frame rate (not shown).

The effect of applying caffeine upon the timecourse of changes of I_{Na-Ca} and $[Ca^{2+}]_i$ is shown in Figures 4 and 5. The application of caffeine produces a transient rise in $[Ca^{2+}]_b$ and an inward current (Figure 4A), the ratio and current traces are shown on an expanded timescale in Figure 4B. In this panel the current trace has again been inverted and scaled with respect to the ratio record. However, in this case the records have been scaled such that the initial rise phases of current and Ca are superimposed. By scaling the traces in this manner it is clear that I_{Na-Ca} initially decays more rapidly than Indo-1 ratio ($[Ca^{2+}]_b$).

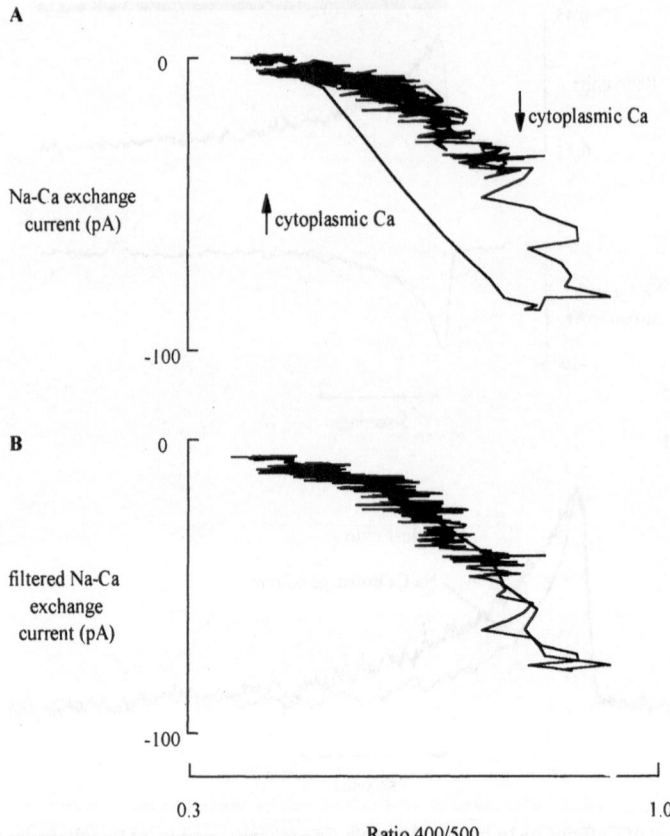

Figure 5. Hysteresis plots between current and ration after caffeine application. Relationship between Na-Ca exchange current and Indo-1 ration during Caffeine application. The hysteresis loop in A) The effect of filtering the current record. A single exponential filter of approximately 100 ms applied to the current record effectively removes the observed hystreresis between current and calcium.

The discrepancy in the rate of decay of the normalised I_{Na-Ca} and $[Ca^{2+}]_b$ observed in Figure 4 once again results in a hysteresis between $[Ca^{2+}]_b$ and I_{Na-Ca}. The hysteresis occurs in the same direction as that shown for spontaneous oscillatory release of Ca from the SR such that the inward current generated by the Na-Ca exchanger is greater when the Ca concentration is increasing in the bulk compartment (as detected by the fluorescent indicator) than when it is falling in this compartment. A similar hysteresis was observed in 11 out of 13 cells (85 %) of I_{Na-Ca} - Indo-1 ratio relationships obtained by caffeine application.

The experimentally observed hysteresis can be removed by applying a low pass single exponential filter to the current record. In Figure 5B such a filter has been applied to the original record shown in the upper panel. It can be seen that the hysteresis between I_{Na-Ca} and Indo-1 ratio has effectively been removed. On average the value of the time constant for the filter required to best remove the hysteresis between the two variables is 132 ± 19 ms (mean $\pm$ s.e.m). This therefore suggests that under these conditions I_{Na-Ca} and Indo-1 ratio are out of phase by approximately 100 ms.

2.1. Does the Hysteresis Result from Macroscopic Non-Uniformities of $[Ca^{2+}]_b$?

The hysteresis between I_{Na-Ca} and $[Ca^{2+}]_b$ (when measured ratiometrically) is seen when Ca release occurs spontaneously from the SR and also when caffeine is used. During

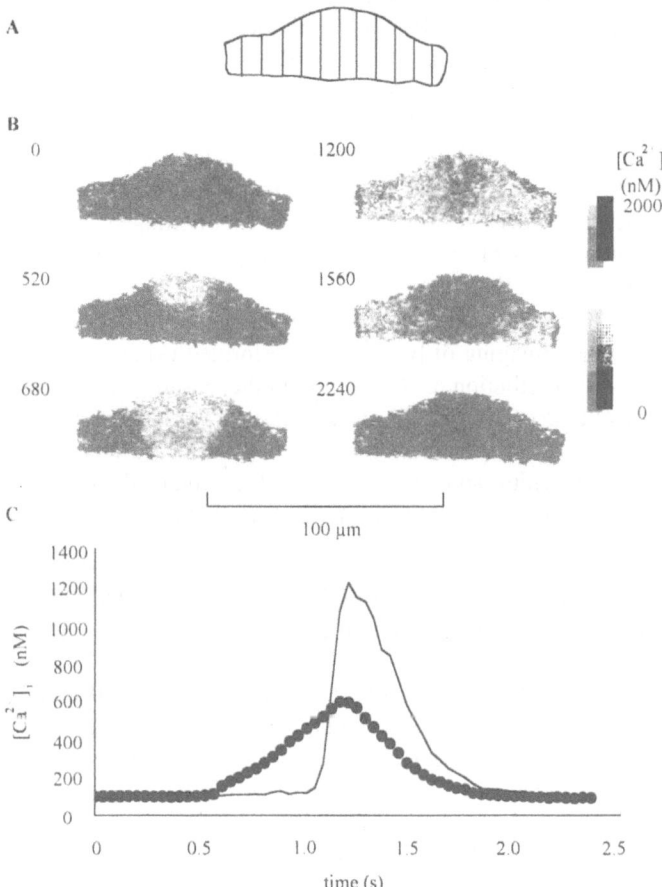

Figure 6. Evaluation of macroscopic Ca gradients. **A)** The windows (~10 µm width) represent the individual regions where the changes of bulk Ca concentration were fitted to the changes in membrane current. **B)** Fluo-3 self ratio pseudocolour images. The changes in Ca concentration are occurring in response to spontaneous oscillatory release of Ca from the SR. The values to the left of the images represent the time (ms) after the start of the image sequence. The cell was voltage clamped and the membrane potential was held at -80mV throughout. **C)** Changes in cytoplasmic Ca concentration during the oscillation shown in B). The filled circles represent the spatially averaged calcium concentration and the solid line the calcium concentration measured from a single window. The calcium concentration in this window rises faster and to a higher level than that averaged across the whole cell.

spontaneous release Ca rises in one region of the cell initially and then propagates as a wave of CICR throughout the cytoplasm (Takamatsu & Wier, 1990; Wier *et al.* 1987; Lipp & Niggli, 1993). Under these conditions the bulk Ca concentration will not be uniform throughout the cell and will also change during the course of the propagating wave of CICR (Figure 6).

Using digital video imaging of $[Ca^{2+}]_b$ the macroscopic distribution of $[Ca^{2+}]_b$ within the cell can only clearly be resolved in the x and y dimensions. This therefore means that Ca gradients occurring in the z direction are still unresolved. However, cardiac myocytes are generally wider than they are deep (Berlin *et al.* 1994). Presumably, if Ca were released from a point at the centre of the z plane and diffused evenly from this point in all directions then by the time the Ca had spread to occupy all of the directly observable y plane of the cell it would also have occupied all of the unresolvable z plane of the cell. As a result, if the hysteresis between I_{Na-Ca} and ratio ($[Ca^{2+}]_b$) is still present when the Ca wave first occupies the resolvable y direction then it is likely that Ca also occupies fully the unresolvable z dimension. As a consequence, if Ca occupies all of the y plane and therefore presumably all of the z plane and the hysteresis is still present it should also be able to exclude the effect of any discrepancy between the distribution of Na-Ca exchanger pumps and SR Ca release sites within the cell. This point has been addressed in Figures 6 and 7, where the arrow in 7A and 7B represents the time when the wave of CICR is first homogeneous in the observable y plane and therefore presumably homogeneous in the z plane. It can clearly be seen that the hysteresis is still present at this time.

To account for the effects of the macroscopic non uniformities of $[Ca^{2+}]_b$ upon ratio and I_{Na-Ca} digital video imaging of $[Ca^{2+}]_b$ was performed using either Fluo-3 or Calcium Green-1 to allow the distribution of $[Ca^{2+}]_b$ within the cell to be monitored during spontaneous oscillatory release or during caffeine application. To calculate the true mean value of $[Ca^{2+}]_b$, or ratio, the cell is divided into a series of measurement windows, each of approximately 10 μm width, and summing the values obtained for $[Ca^{2+}]_b$ (or ratio) and dividing this value by the number of measurement windows.

2.2. Can a Non-Linear Dependence of I_{Na-Ca} or Fluorescence Upon Ca Concentration Account for the Hysteresis?

Since the relationship between fluorescence, or ratio, and $[Ca^{2+}]_i$ is non linear, the apparent $[Ca^{2+}]_b$ in terms of Indo-1 ratio units obtained during spontaneous SR Ca release is likely to be different from the real value of $[Ca^{2+}]_b$. Similarly, as the macroscopic distribution of Ca may change as the wave propagates throughout the cell the discrepancy between the real and apparent values of mean $[Ca^{2+}]_b$ may also alter thereby leading to a hysteresis between I_{Na-Ca} and $[Ca^{2+}]_b$. When caffeine is used to produce SR Ca release the effect of the macroscopic distribution of $[Ca^{2+}]_b$ can be discounted because the rise of Ca produced by caffeine is uniform throughout the cell.

A similar hysteresis would be produced if there were a non linear relationship between I_{Na-Ca} and Ca concentration. However, at least over the physiological range of Ca concentrations, I_{Na-Ca} has been shown to be very nearly linearly related to Ca concentration (Barcenas-Ruiz et al. 1987; Beuckelmann & Wier, 1989; Lipp & Pott, 1988). It is therefore necessary to account for the likely effect of the non uniform distribution of Ca throughout the cytoplasm upon Indo-1 ratio (and possibly I_{Na-Ca}) before suggesting that the hysteresis is due to Ca concentration gradients between the space adjacent to the sarcolemma and the bulk cytoplasmic compartment.

By assuming that I_{Na-Ca} and ratio (or fluorescence) are saturating functions of Ca concentration and are related to Ca concentration in a simple Michaelis fashion it is possible to calculate the expected membrane current for any particular value of $[Ca^{2+}]_b$. With this type of modelling:

$$I = \frac{[Ca^{2+}]_i \cdot I_{max}}{([Ca^{2+}]_i + K_i)} \tag{1}$$

I   –   membrane current (I_{Na-Ca})
I_{max} –   saturating value of I_{Na-Ca}
K_i – $[Ca^{2+}]_i$ at which I_{Na-Ca} is half I_{max}

Similarly ratio can be modelled as:

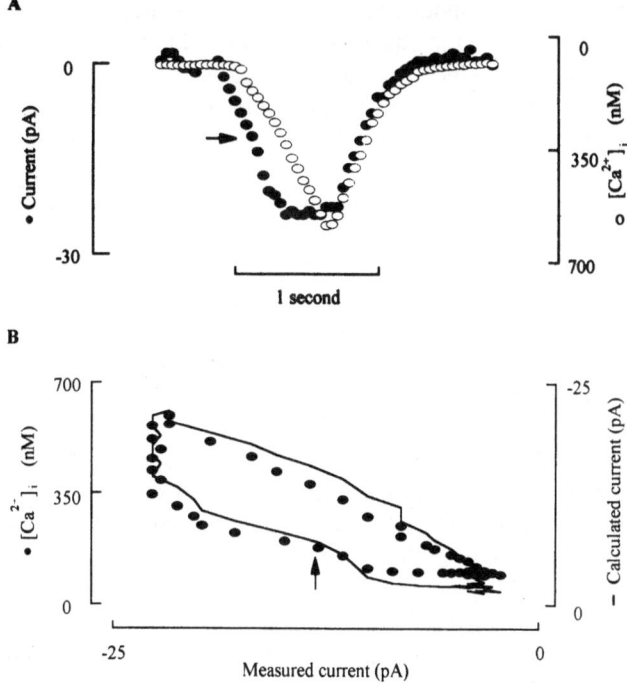

Figure 7. Evaluation of macroscopic non-uniformities of cytoplasmic calcium. Modelling of changes of Na-Ca exchange current and fluorescence as saturating functions of Ca concentration. **A)** Comparison of the timecourse of changes of Na-Ca exchange current (solid circles) and bulk cytoplasmic Ca concentration (open circles and inverted) during a spontaneous propagating wave of **CICR. B)** Hysteresis plots, the solid circles show the experimentally observed hysteresis with Ca concentration (left ordinate) plotted against measured current (abscissa, inverted). The right hand ordinate plots the predicted membrane current (solid line, inverted) against the measured current. The cell was voltage clamped using the perforated patch technique and held at - 80 mV throughout. The arrow in **A)** and **B)** indicates the time when the wave first occupies the full depth of the y plane of the cell.

$$R = \frac{[Ca^{2+}]_i \cdot R_{max}}{([Ca^{2+}]_i + K_R)} \tag{2}$$

R   –   ratio
R_{max} –   saturating value of ratio
K_R – $[Ca^{2+}]_i$ at which ratio is half R_{max}

Equations (1) and (2) can be rearranged and $[Ca^{2+}]_i$ eliminated from the equations such that:

$$I = \frac{I_{max} \cdot K_R \cdot R}{R \cdot (K_R - K_i) + K_i \cdot R_{max}} \tag{3}$$

The equation (3) I_{Na-Ca} can be described as a saturating function of ratio with a K_{app} for ratio of:

$$K_{app} = \frac{K_i \cdot R_{max}}{K_R - K_i} \tag{4}$$

Once again the cell is divided into a series of measurement windows of ~ 10 μm width and the function obtained in equation (3) is applied to the experimentally measured $[Ca^{2+}]_b$ or ratio in each measurement window. The changes in membrane current (I) are then predicted using a computer based curve fitting routine (SigmaPlot, Jandel Scientific). The curve fitting routine attempts to fit the predicted current (I_{pred}) to the experimentally measured current and provides values for the unknown parameters K_R, K_i, I_{max} and R_{max}. The calculated currents for each measurement window are then summed and I_{pred} compared with the experimentally measured current (I_{Na-Ca}).

If it is possible to predict the changes in subsarcolemmal Ca concentration (as measured by I_{Na-Ca}) from the changes of $[Ca^{2+}]_b$ or ratio such that both I_{Na-Ca} and ratio are saturating functions of $[Ca^{2+}]_i$, then, the result of such modelling would be that I_{pred} predicted by the model would be the same as I_{Na-Ca}. Therefore, if I_{pred} were plotted against I_{Na-Ca} there should be no hysteresis. The results of modelling the changes in subsarcolemmal $[Ca^{2+}]_i$ in this way are also summarised in Figures 6 and 7.

The upper panel of Figure 6 shows a schematic representation of the individual measurement windows to which the function in equation (3) was applied. Panel B shows pseudocolour self ratio images of $[Ca^{2+}]_b$ (measured using Fluo-3) following the timecourse of a spontaneous wave of CICR. In this case the wave commences towards the centre of the cell and propagates from this point to both ends of the cell. The lower panel compares the timecourse of the changes of $[Ca^{2+}]_b$ in a single measurement window to the spatially averaged $[Ca^{2+}]_b$ throughout the cell (Takamatsu & Wier, 1990). The discrepancy between the changes of $[Ca^{2+}]_b$, measured in a single window, and the spatially averaged $[Ca^{2+}]_b$ demonstrates that under conditions of spontaneous Ca release it is necessary to apply such modelling to small regions of the cell when considering for the effects of any non linear relationship between ratio, I_{Na-Ca} and $[Ca^{2+}]_b$ and spatial distribution of Ca within the cell.

Figure 7A compares the timecourses of the mean $[Ca^{2+}]_b$ and I_{Na-Ca} for the oscillation shown in the previous figure. The mean $[Ca^{2+}]_b$ (open symbols) clearly lags behind the changes in I_{Na-Ca} (filled symbols) throughout the timecourse of the wave. The temporal discrepancies between I_{Na-Ca} and $[Ca^{2+}]_b$ are shown as the hysteresis plots in panel B of the figure. The function in equation (3) has then been applied to the measured changes of $[Ca^{2+}]_b$ in each individual measurement window. The result of modelling I_{Na-Ca} and fluorescence as saturating functions of $[Ca^{2+}]_b$ in this manner still produces a hysteresis (solid line) which is of similar magnitude and direction to the experimentally observed hysteresis (filled symbols). The arrow in Figure 7B. represents the time when the Ca wave first occupies all of the y axis of the cell and therefore Ca is also occupying all of the unresolvable z dimension, at this time the hysteresis is still clearly present.

Figure 7 shows that the effect of modelling I_{Na-Ca} and fluorescence as saturating functions of $[Ca^{2+}]_i$ and allowing for the macroscopic distribution of $[Ca^{2+}]_i$ within the cytoplasm fails to account for the hysteresis between I_{Na-Ca} and $[Ca^{2+}]_b$. This therefore suggests that macroscopic non uniformities of $[Ca^{2+}]_b$, which may change during the timecourse of a wave of CICR, are not responsible for the hysteresis. In addition, accounting for any non-linear dependence of fluorescence or I_{Na-Ca} fails to account for the experimentally observed hysteresis. Furthermore, the modelling suggests that the $[Ca^{2+}]_m$ detected by the Na-Ca exchanger in the subsarcolemmal space is different to that detected by the fluorescent indicator in the bulk compartment.

2.3. Can Microscopic Ca Gradients Account for the Hysteresis?

The hysteresis between I_{Na-Ca} and $[Ca^{2+}]_b$ or fluorescence suggests that there may be a compartment within the cell adjacent to the surface membrane where the $[Ca^{2+}]_m$ sensed by the Na-Ca exchanger follows a different timecourse from that in the cytoplasm detected by the fluorescent indicator. In order to attempt to account for the temporal discrepancy and possible Ca concentration gradients between the surface membrane and the bulk compartment the simple model shown in Figure 8 has been used.

In this model the cell is assumed to consist of two cellular compartments, firstly, the small subsarcolemmal or 'fuzzy space' compartment where the calcium concentration ($[Ca^{2+}]_m$) is sensed by the Na-Ca exchanger. This space is bound by the surface membrane and a single lumped diffusion barrier. The second compartment is the larger bulk cytoplasmic compartment. In the bulk cytoplasmic compartment the calcium concentration ($[Ca^{2+}]_b$) is detected by the fluorescent indicator. A further assumption is that Ca release from and uptake into the SR and extrusion across the surface membrane only occurs into and from the subsarcolemmal compartment respectively. Finally it is assumed that the two compartments are separated by a single lumped diffusion barrier.

By making these assumptions some properties of Ca movements between the two compartments can be derived. Firstly the flux of Ca from the subsarcolemmal to the bulk compartments is given by:

$$\text{flux} = p \cdot A \cdot ([Ca^{2+}]_m - [Ca^{2+}]_b) \qquad (5)$$

p – effective permeability of the diffusion barrier (cm.s^1)
A – surface area of the barrier (cm^2)

The flux of Ca from the superficial to the bulk compartment can be further defined by the volume, buffering capacity of the bulk compartment and the rate of change of Ca in the bulk compartment such that:

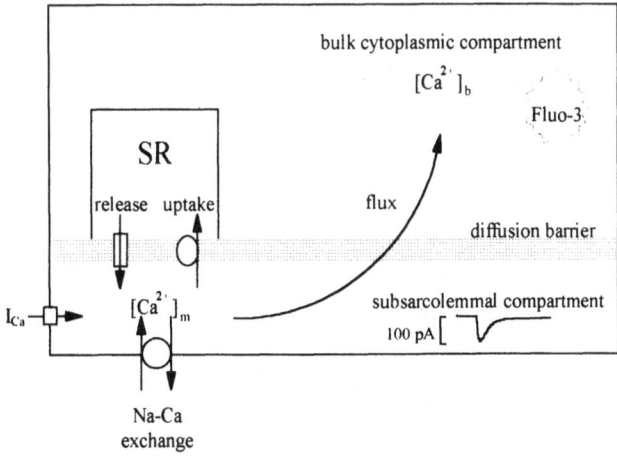

Figure 8. Modelling of microscopic non uniformities in Ca concentration. Schematic representation of the two compartment model used to predict changes in subsarcolemmal Ca concentration from the observed changes in fluorescence. The ventricular cell consists of two compartments, the larger bulk cytoplasmic compartment where the changes in Ca concentration are detected by the fluorescent indicator and the smaller subsarcolemmal compartment adjacent to the surface membrane where changes in Ca concentration are detected by the Na-Ca exchanger. The two compartments are separated by a single lumped diffusion barrier. Ca release from and uptake into the SR occur only from within the subsarcolemmal compartment as does trans-sarcolemmal Ca entry and exit. The flux Ca between the two compartments can be shown to be proportional to the Ca concentration gradient between the subsarcolemmal and bulk compartments.

$$V \cdot \beta \cdot \frac{\delta \, [Ca^{2+}]_b}{\delta \, t} = p \cdot A \cdot (\, [Ca^{2+}]_m - [Ca^{2+}]_b \,)$$

(6)

V	–	volume of the bulk compartment (cm^3)
β	–	buffering capacity of the bulk compartment (dimensionless)
$\delta[Ca^{2+}]_b \, /\delta t$	–	rate of change of $[Ca^{2+}]_b$ (μM.s^{-1})

Berlin *et al.* (1994) calculated the K_D for intrinsic cytoplasmic Ca buffers in rat ventricular myocytes to be 630 nM which suggests that these buffers will tend towards saturation as $[Ca^{2+}]_b$ rises. However for the purposes of this model it is assumed that the buffering capacity for Ca is constant which is therefore a simplification of the real situation. By making this assumption it is possible to define a new constant, γ, with units of seconds (equation (7)). It then follows that the rate of change of $[Ca^{2+}]_b$ can be defined as a function of the concentration gradient between the subsarcolemmal and bulk compartments as given in equation (8):

$$\gamma = \frac{\beta \cdot V}{p \cdot A}$$

(7)

$$\frac{\delta[Ca^{2+}]_b}{\delta \, t} = \frac{(\, [Ca^{2+}]_m - [Ca^{2+}]_b \,)}{\gamma}$$

(8)

The calcium concentration (fluorescence ratio) in the bulk compartment can be measured with the fluorescent indicator and if γ is known it is possible to calculate $[Ca^{2+}]_m$ (equation (9)). Finally by assuming that $I_{Na\text{-}Ca}$ is a saturating function of $[Ca^{2+}]_i$ and can be modelled by the form given in equation (10) it is possible to fit the measured changes in $I_{Na\text{-}Ca}$ to the measured changes of $[Ca^{2+}]_b$ using a computerised curve fitting routine as described previously. The fit will provide the values for the unknown parameters γ and K_m. To model the changes of $I_{Na\text{-}Ca}$ from the measured changes of $[Ca^{2+}]_b$ (as measured using digital video imaging) the cell was again divided into a series of measurement windows (see Figure 9) with the modelling applied to each measurement window and the resulting currents summed to give the predicted membrane current I_{pred}.

$$[Ca^{2+}]_m = [Ca^{2+}]_b + \gamma \cdot \frac{\delta \, [Ca^{2+}]_b}{\delta \, t}$$

(9)

$$I_{Na\text{-}Ca} = \frac{I_{Na\text{-}Camax} \cdot [Ca^{2+}]_m}{[Ca^{2+}]_m + K_m}$$

(10)

$I_{Na\text{-}Camax}$	–	saturating value of $I_{Na\text{-}Ca}$
K_m	–	$[Ca^{2+}]_m$ at which $I_{Na\text{-}Ca}$ is half maximal

The results of modelling the changes in $[Ca^{2+}]_m$ in this way are summarised in Figure 10. In the upper panel of the figure the experimentally measured $I_{Na\text{-}Ca}$ (filled symbols) and $[Ca^{2+}]_b$, expressed as Calcium Green-1 self ratio units, (open symbols) for the spontaneous oscillation shown in the previous figure are plotted as functions of time. It can be seen that once again the changes in $[Ca^{2+}]_b$ lag behind the changes in $I_{Na\text{-}Ca}$ which produces the hysteresis shown in the lower panel of the figure (open symbols). The I_{pred} predicted by the model is shown as the solid line. It can be seen in the upper panel of the figure that I_{pred} closely follows the changes in the experimentally measured $I_{Na\text{-}Ca}$. This therefore suggests

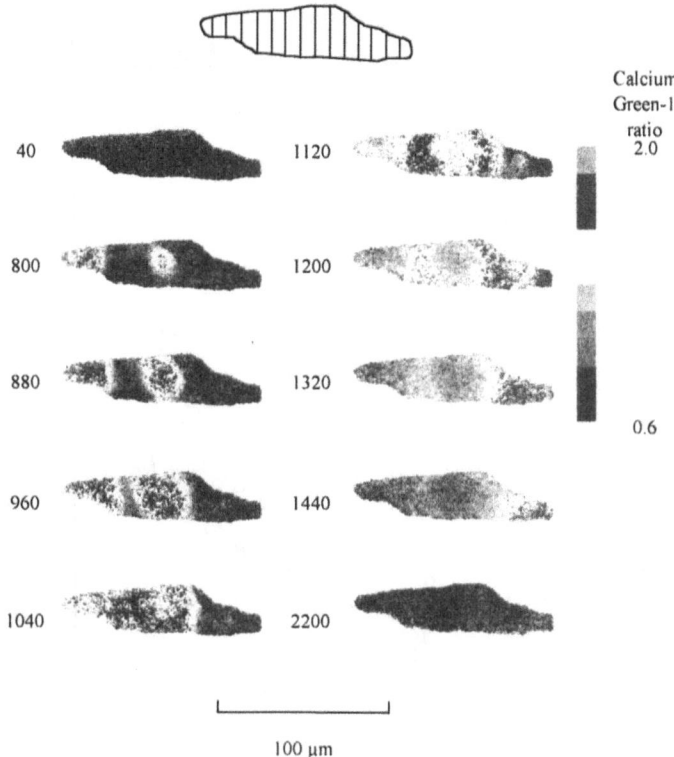

Figure 9. Evaluation of microscopic non-uniformities of Ca. Pseudocolour Calcium Green-1 self ratio images following the timecourse of spontaneous oscillatory release of Ca from the SR. The cell was voltage clamped at-80 mV throughout using the perforated patch and switch clamp techniques. The values to the left of the images represent the time (milliseconds) after the start of the image sequence. The outline sketch represents the individual measurement windows to which the analysis of equations (9) and (10) was applied.

that the changes of $[Ca^{2+}]_m$ and hence I_{Na-Ca} can be predicted from the changes of $[Ca^{2+}]_b$ by using a two compartment model with a barrier to diffusion from the subsarcolemmal to the bulk compartments as illustrated in Figure 8.

The agreement between the timecourses of I_{Na-Ca} and I_{pred} is shown in the hysteresis plot in the lower panel of the figure where it can be seen that when I_{pred} (right hand ordinate, solid line) is plotted as a function of the measured I_{Na-Ca} (abscissa, inverted) the hysteresis is removed.

The mean value obtained for the unknown parameter γ was 133 ± 47 ms (mean ± sem) which is in very close agreement with the value of 132 ± 17 ms required to remove the hysteresis between $[Ca^{2+}]_b$ and I_{Na-Ca} during caffeine application by filtering the current record. This again suggests that the changes of Ca occurring adjacent to the Na-Ca exchanger and in the bulk cytoplasm as detected by the fluorescent indicator are approximately 100 ms out of phase. Obtaining the value for γ also allows the changes in $[Ca^{2+}]_m$ to be calculated from the observed changes in $[Ca^{2+}]_b$ by using equation (9). Figure 11A. compares the observed changes in $[Ca^{2+}]_b$ (solid symbols) and predicted changes in $[Ca^{2+}]_m$ (solid line) for a single spontaneous oscillation in a Fluo-3 loaded cell. The changes of $[Ca^{2+}]_b$ and $[Ca^{2+}]_m$ are measured from a single 10 µm wide measurement window as used in Figures 6 and 9 to account for the spatial non uniformity of the wave within the cell.

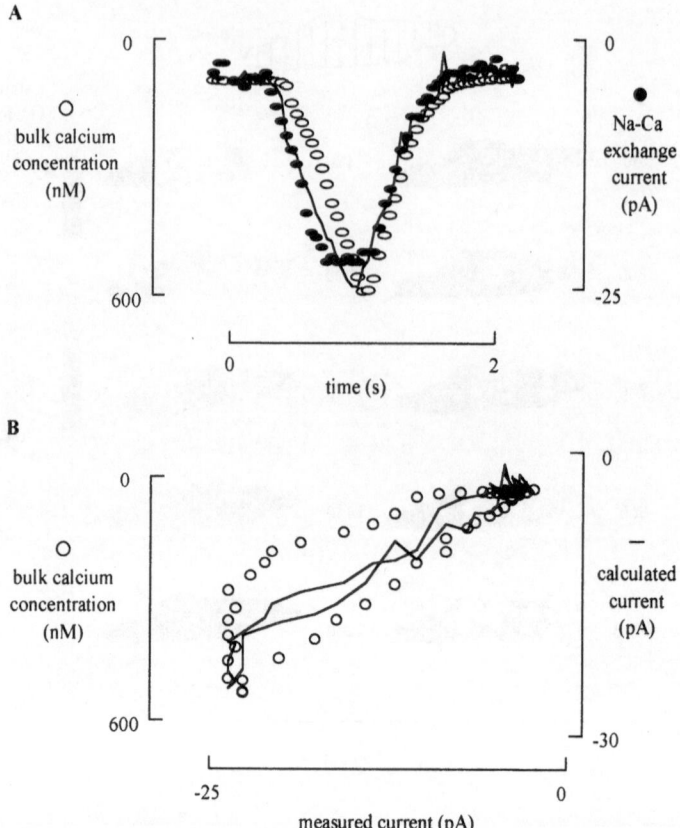

Figure 10. Analysis of relationship between Ca concentration and membrane current in terms of a sub-sarcolemmal Ca pool. **A)** Comparison between experimentally measured bulk cytoplasmic Ca concentration (open symbols, left hand ordinate) and Na-Ca exchange current (filled symbols, right hand ordinate) during a single spontaneous oscillation of Ca. The solid line shows the predicted membrane current obtained using the function given in equation (9). **B)** Hysteresis plots. The experimentally observed hysteresis between bulk Ca concentration and Na-Ca exchange current (open symbols, left hand ordinate) is compared with the hysteresis between the predicted membrane current and bulk Ca concentration (solid line, right hand ordinate). The plot shows that the hysteresis between bulk Ca concentration and current can be removed when changes in Ca concentration adjacent to the sub-sarcolemmal membrane are modelled as described in equation (9).

The calculated change of $[Ca^{2+}]_m$ for a single measurement window rises, and falls, before the change of $[Ca^{2+}]_b$ measured with the fluorescent indicator. In addition the peak value of $[Ca^{2+}]_m$ is greater than the peak value of $[Ca^{2+}]_b$ (2.5 *vs.* 1.2 µM). The difference between the timecourse of $[Ca^{2+}]_m$ and $[Ca^{2+}]_b$ is even more apparent when $[Ca^{2+}]_m$ is first changing as shown in Figure 11B where the initial changes of $[Ca^{2+}]_b$ and $[Ca^{2+}]_m$ are shown on an expanded timescale. Under these conditions when $[Ca^{2+}]_m$ has risen from a resting level of 100 nm to approximately 1.5 µM (rate of rise of $[Ca^{2+}]_m \sim 15.6$ µM.s^{-1}) $[Ca^{2+}]_b$ has only risen to 550 nM (rate of rise of $[Ca^{2+}]_b$ 4.75 µM.s^{-1}). Similarly as $[Ca^{2+}]_m$ is proportional to the rate of change of $[Ca^{2+}]_b$ and the Ca concentration gradient between the two compartments it would be expected that $[Ca^{2+}]_m$ will also tend to decay before $[Ca^{2+}]_b$. This latter point is again shown in the upper panel of Figure 11.

By modelling a lumped diffusion barrier between the putative subsarcolemmal and bulk compartments the above results suggest that it is possible to account for the difference in the timecourse of I_{Na-Ca} and $[Ca^{2+}]_b$ observed during spontaneous oscillatory release of Ca

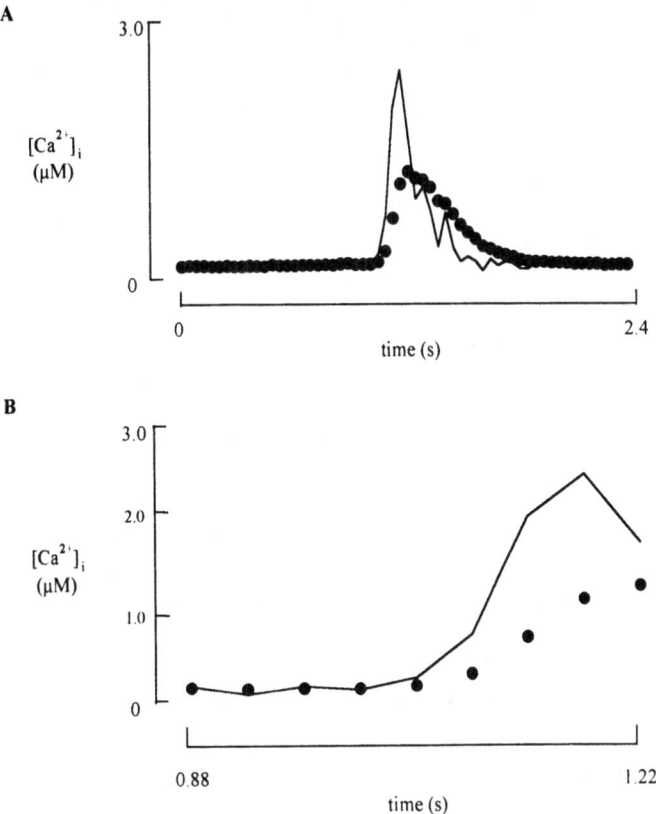

Figure 11. Differing rates of rise of Ca in the two cellular compartments. Comparison of bulk cytoplasmic and subsarcolemmal calcium concentration during spontaneous oscillatory release of calcium from the SR. **A)** filled circles show the changes of bulk cytoplasmic calcium concentration in a single measurement window (~ 10 m width) estimated using the fluorescent indicator Fluo-3, the solid line shows the calculated changes in subsarcolemmal calcium concentration for the same measurement window using equation (9). **B)** expanded portion of A) showing the more rapid rise of calcium in the subsarcolemmal space compared with that in the bulk space.

from the SR. The modelling also suggests that the Ca concentration in the subsarcolemmal space which is 'sensed' by the Na-Ca exchanger is different from that in the bulk cytoplasmic compartment as detected by the fluorescent indicator and that this results in the current carried by the Na-Ca exchanger rising (and falling) faster and peaking earlier than the bulk cytoplasmic Ca concentration.

3. DISCUSSION

In this paper the timecourses of $[Ca^{2+}]_b$ and I_{Na-Ca} have been compared during spontaneous oscillatory releases of Ca from the SR of adult rat ventricular myocytes. The changes of I_{Na-Ca} were observed to occur before the changes of $[Ca^{2+}]_b$ as detected using Ca sensitive fluorescent indicators in the cytoplasm. However, it is important to account for the kinetics of the indicator being used to measure $[Ca^{2+}]_b$ since any discrepancy between the actual rise of $[Ca^{2+}]_b$ and fluorescence could produce a hysteresis especially if similar delays were not present in the measurements of I_{Na-Ca}. Sipido and Wier (1991) analysed this effect when measuring SR Ca fluxes using Indo-1. In conclusion the authors estimated that the kinetics of Ca binding to Indo-1 would produce a delay in the upstroke of the systolic Ca transient of less than 20 ms. This is much less than the delay (133 ms) measured in these experiments between $[Ca^{2+}]_b$ and I_{Na-Ca}. In addition the majority of the experiments described were performed using either Fluo-3 or Calcium Green-1. Both of these indicators have a higher K_D and therefore faster off rate for Ca than either Indo-1 or Fura-2. As a consequence they are even less likely to produce a 'false' hysteresis.

Other possible artifacts which are accounted for are the limited resolution of the imaging system in the z plane of the cell and the response of the camera to changes in fluorescence. The lack of resolution in the z plane using conventional imaging techniques would serve to underestimate the changes of mean $[Ca^{2+}]_i$ and thereby produce a greater discrepancy between $[Ca^{2+}]_b$ and I_{Na-Ca}. The possible discrepancy caused by imaging $[Ca^{2+}]_i$ in this way is made less likely by the fact that the hysteresis is still present even if the measurements are made when the wave occupies the full y plane and therefore presumably the full z plane.

A further artifact may arise due to a delay between the increase in fluorescence and the response of the intensified camera. This point has been addressed previously and the response time of the camera is small in comparison to the delay between I_{Na-Ca} and $[Ca^{2+}]_b$ (Trafford *et al.* 1995). For those experiments where $[Ca^{2+}]_i$ was measured using photomultiplier tubes and Indo-1 the delay between $[Ca^{2+}]_b$ and I_{Na-Ca} was very similar to that measured using digital video imaging. The PMT's response to changes in illumination intensity is extremely rapid and therefore the hysteresis cannot be attributed to the delays in the imaging system.

Both the current and $[Ca^{2+}]_b$ records are sampled and filtered at the same frequency. A final possible artifact of the imaging system is the synchronization of the video frame rate and the current records. To address this problem the video frame counter and the current acquisition hardware and software were triggered simultaneously.

The hysteresis between $[Ca^{2+}]_b$ and I_{Na-Ca} can be accounted for if a single lumped diffusion barrier is modelled between two cellular compartments, the larger bulk cytoplasmic compartment and the smaller surface membrane bound subsarcolemmal compartment. The predicted changes in $[Ca^{2+}]_m$ occur faster than the experimentally observed changes in $[Ca^{2+}]_b$ and the delay introduced between the changes of Ca in the two compartments is estimated to be 133 ms during spontaneous oscillatory release of Ca from the SR. During caffeine application the hysteresis can be removed by filtering the current record with a single exponential filter of time constant 132 ms. This similarity to the time constant predicted by the model indicates that the model provides a first step to accounting for the timecourses of calcium and current.

The model does not provide an indication of what the diffusion barrier between the two cellular compartments may be. It seems plausible that the SR itself may contribute to forming the barrier. In addition, the mitochondria occupy approximately 35 % of the total ventricular volume with a layer of mitochondria also seen adjacent to the surface membrane on electron micrographs (Page, 1978; Page *et al.* 1971) and therefore it is likely that the mitochondria may also form part of the diffusion barrier.

The results presented above are supportive of substantial gradients for Ca within the subsarcolemmal space during normal excitation contraction coupling. By modelling the changes in subsarcolemmal Na and Ca concentration the direction of action of the Na-Ca exchanger during systole can also be predicted. These results, although achieved by making assumptions as to the time course of changes in subsarcolemmal Na concentration, also predict that during the action potential the Na-Ca exchanger would be predominantly involved in Ca extrusion from the cell.

In summary, the work presented in this paper provides evidence for the existence of a subsarcolemmal compartment for Ca in cardiac ventricular myocytes. Substantial gradients of Ca concentration are predicted to arise between the subsarcolemmal compartment and the bulk cytoplasmic compartment of the cell. The timecourse of changes of Ca concentration in the two compartments has also been shown to be different during Ca release from the SR (at least during spontaneous oscillatory release of Ca from the SR). By modelling the changes of Ca concentration within the subsarcolemmal space the results of this chapter also predict that Ca entry via Na-Ca exchange could occur early in the action potential. The results also provide a characterisation of the diffusion properties of the 'fuzzy space' thought to be important in excitation contraction coupling.

REFERENCES

Ashley, C. C., Ellory, J. C. & Griffiths, P. J. (1977). Caffeine and the contractility of single muscle fibres from the barnacle balanus nubilis. *Journal of Physiology* 269, 421-439.

Barcenas-ruiz, L., Beuckelmann, D. J. & Wier, W. G. (1987). Sodium-calcium exchange in heart: Membrane currents and changes in [Ca2+]i. *Science* 238, 1720-1722.

Berlin, J. R., Bassani, J. W. & Bers, D. M. (1994). Intrinsic cytosolic calcium buffering properties of single rat cardiac myocytes. *Biophysical Journal* 67, 1775-1787.

Berlin, J. R., Cannell, M. B. & Lederer, W. J. (1989). Cellular origins of the transient inward current in cardiac myocytes Role of fluctuations and waves of elevated intracellular calcium. *Circulation Research* 65, 115-126.

Beuckelmann, D. J. & Wier, W. G. (1989). Sodium-calcium exchange in guinea-pig cardiac cells: Exchange current and changes in intracellular Ca2 l. *Journal of Physiology* 414, 499-520.

Bond, M., Shuman, H., Somlyo, A. P. & Somlyo, A. V. (1984). Total cytoplasmic calcium in relaxed and maximally contracted rabbit portal vein smooth muscle. *Journal of Physiology* 357, 185-201.

Brown, A. M., Lee, K. S. & Powell, T. (1981). Sodium current in single rat heart muscle cells. *Journal of Physiology* 318, 479-500.

Callewaert, G., Cleemann, L. & Morad, M. (1989). Caffeine-induced Ca2+ release activates Ca2+ extrusion via Na+-Ca2+ exchanger in cardiac myocytes. *American Journal of Physiology* 257, C147-C152.

Cannell, M. B., Berlin, J. R. & Lederer, W. J. (1987). Effect of membrane potential changes on the calcium transient in single rat cardiac muscle cells. *Science* 238, 1419-1423.

Cheng, H., Lederer, W. J. & Cannell, M. B. (1993). Calcium sparks: elementary events underlying excitation-contraction coupling in heart muscle. *Science* 262, 740-744.

Eisner, D. A. & Lederer, W. J. (1979). The role of the sodium pump in the effects of potassium-depleted solutions on mammalian cardiac muscle. *Journal of Physiology* 294, 279-301.

Fabiato, A. (1983). Calcium-induced release of calcium from the cardiac sarcoplasmic reticulum. *American Journal of Physiology* 245, C1-C14.

Fabiato, A. (1985). Simulated calcium current can both cause calcium loading in and trigger calcium release from the sarcoplasmic reticulum of a skinned canine cardiac Purkinje cell. *Journal of general Physiology* 85, 291-320.

Fabiato, A. & Fabiato, F. (1975). Contractions induced by a calcium-triggered release of calcium from the sarcoplasmic reticulum of single skinned cardiac cells. *Journal of Physiology* 249, 469-495.

Fedida, D., Noble, D., Rankin, A. C. & Spindler, A. J. (1987). The arrhythmogenic transient inward current iTI and related contraction in isolated guinea-pig ventricular myocytes. *Journal of Physiology* 392, 523-542.

Ganitkevich, V. Y. & Isenberg, G. (1996). Dissociation of subsarcolemmal from global cytosolic $[Ca^{2+}]$ in myocytes isolated from guinea-pig coronary artery. *J.Physiol.* 490, 305-318.

Isenberg, G. & Wendt-Gallitelli, M. F. (1990). X-ray microprobe analysis of sodium concentration reveals large transverse gradients from the sarcolemma to the centre of voltage-clamped guinea-pig ventricular myocytes. *Journal of Physiology* 420, 86P

Kass, R. S., Lederer, W. J., Tsien, R. W. & Weingart, R. (1978). Role of calcium ions in transient inward currents and aftercontractions induced by strophanthidin in cardiac Purkinje fibres. *Journal of Physiology* 281, 187-208.

Kass, R. S., Tsien, R. W. & Weingart, R. (1978). Ionic basis of transient inward current induced by strophanthidin in cardiac Purkinje fibres. *Journal of Physiology* 281, 209-226.

Kimura, J., Noma, A. & Irisawa, H. (1986). Na-Ca exchange current in mammalian heart cells. *Nature* 319, 596-597.

Leblanc, N. & Hume, J. R. (1990). Sodium current-induced release of calcium from cardiac sarcoplasmic reticulum. *Science* 248, 372-376.

Lederer, W. J., Niggli, E. & Hadley, R. J. (1992). Sodium-calcium exchange in excitable cells: Fuzzy Space. *Science* 248, 283

Lipp, P. & Niggli, E. (1993). Microscopic spiral waves reveal positive feedback in subcellular calcium signalling. *Biophysical Journal* 65, 2272-2276.

Lipp, P. & Pott, L. (1988). Transient inward current in guinea-pig atrial myocytes reflects a change of sodium-calcium exchange current. *Journal of Physiology* 397, 601-630.

Lipp, P., Pott, L., Callewaert, G. & Carmeliet, E. (1990). Simultaneous recording of Indo-1 fluorescence and Na^+/Ca^{2+} exchange current reveals two components of Ca^{2+} release from sarcoplasmic reticulum of cardiac atrial myocytes. *FEBS.Letters.* 275, 181-184.

Mechmann, S. & Pott, L. (1986). Identification of Na-Ca exchange current in single cardiac myocytes. *Nature* 319, 597-599.

Miura, Y. & Kimura, J. (1989). Sodium-calcium exchange current. Dependence on internal Ca and Na and competitive binding of external Na and Ca. *Journal of general Physiology* 93, 1129-1145.

Nabauer, M., Callewaert, G., Cleemann, L. & Morad, M. (1989). Regulation of calcium release is gated by calcium current, not gating charge, in cardiac myocytes. *Science* 244, 800-803.

O'neill, S. C., Donoso, P. & Eisner, D. A. (1990). The role of $[Ca^{2+}]_i$ and $[Ca^{2+}]_i$-sensitization in the caffeine contracture of rat myocytes: measurement of $[Ca^{2+}]_i$ and $[caffeine]_i$. *Journal of Physiology* 425, 55-70.

Osipchuk, Y. V., Wakui, M., Yule, D. I., Gallacher, D. V. & Petersen, O. H. (1990). Cytoplasmic Ca2+ oscillations evoked by receptor stimulation, G- protein activation, internal application of inositol trisphosphate or Ca2+: simultaneous microfluorimetry and Ca2+ dependent Cl- current recording in single pancreatic acinar cells. *EMBO.J.* 9, 697-704.

Page, E. (1978). Quantitative ultrastructural analysis in cardiac membrane physiology. *American Journal of Physiology* 235, C147-C158.

Page, E., Mccallister, L. P. & Power, B. (1971). Stereological measurements of cardiac ultrastructures implicated in excitation-contraction coupling (sarcotubules and t-system). *Proc.Natl.Acad.Sci.U.S.A.* 68, 1465-1466.

Sipido, K. R. & Wier, W. G. (1991). Flux of Ca^{2+} across the sarcoplasmic reticulum of guinea-pig cardiac cells during excitation-contraction coupling. *Journal of Physiology* 435, 605-630.

Stehno-Bittel, L. & Sturek, M. (1992). Spontaneous sarcoplasmic reticulum calcium release and extrusion from bovine, not porcine, coronary artery smooth muscle. *Journal of Physiology* 451, 49-78.

Stern, M. D. (1992). Theory of excitation-contraction coupling in cardiac muscle. *Biophysical Journal* 63, 497-517.

Stern, M. D., Capogrossi, M. C. & Lakatta, E. G. (1988). Spontaneous calcium release from the sarcoplasmic reticulum in myocardial cells: mechanisms and consequences. *Cell Calcium.* 9, 247-256.

Stern, M. D. & Lakatta, E. G. (1992). Excitation-contraction coupling in the heart: the state of the question. *FASEB* 6, 3092-3100.

Takamatsu, T. & Wier, W. G. (1990). Calcium waves in mammalian heart: quantification of origin, magnitude, waveform, and velocity. *FASEB.J.* 4, 1519-1525.

Trafford, A. W., Diaz, M. E., O'neill, S. C. & Eisner, D. A. (1995). Comparison of subsarcolemmal and bulk calcium concentration during spontaneous calcium release in rat ventricular myocytes. *J.Physiol* 488, 577-586.

Trafford, A. W., O'Neill, S. C. & Eisner, D. A. (1993). Factors affecting the propagation of locally activated systolic Ca transients in rat ventricular myocytes. *Pflugers Archiv European Journal of Physiology* 425, 181-183.

Varro, A., Negretti, N., Hester, S. B. & Eisner, D. A. (1993). An estimate of the calcium content of the sarcoplasmic reticulum in rat ventricular myocytes. *Pflugers Archiv European Journal of Physiology* 423, 158-160.

Wendt-Gallitelli, M. F. & Isenberg, G. (1991). Total and free myoplasmic calcium during a contraction cycle: x-ray microanalysis in guinea-pig ventricular myocytes. *Journal of Physiology* 435, 349-372.

Wendt-Gallitelli, M. F., Voigt, T. & Isenberg, G. (1993). Microheterogeneity of subsarcolemmal sodium gradients, electron probe microanalysis in guinea-pig ventricular myocytes. *Journal of Physiology* 472, 33-44.

Wier, W. G., Cannell, M. B., Berlin, J. R., Marban, E. & Lederer, W. J. (1987). Cellular and subcellular heterogeneity of $[Ca^{2+}]_i$ in single heart cells revealed by fura-2. *Science* 235, 325-328.

10

FACTS AND FANTASIES ABOUT HAIR CELLS

Jonathan F. Ashmore

Department of Physiology
School of Medical Sciences
University Walk
Bristol BS8 1TD, United Kingdom

1. INTRODUCTION

The sensory cell of the auditory, vestibular and lateral line organs is the hair cell, a highly specialised neuroepithelial cell designed to detect small perturbations applied to the stereocilia projecting from its apical membrane. The mechanical displacements induced by a sound stimulus may be only a few nanometres and this presents technical problems in both measurement and conceptualisation. In the case of the inner ear, the threshold for hearing has been measured to correspond to distortions of cochlear structures of about 0.3nm. How these small displacements are sensed, from the structure of the mechanosensing channel to the mechanics of the embedding tissue within the cochlea, to the neural encoding employed by the auditory pathway is the subject of auditory research. The purpose of this chapter is to highlight some topics which recur in this field. These are the facts. The fantasies are the models which attempt to summarize some of the experimental observations.

2. IS THERE A SINGLE TYPE OF HAIR CELL?

Most of our present knowledge about hair cells and how they work is derived from relatively few preparations. These have tended to be lower vertebrate systems as the cells are large and relatively easy to maintain at room temperatures (Hudspeth & Gillespie, 1994). Of particular value have been cells from the frog sacculus, which is an organ specialised for the detection of sound carried through the ground on which the animal stands. The cells are well organised in an epithelium and the stereocilia are organised, like many vestibular hair cell types, in a tight bundle.

Another preparation which has provided considerable information is the turtle auditory papilla (Art & Fettiplace, 1987; Crawford, Evans & Fettiplace, 1989) where the cells are tuned to a range of frequencies up to about 700Hz. As with the frog, the cells are robust, but contain on the basolateral membrane the ionic machinery for electrical tuning, ensuring that each cell behaves like a narrow bandpass filter. The result is that each cell produces the largest receptor potential at one frequency even though all hair cells receive the same input.

In mammalian inner ear organs there are further questions which arise from the frequency above a few kilohertz. Such high frequencies are beyond the range at which electrical tuning as found in frogs and turtles operates. Mammals have been subject to considerable selection pressure to evolve high frequency hearing both as a means of exploiting a small acoustic head shadow and, in the case of auditory specialists such as bats, to use echolocation. The questions include whether transduction mechanisms are specialised to operate at frequencies above 10 kHz, how frequency selectivity operates at high frequencies, whether high frequency systems adapt and how cochlear sensitivity is determined.

3. THE HEARING RANGE

The hearing range in the mammalian auditory system varies probably less than expected (Fay, 1988; Greenwood, 1990). The design principle of the cochlea ensures that sound frequency is encoded logarithmically. Thus each octave of the auditory range has approximately equal numbers of hair cells devoted to the separation and encoding of each frequency range. Most mammals are capable of detecting frequencies over 7 to 10 octaves. Thus the total length of the guinea pig organ of Corti is 18mm, whereas that of a human is 34mm, and even the total length of the cochlea of a whale is only about 60mm long and most of the additional length in the latter case is associated with the additional infrasound processing capability.

The main simplification in modelling cochlear processing is to assume that the cross section of the cochlea is relatively invariant and that the graded properties of the mechanical parameters are solely responsible for the frequency-place map. Echolocating bats are often quoted in support of the extreme hearing range which mammals can exploit. However, it is clear that some modifications to a uniform cochlear structure are required to encompass all mammals. The cochlea of the bat has what has been termed an acoustic fovea, where the frequency map is distorted, so that a considerable fraction of the cochlea is devoted to a narrow region of frequencies around the echolocating frequency. The cells within this region are also modified and, while maintaining recognisable form as hair cells or supporting cells, have their dimensions significantly altered (Kossl & Vater, 1985). Recent measurements of the mechanics of the bat cochlea suggest that there are differences in mechanical tuning properties (Kossl & Russell, 1995). For example, at present it is not known whether there is a travelling wave in the bat cochlea.

The threshold displacement of the basilar membrane in mammals is about 0.3nm, (Sellick et al, 1982). This is not necessarily the deflection which will be experienced by the stereocilia at threshold because of the geometrical coupling factor between basilar membrane and moving of the overlying tectorial membrane. The deflection that should be applied to the stereocilia may therefore be about 6 times greater (Mammano & Ashmore, 1993), and therefore at threshold the deflection of the bundle may be 2 nm applied to the top of the bundle. Since in the basal cochlea, where these measurements are made, the bundle height is about 2 μm, the angular deflection, at a sound level of 40dB above threshold would be 0.1 rad. This is near the range at which the non-linearity of the transduction process becomes apparent.

4. THE SITE OF MECHANOTRANSDUCTION

The apical surface of hair cells is specialised. It contains a group of about 100 modified actin filled villi which project from the surface and which pivot as a bundle around their base. A variety of experiments now indicate that the top surface of this bundle contains the transducing channels. It can be shown that this is where current enters the cell during bundle deflection (Hudspeth & Gillespie, 1994), where specific blockers of mechanotransduction act (Jaramillo & Hudspeth, 1991), and where calcium enters the cell when the bundle is deflected.

There is also ultrastructural specialisation of the tips of the stereocilia. Running between the tip of all but the tallest stereocilium and the adjacent one is a fine extracellular linking protein (Pickles et al, 1984). These tip links form the structural basis for one model for mechanotransduction which has for the past decade. The model is that deflection of the bundle leads to the link being extended by x. Therefore an energy $1/2\ k_s\ x^2$ where k_s is the stiffness of the tip link, is transferred to gating the channel (s) which may be at either end (or both) of the link. The molecular identity of the link is not known, but it can be removed by enzyme digestion and by low (i.e. < 1 μM) calcium levels. The best candidate at present is an elastase-related protein (Gillespie, 1995). Individual high-resolution images of the link suggest that the protein may be in a dimeric form.

There have been other proposals for the position of the mechanotransducer. In one of the first imaging experiments (using chick hair cells), the permeability of the channel to manganese was used to investigate entry sites around the bundle. These results suggested that the transducer was located near the base of the stereocilia. Although this has not been reduplicated elsewhere it acted as a spur to investigate the precise localisation, and there are now several reports using confocal microscopy of the site of calcium entry in stereocilia.

A second alternative proposal is that the transducer channels are not localised at one or both end of the tip link, but are placed at the apposition point between the two stereocilia, (Hackney and Furness, 1992). The evidence for this is immunohistochemistry. The transducer channel has antigenic similarities with the Na channels in the epithelial cells of kidney, and at least one of the antibodies raised against this channel labels the apical portions of mammalian hair cell stereocilia. Using electronmicroscopy for better resolution, it has been found that the label adheres not to the ends of the tip links but to the region of membrane which are opposed at the top of the stereocilia. It is not too hard to see that any movement of the stereocilia would also bring about a shearing force between the surfaces, which if suitably coupled to a stretch activated channel could result in mechanotransduction.

5. ADAPTATION OF THE MECHANO-TRANSDUCER

Adaptation is the process where the sensitivity of the sensory system is reset and the dynamic range extended. In the case of hair cells, this amounts to a continuous mechanical redefining of the gain of the cell, and the mechanisms involved have been the subject of several studies. There are two mechanisms proposed: 1) where adaptation arises because the coupling between the stimulus displacement and transducing channel is altered and 2) where adaptation arises from a direct modification of the transducing channel.

Do all hair cells adapt? Adaptation has been most extensively studied in frog saccular hair cells and turtle hair cells. There is there ample evidence which shows that when a steady displacement is applied to the bundle, then the input output curve moves along the X-axis to ensure that the maximum sensitivity around the new set point is maintained. Adaptation, observed as a deactivation of the transducer current during step displacements, is also seen in organotypic mouse cochleas (Rusch et al., 1994).

However, hair cells recorded *in vivo* show relatively little adaptation (Dallos, 1992) This can be shown by presenting either a low frequency tone to the cochlea and showing that the receptor potential at each cycle does not diminish or by observing that the pedestal of the hair cell receptor potential (the DC component) does not change with time even during

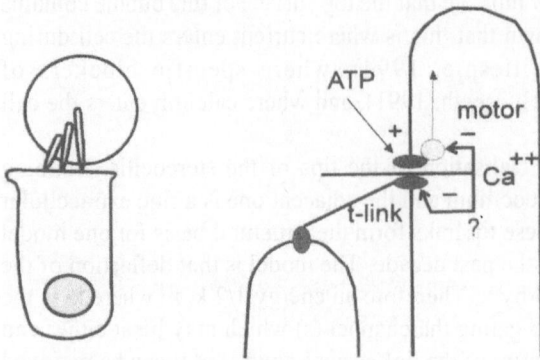

Figure 1. Model for the hair cell transducer. A scheme which summarises data for the hair cell transducer, located at the tip (L) of the hair cell stereocilia. The tip-link runs between stereocilia and gates a channel at one (or both) ends. The influx of calcium may either reduce the upward motion of the motor (a myosin 1b / actin system) or act directly to close the transducer channel. The action of extracellular ATP is shown here as increasing the channel open probability. Not shown is the intracellular requirement for ATP of the motor.

a prolonged tone burst. This is in contrast with receptor potentials (Preyer et al, 1994) or whole cell currents (Ashmore et al, 1993) recorded from isolated outer hair cells.

The mechanisms of adaptation all involve calcium. The best studied system is the frog saccular hair cell (reviewed by Gillespie, 1995). The current ideas about the mechanism of adaptation condense a combination of both physiological and structural observations, (Figure 1). The location of the transducer at the top of the stereocilial bundle (e.g. Jaramillo & Hudspeth, 1991), its permeability to calcium (Ohmori, 1985), the presence of myosin Ib at the stereocilial tips (Solc et al, 1994) and the close association with the tip link suggest that the tip link is continuously retensioned by a steady influx of calcium through the transducer. Thus the normal situation is where the link is continuously sliding up towards the top of the stereocilium. Experiments with non-hydrolysable forms of ATP are consistent with the idea that the myosin-dependent adaptation motor involves ATP hydrolysis.

One consequence of this model is that an elevation of intracellular calcium will cause the motor to generate less force and allow the link to slip down the stereocilium. Conversely, a reduction of the calcium influx, caused either by a closure of the transducer channel as the stereocilial bundle moves in away from the tallest stereocilium or by removal of the external calcium, should cause the link to slide up the bundle. There may well be implications for mammalian hair cells. In the mammalian cochlea the endolymph which faces the transducer contains low calcium (30μM in mammals). In other vertebrate hearing organs the endolymph calcium can be an order of magnitude higher (230μM is estimated for the frog (Eatock et al., 1987)). Thus in mammals, the changes of intracellular calcium in the stereocilia which result from the opening and closing of the transducer channel would be smaller. There is also the possibility that the tension in the tip link may be larger than in the lower vertebrate preparations.

Although the apical membrane is specialised for mechanotransduction, there is good evidence that the apical membrane also expresses ion channels which are modulated by ATP. These receptors are mainly of the P_{2X} subtype (Housley et al., 1992). The effect is that extracellular ATP produces inward currents when applied locally to the apical membrane. Since the P_{2X} ionotropic receptors are calcium permeable, the inward flux of calcium can be measured using indicator dyes. The precise role of the ATP receptors is not clear, although the presence of low (50nM) levels of ATP in scala media facing the apical membranes (quoted in Housley, 1995) suggests that calcium levels in the stereocilia may be subject to long term modulation. Perhaps more surprising is the disparity between the ATP-modulated conductance (approx 25nS for a 20 μm long basal turn cell) and the transducer conductance which is likely not to exceed 10nS (Kros et al, 1995). Thus it seems unlikely that the P_{2X} receptor and mechanotransducer channels are identical, although some association between the two cannot be ruled out.

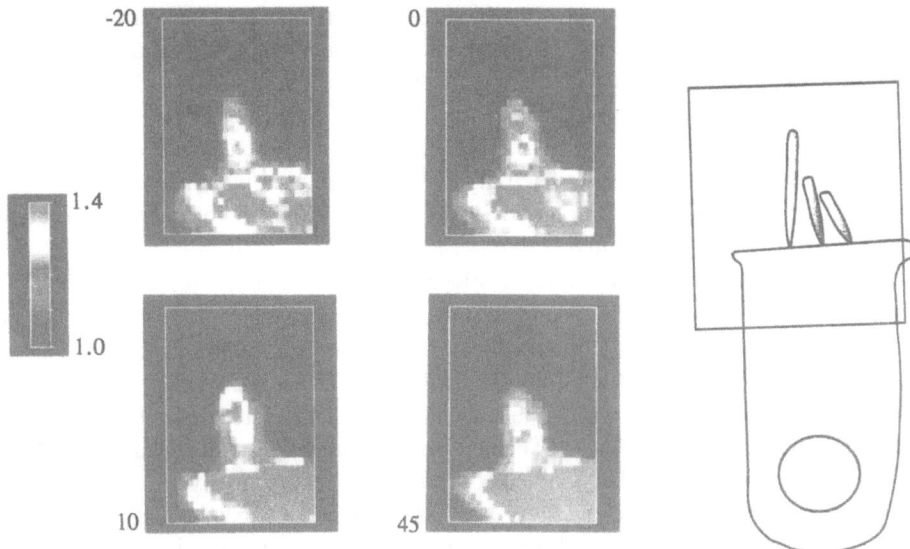

Figure 2. Action of ATP on the outer hair cell stereocilia. A fura-2 ratio image of an outer hair cell during the application of extracellular 30 μM ATP. Time from the application of the solution indicated in seconds by each frame. Pseudo colour scale marked (left) as an arbitrary ratio, with highest values corresponding to 140nM approximately. .Stereocilia only imaged, as in imaging box (left). The ratio image shows an elevated calcium region near the midpoint of the stereocilial bundle, the calcium signal eventually growing throughout cell body. (Unpublished data from Kolston and Ashmore).

Some support for this proposal comes from imaging experiments to investigate hair cell calcium when extracellular ATP is applied around the stereocilia. Figure 2 shows an experiment using fura-2 as the indicator dye in an outer hair cell. The experiment shows that the effect of micromolar levels of ATP applied to the cell is to produce a small increase in calcium near the midpoint of the tallest stereocilia. This is close to the site of where the transducer is located. There is, in addition, a movement of the calcium 'hotspot' towards the top of the stereocilium which may be explained most simply by the involvement of an active stereocilial motor.

6. REVERSE TRANSDUCTION

A property of hair cells in the mammalian cochlea is that transduction can be reversed, in the sense that electrical to mechanical energy conversion can be brought about. The cochlea contains two types of cell which can be distinguished by their position and innervation pattern. There are also differences in cell morphology, with inner hair cells, which act as the sensory input to the auditory pathways, having an ellipsoidal cell body whereas the outer hair cells are cylindrical. Both inner and outer hair cells are mechanosensory cells and this can be determined both from their intracellular responses to sound and by the extracellular current flow produced. However, whereas inner hair cells are exclusively sensory cells, outer hair cells can act like motor cells by to convert membrane potential in an axial change of length of the cell, and by implication force elements in the cochlea.

The process of reverse transduction lies at the basis of preprocessing of sound by the mammalian cochlea (Ashmore, 1994; Ashmore & Kolston, 1994). The macroscopic consequences of this force generation now appear to be compatible with the idea of a cochlear amplifier, a term given to the processes which result in an enhanced motion of the basilar membrane to produce the frequency selectivity of the auditory nerve. The sequence of events can be described by: deflection of the stereocilia by the overlying tectorial membrane; generation of a receptor potential by the outer hair cells; sensing of the potential by a membrane bound motor in the lateral membrane; and generation of force by the outer hair cells along the cell axis as a result of activity by the motor. Although this is a motor which is found only in outer hair cells, the activity of the motor is sufficient to distort the structures of the organ of Corti, (Mammano & Ashmore, 1993).

Most of the current interest has revolved around the identity of the outer hair cell motor. The main requirement has been that it operates at rates which allow it to participate in acoustic events and that it is a motor which extracts energy from the electric field across the membrane. Ultrastructure of the basolateral membrane of hair cells shows a dense packing of particles about 12nm in diameter and these are the candidates for the motor itself. There is no essential requirement for ATP nor is the operation of the motor calcium dependent (Holley & Ashmore, 1988). The mode of operation appears to involve a conformational change of area of the motor protein, the area increasing when the membrane hyperpolarizes and decreasing when it depolarizes. The maximum area change need only be about 8% which is probably below the resolution of electron-microscopy techniques, but would be large enough to produce the 4% length increase (or decrease, respectively) in the length of the cell which can be detected by light microscopy.

When the motor undergoes a conformation change, the electrophysiological event is a movement of charge across the membrane (reviewed, Ashmore, 1994). A favoured model for the motor is that it is a protein (or a cluster of subunits) which has a deep access pore allowing the movement of ions (probably cations) across a fraction of the membrane electric field. It is this ion movement which is the major contributor to the observed charge movement. Figure 3 shows such a scheme. This model has affinities to current models for the sodium/calcium exchanger and the sodium/potassium exchanger, both of which can be studied in macropatches (Hilgemann, 1994), which are both considered to have a deep access pore. There is some preliminary evidence such a scheme in hair cells, (Gale & Ashmore, 1995) for experiments show that the voltage sensitivity, but not the charge moved, is sensitive to the level of extracellular cations. The kinetics of the movement is as fast as the patch clamp can readily record and is over within 15 microseconds. This distinguishes the conformational change from one such as occurs in the sodium channel where the S4 region

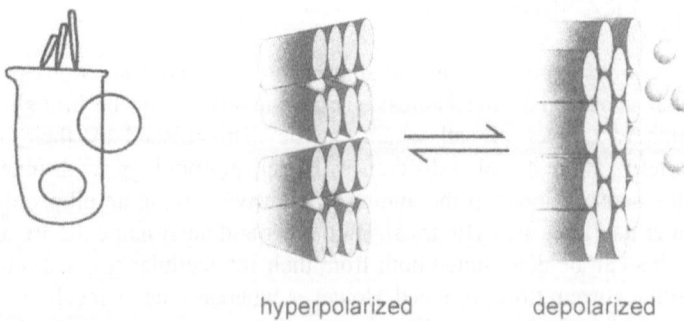

                    hyperpolarized   depolarized

Figure 3. Model for the outer hair cell motor. The scheme depicted shows a region of the basolateral membrane. The dense motor array reorders in the presence of cations: on hyperpolarization, cations being drawn into the pore and the unit area increases; on depolarisation they are expelled from pore and the area decreases. The effect results in a change the length of the cell.

moves as a voltage sensor and gives rise to a (much slower) movement of charge, (Heine-mann et al., 1992).

7. THE TRANSFER FUNCTION OF COCHLEAR HAIR CELLS

The relation between bundle displacement and intracellular current flow can be determined in isolated cell preparations. This transfer function is a curve with a sigmoidal function. one, where the inward current is given as a function of the tip displacement X and a neutral point X_o in the following form:

$$I(X) = I_{max} /(1+\exp(-a(X-Xo) (1 + \exp(-b(X-X_1))))$$

where a,b and X_o and X_i are constants used to fit the data. Data from the turtle can be fitted with a = 11 μm^{-1} and b = 4 μm^{-1} although mammalian data requires larger values. The values of X_i vary with the adaptation state of the transducer. Although these curves describe the transfer function of isolated hair cells, there are discrepancies between inner and outer hair cells of the mammalian cochlea *in vivo*. Inner hair cells appear to be described by transfer curves where approximately 20% of the channels are open at rest (Russell et al, 1986). The consequence is that for more intense sounds and /or those which exceed threshold, the receptor potential is asymmetric. Outer hair cells *in vivo* however, can be described by a symmetric transfer function: at rest approximately 50% of the transducer channels are open (Dallos, 1992). Thus for outer hair cells, the receptor potentials and the cochlear microphonic (the extracellularly recorded transduction current) are symmetric, and there is no offset produced by the harmonic components of the currents.

Why should there be a difference? One possibility is that the transducer channel differs between inner and outer hair cells. Records from isolated preparations (e.g. Russell et al, 1986; Rüsch et al., 1994) do not readily support this idea. These authors find both outer and inner hair cells have comparable transfer curves. Since these recordings were made in elevated calcium solution (rather than at the low levels which correspond to scala media) the conclusions still need to be interpreted with caution. The other possibility is that in the intact cochlea, the outer hair cell stereocilia are subject to a steady bias which ensures that nearly 50% of the transducer channels are open. This hypothesis requires either that there really is no adaptation in mammalian hair cells or that the hair cell mechanics is under control from local or long loop neural control. In this case the open the transducer channel could be maintained near the observed level at rest.

8. DISCUSSION

The specialised mechanisms found in hair cells exert a fascination because the mechanisms are built around molecular scale structures — nanostructures — which are particularly elegant in their design and assembly. As emphasised above, however, it is the small scale of the transducing machinery which has posed particular problems in the identification of the molecular components. There are as yet no clear, sequenced candidates for the channel and its coupling proteins which act as the transducer in hair cells, although there has been very promising progress made recently in the identification of the molecular biology of mechanotransduction in the nematode worm *C.elegans* in the form of the identifiucation of components of the *mec* gene family. There are also no positively identified candidates for the motor which operates in outer hair cells to drive the process of cochlea amplification. One reason for both situations is clearly that the amount of starting tissue is small. A second reason, however, is that assays for the proteins are functional assays: to determine whether a candidate protein acts as a mechanotransducing channel, it is necessary to perturb it mechanically. The development of the techniques to pursue these strategies present biophysical challenges for the immediate future.

ACKNOWLEDGMENTS

This work was carried out with support of EC grant SSS 6961, the Wellcome Trust and the Medical Research Council. I thank Drs. Matthew Holley, Paul Kolston, Fabio Mammano and Jonathan Gale for helpful comments.

REFERENCES

Art, J.J. and Fettiplace, R. (1987) Variation of membrane properties in hair cells isolated from the turtle cochlea. *Journal of Physiology,* 385, 207-242.

Ashmore, J.F. (1994) The cellular machinery of the cochlea. *Experimental Physiology*, 79, 113-134.

Ashmore, J.F. and Kolston P.J. (1994) Hair cell based amplification in the cochlea. *Current Opinions in Neurobiology*, 4, 503-508.

Ashmore, J.F., Kolston, P.J. and Mammano, F. (1993) Dissecting the outer hair cell feedback loop. In: *Biophysics of Hair Cell Sensory Systems* (H.Duifhuis, J.W.Horst, P. van Dijk, S.M. van Netten, Eds) pp 151-158. World Scientific, Singapore.

Crawford, A.C., Evans, M.G. and Fettiplace,R. (1989) Activation and adaptation of transducer currents in turtle hair cells. *Journal of Physiology,* 419, 405-434.

Dallos, P. (1992). The active cochlea. *Journal of Neuroscience* 12, 4575-4585.

Eatock, R., Corey, D.P. and Hudspeth, A.J. (1987). Adaptation of mechanoelectrical transduction in hair cells of the bullfrog sacculus. *Journal of Neuroscience*, 7, 2821-2836.

Fay, R.R. (1988) Hearing in vertebrates: a psychophysics data book. Hill-Fay Associates, Chicago.

Gale, J.E. and Ashmore, J.F. (1995) Effect of extracellular cations on charge movement associated with the motor of outer hair cells isolated from the guinea pig cochlea. *Journal of Physiology,* 485.P. 15-16P.

Gillespie, P.G. (1995) Molecular machinery of auditory and vestibular transduction. *Current of opinions in Neurobiology.*5, 449-455.

Gillespie, P.G. and Hudspeth, A.J (1991) Adenine nucleoside diphosphates block adaptation of mechanoelectric transduction in hair cells. *Proceeding of the national Academy of Sciences, USA* 90, 2710-2714.

Greenwood, D.D. (1990) A cochlear frequency-position function for several species - 29 years later. *Journal of the Acoustical Society of America*, 87: 2592-2605.

Hackney, C.M., Furness, D.N., Benos, D.J., Woodley, J.F. and Barratt, J.(1992) Putative immunolocalization of the mechano-electrical transduction channels in cochlear hair cells. *Proceedings of the Royal Society of London B.* 248,215-221.

Heinemann, S.H., Conti , F. and Stuhmer, W. (1992). Recording gating currents from *Xenopus* oocytes and gating noise analysis. *Methods in Enzymology* 207, 353-368.

Hilgemann, D.W. (1994) Channel-like function of the Na,K pump probed at microsecond resolution. *Science,* 263, 1429-1432.

Holley, M.C. and Ashmore, J.F. (1988) On the mechanism of a high frequency force generator in outer hair cells isolated from the guinea pig cochlea. *Proceedings of the Royal Society of London B.* 232, 413-429.

Housley, G.D.., Greenwood,D., and Ashmore, J.F. (1992) Localisation of the cholinergic and purinergic receptors on outer hair cells isolated fro the guinea pig cochlea. *Proceedings of the Royal Society of London B.* 249, 265-273.

Housley, G.D., Connor, B.J. & Raybould, N.P. (1995) Purinergic modulation of outer hair cell electromotility. In: *Active Hearing*, Flock, A., Ottoson, D. and Ulfendahl, M. Eds., pp. 113-125. Elsevier Science, Amsterdam.

Hudspeth, A.J. & Gillespie, P.G. (1994) Pulling strings to tune transduction: Adaptation by hair cells. *Neuron* 12, 1-9.

Jaramillo, F. & Hudspeth, A.J, (1991). Localisation of the hair cell's transduction channel at the hair cell's top by iontophoretic application of a channel blocker. *Neuron* 7, 409-420.

Kros, C.J., Rüsh, A., Lennan, G.W.T. and Richardson, G.P. (1995) Transducer currents and bundle movements in outer hair cells of neonatal mice. In: *Active Hearing*, Flock, A., Ottoson, D. and Ulfendahl, M. Eds.,pp. 113-125. Elsevier Science, Amsterdam.

Kossl, M. and Russell, I.J. (1995) Basilar membrane resonance in the cochlea of the mustached bat. *Proceeding of the national Academy of Sciences, USA.* 92, 276-280.

Kossl, M. and Vater,M. (1985) The cochlear frequency map of the mustache bat, *Pteronotus parnelli. Hearing Research* 19, 157-170

Mammano, F. and Ashmore, J.F. (1993). Reverse transduction measured in the isolated cochlea by laser Michelson interferometry. *Nature*, 365, 838-841.

Ohmori, H. (1985) Mechano-electrical transduction currents in isolated vestibular hair cells of the chick. *Journal of Physiology*, 359, 189-217

Preyer, S, Hemmert, W., Pfister, M, Zenner, H-P, and Gummer, A.W. (1994). Frequency response of mature guinea pig outer hair cells to stereociliary displacement. *Hearing Research* 77, 116-124.

Rusch, A. Kros, C.J. and Richardson, G.P. (1994) Block by amiloride and its derivatives of mechano-electrical transduction in outer hair cells isolated from the guinea pig cochlea. *Journal of Physiology* 474, 75-86

Russell, I.J., Richardson, G and Cody, A. (1986) The responses of inner and outer hair cells in the basal turn of the guinea pig and the mouse cochlea grown in vitro. *Hearing Research*, 22 196-216.

Russell, I.J. and Sellick, P.M. (1983) Low frequency characteristics of intracellularly recorded receptor potentials in guinea pig cochlear hair cells. *Journal of Physiology* 338, 179-206.

Sellick, P.M. , Patuzzi, R and Johnstone, B.M. (1982). Measurements of the basilar membrane motion in the guinea-pig using Mossbauer technique. *Journal of the acoustical Society of America.* 72, 131-141.

Solc, C.F., Derfler, B.H., Duyk, G.M. and Corey, D.P. (1994) Molecular cloning of myosins from bullfrog saccular macula: a candidate for the hair-cell adaptation motor. *Auditory Neuroscience*, 1, 63-75.

11

PRECISE AND PERCEPTUALLY RELEVANT PROCESSING OF AMPLITUDE MODULATION IN THE AUDITORY SYSTEM

Physiological and Functional Models

Frédéric Berthommier[*] and Christian Lorenzi

Institut de la Communication Parlée/INPG
46, Avenue Félix Viallet
38031 Grenoble Cedex

1. INTRODUCTION

The peripheral representation of sounds is transmitted by the auditory nerve to intermediate stages of the nervous system, preceding identification. We suppose that main features of sounds corresponding to perceptual dimensions like pitch and timbre are already extracted in these levels, complementing richer and extensive representations localised in cortical areas. This approach allows us to consider a biological system having a rather simple architecture, known inputs and outputs. Understanding physiology allows us to perform auditory scene analysis (ASA) grounded on plausible basis. ASA is an emerging concept integrating particular properties of the auditory system working together in order to deal with a complex environment (Bregman, 1990). From the neurobiological point of view, the goals of such a modelling work are: (1) better approaching the way of coding and controlling information in the nervous system, (2) comparing with physiological data, and finally (3) reducing the computational processes.

The sound waves are converted into a spatio-temporal representation by the peripheral auditory system (PAS). The spatial dimension is the tonotopic axis. In the literature, this peripheral auditory image is currently approximated by filter banks, rectification and adaptation in order to recover the effective input shapes of the mid-brain levels of the auditory system. For simplicity, we neglect parts of this complete model of the PAS in the case of periodic-stationary signals.

The first question addressed here concerns the ability of single neurons receiving this information to be sensitive and to represent periodicity. The temporal axis of the neural

---

[*] Address correspondence to: Frédéric Berthommier, Institut de la Communication Parlée/INPG, 46, av. Félix Viallet, 38031 Grenoble Cedex, France. email: bertho@icp.grenet.fr; Tel: 76 57 48 28; Fax: 76 57 47 10.

auditory image contains perceptually relevant information, periodic and aperiodic. After rectification, only the positive part of resolved periodic components remains in the low frequency domain (< 500 Hz). For higher frequencies, beating due to the summation of unresolved harmonics appears to produce an amplitude modulation (AM) of the neural activity, synchronised on the acoustic stimulus fundamental frequency (f0). We use the same terminology (AM) for the resultant periodic signals of the two domains, because similar variations of auditory nerve fibres discharge rates are observed.

Depending on their membrane properties, neurons can be integrators or coincidence detectors (Abeles, 1981), but models and experiments show other dynamical properties at the single neuron level. Having a neuron model with proper oscillatory characteristic is a critical point for well understanding build up of neuronal assemblies and object recognition according to the correlational point of view. This was the first reason for developing a time-dependent neuron model and we then connect a rather similar neuronal unit for building a model of olfactory integration (Berthommier, Buonviso and Chaput, 1995). We explain here the second motivation of this model.

2. MODELLING OF PHYSIOLOGICAL AM PROCESSING

Our auditory models are based on dynamical Hodgkin-Huxley (HH) equations (1952), or on the equivalent probabilistic form. This directly provides instantaneous discharge rates of stellate cells. This population of cells, connected to auditory nerve fibres, is localised in the CN. For auditory modelling, a complex model including both stochastic and non-linear dynamical aspects - the HH system - can be reduced to a simple probabilistic formulation. We present the characteristics of the probabilistic model, and we apply it on AM signals. This gives a representation of the AM characteristic in intermediate stages of the auditory system, which we assume would be precise and useful for processing complex sounds.

While several models of AM processing are based on an explicit autocorrelation of signals at the filterbank output (Slaney and Lyon, 1991), a quantisation of periodicity (Patterson and Holdsworth, 1991), or a characterisation of synchronicity (Cooke, 1992), physiological data favour a theory of resonance based on a selective enhancement of amplitude modulation depth (Langner and Schreiner, 1988; Frisina, Smith and Chamberlain, 1990). Many modellers have simulated the neuronal responses observed in the ventral CN, using simple neuronal models derived from "integrate and fire" units (Berthommier, 1991), or from the classical HH equations (Banks and Sachs, 1991; Hewitt, Meddis and Shackleton, 1992; Lorenzi, Berthommier and Tirandaz, 1993). A probabilistic formalisation of these stochastic simulations clarifies the mechanisms of neuronal integration and we obtain a particular driven oscillator having an implicit delay line providing to the unit a rather large memory.

The first model we present — Model I — has two main stages: the peripheral signal is directly synthesised and then processed by the units (fig. 1). With the probabilistic model, each CN cell receives only one channel and this output is transmitted to only one IC cell. On the contrary, a great convergence is necessary when we use pulse coding, in order to get a redundancy. Hence, the configuration of probabilistic model could be modified in order to add some convergence and to obtain a cross-channel associativity. The function of the present implementation is not to associate and to categorise, but to extract and to map components of the signal.

The temporal/periodic information observed at the PAS output is rate-place coded at the IC level, where a great proportion of the units has a mean discharge rate depending directly on AM frequency and modulation depth.

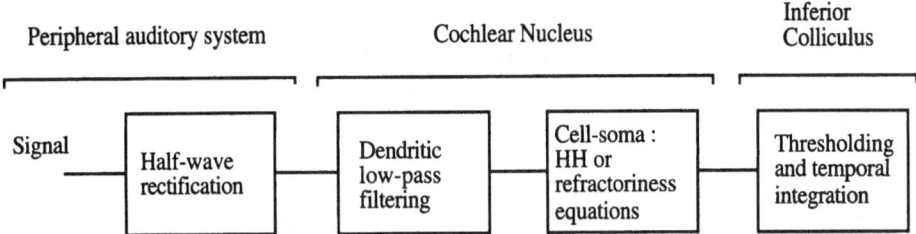

Figure 1. Diagram of physiological AM processing in the intermediate auditory system (Model I).

2.1. Probabilistic Model of Neural Integration

This single-neuron model represents the instantaneous discharge rate of a single neuron or the mass activity of a group of independent cells by a continuous real-valued signal. The validity of this model has been well tested for auditory modelling, where the information directly arrives from the auditory nerve (Berthommier, 1992; Berthommier, 1993; Lorenzi et al., 1993). Our conclusion was that probabilistic form and the classical models of neural integration like the HH system or "integrate and fire" units with refractory period show very similar properties despite very different computational costs due to the integration of a differential system or to the stochastic generation of inputs. Such unit model has a great degree of generality because it is plausible that principal cells of the nets observed elsewhere in the brain have their main properties determined firstly by excitation parameters: weights, time constant(s) of the excitatory post-synaptic potential (EPSP) or the inhibitory one (IPSP); and secondly by refractoriness parameters: time constant(s) of the refractory period. The refractory period can be relative or absolute and when the neuron discharges, no spikes are emitted during the absolute refractory period.

The probabilistic model evaluates at each time the neuronal discharge rate when the input is itself the time varying probability of a single point process corresponding to the sum of point processes received by an equivalent "integrate and fire" unit. We assume that each stellate cell integrates spike trains transmitted by about fifty auditory nerve fibres (over thirty thousands), and consequently responds to a weakly spatially-distributed signal. Similarly, in the temporal domain, the signal is integrated by cells with a weak lowpass filtering of about 0.5 to 1ms. Despite these integrative properties, the global effect of neuronal process is a selective and rather sharp enhancement of the input AM when the AM frequency is related to the absolute refractory period duration: neurons have an intrinsic resonance property and can integrate the signal over consecutive periods. Consequently, isolated neurons can filter the signal to extract periodic variations. The mathematical basis of this filtering property are explained in the following paragraphs.

A real neuron having an absolute refractory period exhibits a PSTH (Post Stimulus Time Histogram) with a decreasing oscillatory response when input is a burst (a step) of noise. Stellate cells have a so-called "chopper" type of response. This damping is due to a progressive loss of memory and a decorrelation of responses with the beginning of the step variation. On the contrary, neuronal activity remains well phase-locked with low frequency periodic inputs. The best model is not a second-order oscillator if we assume the stochastic structure of inputs. In fact, the response is damped because inputs are stochastic. The formal neuron has a variable representing the probability to be outside the absolute refractory state, knowing the previous probabilities of discharge. The recurrent evaluation of this refractory variable $R1(t)$ is:

$$R_1(t+1) = (1 - \sum_{\theta=t-\text{ref1}}^{t} P(\theta)) \tag{1}$$

= Probability of having no spike since t-ref1, and ref1 the duration of the absolute refractory period.

$P(t)$ is the solution of a differential system and this is the integral form of the delayed differential equation. Neuronal output $P(t)$ can be described by an analytical solution only if input is stationary or is an Heaviside function. Otherwise, we integrate numerically this differential equation, and we compare $P(t)$ with simulations of the equivalent stochastic system: a unit having an absolute refractory period and receiving spikes. From the signal processing point of view, this is a feedback filter because we compute outputs at a given time using previous outputs. Contrasting with the autocorrelation which results from a product between current and delayed inputs, this depends on a product between inputs - represented by $E(t)$ - and a function of delayed outputs stored in delay lines.

The relative refractory variable R2(t) takes into account the exponential weighting in time and the probability of having a discharge at a given time and no discharge since:

$$R_2(t+1) = 1 - \sum_{\theta=-\infty}^{t-\text{ref1}} P(\theta) \left[\prod_{\theta'=\theta+\text{ref1}}^{t} (1 - P(\theta')) \right] e^{-(t-\theta-\text{ref1})/\text{ref2}}, \tag{2}$$

where ref2 is the time constant of the relative refractory period

$$P(t) = E(t)R_1(t)R_2(t) \tag{3}$$

is the probability of discharge at t of a formal neuron having both refractory variables, where $E(t)$ is the excitation variable.

Two other operators are necessary for retrieving the precise physiological shape of the stellate cell response and for well describing the excitation source $E(t)$: blocking and reset (fig. 2). When the neuron discharges, both excitation and refractory variables are reset. The excitation is blocked during the absolute refractory period. Blocking the arrival of afferent impulses also modifies excitation value during the absolute refractory period. This is evaluated by multiplying at each time the absolute refractority R1(t) by the input value I(t). Since reset drives the membrane potential to zero, we also take into account the probability of having no second discharge since the reception of the last input:

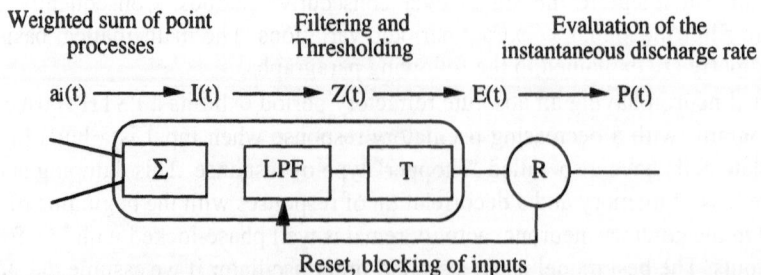

Figure 2. Block diagram of the probabilistic model. Steps of computation are: (1) linear input summation (2) Lowpass Filtering (3) thresholding (4) refractority. There are two main variables: excitation E(t) — at the output of the main box - and refractory R(t) — at the output of box R. This corresponds to physiological steps of integration and relevant parameters. All variables have real/continuous values.

$$Z(t+1) = (1-e^{-1/\tau}) \sum_{\theta=-\infty}^{t+1} I(\theta)R_1(\theta) \left| \prod_{\theta'=\theta}^{t}(1-P(\theta')) \right| e^{-(t+1-\theta)/\tau}$$

(4)

$$\text{with } P(t) = T(Z(t))R_2(t) = E(t)R_2(t)$$

(5)

$$\text{and } T(x) = \frac{x}{S}(1-e^{-kx/S}), \ k=1, \text{ depending on FS}$$

(6)

$T(x)$ is a non-linear function having an asymptotic linear branch. The saturation of the response to increasing intensity levels is only due to the refractory components (fig. 3). The

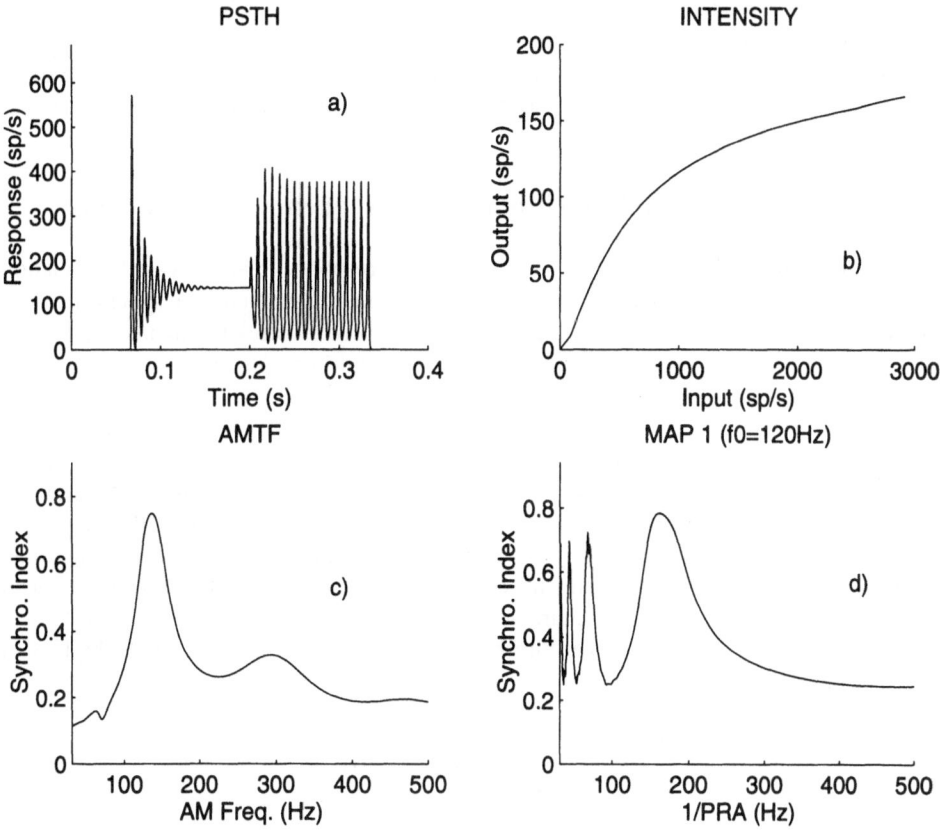

Figure 3. Resonance and transfer function of stellate cells at the cochlear nucleus level. This is computed with the probabilistic model described in text. The mean input rate is a sampled signal (10 KHz) corresponding to a Poisson point process. The mean discharge rate is always 1500 spikes/sec. For an AM/rectified signal, the modulation depth is 0.5. Parameters of the model: S=4, τ=1ms, refl=5ms, ref2=6ms. a) PSTH of this cell for a constant signal followed by an AM signal modulated at 120 Hz, near the best frequency of the cell. We show that the modulation depth is greatly increased. b) Saturation of the mean discharge rate. Input is a constant signal. This model has a weak threshold effect which can be easily increased. c) Amplification of the synchrony index (input = 0.25) relative to frequency, showing a best frequency and a little near-harmonic rebound. Signal is filtered for other AM frequencies. d) Synchrony index on the AM freq of the stimulus (120 Hz), computed for an array of stellate cells. Hence, this is relative to the driving signal. The variable is the absolute refractory parameter refl. Other parameters remain the same. Cells with refl~6ms are better synchronised. Secondary peaks are observed for very long (>10 ms) refractory periods, probably not realistic for stellate cells. This resonance of the cells is suitable for building a rate place map at the inferior colliculus level.

refractory component R1(t) is already introduced in the term Z(t). The reset concerns both the unit excitation variable Z(t) and its recovery function R(t). Reset has the same effect as an inhibition of the units, and it can also be used for desynchronising a group of cells and enhancing the short-term correlation of their activity (Berthommier et al., 1995).

The amplitude modulation transfer function (AMTF) of the formal unit has a first peak of synchrony rate around the frequency 1/ref1 (fig. 3). This is a non-linear filter and we remark that AMTF shape varies with the intensity level of the input signal. Synchrony rates increase for higher intensities, with a better phase-locking on the modulation. Peaks disappear for low intensity levels, and we obtain a lowpass characteristic. All these properties are also retrieved with HH and "integrate and fire" models (Hewitt et al., 1992, Lorenzi et al., 1993).

Practical algorithms have a low, linear in time, complexity because the filtering stages related to the dendritic and refractory processes are computed recursively. They run fast, even for a great number of neurons (used for fig. 3 and 4). Such a probabilistic model is an efficient tool for modelling biological information processing, when periodicity is the significant information. We have planned out the VLSI implementation of a set of probabilistic units in order to show the interest of a numerical solution in comparison to other ones (Gautier and Pouillot, 1995). We hope to use this circuit for hearing aids and cochlear nucleus implants.

2.2 Mapping of the Amplitude Modulation Frequency at the IC Level

For CN units, an increased synchrony index corresponds to an increased correlation of their outputs. Consequently, a first mechanism allowing a dependence on synchrony rates is coincidence detection. A unit discharging when it receives correlated inputs is a coincidence detector. We can suppose that, physiologically, dendrites of IC units compute coincidence detection while soma performs an integration of resultant activity. At the modelling point of view, a coincidence detection operator can be a sigmoidal shaped function applied after a short term temporal integration (~0.5 ms) and a rather large convergence of input channels. Our model is a cascade between two units — one CN and one IC — without any convergence. This is directly motivated by the use of probabilistic values. IC integrative process is based on an adaptive threshold in order to capture both onsets and periodic variations. It removes at the same time the constant signals independently of the input mean discharge rate: IC unit does not discharge when input synchrony index is zero.

The mean of the residue over threshold during a period of time T gives the IC cell discharge rate:

$$\text{MOT} = \frac{1}{T} \int_0^T R_+(P(\theta) - c\, Q(\theta))\, d\theta,$$

(7)

where c is a constant value, and $R_+()$ the rectification function

$$\text{with } Q(t) = \frac{1}{\tau 2} \int_{-\infty}^t P(\theta) e^{-(t-\theta)/\tau 2} d\theta$$

(8)

Then, we obtain a rate-place code of the variations of synchrony index observed at the CN level (fig. 4).

Finally, the transfer function of CN cell quantified on the basis of synchrony index and the rate-place coded IC response are close (fig. 4). But, at the IC level, we evaluate directly the response rates to AM signals without the (period) reference required for synchrony index computation. The resolution of peaks is sufficient for separating two components having a 20% difference in frequency (100Hz + 120Hz) and 0.5 modulation depth. These components come from the same channel(s) of the PAS filterbank and are not resolved in the tonotopic representation, but they appear well separated in this new representation.

2.3. Comparison Between Model I Properties and Psychoacoustical Detection Performance

A comparison has been done between responses of models and psycho-acoustical detection performances of subjects. A first experiment evaluates the detection threshold of modulation depth for AM noise. The result indicates an equivalence at the neuronal level:

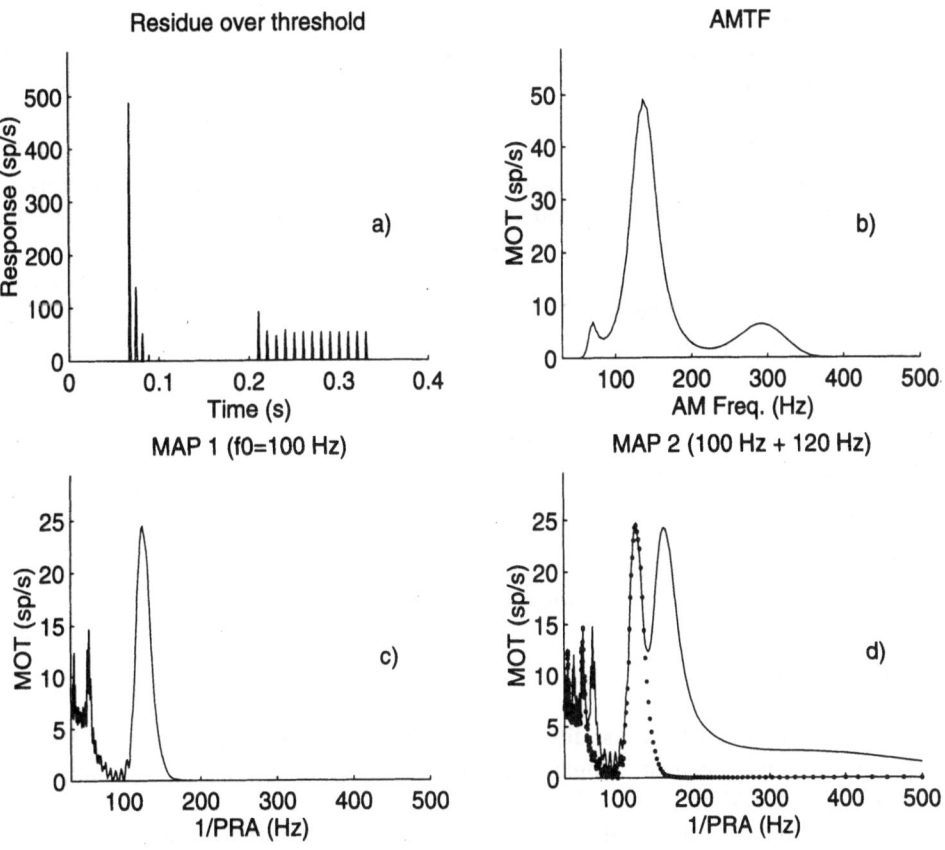

Figure 4. Mapping at the IC level. Responses are directly computed with outputs of CN units having the parameters used for fig. 3. Threshold values are given by a first order lowpass filtered signal Q(t) (time cst=5ms) multiplied by a constant value (c=1.45). a) Residue of excitation with AM freq=100 Hz. This residue is integrated over the period of time T which is the second periodic part of the excitation. The mean discharge rate is the MOT (Mean over threshold) corresponding to the activity of IC cells. b) Transfer function of the two stages. Harmonic and sub-harmonic peaks persist. c) Response of the array of 250 cells connected to the preceding set (fig. 3). Only the ref1 parameter of CN cells varies among the map. The AM freq=100 Hz is well mapped by cells connected to 8ms PRA stellate cells. We also notice a great response for IC cells related to CN cells having a refractory period which is a multiple of the best cell ref1 refractory period. d) Mapping of mixed signals with AM freq = 100 Hz + 120 Hz. A 20% difference is well resolved. The isolated response c) is superimposed, showing the independence of the two peaks. Peak position of the 120 Hz component remains the same, compared with fig. 3d.

neurometric curve of an IC neuron receiving stellate cells afferences and psychometric responses of subjects are similar. The authors used the same detection task to compare the performance of the models of inferior colliculus neurons with that of four human observers. The activity of a relatively small number of IC neurons seems to provide sufficient information to account for psychoacoustical judgements dealing with amplitude modulation detection (Lorenzi, Micheyl and Berthommier, 1995).

3. SPECTRAL SEGREGATION BASED ON A BI-DIMENSIONAL REPRESENTATION

From a functional point of view, separating "auditory objects" like vowels requires a precise representation of AM frequencies easy to get through DFT at the output of a simple pre-processing based on half-wave rectification and bandpass filtering. Following the physiological principle of mapping of periodic components, we propose a functional method (functional Model II, fig. 5). This represents quasi-stationary periodic sound components in a <carrier frequency/AM frequency> plane by using rectification, bandpass filtering and Fourier transform. This allows us to separate clearly respective spectra of two superimposed complex sounds when they have different f0s. Separated spectra can be sent to a recogniser. A separation power equivalent to previous models based on autocorrelation (Assmann and Summerfield, 1990) has been shown for superimposed double vowels (Berthommier and Meyer, 1995; Meyer and Berthommier, 1995). This argues for the physiological plausibility of separation mechanisms not directly based on autocorrelation and operating at intermediate levels of the auditory pathway. After the proposal of Licklider's theories (Licklider, 1959), autocorrelation was the unique effective method developed in the literature for separating and identifying spectra of complex sounds.

The functional model II can be applied to vowels segregation while Model I is not currently appropriated in this way because, even if single cell computation is very fast, the final cost is heavy: we must evaluate, during at least 100 ms, outputs of a map of cells having dozens of units connected to each peripheral channel. Furthermore, tuning the physiological model for this task is not so easy. This was a great motivation for developing a practical model sharing common properties with the physiological one.

3.1 Design of Bandpass Filters

Rectification and bandpass filtering stages convert AM into cosine-demodulated signals. In the high frequency domain, carrier frequency is filtered out and envelope corresponding to AM is preserved. A simple method for building temporal recursive bandpass filters defines a temporal kernel f(t) as a weighted sum of pairs of first order filters:

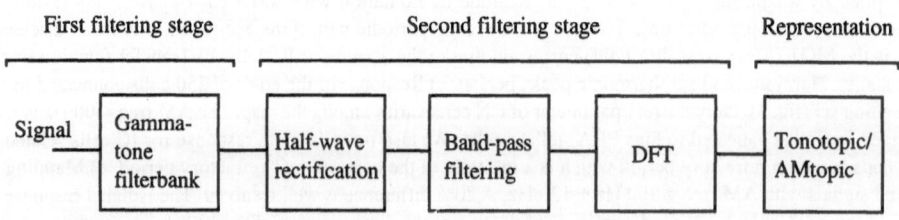

Figure 5. Diagram of functional processing for complex sounds segregation (Model II).

$$f(t) = \sum_i \alpha_i F(k_i, b_i, t) \tag{9}$$

Members of these pairs are exponentials, a positive and a negative one, weighted to obtain a null integral:

$$F(k_i, b_i, t) = \frac{1}{b_i} e^{-k_i b_i t} - \frac{1}{k_i b_i} e^{-b_i t} = F_{i+}(t) - F_{i-}(t), \tag{10}$$

where b is the time constant and k tunes the ratio between the positive and the negative exponentials.

When the DFT is applied on the resultant signal, DC peak is abolished because:

$$\int_0^{+\infty} F(k_i, b_i, \theta) d\theta = 0 \tag{11}$$

Such difference of exponentials (DOE) performs a bandpass filtering in the frequency domain. In practice, we only use one pair of first order filters running in parallel:

$$f(t) = e^{-2at} - \frac{1}{2} e^{-at} \tag{12}$$

This is an extension in the temporal domain of the classical DOG (Difference Of Gaussians) used for modelling visual filtering:

$$G(k, b, x) = \frac{1}{b} e^{-k^2 b x^2} - \frac{1}{kb} e^{-bx^2} = G_+(x) - G_-(x), \tag{13}$$

where b is here the space constant.

We easily verify that:

$$\int_{-\infty}^{+\infty} G(k, b, \chi) d\chi = 0 \tag{14}$$

This filtering method is actually used in order to extract auditory speech events (Piquemal, Schwartz and Berthommier, 1995) and to model the first layer of the olfactory bulb (Berthommier et al., 1995).

3.2. Principle of Double Vowels Segregation with AM Maps

The example we show is the AM map representation of two superimposed stationary vowels /a/ + /i/ having close f0s (fig. 7 and 8). They are synthesised with a standard Klatt formant synthesiser with similar intensity levels. The vertical axis of the bi-dimensional representation is the tonotopic one, quasi-logarithmic, whereas the horizontal AM frequencies are linearly distributed. The vertical axis corresponds to gammatone filterbank channels, and the horizontal axis to outputs of the DFT. In the low frequency domain of the tonotopic axis, the harmonics are well resolved among the two dimensions while they produce an AM signal due to beating. This AM signal is synchronised on the fundamental frequency, among the medium and high frequency domains. The DOE bandpass filter, identical for all vertical

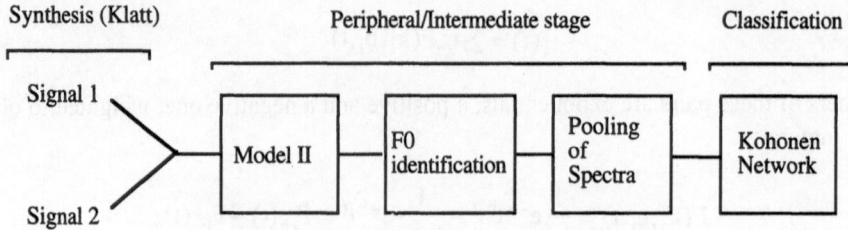

Figure 6. Double vowels segregation based on model II.

channels, has a wide selectivity centred on the low frequency domain (50-500 Hz). Hence, lower harmonics are well represented by both frequency and place, but it is necessary to pool these components for recovering the whole spectrum. Simple rules for implementing a grouping process are based on the distribution of peaks in this representation.

Our algorithm searches the fundamental frequencies before pooling the spectra (fig. 6). This is not an absolute constraint, because the distribution of peaks in AM maps is highly structured: a point related to the fundamental frequency is the root of two lines (fig. 7). The vertical one is centred on f0 and we can directly keep the energy of this section. The second line is a curve passing through the lower resolved harmonics. This is defined from both the tonotopic distribution of frequencies and temporal coding. It is then necessary to know the places of these peaks in order to pool their energies. Hence, we could map these relations directly, but pooling after f0 identification is the method we use. Grouping of AM harmonics is optional in the medium frequency (MF) and high frequency (HF) domains. The resultant spectra are identified with a Kohonen net (1982) trained with a set of isolated vowels. This evaluates a distance between a given spectrum and stored prototypes. A normalisation of the spectra is preferable in order to be insensitive to intensity variations.

Other beating components appearing in the very low frequency domain (0-100 Hz) are currently not appropriated, but they carry useful information to detect superimposition of components and to give a criteria for a better partition of energies between them.

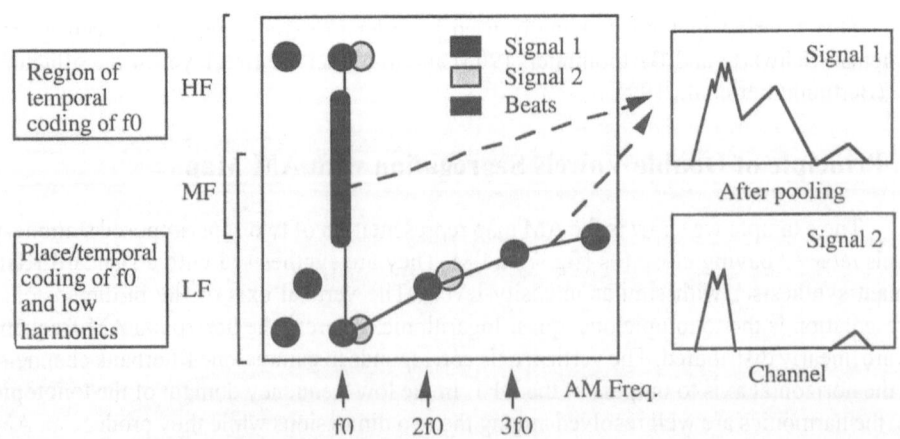

Figure 7. Method for pooling spectra from AM maps.

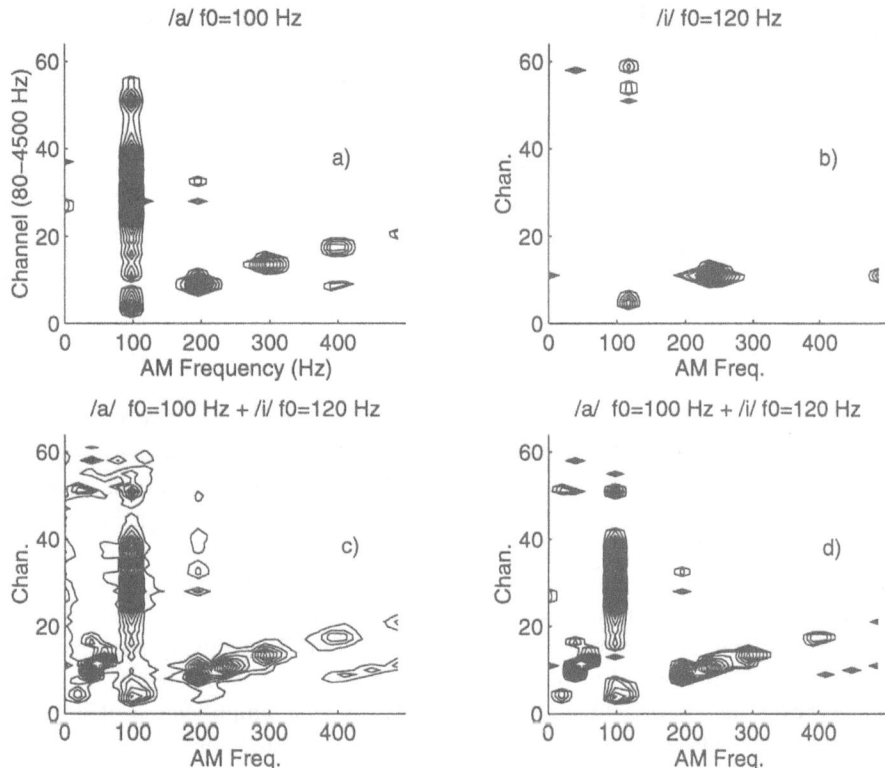

Figure 8. Tonotopic/AMtopic mapping of complex sounds as vowels. The signal is processed with a middle ear model and the Patterson/Holdsworth gammatone filterbank. Set-up with standard parameters. Outputs are half-wave rectified and DOE bandpass filtered, time const.=3ms. Signal duration: 100 ms. A 512 points DFT is computed for each channel only during the stationary second part of the response, giving the AM frequency axis. Zero peaks are weak. Images a), b), d) are thresholded in order to isolate the peaks (0.15 of maxima). Own components of each sound and interacting ones are well displayed by this representation. a) Stationary synthetic /a/ isolated, b) synthetic /i/ having a close f0 frequency c) non thresholded representation for /a/+/i/. d) Own components displayed by a), b) are retrieved, but beating components appears (freq. < 100 Hz here). These could improve the analysis of non resolved components.

3.3. Identification of the Fundamental Frequency

In order to conclude by fitting physiology and signal processing, we propose a method derived from AM maps, able to represent the pitch of a complex sound like the brain does. This method evaluates pitches of sounds in different situations by place recoding of the AM frequency. We have shown that the precise knowledge of f0s is a key point for recovering spectra from AM maps, and amplifying f0 place-coded energy is an appealing pre-processing strategy for tracking simultaneous pitches of superimposed signals. Furthermore, we have recently developed a new algorithm using f0s and working directly with sound waves (Berthommier and Tessier, 1996). This is able to separate mixed complex sounds and to produce satisfactory audible outputs without resynthesis.

The AM map has two domains - LF and HF - among the tonotopic axis, because the lack of resolution of peripheral auditory filters. In this map, the modulations produced in the HF domain of the tonotopic axis are projected in the low frequency domain of the AMtopic axis after rectification and bandpass filtering (i.e. after demodulation). When the DFT is

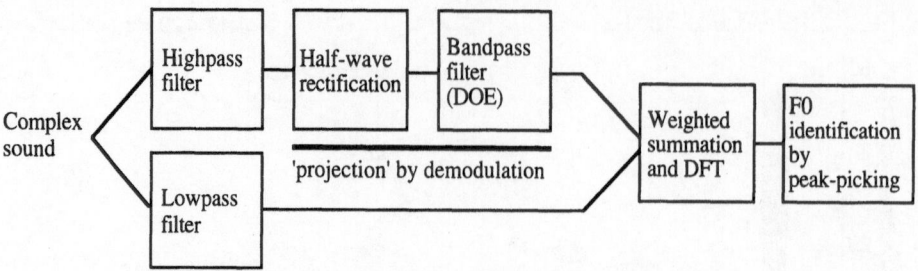

Figure 9. Fundamental frequency identification.

applied on a demodulated signal, we obtain an AM spectrum containing low frequency components of the envelope. Consequently, a straightforward procedure can be proposed to place-recode the temporal information and to summarise the contributions of the two domains (fig. 9) : (1) we cut the spectrum in two parts by filtering the signal directly with two complementary, lowpass and highpass, filters (2) the resultant waves are added together after demodulation of the HF components (3) the DFT produces a composite spectrum which is the conventional one, representing the place-coded information, added together with the AM spectrum which is an image of the temporal information. This roughly corresponds with the vertical summation of AM maps components, which is used for f0 identification before pooling (Meyer and Berthommier, 1995). We obtain a great amplification of f0 and low frequency components in the new summarised spectrum (fig. 10), or the creation of a f0 peak coming from the AM spectrum even if it is absent in the conventional spectrum. A simple peak-peaking of the greatest component performs the f0 identification when amplification of the f0 peak is sufficient.

Psychophysically, the fundamental frequency of a harmonic complex — a set of components having the same f0 — is heard even if it is not an effective component of the input stimulus. Recent recordings done at the cortical level show neuronal responses related to the absent component (Riquimaroux and Hashikawa, 1994). Place recoding of temporal information at a given level of the auditory pathway is a plausible explanation of this strong psychoacoustical effect.

4. DISCUSSION

Physiological and functional points of view, and their underlying mechanisms, are apparently similar. However, let us address the key points of their differences in order to come back to experimentation.

The functional process (model II) exhibits a very high selectivity thanks to DFT and it is probably an optimal process from the signal processing point of view. Let us remark that extraction of the AM frequency by resonators corresponds to the Helmholtz' hypothesis of place coding of frequency components extended to the AM encoding. This could be physiologically implemented with large banks of selective neural filters, ordered in computational maps as observed in the IC. But such physiological mechanism can be replaced by the DFT for modelling, when the task consists in obtaining a fine representation of the frequency of the components. Then, when a rule-based grouping process is applied on this map to recover the "auditory objects," we obtain a clear bottom-up procedure.

But what remains of the interest of physiological modelling? A classical argumentation against the Helmholtz' hypothesis of place coding of the frequency components is the

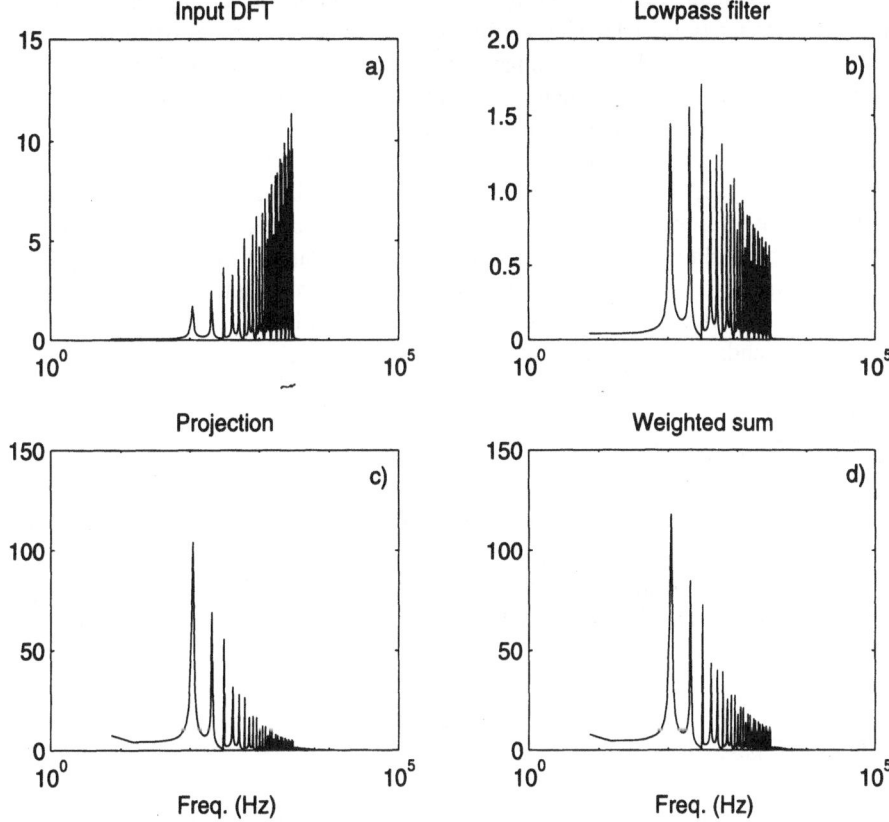

Figure 10. An example of f0 identification. The stimulus is a ramped 30 harmonic complex having a weak f0=100 Hz component and energy mainly concentrated in the high frequency domain. a) Conventional spectrum of this stimulus. b) Lowpass filtered signal showing an incomplete inversion of the ramp (time cst=1 ms) c) AM spectrum after highpass filtering (time cst=1 ms) and demodulation. We see the complete inversion of the ramp. In this representation, f0 is the greatest component. d) Summary spectrum where f0 is identified with a simple peak-peaking of the greatest component.

sensitivity to dephasing. The DFT and the autocorrelation loses phase information whereas model I is sensitive to the modulation envelope shape which depends on the relative phase of the components. Strickland and Viemeister (1994) have proposed a new experimental paradigm using AM harmonics in order to reveal a sensitivity to AM dephasing: we can observe a perceptual sensitivity between a fundamental AM frequency and his dephased first harmonic when the modulated stimulus recruits the same channels of the PAS. This is the case when stimuli is amplitude-modulated white noise. In collaboration with Laurent Demany (lab. de psychoacoustique, Bordeaux), we have developed an experimentation based on this principle for demonstrating the validity of Model I.

Moreover, the harmonicity of components of complex signals is directly detected by the neural filters in the CN and mapped at the IC level, because AMTFs have at least two peaks, one for their stimulus best frequency, and another one for the first harmonic (fig. 3 and 4). Stellate cells, together with the model we propose, are not simple bandpass filters like gammatone filters, but rather non-linear dynamical systems. Furthermore, models show a multi-bandpass property: they are also sensitive to the harmonics of their best frequency, and more generally to the consonance of the components. This suggests a physiological mechanism of pooling of harmonic components.

The third main difference comes from the potential adaptivity of physiological auditory processing (model I). This could support cooperativity and descending control because neural selectivity depends on parameters like refractory period duration. It is then possible to tune neural filters adaptively and cooperatively without computing a fine exhaustive map for finding the stimulus modulation frequency. The mechanism can be an adaptive addition of a refractory-like effect: the inhibition by neighbouring neurons or by upper centers. We have described a two-steps mechanism of integration. The signal is firstly gated by selective resonators - the CN stellate cells - and non-enhanced modulations are blocked by a threshold which is located at the IC level. A lack of exhaustivity of the set of resonators could be compensated by a dynamical short-term adaptivity. A bottom-up/top-down strategy of information processing becomes possible by taking into account the retroactive descending controls and the attentional processes.

ACKNOWLEDGMENTS

This work was partly supported by the HCM Network SPHERE (Contract No. CHRX-CT93-0098) and was presented 5th January 1995, in Sheffield, during the workshop "Links between speech technology, Speech science and Hearing" (excepted the paragraph on f0 identification). Thanks to J.L. Schwartz and G. Meyer for helpful discussions and useful criticism.

REFERENCES

1. Abeles, M. (1982) Role of the cortical neuron: Integrator or coincidence detector?, Isr. J. Med. Sci., 18, 83-92.
2. Assmann, P.F. & Summerfield, Q. (1990) Modelling perception of concurrent vowels: vowels with different fundamental frequencies, JASA, 88:2, 680-697.
3. Banks, M.I. & Sachs, M.B. (1991) Regularity analysis in a compartmental model of chopper units in the anteroventral cochlear nucleus neurons, Biol. Cybern., 64, 273-283.
4. Berthommier F. (1991) Neural mapping of sensory inputs in the auditory system, in Cognitiva 90, Kohonen, T. & Fogelman-Soulie, F. (Eds.), Amsterdam North Holland, 25-34.
5. Berthommier, F. (1992) Intégration neuronale dans le système auditif; Modélisation de réseaux neuronaux temporo-dépendants, Thèse GBM-USMG.
6. Berthommier, F. (1993) A probabilistic model of neuronal integration, 2d Conf. in Math. Applied to Biol. and Med., Lyon, to appear in J. of Biol. Systems, vol. 3, no. 4.
7. Berthommier, F., Buonviso, N. & Chaput, M. (1995) A probabilistic model of temporal processing in the olfactory bulb, AIDRI, Hermès, Paris.
8. Berthommier, F. & Meyer, G. (1995) Source separation by a functional model of amplitude demodulation, Eurospeech, Madrid, Vol. 1, 135-138.
9. Berthommier, F. & Tessier, E. (1996) Source segregation based on knowledge of physical properties of sources and sounds, ESCA workshop, Keele (submitted).
10. Bregman, A.S (1990) Auditory scene analysis, MIT Press, London.
11. Cooke, M.P. (1992) An explicit time-frequency characterization of synchrony in an auditory model, Comp. Speech and Lang., 44, 99-122.
12. Frisina, R.D., Smith, R.L. & Chamberlain, S.C. (1990) Encoding of amplitude modulation in the gerbil cochlear nucleus: I. A hierarchy of enhancement, Hear. res., 44, 99-122.
13. Gautier, G. & Pouillot, M. (1995) Etude et conception d'un circuit intégré neuronal à dynamique temporelle, Project supervised by Berthommier, F. & Castelli, E., Tech. Report ENSERG.
14. Hewitt, M.J., Meddis, R. & Shackleton, T.M. (1992) A computer-model of cochlear nucleus stellate cell: responses to amplitude modulated and pure tone stimuli, JASA, 91, 2096-2109.
15. Hodgkin, A.L. & Huxley, A.F. (1952) Currents carried by sodium and potassium ions through membranes of the giant axon of Loligio, J. Physiol., London, 116, 500-544.

16. Kohonen, T. (1982) Self-organized formation of topologically correct feature maps, Biol. Cybern., 43, 59-69.
17. Langner, G. & Schreiner, C.E. (1988) Periodicity coding in the inferior colliculus of the cat. I: Neuronal mechanisms, J. Neurophysiol., 60, 1799-1822.
18. Licklider, J.C. (1959) Three auditory theories, In Psychology: A study of a science, Koch, S. (Ed.), New york: McGraw-Hill, Vol. 1, 41-144.
19. Lorenzi, C., Berthommier, F. & Tirandaz, N. (1993) Physiological Modeling of Cochlear Nucleus Responses: Perception of complex sounds, ESANN 93, Bruxelles.
20. Lorenzi, C., Micheyl, C. & Berthommier, F. (1995) Neuronal correlates of perceptual amplitude-modulation detection, Hear. Res. (in press)
21. Meyer, G. & Berthommier, F. (1995) Vowel segregation with amplitude modulation maps: modelling studies, SPHERE Tech. Report.
22. Patterson, R.D. & Holdsworth, J. (1991) A functional model of neural patterns and auditory images, in: Advance in speech, Hearing and Langage processing, Vol. 3, Ainsworth, W.A. (Ed.), JAI Press, London.
23. Piquemal, M., Schwartz, J.L. & Berthommier, F. (1995) Auditory and visual detection of plosive bursts in noisy CVC sequences, Neuronime, Marseille.
24. Riquimaroux, H. & Hashikawa, T. (1994) Units in the primary auditory cortex of the Japanese monkey can demonstrate a conversion of temporal and place pitch in the central auditory system, J. de Phys., Suppl. au no. III, Vol. 4, C5-419-425.
25. Slaney, M. & Lyon, R.F. (1991) Apple Hearing Demo Reel, Apple Comp. Inc.Apple, Tech. Rep. no. 25.
26. Strickland, E.A. & Viemeister, N.F. (1994) What aspects of the enveloppe are relevant for detection of amplitude modulation?, JASA, 95, 2964.

12

OLFACTORY SENSORY TRANSDUCTION

Marie-Christine Broillet and Stuart Firestein

Department of Biological Sciences
Columbia University
New York, New York 10027

1. INTRODUCTION

Organisms continuously monitor their environment, employing highly specialized systems such as vision, hearing and olfaction to extract pertinent information from an array of physical stimuli. These physical sensations are then transformed into transduction events.

The olfactory perception, which is the topic of this short review, is the result of a complex cascade of biochemical and electrophysiological processes. Recent evidence suggests that for olfaction similar mechanisms are utilized by all vertebrates. Chemical signals, e.g. odorants or pheromones, are perceived by the remarkably specific and highly sensitive olfactory systems (the olfactory epithelia of vertebrates or the antennae of insects). Although, with evolution, there has been reduced dependence on this modality for critical behaviour, such as food finding or mating, the human nose is still, for example, able to detect between several thousand odorants molecules. Only the immune system surpasses the capacity of the olfactory system to recognize as many ligands.

The focus of this review will be restricted to events occurring at the level of the olfactory receptor neuron.

2. THE OLFACTORY RECEPTOR NEURON

The remarkable capacity to discriminate among a wide range of odour molecules begins at the level of the olfactory receptor neurons (ORNs). These particular neurons perform the complex task of converting the chemical information contained in the odour molecules into information contained in membrane signals and neural space (1).

In most vertebrates, the ORNs form a sensory epithelium within the nasal cavity. They are true neurons, sending an axon to the central nervous system. They have a bipolar morphology (Figure 1) with a soma diameter of 5-25 μm and a single dendrite extended to the epithelial surface. From the knob-like dendritic terminus, 10 to 12 cilia having a diameter of 0.5 μm and a length up to 100 μm are extended into the layer of mucus that lines the nasal lumen. The composition of the mucus layer, secreted by the Bowman glands and the supporting cells, has not yet been identified, but it contains proteins which are able to bind

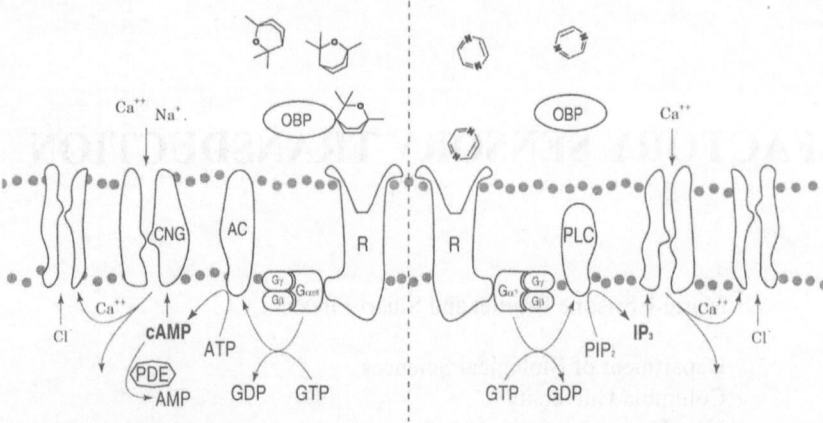

Figure 1. Schematic representation of a bipolar shaped olfactory receptor neuron. The transduction pathway (odor receptor-coupled second messenger system) occurs in the ciliary membrane resulting in the opening of cation selective channels. The influx of cations depolarizes the cell membrane, the resting potential dropping from -65 mV to -45 mV. This depolarization spreads by passive current flow through the dendrite to the soma where it activates voltage-gated Na^+ channels initiating impulse generation. The combination of the Na^+ current, voltage-dependent K^+ currents and a small Ca^{2+} current acts to produce one or more action potentials that can propagate via the axon to the olfactory bulb.

hydrophilic molecules like odours, mitogens or noxious substances, and/or to remove these molecules from the area (2).

The sensory transduction occurs at the level of the ciliary membrane. This cascade of events results in the opening of cation selective channels. The influx of cations depolarizes the cell membrane shifting the resting potential from -65 mV to -45 mV. This depolarization spreads by passive current flow through the dendrite to the soma where it activates voltage-gated Na^+ channels initiating impulse generation. The combination of Na^+ current, voltage-dependent K^+ currents and a small Ca^{2+} current acts to produce one or more action potentials that can propagate via the axon to the olfactory bulb of the brain.

3. THE SECOND MESSENGER TRANSDUCTION CASCADE

Odorant recognition involves membrane protein receptors and transduction components analogous to those which mediate the specific responses to hormones, growth factors and neurotransmitters (3). Every molecular element of the olfactory transduction cascade has been isolated, cloned and expressed allowing the establishment of the scheme presented in Figure 2. We will come back and detail some of the most important components of this pathway in the following paragraphs, but the different steps of the transduction cascade can be first summarized as follows: when a receptor molecule is occupied by an odorant, it activates a specific GTP-binding protein (G_{olf}), which modulates the activity of an adenylyl cyclase (AC type III), an enzyme producing the second messenger cAMP. cAMP directly activates a cyclic nucleotide-gated (CNG) channel representing the final step in the biochemical cascade and the first step in the generation of the electrical response. An additional, unique membrane conductance, a Ca^{2+}-activated chloride current, is also involved in the electrical response to odours. Thus, olfactory neurons use the concentration of an intracellular ion (Cl^-) to control their membrane

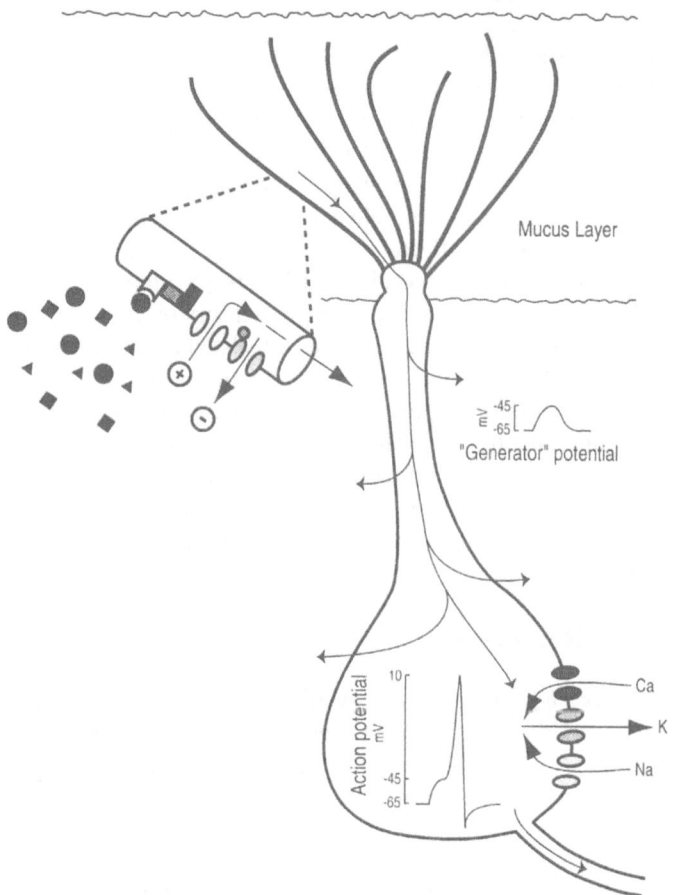

Figure 2. Olfactory transduction pathway. In this pathway, the binding of an odorant molecule (cineole is represented here) carried through the mucus layer via an odour binding protein (OBP) to the odorant receptor leads to the interaction of the receptor to a GTP-binding protein (G protein). This interaction in turn leads to the release of the GTP-coupled alpha subunit of the G protein, which then stimulates the adenylyl cyclase (AC) to produce elevated levels of cAMP. The increase of cAMP opens cyclic nucleotide gated channels (CNG) causing an alteration of the membrane potential. In some species or with some types of odorants (pyrazine is represented here), a phospholipase C-IP3 pathway is activated acting on a Ca^{2+} channel in the plasma membrane. In some species also, Ca^{2+} entering via the CNG channel or via the Ca^{2+} channel of the IP3 pathway gates a Cl⁻ channel that contributes substantially to the sensory response.

potential. Indeed, the transduction machinery of the olfactory neuron is rather peculiar among sensory neurons as a consequence of its direct contact with the external environment. Because of this direct contact, the concentration of ions in the mucus will vary from a dry day to a humid day, resulting in variations in the extracellular Na^+ concentration. These cells have therefore developed an alternative mechanism involving their unusually high intracellular Cl⁻ concentration. Cl⁻ leaves the cells via Ca^{2+}-activated Cl⁻ channels, creating a negative outward current in the presence of an elevated intracellular Ca^{2+} concentration. This anionic current, which can be as large as the cationic current through the CNG channels, causes an additional membrane depolarization (4) and is also voltage dependent (5).

3.1. The Odorant Molecules

Odorant molecules are hydro- or lipophilic molecules of low molecular weight and varied chemical structures. A slight alteration in their structure can lead to profound changes in the perceived odour quality. One commonly cited example is carvone whose L- and D-stereo isomers are perceived as spearmint and caraway, respectively. Other similarly subtle modifications can also generate striking changes in perception.

Aquatic animals smell water-soluble odorants, such as amino-acids and nucleotides which have direct access to the olfactory receptor neurons. Terrestrial animals smell, in addition, small volatile lipophilic molecules which must pass through an aqueous medium (mucus, sensilla lymph) in order to reach their chemosensory receptors (6, 7). The discovery of small, water soluble, odorant-binding proteins (OBPs) abundant in the sensillum lymph of various insect species (8, 9), as well as in the nasal mucus of cow (10), rat (11) and frog (12), led to the concept that OBPs might enhance the capture rate of volatile odour molecules by helping the partitioning of the hydrophilic odorants into the aqueous environment surrounding the sensory neurons. Furthermore, OBPs are supposed to keep the lipophilic odorants in solution and shuttle them to their receptors. It has been suggested that the acquisition of OBPs may represent one of the molecular adaptations that animals evolved to deal with terrestrial life (13).

3.2. The Odorant Receptors

In 1991, Buck and Axel (14) discovered a multigene family coding for odorant receptors in rat ORNs. This family of genes is enormous, containing hundreds of individual genes and it is also extremely diverse. Although odorant receptors share some common motifs, they are very heterogeneous in amino acid sequence. This diversity is consistent with the task of recognizing a wide variety of structurally diverse odour ligands (15). The human and mouse odorant receptor gene families contain 500-1000 genes while that of the catfish, for example, has only a 100 genes (16). The odour receptors do not contain introns and exist separately on the genome. The expression of these genes is restricted to the olfactory epithelium and testis (17).

The receptors resemble other G-protein coupled receptors which are characterized by having seven transmembrane spanning domains (14). The coupling with G-proteins occurs in the intracellular loop between the transmembrane domains TM5 and TM6, whereas odour-ligand binding takes place in a highly variable region formed, in the plane of the membrane, by the combination of transmembrane domains TM3, TM4 and TM5. The huge size and diversity of the odorant receptor family suggest that each odorant receptor might be specific for one or few odorants. A rapid comparison of the total number of odorant receptor genes in the human genome (about 500) with the estimated number of odours that can be discriminated (about 10000) also implies that each odorant receptor should be able to interact with a small number of different odorants.

3.3. The Enzymatic Activities

Experiments using an isolated cilia preparation and, later, rapid kinetic measurements have demonstrated an increase in adenylyl cyclase activity (18, 19) and a linear increase in cAMP concentration (20, 21) immediately after the application of odours. The maximum increase in cAMP level occurs 50 ms after the beginning of the odour exposure and returns to baseline within 300 ms, suggesting a pulse-like production of cAMP.

These results implicate a G protein-coupled pathway which is able to generate the second messenger cAMP. A G protein (G_{olf}) coupled to the odorant receptor has been identified (22). It is an isoform of G_s which is specific to olfactory neurons and present in very high concentration in the olfactory cilia.

The adenylyl cyclase, called adenylyl cyclase type III (23), is also specific to olfactory neurons. It has a very low basal activity in the absence of the odour stimulus, but can very rapidly generate large concentrations of cAMP upon stimulation (24).

3.4. The Odour-Induced Current

Nakamura and Gold (1987,(25)) using excised patches of cilia membrane, have recorded macroscopic ionic currents that are directly activated by cAMP. The channel responsible for this conductance has been cloned and identified in several species (26-28). Cyclic nucleotide-gated channels (CNG channels) form a family of ion channels that are structurally related to voltage gated channels. They require the binding of at least 3 cyclic nucleotide molecules for activation (29). Although they have recently been identified in an assortment of cell types and tissues (30), they are most prevalent in the peripheral sensory receptor cells of the visual and olfactory systems where they play a crucial role in transducing sensory information into changes in membrane potential (31). In olfactory neurons, these channels can be activated by either cAMP (K_d = 20 μM) or cGMP (K_d = 5 μM), although it is generally believed that under normal physiological conditions it is a rise in intracellular cAMP that is responsible for channel activation (1, 32). A high density of CNG channels is present on the ciliary membrane. These channels are selective for cations and their activation would lead to cell membrane depolarization (Figure 3).

The odour-induced current possesses the following characteristics : a long (150-450 ms) concentration dependent latency between the binding of the odour molecule and the activation of the current; a peak current amplitude which is sigmoidal and can be fitted with the Hill equation giving a Hill coefficient of 2-4; a narrow operating range; the ability to integrate stimulus detection over time, and a response that adapts to maintained stimulation (33, 34).

Recently it has been shown that the olfactory system, as many sensory systems, has evolved a signal detection sensitivity of single events in analogy with the single photon for photoreceptors, or the single pheromone molecule for insects. Quantal event types of responses of about 0.3 to 1 pA of current have been recorded in salamander olfactory neurons (35) possibly representing the binding of a single odorant molecule (Figure 4).

3.5. Adaptation of the Odour-Induced Response

The macroscopic current measured during application of sustained odour stimulus returns to baseline in 4 to 5 seconds despite the continuous presence of the stimulus. This phenomenon is characteristic of many signalling systems and is usually called adaptation or desensitization depending on the site of the off mechanism.

In the olfactory neurons, a feed-back meachanism exists in which Ca^{2+} ions and cAMP might play a role. The CNG cations channels are Ca^{2+} permeant (36). Thus, an increase in channel activity results in an influx of calcium ions and a transient rise in the intracellular concentration of calcium. As there are no organelles in the cilia, there is no storage compartment for Ca^{2+}. CNG channels have different open probabilities in the presence of a low or high intracellular calcium concentration. The 0.65 open probability generally recorded in low calcium decreases to 0.1 in a high calcium concentration (1-3 μM) (37). The mechanism could be a direct effect of calcium ions on the channel or could be mediated by a calcium binding protein like Ca-Calmodulin (Ca-CaM); these results are still controversial. Kramer and Siegelbaum (1992, (38)), working on catfish channels reconstituted in *Xenopus* oocytes, have found that intracellular Ca^{2+} has a marked inhibitory effect on the CNG

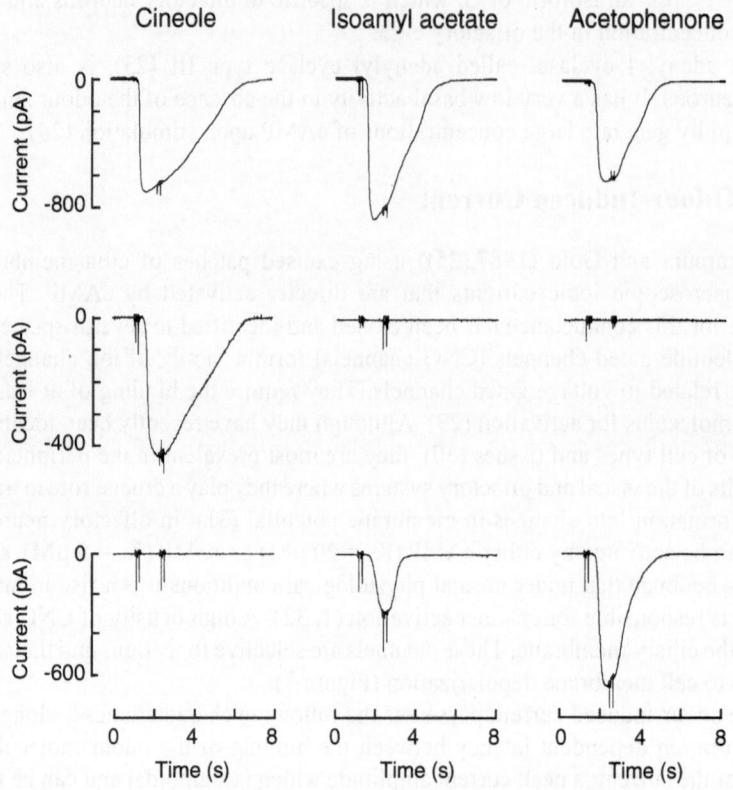

Figure 3. Current responses of three different olfactory neurons from salamander (*Ambystoma tigrinum*) to 5 x 10^{-4} M of the indicated odorants. Downward deflections of the current traces denote positive current flowing into the cell. An electrical artifact, caused by the stepping motor of the perfusion system, is indicative of the on- and offset of the stimulus. Holding potential: -55 mV. From (34) with permission.

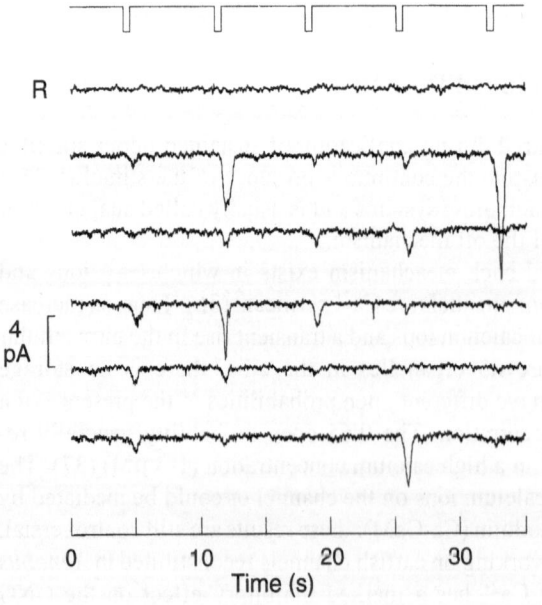

Figure 4. Responses of repeated exposures to short pulses of odorant illustrating the quantal nature of odorant response. Holding potential : -50 mV. Details of the procedures can be found in Menini et al. (1995, (35)). From (35) with permission.

channels distinct from the one described earlier on the voltage dependent open channel block (25, 37). It reduces the sensitivity of CNG channels for cAMP, but, instead of acting directly on the channel, Ca^{2+} appears to act via a protein intermediate. This was confirmed by Liu and Yau in 1994, (39) with their demonstration that the modulation of the CNG channel activity by Ca-CaM is caused by the direct binding of Ca-CaM to a specific domain at the N-terminus of the channel. This binding reduces the apparent affinity of the channel for cyclic nucleotides (cAMP affinity decreases 20 times in the presence of 235 nM Ca-CaM).

Other actions of Ca^{2+} that would lead to an adapted state include activation of the phosphodiesterase (PDE) (40) and possibly inhibition of adenylyl cyclase. However, calcium also activates the Ca^{2+}-activated Cl^- channels creating the above mentioned depolarizing Cl^- current. The significance of these apparently opposite actions of Ca^{2+} still needs to be resolved.

A second feed-back pathway for adaptation may utilize cAMP and a phosphorylation step, as it has been described for the β-adrenergic receptor or rhodopsin (41, 42). cAMP not only activates the CNG channels but also a kinase (PKA) that inhibits some early steps of the transduction cascade. The addition of a fragment of the PKA antagonist, WIPTIDE, prolongs the odour-induced production of cAMP (43). Recently an isoform of βARK (β adrenergic receptor kinase) showing great homology to the rhodopsin kinase has been identified in olfactory neurons (42). It phosphorylates the receptor only when the receptor is occupied by the ligand. Odours also induce phosphorylation of ciliary proteins using both the protein kinases A and C (43), and it is therefore possible that cAMP generated in response to odour stimulation may itself activate a negative feed-back pathway to stop the electrical response.

3.6. Alternative Second Messenger Pathway

In addition to the cAMP pathway, alternative second messenger pathways and ionic conductances might be involved in odour transduction. Inositol-1,4,5-triphosphate (IP_3) is produced by certain classes of odours (44, 45). In catfish (46), crustaceans (47) and recently in rat (48), IP_3 receptors and phospholipase C-IP_3-Ca^{2+} channels have been described, but this IP_3 gated conductance may be species specific. IP_3 sensitive channels are normally found on the membrane of intracellular compartments but as there are no organelles in the cilia, IP_3 must act at the plasma membrane. This has been shown in cultured olfactory neurons from lobster (47). In these cells, injection of IP_3 produced a current similar to the odour induced current and it was possible to obtain single channel recordings of IP_3 gated channels from plasma membrane patches.

Different odorants can selectively activate different second messengers. Odorants having the same fruity flavour like citralva or lyral induce an increase in different second messengers, so that an increase in cAMP is observed with citralva whereas lyral activates IP_3 without affecting the cAMP level (44, 45). This principle has been confirmed for a number of compounds with different odour quality, including floral, herbaceous and putrid odorants. IP_3 may be involved in signalling specific odour types, particularly non volatiles, and may be more important in aquatic animals where odours must be soluble substances.

4. DISCUSSION

4.1. Critical Issues Still to Be Resolved

Significant advances in our understanding of the olfactory system have occurred at a rapid pace over the last several years. While this knowledge has provided some very critical insights, it also serves to further focus the remaining critical questions. Among the most immediate are:

- How many receptors are expressed in one olfactory neuron?
- How many odours are recognized by one odour receptor?
- What part of an odour molecule is recognized by the receptor?
- What would be the role and physiological importance of other second messengers such as the gaseous NO or CO or even cGMP in the transduction cascade?
- What about the presence and importance of the IP_3 pathway in non aquatic vertebrates olfactory neurons?
- What are the mechanisms responsible for adaptation?
- What are the respective roles of Ca^{2+}, cAMP and phosphorylation?

These few issues are only restricted to the level of the olfactory neurons, higher processing in the brain, organization in the cortex or plasticity being other of the many unresolved issues in olfaction.

ACKNOWLEDGMENTS

We gratefully thank Nathan Tableman for assistance in producing the figures.

REFERENCES

1. Shepherd GM (1994) Discrimination of molecular signals by the olfactory receptor neuron. Neuron 13:771-790
2. Dahl A (1988) The effect of cytochrome P450-dependent metabolism and other enzyme activites on olfaction. In: Margolis TV (ed) Molecular Neurobioloy of the Olfactory System. Plenum Press, New York, pp 51-70
3. Hibert MF, Trump-Kallmeyer S, Bruinvels A, Hoflack J (1991) Three-dimensional models of neurotransmitter G-binding protein coupled receptors. Molecular Pharmacology 40:8-15
4. Kurahashi T, Yau K-W (1993) Co-existence of cationic and chloride components in odorant-induced current of vertebrate olfactory receptor cells. Nature 363:71-74
5. Firestein S, Shepherd GM (1995) Interaction of anionic and cationic currents leads to a voltage dependence in the odor response of olfactory receptor neurons. Journal of Neurophysiology 73:562-567
6. Kaissling K-E (1986) Chemo-electrical transduction in insect olfactory receptors. Annual Review of Neuroscience 9:121-145
7. Lancet D (1986) Vertebrate olfactory reception. Annual Review of Neuroscience 9:329-355
8. Kaissling K-E, Thorson J (1980) Insect olfactory sensilla:structural, chemical and electrical aspects of the functional organization. In: Satelle DB, Hall ML, Hildebrand JG (ed) Receptors for neurotransmitters, hormones and pheromones in insects. Elsevier/North-Holland, Amsterdam, pp 261-282
9. Vogt RG, Riddiford LM (1981) Pheromone binding and inactivation by moth antennae. Nature 293:161-163
10. Bignetti E, Cavaggioni A, Pelosi P, Persaud KC, Sorbi RT, Tirindelli R (1985) Purification and characterization of an odorant-binding protein from cow nasal tissue. European Journal of Biochemistry 149:227-231
11. Pevners J, Sklar P, Snyder SH (1985) Characterization of an odorant-binding protein from bovine and rat nasal mucosa. Proc. Natl Acad. Sci. 83:4942-4946
12. Lee KH, Wells RG, Reed RR (1987) Isolation of an olfactory cDNA: similarity to retinal-binding protein suggest a role in olfaction. Science 1053-1056
13. Vogt RG, Rybczynski R, Lerner MR (1990) The biochemistry of odorant reception and transduction. In: Schild D (ed) Chemosensory Information Processing. Springer Verlag, Berlin, pp 33-76
14. Buck LB, Axel R (1991) A novel multigene family may encode odorant receptors: A molecular basis for odor recognition. Cell 65:175-187

15. Ressler KJ, Sullivan SL, Buck LB (1994) A molecular dissection of spatial patterning in the olfactory system. Current Opinion in Neurobiology 4:588-596

16. Ngai J, Dowling MM, Buck LB, Axel R, Chess A (1993) The family of genes encoding odorant receptors in the channel catfish. Cell 72:657-666

17. Parmentier M, Libert F, Schurmans S, Schiffmann S, Lefort A, Eggerickx D, Ledent C, Mollereau C, Gerard C, Grootegoed A, Vassart G (1992) Expression of members of the putative olfactory receptor gene family in mammalian germ cells. Nature 355:453-455

18. Pace U, Hanski E, Salomon Y, Lancet D (1985) Odorant-sensitive adenylate cyclase may mediate olfactory reception. Nature 316:255-258

19. Sklar PB, Anholt RRH, Snyder SH (1986) The odorant sensitive adenylate cyclase of olfactory receptor cells: Differential stimulation by distinct classes of odorants. Journal of Biological Chemistry 261:15538-15543

20. Breer H, Boekhoff I, Tareilus E (1990) Rapid kinetics of second messenger formation in olfactory transduction. Nature 345:65-68

21. Restrepo D, Boekhoff I, Breer H (1993) Rapid kinetic measurements of second messenger formation in olfactory cilia from channel catfish. The American Journal of Physiology 264:c906-c911

22. Jones DT, Reed RR (1989) G_{olf}: An olfactory neuron specific-G protein involved in odorant signal transduction. Science 244:790-795

23. Krupinski J, Coussen F, Bakalyar HA, Tang W-J, Feinstein PG, Orth K, Slaughter C, Reed RR, Gilman AG (1989) Adenylyl cyclase amino acid sequence: Possible channel- or transporter-like structure. Science 244:1558-1564

24. Bakalyar HA, Reed RR (1990) Identification of a specialized adenylyl cyclase that may mediate odorant detection. Science 250:1403-1406

25. Nakamura T, Gold GH (1987) A cyclic-nucleotide gated conductance in olfactory receptor cilia. Nature 325:442-444

26. Dhallan RS, Yau K-W, Schrader KA, Reed RR (1990) Primary structure and functional expression of a cyclic nucleotide-activated channel from olfactory neurons. Nature 347:184-187

27. Ludwig J, Margalit T, Eismann E, Lancet D, Kaupp UB (1990) Primary structure of cAMP-gated channel from bovine olfactory epithelium. FEBS Letters 270:24-29

28. Goulding E, Ngai J, Kramer R, Colicos S, Axel R, Siegelbaum S, Chess A (1992) Molecular cloning and single-channel properties of the cyclic nucleotide-gated channel from catfish olfactory neurons. Neuron 8:45-58

29. Zufall F, Firestein S, Shepherd GM (1991) Analysis of single cyclic nucleotide-gated channels in olfactory receptor cells. Journal of Neuroscience 11:3573-3580

30. Yau K-W (1994) Cyclic nucleotide-gated channels: An expanding new family of ion channels. Proc. Natl. Acad.Sci. 91:3481-3483

31. Kaupp UB, Altenhofen W (1992) Cyclic nucleotide-gated channels of vertebrate photoreceptor cells and olfactory epithelium. In: Corey DP, Roper SD (ed) Sensory Transduction. Rockefeller Univ Press, New York, pp 133-150

32. Firestein S, Zufall F (1994) The cyclic nucleotide-gated channel of olfactory receptor neurons. Seminars in Cell Biology 5:39-46

33. Firestein S (1992) Physiology of transduction in the single olfactory sensory neuron. In: Corey DP, Roper SD (ed) Sensory Transduction. Rockefeller Univ Press, New York. pp 61-71

34. Firestein S, Picco C, Menini A (1993) The relation between stimulus and response in olfactory receptor cells of the tiger salamander. Journal of Physiology 468:1-10

35. Menini A, Picco, C, and Firestein, S (1995) Quantal-like current fluctuations induced by odorants in olfactory receptor cells. Nature 373:435-437

36. Kurahashi T (1989) Activation by odorants of cation-selective conductance in the olfactory receptor cell isolated from the newt. Journal of Physiology 419:177-192

37. Zufall F, Shepherd GM, Firestein S (1991) Inhibition of the olfactory cyclic nucleotide-gated ion channel by intracellular calcium. Proc Roy Soc Ser B 246:225-230

38. Kramer RH, Siegelbaum SA (1992) Intracellular Ca^{2+} regulates the sensitivity of cyclic nucleotide-gated channels in olfactory receptor neurons. Neuron 9:897-906

39. Liu M, Chen T-Y, Basheer A, Li J, Yau K-W (1994) Calcium-calmodulin modulation of the olfactory cyclic nucleotide-gated cation channel. Science 266:1348-1354

40. Borisy FF, Ronnett GV, Cunningham AM, Julifs D, Beavo J, Snyder SH (1992) Calcium/calmodulin-activated phosphodiesterase expressed in olfactory receptor neurons. Journal of Neuroscience 12:915-923

41. Dohlman HG, Torner J, Caron MG, Lefkowitz RJ (1991) Model systems for the study of seven-trans-membrane-segment receptors. Annual Review of Biochemistry 60:653-688

42. Dawson TM, Arriza JL, Jaworsky DE, Borisy FF, Attramadal H, Lefkowitz RJ, Ronnett GV (1993) B-adrenergic receptor kinase-2 and B-arrestin-2 as mediators of odorant-induced desensitization. Science 259:825-829

43. Breer H, Boekhoff I, Krieger J, Raming K, Strotmann J, Tareilus E (1992) Molecular mechanisms of olfactory signal transduction. In: Corey DP, Roper SD (ed) Sensory Transduction. Rockefeller University Press, New York. pp 93-108

44. Boekhoff I, Breer H (1990) Differential stimulation of second messengers pathways by distinct classes of odorants. Neurochemistry International 17:553-557

45. Boekhoff I, Tareilus E, Strotmann J, Breer H (1990) Rapid activation of alternative second messenger pathways in olfactory cilia from rats by different odorants. EMBO Journal 2453-2458

46. Kalinoski DL, Aldinger SB, Boyle AG, Huque T, Marecek JF, Prestwich GD, Restrepo D (1992) Characterization of a novel inositol 1,4,5-triphosphate receptor in isolated olfactory cilia. Biochemical Journal 281:449-456

47. Fadool DA, Ache BW (1992) Plasma membrane inositol 1,4,5-trisphosphate-activated channels mediate signal transduction in lobster olfactory receptor neurons. Neuron 9:907-918

48. Okada Y, Teeter JH, Restrepo D (1994) Inositol 1,4,5-trisphosphate-gated conductance in isolated rat olfactory neurons. Journal of Neurophysiology. 71:595-602

13

THE VOMERONASAL ORGAN

A. Cavaggioni,[1] Carla Mucignat-Caretta,[2] G. Sartor,[3] and R. Tirindelli[2]

[1] Istituto di Fisiologia Umana
Università di Padova
53131 Padova, Italy
[2] Istituto di Fisiologia Umana
[3] Istituto di Chimica Biologica
Università di Parma
43100 Parma, Italy

1. INTRODUCTION

The vomeronasal (VN) organ of Jacobson is the receptor organ of the accessory olfactory system (AOS). The VN organ consists mainly of an olfactory epithelium (OE) and a vascular pump. The OE is characterised by the presence of primary receptor neurons whose axons form the VN nerve and contact the secondary neurons in the AO bulb (AOB). Hence the neural outflow goes to brain structures controlling the pituitary-gonadal axis and the reproductive behaviour. The adequate stimuli for the VN organ are volatile as well as non-volatile substances. They are driven toward the primary VN neurons through a blind ended canal, the VN canal, by the vascular pump. The VN system is important in chemical communication within the species and the majority of the effects driven by the activity of this system in mammals gather under the umbrella of reproduction. The evidence from molecular biology, biochemistry, histology, physiology and behaviour studies singles out the VN system as a distinct system when compared to the main olfactory system.

The VN system is present in almost all terrestrial vertebrates (Stoddart, 1980, Wysocki and Meredith, 1987, Halpern, 1987, Harrison,1987, Eisthen, 1992). Absent in birds it is well developed in many reptiles. Scleroglossa reptilia have evolved a flicking tongue which brings the stimuli from outside to the VN organ in the mouth cavity (Halpern and Kubie, 1984, Halpern, 1987, Wang, Chen, Inouchi, 1993). The VN system is absent in some Iguana reptilia in which the tongue is used for food prehension, e.g., chamaeleonids. It is present in most mammal species to the exception of cetaceans and some bat species (Cooper and Bhatnagar, 1976). Well developed in the "flying lemur" *Cynocephalus* (Bhatnagar and Wible, 1994), in the prosimian *Tupaia glis* and shown also to be active in the primitive primate *Microcebus murinus*, it is less developed in platirrhine and vestigial or rudimentary in catarrhine monkeys (Schilling, Serviere, Gendrot and Perret, 1990). Its presence as a functioning entity in man is a matter of debate (Garcia-Velasco and Mondragon,1991, Moran, Jafek and Rowley, 1991, Stensaas, Lavker, Monti-Bloch, Grosser and Berliner, 1991).

In phylogenesis it can be traced back to amphibians as early as in the larval aquatic stage (Schmidt and Roth,1990, Burton, Cogan and Borror, 1990, Franceschini, Sbarbati and Zancanaro, 1991, Jones, Pfeiffer and Asashima, 1944) suggesting that the VN system evolved before adaptation to the terrestrial ecological niche. No VN system, however, has been so far recognised as distinct from the main olfactory system in fishes nor even in African dipnoi pulmonate fishes (von Berthold, Claas, Munz and Meyer, 1988).

The primary neurons of the VN organ originate in the embryo from the VN olfactory placode. This placode gives also origin to the terminal nerve. No sensory function has been ascertained for the terminal nerve in adult life, but cells containing the releasing factor for the luteinizing hormone (LHRH) originate in this placode and migrate to the septal area of the brain and hypothalamus during embryo genesis. This migration occurs along a track of cell adhesion molecules, namely NCAM molecules (Schwanzel-Fukuda, Zheng, Bergen, Weesner and Pfaff, 1992) and is probably driven by chemical factors released by target brain areas (Daikoku, Daikoku-Ishida, Okamura, Chikumori, Aoyama and Yokote, 1991). The origin of LHRH expressing neurons is an early commitment of this placode to the regulation of the reproductive process. During embryo genesis the VN organ originates first as a pit in the placode which then assumes the form of a closed end narrow tube whose neuroepithelium ends up by lining the medial side of the VN canal in many mammals (Mendoza and Szabo, 1988). Although the VN organ is probably a functional organ at birth, its maturation goes on for the first few weeks after birth, a longer period as compared to the main olfactory system (Coppola, Budde and Millar, 1933). There is a remarkable conservation of the neuroepithelium structure among different species. Sensory, supporting and basal cells are present as in the main olfactory neuroepithelium. There are differences, however (Bannister and Dodson, 1992). The primary sensory neurons are characterised by microvillar extensions at their apical pole which extend in the fluid filling the VN canal and do not have cilia. In addition, the cytoplasm is rich in agranular endoplasm reticulum suggesting a rapid membrane turnover and calcium stores. The olfactory glands of Bowmann which are a major element of the main olfactory mucosa are not present under the VN epithelium, although a few glands are present in the organ close to the vascular pump. Olfactory neurons of the VN olfactory epithelium are renewed throughout life, the dead cells being replaced by new neurons which originate by division of stem cells. After transection of the VN nerve and retrograde degeneration of the neurons, the olfactory epithelium is regenerated in about 30 days (Barber and Raisman, 1978). This property, shared by all olfactory epithelia, is due to neurons that do not enter the post mitotic pause as other neurons do. This property is remarkable in that the newly formed axons diligently find their way to make adequate synaptic contacts in the AOB.

The secondary neurons of the VN system are in the AOB. The segregation of secondary neurons that receive signals related to sex of conspecifics is a feature common to vertebrate species with a VN organ as well as to insects that have sensilla specialised for the detection of pheromones (Masson and Mustaparta, 1990).

The structure of the AOB is characterised by layers: The fiber layer, the glomerular, outer plexiform, mitral, inner plexiform, and granular layer. The internal connections of the AOB are similar to the main olfactory bulb. Briefly, there is a monosynaptic path from olfactory afferent fibre to the dendrites of mitral cells, and a major inhibitory feedback via dendro-dendrite reciprocal synapses between mitral and granule interneurons. Glutamate is probably the excitatory mediator of the olfactory afferent (Dudley and Moss, 1995) and GABA the inhibitory mediator at the dendro-dendrite synapse between granule and mitral cell (Kaba, Hayashi, Higuchi and Nakarashi, 1994). It should be realised that the inhibitory feedback may confer very complex dynamic properties to the olfactory bulb. The absence of a clear cut topological relation between the epithelium and its representation in the AOB as well as the great convergence of signals from many primary neurons into a single mitral cell make unlikely that the AOB operates an analysis of the olfactory signals according to the principles of parallel and hierarchical organisation as described in the visual cortex. The AOB is also characterised by the presence of hypothalamic releasing hormones (LHRH and thyrotropin releasing hormone) as well as dopamine and norepinephrine (NE). Interestingly the levels of LHRH and dopamine are higher in rodents exposed to sex-related odours (Wang and Tsai, 1991). Clearly the AOB is not a simple relay station for olfactory signals but a complex neurochemical integration centre.

In mammals the outflow of the AOB is directed mainly to the medial and posteromedial cortical amygdaloid nuclei. These nuclei project to lower structures, the bed nucleus of the stria terminalis, the medial preoptic and ventromedial hypothalamic nuclei. The tubero-infundibular neurons of the arcuate nucleus are also a target of the olfactory outflow of the VN system and they play a key role in controlling the release of hypothalamic factors and hormones.

It thus appears that the VN outflow in the AO system is well placed to control the reproductive behaviour and the brain-pituitary-gonadal axis. In turn, the AO system is under the influence of sex hormones; the VN organ, the AOB and AO system are sexually dimorphic in size in adult rodents, having male conspecifics greater structures with more neurons (Segovia and Guillamon, 1982, 1993, Collado, Segovia, Cales, Perez-Laso, Rodriguez-Za-fra, Guillamon and Valencia, 1992). The dimorphism can be eliminated either by male castration or female treatment with androgens on the day of birth. Androgens are probably converted to estradiol as estradiol receptors are present in the neurons of the AO system (Koch, 1990). These receptors probably induce the sexual dimorphism in the critical period after birth but the mechanism is not known. Estrogens have also short term effects. They may enhance the excitatory transmission in the amygdala from AOB to tubero-infundibular neurons (Kaba, Saito and Seto, 1992). Estradiol receptors in pup rats are upregulated by parental care (Modney, Yang and Hatton, 1990) in the AO system: An observation which highlights the complexity of this regulation. Furthermore, only in lactating rats, stimulation of the AOB enhances the dendro-dendrite coupling between neurons of the supraoptic nucleus of the hypothalamus (Hatton and Yang, 1990), showing that olfactory stimuli may have different effects depending on the endocrine condition of the animal.

In conclusion there seems to be a complex reciprocal relationship between the AO system and reproduction.

2. THE PUMPING MECHANISM

The VN organ is open either in the mouth as in snakes or in the nasal cavity in proximity of the nasopalatine duct as in most mammals. Non volatile substances as fluorescent dyes dispersed with conspecific urine have access to the VN canal of rodents (Wysocki, Wellington and Beauchamp, 1980). This may occur through licking and carrying the substances in close proximity of the duct opening when they are pumped within the canal. The lateral aspect of the canal, which is not lined by neuroepithelium, covers a cavernous tissue which undergoes cycles of swelling and emptying with blood thereby compressing and decompressing the lumen. This moves back and forth the lumen fluid helping substances to harbour close to receptors. This vascular pump is activated in a reflex way by olfactory stimuli as well as by any stimulus that arouses the animal interest (Meredith and Fernandez-Fewell, 1994, Meredith, 1994).

3. BIOCHEMISTRY

There are differences in the protein asset of the VN neurons as compared to neurons of the main olfactory system (Ichikawa, Osada and Kai, 1992, Abe, Watanabe and Kondo,

1992, Krishna, Getchell and Getchell, 1992, Kishimoto, Keverne and Emson, 1993). In *Xenopus laevis*, monoclonal antibodies raised against plexin react with an antigen in VN neurons (Satoda, Takagi, Ohta, Hirata and Fujiama, 1995), and so do anti-calbindin antibodies in the rat (Johnson, Eller, Jafek and Norman, 1992). In the rat, different agglutinins bind to VN cells showing a peculiar glycoprotein expression. The putative olfactory molecular receptors of the VN primary neurons have not been identified as yet. It would be interesting to know whether they are similar to those described in fish (Ngai, Dowling, Buck, Axel and Chen, 1993) or in the main olfactory epithelium of airborne species (Danty, Cornuet and Masson, 1994). The sensory transduction of the VN organ has not been studied as yet, there is no reason to assume that it is identical to the main olfactory neuroepithelium and we have experimental data heading in this direction. Also the fluid which fills the VN canal has a different protein composition as compared to the mucus which streams over the main olfactory epithelium. A protein, vomeromodulin and two lipocalins have been shown to be particularly abundant in the fluid of the VN canal of rodents (Khew-Goodall, Grillo, Getchell, Danho, Getchell and Margolis, 1991, Miyavaki, Matsushita, Ryo and Mikoshiba, 1994). The VN perireceptor ambient and possibly the events there occurring are likely to be peculiar of the VN organ and functional to the detection of water soluble pheromones (Gaupp, 1902, Broman, 1920, Getchell, Margolis and Getchell, 1984).

4. VN PRIMER EFFECTS

The VN organ detects priming pheromones, namely chemicals released by conspecifics that elicit endocrine responses (Milligan, 1980, Keverne, 1983, Brennan, Kaba and Keverne, 1990, Wysocki and Lepri, 1991, Hatanaka, 1992, Meredith and Fernandez-Fewell, 1994). The best known primer effects in mice are the suppression of oestrus in group-housed females (Lee-Boot effect), the induction of oestrus and oestrus synchrony by male urine of conspecifics (Whitten effect), the acceleration of puberty in female mice by adult male urine (Vendenbergh effect) and the pregnancy block by the odour of a "strange" male (Bruce effect). These four effects seem mediated by the hypothalamic factors, LHRH and prolactin inhibiting factor, as suggested by LH and prolactin levels in blood (Meredith and Howard, 1992). Their release is driven by the dopaminergic terminals of tubero-infundibular neurons, which are therefore central in this regulation. So the Lee-Boot effect is caused by a fall of prolactin induced by odorant stimuli of females that act on female mice giving suspension of the oestrus cycle; the Whitten effect is caused by a surge of LH by odorant stimuli of male mice that induce oestrus; the Vandenbergh effect is due to an LH surge in prepuberal female mice which triggers early puberty. The adequate stimuli for this latter effect have been studied in some detail. The stimuli are present in adult male urine, are non-volatile and their apparent molecular weight ranges from 18000 Da to 860 Da (Vandenbergh, Finlayson, Dobrogosz, Dills and Kost, 1776). The most parsimonious interpretation of a series of biochemical characterisations is that proteins and peptides are in cause. Adult male urine has an unusually high concentration of proteins, the Major Urinary Protein (MUP) complex Fig.1.

This belongs to the lipocalin superfamily (Samson, North and Sawyer, 1994) a phylogenetically old family of secretory proteins whose members share a similar hydrophobic binding pocket, but are otherwise diversified as to the function. In mice the MUP is expressed in the liver under androgen induction by a large polymorphic multigene family located in chromosome 4 (Clark, Ghazal, Bingham, Barrett and Bishop, 1985). They can be filtered by the kidney and be present in urine because of their relatively small molecular weight. Also a MUP-related peptide is probably secreted in urine as some of the MUP genes have a stop codon in the DNA coding sequence and could express a short peptide only. It is likely that the N-terminal part of MUP and the related peptide are recognised by a VN

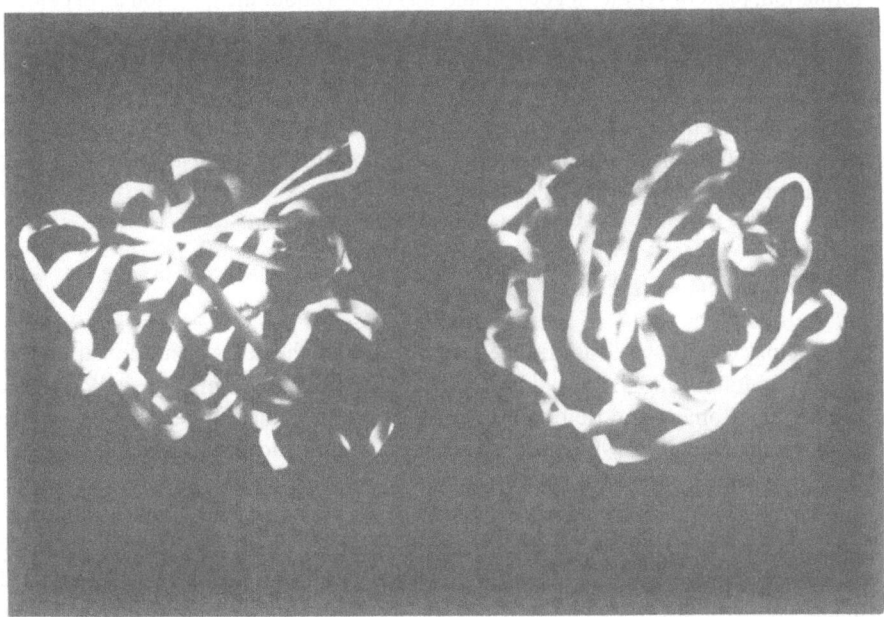

Figure 1. Peptide backbone of MUP with a bound pheromone. The peptide backbone is shown as a yellow ribbon. The barrel like structure formed by eight beta-sheets, typical of lipocalins, and a short alpha-helix are visible. In the left hand image the N-terminal is in the lower part of the molecule. The right hand image is taken facing the open end of the hydrophobic cavity or pocket of the barrel structure.The bound pheromone (2-*sec*-butyl-thiazoline) is shown within the pocket with C-atoms in white, S-atom in yellow and N-atom in blue color. Obtained from X-ray diffraction data (Bocskei, Groom, Flower, Wright, Phillips, Cavaggioni, Findlay and North, 1992).

molecular receptor protein (Mucignat-Caretta, Caretta and Cavaggioni, 1995). MUP binds volatile odorants in the mouse body (Bacchini, Gaetani and Cavaggioni, 1992, Robertson, Beynon and Evershed, 1993, Pelosi, 1994) and slowly releases them in air as urine is in the field, giving the peculiar odour of mice urine. In this way the attention and the search of conspecifics is attracted toward the protein. In conclusion MUP seems to have a dual role: It stores and releases volatiles that induce searching behaviour, and it stimulates directly the vomeronasal organ triggering endocrine modifications.

The Bruce effect is a very complex VN mediated phenomenon whose nervous mechanism has been studied to some extent. This effect is the failure of fertilised eggs to implant in uterus (pregnancy block) caused by the odour of a "strange" male whose urine odour is different from the "stud" male (Brennan et al., 1990, Kaba, Rosser and Keverne, 1988). The "stud" male urine is ineffective provided the female is allowed to be imprinted by the "stud" male urine for a few hours after mating. This implies that in the critical period after fecundating a memory trace of the odour of "stud" male urine forms which prevents the pregnancy block. The memory trace takes place in the AO bulb and requires the coincidence of "stud" olfactory cues and the activity of a noradrenergic input over the critical period. At the synaptic level, pharmacological as well as lesion experiments make likely that the noradrenergic input reduces the efficacy of the GABAergic inhibitory synapse of the granule cell on the mitral cell. This effect results in a higher activity of the mitral cells and finally the formation of a trace that lasts for a few months, by means of an unknown mechanism. This trace can be understood as a selective filter which cuts off the blocking action of "stud" urine but not the "strange" urine. During the few hours necessary for the trace formation, protein synthesis goes on and this brings about the question as to control of expression in this period. Interestingly, immediate-early genes c-fos and egr-1 are expressed in the AO system during the period over which the trace forms (Baum and Everitt, 1992, Brennan, Hancock and Keverne, 1993).

The final event responsible for pregnancy block is a fall in prolactin levels with consequent reduction of corpus luteum, the endometrium does not undergo the normal changes and the egg fails to implant and develop.

5. VN EFFECTS ON BEHAVIOUR

Olfactory cues may release fast behaviours without the delay of hormonal responses (Mennella and Moltz, 1988, Bean and Wysocki, 1989, Saito Higarashi, Hokao, Wakafuji and Takahashi, 1990, Romero and Beltramino, 1990, Fleming, Gavorth and Sarker, 1992, Meek, Lee, Rogers and Hernandez, 1994). Several behaviours related to reproductive life are known to depend on the VN system. The behaviours range from aggression, pup defence, maternal care, pup recognition, androtropism and lordosis of female mice, ultrasound vocalisation of male mice. The sexual behaviour of male hamster has been studied to some depth (Singer, Clancy, Macrides, Agosta and Branson, 1988, Henzel, Rodriguez, Singer, Stults, Macrides, Agosta and Niall, 1988). Male hamsters are attracted by vaginal smears of conspecifics in oestrus. A protein, aphrodisin, has been found in the vaginal discharge which induces mating behaviour in normal hamsters but not in hamsters that had the VN organ removed. Aphrodisin belongs to the lipocalin family and is probably recognised by VN receptors. It induces mating behaviour in male hamster as a consequence of LHRH released in the hypothalamus.

6. DISCUSSION

Two lipocalins, MUP and aphrodisin are recognised by the vomeronasal organ and carry information related to sex in rodents and hamsters respectively. The anatomical position as well as the structure of this organ is well suited to detect water-bound molecules, peptides and proteins (Broman, 1920). Lipocalins, in turn, are well suited to act as a complex pheromonal system with both long distance signalling by means of the volatiles released from the hydrophobic pocket as well as for direct stimulation of VN receptor neurons. Volatile presentation by lipocalins to vomeronasal receptors is an interesting possibility. It has the advantage of the high specificity of protein-receptor interaction endorsed by the chemical signature of the volatile. Borrowing the words of immunology, MUP and aphrodisin may be ranked among the compatibility proteins, namely proteins that evolved in defence of the genetic identity of the species, and the vomeronasal organ as the response element of the reproductive system.

REFERENCES

Abe, H., Watanabe, M. and Kondo, H. (1992). Developmental changes in expression of a calcium-binding protein (spot 35 - calbindin) in the Nervus terminalis and the vomeronasal and olfactory receptor cells. Acta Otolaryngol Stockh. 112, 862-871.

Bacchini, A., Gaetani, E. and Cavaggioni, A. (1992). Pheromone binding proteins of the mouse, *Mus musculus*. Experientia 48, 419-421.

Bannister, L.H. and Dodson, H.C. (1992). Endocytic pathways in the olfactory and vomeronasal epithelia of the mouse: ultrastructure and uptake of tracers. Microsc. Res. Tech. 23, 128-141.

Barber, P.C. and Raisman , G. (1978). Replacement of receptor neurones after transection of the vomeronasal nerves in the adult mouse. Brain Res. 147, 297-313.

Baum, M.J. and Everitt, B.J. (1992). Increased expression of c-fos in the medial preoptic area after mating in male rats: role of afferent inputs from the medial amygdala and midbrain central tegmental field. Neuroscience 50, 627-646.

Bean, N.J. and Wysocki, C.J. (1989). Vomeronasal organ removal and female mouse aggression: the role of experience. Physiol. Behav. 45, 875-882.

Bhatnagar, K.P. and Wible, J.R. (1994). Observations on the vomeronasal organ of the *Colugo cynocephalus*. Acta Anatomica 151, 43-48.

Bocskei, Z., Groom, C.R., Flower, D.R., Wright, C.E., Phillips, S.E.V., Cavaggioni, A., Findlay, J.B.C. and North, A.C.T. (1992). Pheromone binding to two rodent urinary proteins revealed by X-ray crystal-lography. Nature 360, 186-188.

Brennan, P., Kaba, H. and Keverne, E.B. (1990). Olfactory recognition: a simple memory system. Science 250, 1223-1226.

Brennan, P.A., Hancock, D. and Keverne, E.B. (1992). The expression of the immediate-early genes c-fos, egr-1 and c-jun in the accessory olfactory bulb during formation of an olfactory memory in mice. Neuroscience 49, 277-284.

Broman, I. (1920). Das Organon vomero-nasale Jacobsoni - ein Wassergeruchorgan! Anat. Hefte,1 Abt., H.174, 137-191.

Burton, P.R., Coogan, M.M. and Borror, C.A. (1990). Vomeronasal and olfactory nerves of adult and larval bull frogs: I. Axons and the distribution of their glomeruli. J. Comp. Neurol. 292, 614-623.

Clark, A.J., Ghazal., P., Bingham, R.W., Barrett, D. and Bishop J.O. (1985). Sequence structures of a mouse major urinary protein gene and pseudogene compared. The EMBO J. 4, 3159-3165.

Collado, P., Segovia, S., Cales, J. M., Perez-Laso, C., Rodriguez-Zafra, M., Guillamon, A. and Valencia, A. (1992). Female's DHT controls sex differences in the rat bed nucleus of the accessory olfactory tract. Neuroreport. 3, 327-329.

Cooper, G.J. and Bhatnagar, K.P (1976). Comparative anatomy of the vomeronasal complex in bats. J.Anat. 122, 571-601.

Coppola, D.M., Budde, J. and Millar L. (1993). Vomeronasal duct has a protracted postnatal development in the mouse. J. Morphol. 218, 59-64.

Daikoku, S., Daikoku-Ishido, H., Okamura, Y., Chikamori-Aoyama, M. and Yokote, R. (1991). Further evidence of the presence of rat embryonic hypothalamic factors that induce the differentiation of gonadotropic hormone-releasing hormone-containing secretory neurons. Anat. Rec. 230, 539-550.

Danty, E., Cornuet, J.M. and Masson C. (1994). Honeybees have putative olfactory receptor proteins similar to those of vertebrates. C.R.Acad.Sci.Paris 317, 1073-1079.

Doty, R.L. (1986). Odor-guided behavior in mammals. Experientia 42, 257-271.

Dudley, C.A. and Moss, R.L. (1995). Electrophysiological evidence for glutamate as a vomeronasal receptor cell neurotransmitter. Brain Res 675 , 208-214.

Eisthen, H.L. (1992). Phylogeny of the vomeronasal system and of receptor cell types in the olfactory and vomeronasal epithelia of vertebrates. Microsc. Res. Tech. 23, 1-21.

Fleming, A.S., Gavorth, K. and Sarker, J. (1992). Effects of transections to the vomeronasal nerves or the main olfactory bulbs on the initiation and long-term retention of maternal behavior in primiparous rats. Behav. Neural. Biol. 57, 177-188.

Franceschini, F., Sbarbati, A. and Zancanaro, C. (1991). The vomeronasal organ in the frog, *Rana esculenta*. An electron microscopy study. J. Submicrosc. Cytol. Pathol. 23, 221-231.

Garcia-Velasco, J. and Mondragon M. (1991). The incidence of the vomeronasal organ in 1000 human subjects and its possible clinical significance. J. Steroid Biochem. Mol. Biol. 39, 561-563.

Gaupp, E. (1904). Anatomie des Frosches. Vieweg Verlag, Braunschweig, 673.

Getchell, T.V., Margolis, F.L. and Getchell M.L. (1984). Perireceptor and receptor events in vertebrate olfaction. Progress in Neurobiology 23, 317-345.

Halpern, M. and Kubie, J.L. (1984). The role of the ophidian vomeronasal system in species-typical behavior. TINS December 1984, 472-477.

Halpern, M. (1987). The organization and function of the vomeronasal system. Am. Rev. Neurosci. 10,325-365.

Harrison, D.(1987). Preliminary thoughts on the incidence, structure and function of the mammalian vomero-nasal organ. Acta Otolaryngol. (Stockholm) 103, 489-495.

Hatanaka, T. (1992). The mouse vomeronasal organ a sex pheromone receptor? Chemical Signals in Vertebrates VI, edited by R.L. Doty and D. Muller-Schwarze, Plenum Press, New York, 27-31.

Hatton, G.I. and Yang, Q.Z. (1990). Activation of excitatory amino acid inputs to supraoptic neurons. I. Induced increases in dye-compling in lactating, but not virgin or male rats. Brain Res. 513, 264-269.

Henzel, W.J., Rodriguez, H., Singer, A.G. Stults, J. T., Macrides, F., Agosta, W.C. and Niall, H. J. (1988). The primary structure of aphrodisin. Biol. Chem. 263, 16682-16687.

Ichikawa, M., Osada, T. and Ikai, A. (1992). *Bandeiraea simplicifolia* lectin I and *Vicia villosa* agglutinin bind specifically to the vomeronasal axons in the accessory olfactory bulb of the rat. Neurosci. Res. 13, 73-79.

Johnson, E.W., Eller, P.M., Jafek, B.W. and Norman, A.W. (1992). Calbindin-like immuno-reactivity in two peripheral chemosensory tissues of the rat: taste buds and the vomeronasal organ. Brain Res. 572, 319-324.

Jones, F.M., Pfeiffer, C. and Asashima, M. (1994). Ultrastructure of the olfactory organ of the newt Cynops-pirrhogaster. Annals of Anatomy 176, 269-275.

Kaba, H., Hayashi, Y., Higuchi, T. and Nakanishi S. (1994). Induction of an olfactory memory by the activation of a metabotropic glutamate receptor. Science 265, 262-264.

Kaba, H., Rosser, A. E. and Keverne, E.B. (1988). Hormonal enhancement of neurogenesis and its relationsship to the duration of olfactory memory. Neuroscience 24, 93-98.

Keverne, E. (1983). Pheromonal influences on the endocrine regulation of reproduction. TINS, September 1983, 381-384.

Khew -Goodall, Y., Grillo, M., Getchell, M.L., Danho, W., Getchell, T.V. and Margolis, F.L. (1991). Vomero-modulin, a putative pheromone transporter: cloning, characterization, and cellular localization of a novel glycoprotein of lateral nasal gland. FASEB J. 5, 2976-2982.

Kishimoto, J., Keverne, E.B. and Emson, P.C. (1993). Calretinin, calbindin-D28k and parvalbumin - like immunoreactivity in mouse chemoreceptor neurons. Brain Res. 610, 325-329.

Koch, M. (1990). Effects of tratment with estradiol and parental experience on the number and distribution of estrogen-binding neurons in the ovariectomized mouse brain. Neuroendocrinology 51, 505-514.

Krishna, N.S., Getchell, M.L. and Getchell, T.V. (1992). Differential distribution of gamma-glutamyl cycle molecules in the vomeronasal organ of rats. Neuroreport 3, 551-554.

Krishna, N.S.R., Getchell, M.L. and Getchell, T.V. (1994). Expression of the putative pheromone and odorant transporter vomeromodulin messenger-RNA and protein in nasal chemosensory mucosae. J. Neurosci. Res. 39, 243-259.

Li, C.S., Kaba, H., Saito, H. and Seto, K. (1992). Oestrogen infusions into the amygdala potentiate excitatory transmission from the accessory olfactory bulb to tuberoinfundibular arcuate neurones in the mouse. Neurosci. Lett. 143, 48-50.

Masson, C. and Mustaparta, H. (1990). Chemical information processing in the olfactory system of insects. Physiological Reviews 70, 199-245.

Meek, L.R., Lee, T.M., Rogers, E.A. and Hernandez, R.R. (1994). Effect of vomeronasal organ removal on behavioural estrus and mating latency. Biology of Reproduction 51, 400-404.

Mendoza, A.S. and Szabo, K. (1988). Developmental studies on the rat vomeronasal organ: vascular pattern and neuroepithelial differentiation. II . Electon microscopy. Brain Res. 467, 259-268.

Mennella, J. A. and Moltz, H. (1988). Infanticide in the male rat: the role of vomeronasal organ. Physiol. Behav. 42, 303-306.

Meredith, M. (1994). Chronic recording of vomeronasal pump activation in awake behaving hamsters. Physiol. Behav. 56, 345-354.

Meredith M. and Howard, G. (1992). Intracerebro-ventricular LHRH relieves deficits due to vomeronasal organ removal. Brain Res. Bull. 29, 75-79.

Meredith, M. and Fernandez-Fewell, G. (1994). Vomeronasal system, LHRH and sex behavior. Psychoneuroen-docrinology 19, 657-672.

Milligan, S.R. (1980). Pheromones and rodent reproductive physiology. Symp. Zool. Soc. Lond. 45, 251-275.

Miyawaki, A., Matsushita, F., Ryo, F. and Mikoshiba, K. (1994). Possible pheromone-carrier function of two lipocalin proteins in the vomeronasal organ. The EMBO J. 13, 5835-5842.

Modney, B.K., Yang Q.Z. and Hatton, G.I. (1990). Activation of excitatory amino acid inputs to supraoptic neurons. II. Increased dye-coupling in maternally behaving virgin rats. Brain Res. 513, 270-273.

Moran, D.T., Jafek, B.W. and Rowley, J. C. (1991). The vomeronasal (Jacobson's) organ in man: ultrastructure and frequency of occurrence. J. Steroid Biochem. Mol. Biol. 39, 545-552.

Mucignat-Caretta, C., Caretta, A. and Cavaggioni, A. (1995). Acceleration of puberty onset in female mice by male urinary proteins. J. Physiol. (London) 486.2, 517-522.

Ngai, J., Dowling, M.M., Buck, L., Axel, R. and Chess A. (1993). The family of genes coding odorant receptors in the channel catfish. Cell 72, 657-666.

Pelosi P. (1994). Odorant binding proteins.Critical Reviews in Biochemistry and Molecular Biology 29,199-228.

Robertson, D.H.L., Beynon, R.J. and Evershed, R.P. (1993). Extraction, characterisation, and binding analysis of two pheromonally active ligands associated with the major urinary protein of the house mouse (*Mus musculus*) .J. Chem. Ecol. 19, 1405-1414.

Romero, P.R., Beltramino, C.A. and Carrer, H.F. (1990). Participation of the olfactory system in the control of approach behavior of the female rat to the male. Physiol. Behav. 47, 685-690.

Saito, T.R. Igarashi, N., Hokao, R., Wakafuji, Y. and Takahashi, K.W. (1990). Nursing behavior in lactating rats-the role of the vomeronasal organ. Jikken-Dobutsu 39, 109-111.

Sansom, C.E., North, A.C.T. and Sawyer L. (1994). Structural analysis and classification of lipocalins and related proteins using a profile-search method. Biochimica et Biophysica Acta 1208, 247-255.

Satoda, M., Takagi, S., Ohta, K., Hirata, T. and Fujiama, T. (1995). Differential expression of 2 cell-surface proteins, neuropilin and plexin in Xenopus olfactory axon subclasses. J. Neurosci. 15, 942-955.

Schilling, A., Serviere, J. Gendrot, G. and Perret, M. (1990). Vomeronasal activation by urine in the primate *Microcebus murinus*: a 2 DG study. Exp. Brain Res. 81, 609-618.

Schmidt, A. and Roth G. (1990). Central olfactory and vomeronasal pathways in salamanders. J. Hirnforsh. 31, 543-553.

Schwanzel-Fukuda, M., Zheng, L.M., Bergen, H.,Weesner, G. and Pfaff, D.W. (1992). LHRH neurons: functions and development. Prog. Brain Res. 93, 189-201.

Segovia, S. and Guillamon, A. (1982). Effect of sex steroids on the development of the vomeronasal organ in the rat. Developmental Brain Res. 5, 209-212.

Segovia, S. and Guillamon, A. (1993). Sexual dimorphism in the vomeronasal pathway and sex differences in reproductive behaviors. Brain Res. Rev. 18, 51-74.

Singer, A.G., Clancy, A.N., Macrides, F., Agosta, W. C. and Bronson, F.H. (1988). Chemical properties of a female mouse pheromone that stimulates gonadotropin secretion in males. Biol. Reprod. 38, 193-199.

Stensaas, L.J., Lavker, R.M., Monti-Bloch, L., Grosser, B.I. and Berliner, D.L. (1991). Ultrastructure of the human vomeronasal organ. J. Steroid Biochem. Mol. Biol. 39, 553-560.

Stoddart, D.M. (1980). The ecology of vertebrate olfaction. Champman and Hall, London and New York.

Vandenbergh, J.G., Finlayson, J.S., Dobrogosz, W.J., Dells, S.S. and Kost, T.A. (1976). Chromatographic separation of puberty accelerating pheromone from male mouse urine. Biology of reproduction 15, 260-265.

Von Bartheld, C.S.,Claas, B., Munz, H. and Meyer, D.L. (1988). Primary olfactory projections and the Nervus terminalis in the African lung fish: implications for the phylogeny of cranial nerves. Am. J. Anat. 182, 325-334.

Wang, D., Jiang, X.C., Chen, P., Inouchi, J. and Halpern, M. (1993). Chemical and immunological analysis of prey-derived vomeronasal stimulants. Brain. Behav. Evol. 41, 246-254.

Wang, H.J. and Tsai, Y.F. (1991). Effects of male rat urine on norepinerphine levels in the accessory olfactory bulb of young and aged female rats. Proc. Natl. Sci. Counc. Repub. China 15,160-164.

Wysocki, C.J. and Lepri, J.J. (1991). Consequences of removing the vomeronasal organ. J. Steroid Biochem. Mol. Biol. 39, 661-669.

Wysocki, C. J., Wellington, J. L. and Beauchamp, G.K., (1980). Access of urinary nonvolatiles to the mammalian vomeronasal organ. Science 207, 781-783.

Wysocki, C.J. and Meredith M. (1987). The vomeronasal system. In: "Neurobiology of taste and Smell." Finger T.E. and Silver W.L. eds, John Wiley and Sons, 125-151.

14

NEUROGENIC INFLAMMATION OF THE SKIN INDUCES PLASTIC CHANGES IN α2-ADRENERGIC PAIN MODULATION IN RATS

Heikki Mansikka[1*] and Antti Pertovaara[1,2]

[1] Department of Physiology
Institute of Biomedicine
University of Helsinki, Helsinki, Finland
[2] Department of Physiology
Institute of Biomedicine
University of Turku, Turku, Finland

1. INTRODUCTION

The function of the pain modulating system can change because of nerve trauma or disease affecting nerves (Bennett, 1994; Willis, 1994). The pain may develop to a chronic, persistent symptom independent of the original etiologic factor. Chronic pain conditions have been shown to induce changes in the functions of the peripheral (Culp et al., 1989; LaMotte et al., 1992) and the central (Treede et al., 1992; Woolf et al., 1992; Palacek et al., 1992) sensory nerve cells.

Topical application of mustard oil or capsaicin to the skin of humans or animals produces neurogenic inflammation and activates nociceptive fibers. This leads to spontaneous pain and hyperalgesia to mechanical and thermal stimuli (Reeh et al., 1986; Szolcsanyi et al., 1988), corresponding to the symptoms seen in the chronic pain patients. Hyperalgesia in the treated region can, at least partly, be explained by sensitization of primary afferents (i.e. primary hyperalgesia; Culp et al., 1989; LaMotte et al., 1992). Central mechanisms have an important contribution to the secondary hyperalgesia outside the treated region (Woolf and Wall, 1986) because the response properties of the primary afferent fibers are not changed (Torebjörk et al., 1992). The application of chemical irritants such as mustard oil or capsaicin to the skin of animals or humans and the determination of sensory or nocifensive

[*] Correspondence: H. Mansikka, Institute of Biomedicine, Department of Physiology, P.O.B. 9, FIN-00014 University of Helsinki, Finland. FAX: + 358 - 0 - 1918681; E-mail: hmansikk@helsinki.fi

responses to mechanical stimulation outside the treated region provide a model to study secondary hyperalgesia and the underlying central changes in the pain system.

It is well established that α2-adrenoceptors of the endogenous noradrenergic system play an important role in the modulation of pain signals. Systemic or intrathecal administration of α2-adrenoceptor agonists induces antinociception which is due to activation of α2-adrenoceptors at the spinal cord level (Pertovaara, 1993). In the caudal medulla, the lateral reticular nucleus (LRN) is a site, which has a significant role in descending control of nociceptive signals (Gebhart and Ossipov, 1986). The LRN is richly innervated by neurons with α2-adrenergic receptors (Bousquet et al., 1981; Scheinin et al., 1994), and the LRN-induced antinociception can be reversed by α2-adrenoceptor antagonist at the spinal cord level (Janss and Gebhart, 1987). The nucleus raphe magnus (RMG) in the rostroventromedial medulla is also considered an important source of descending control of spinal nociceptive neurons (Basbaum and Fields, 1984).

In the present study we studied the effects of a selective α2-adrenoceptor agonist, medetomidine (Virtanen et al., 1988), and an α2-adrenoceptor antagonist, atipamezole (Scheinin et al, 1988) on the mustard oil induced secondary hyperalgesia to mechanical stimuli. First we wanted to study if secondary hyperalgesia can be attenuated by a systemically administered α2-adrenergic agent at doses which do not influence the nocifensive response in the intact limb. Secondly we wished to evaluate the roles of spinal versus medullary α2-adrenoceptors in the modulation of secondary hyperalgesia by determining hindlimb withdrawal thresholds to mechanical stimulation of the rat paw following microinjection of medetomidine, atipamezole or saline control in the spinal cord, the LRN or the RMG.

2. MATERIALS AND METHODS

The experiments were performed with adult male Hannover-Wistar rats (The Finnish National Laboratory Animal Center; weight range: 250-400 g). The experiments were approved by the Institutional Ethics Committee of the University of Helsinki.

To produce central hyperalgesia, mustard oil (50% in ethanol, Merck, Darmstadt, Germany) was applied for 2 minutes on a piece of filter paper (2 cm^2) on the skin of the ankle of the rat. During testing the rat was freely moving on a metal grid and the hindpaw ipsilateral or contralateral to the mustard oil treatment was stimulated with von Frey hairs (Stoelting, Wood Dale, IL; Chaplan et al., 1994). The stimulus site in the treated hindpaw was at least 2 cm distal from the area treated with mustard oil, because the focus of this study was on the secondary hyperalgesia. The hairs used in this experiment produced forces ranging from 0.445 to 84.96 g. At each time point monofilaments were applied to the foot pad in a series of increasing forces until the rat withdrew its hindlimb. The lowest force producing a withdrawal response was considered nociceptive threshold. The left and right hinpaws were consecutively tested, and at each time point the threshold for each hindpaw is based on two separate measurements. The withdrawal thresholds were determined before the administration of mustard oil, and at various time points following mustard oil application. Medetomidine, atipamezole or saline were applied 12 min before the application of mustard oil.

There were fourteen experimental groups. In seven of the groups the drugs were administered subcutaneously: a control group with saline, three medetomidine groups (3-30 μg/kg), three atipamezole groups (10-1000 μg/kg). In seven of the groups the drugs were administered intramedullary or intrathecally: two control groups with saline into the LRN or intrathecally, medetomidine (1 μg) into the LRN or intrathecally, atipamezole (2.5 μg) into the LRN, RMG or intrathecally. The intramedullary drugs were delivered in a volume of

0.5 µl and intrathecal injections to the lumbar spinal cord in a volume of 10 µl. For more detailed description of the surgical preparation of the rats and the microinjection procedures see Yaksh and Rudy (1976) and Mansikka and Pertovaara (1995). The LRN and the RMG were localized according to the stereotaxic atlas of Paxinos and Watson (1986).

The withdrawal thresholds were determined for both hindpaws before drug applications and at 5 minutes intervals for 40 minutes following mustard oil application. Medetomidine and atipamezole were provided by the Farmos Group, Orion, Turku, Finland.

Statistical evaluation was done with two- and three-way analysis of variance (ANOVA) and t-test. P<0.05 was considered to represent a significant difference.

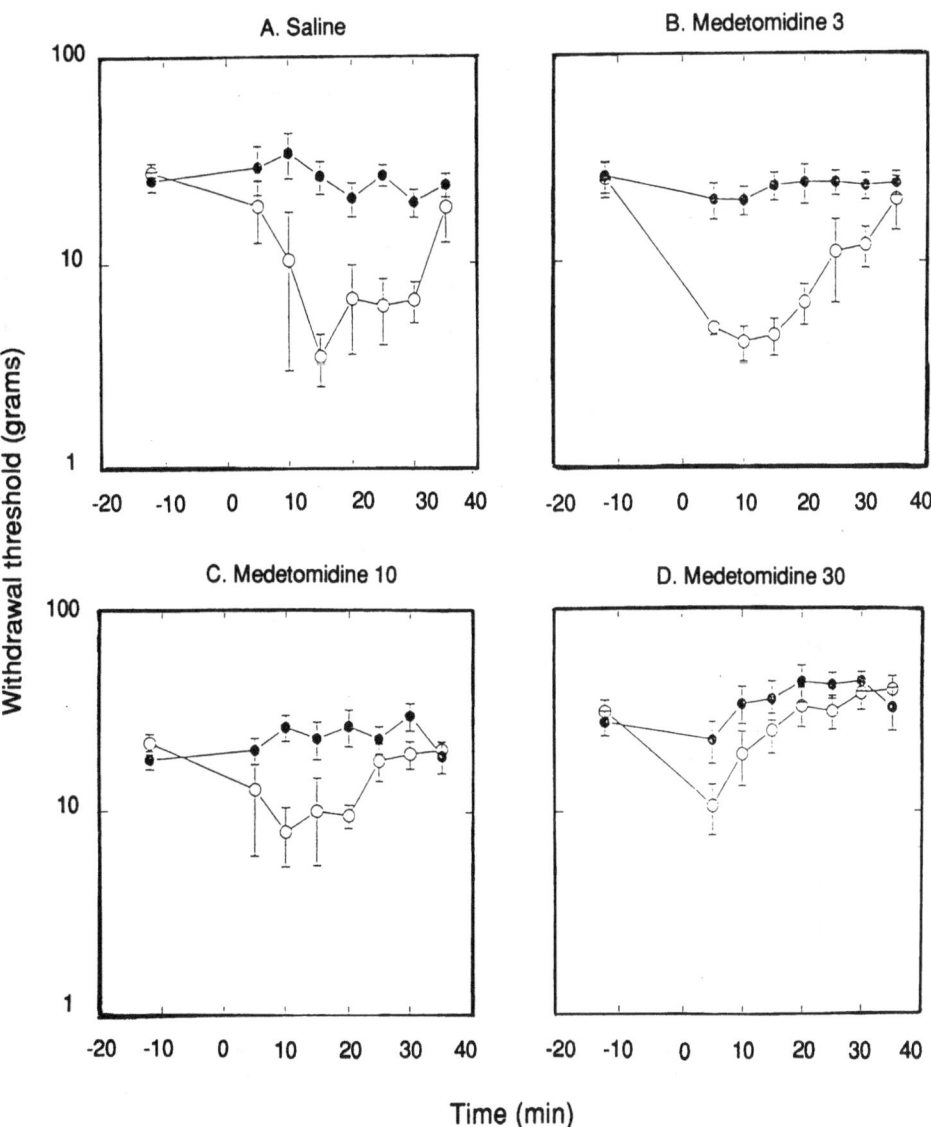

Figure 1. Effect of medetomidine on the threshold of the mechanically induced hindlimb withdrawal response. Dose is given above each graph (µg/kg, s.c.). Mustard oil was applied at time point 0 and the drug/saline control 12 min prior to mustard oil. The open symbols indicate the threshold in the mustard oil treated limb, and the filled symbols in the contralateral limb.

3. RESULTS

3.1. Systemic Administration of the Drugs

Mustard oil produced hyperalgesia in the treated hindlimb but not contralaterally. This was shown by the decrease of the mechanically induced withdrawal threshold in the treated hindlimb but not in the contralateral limb (Fig.1A). Mustard oil-induced hyperalgesia lasted about 30 minutes after which the threshold recovered to the pre-mustard oil level. Medetomidine, an α2-adrenoceptor agonist, dose-dependently (3-30 μg/kg) attenuated the mustard oil-induced mechanical hyperalgesia at doses which did not influence the threshold in the contralateral limb (Fig.1B-1D). According to ANOVA this effect was highly significant (P=0.0001).

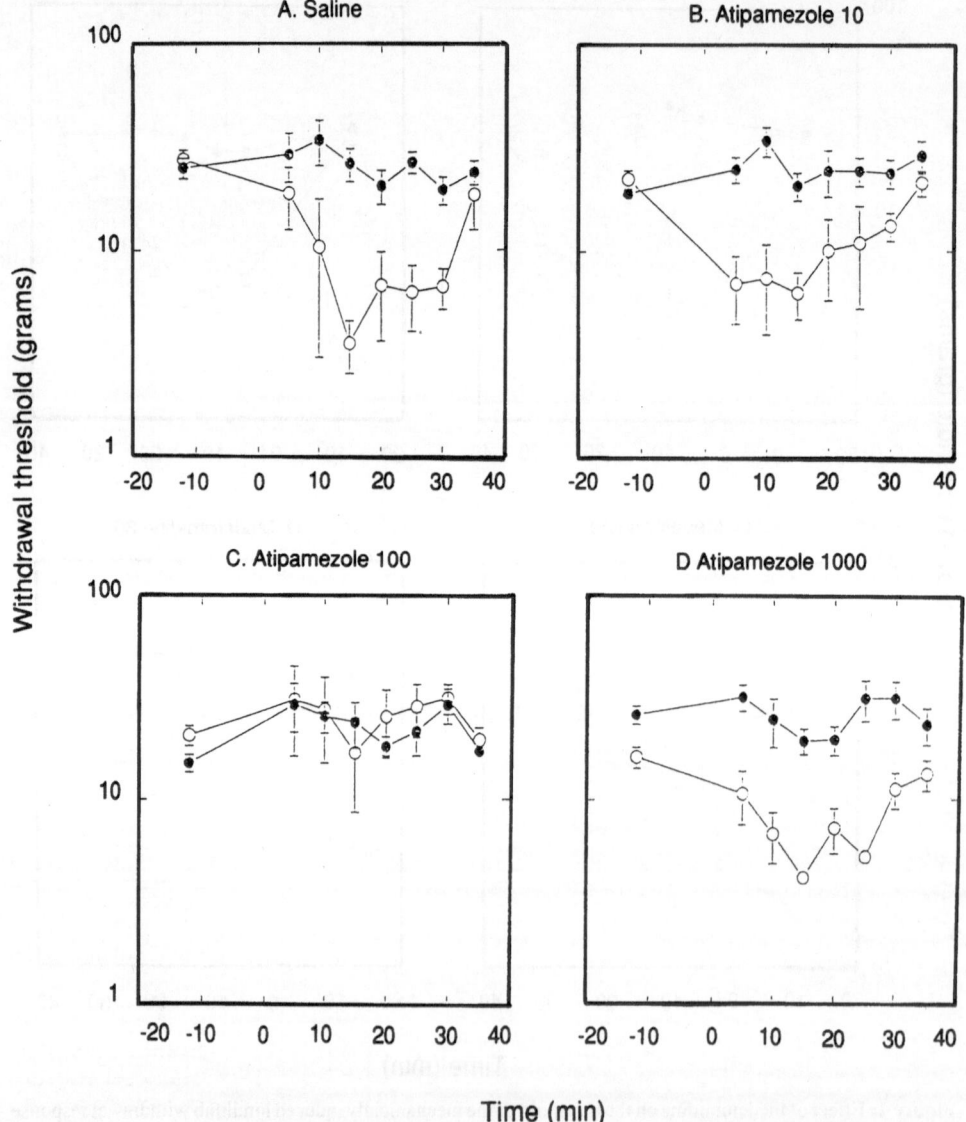

Figure 2. Effect of atipamezole on the threshold of the hindlimb withdrawal response. Dose is given above each graph (μg/kg, s.c.). Atipamezole was administered 12 min prior to mustard oil treatment.

Atipamezole, an α2-adrenoceptor antagonist, also had a significant attenuating effect on the mustard oil-induced hyperalgesia (P=0.0221), at an intermediate dose (100 µg/kg, Fig.2C). At low or high doses (10 µg/kg or 1000 µg/kg, Fig.2B and Fig.2D) atipamezole was without an effect.

3.2. Intramedullary and Intrathecal Administration of the Drugs

The drug administration into the medulla had a highly significant effect on the hindlimb withdrawal threshold (P=0.0003). Saline (Fig.3A) or medetomidine (1 µg, Fig.3B) admistered into the LRN or atipamezole (2.5 µg) administered into the RMG (Fig.3D) did

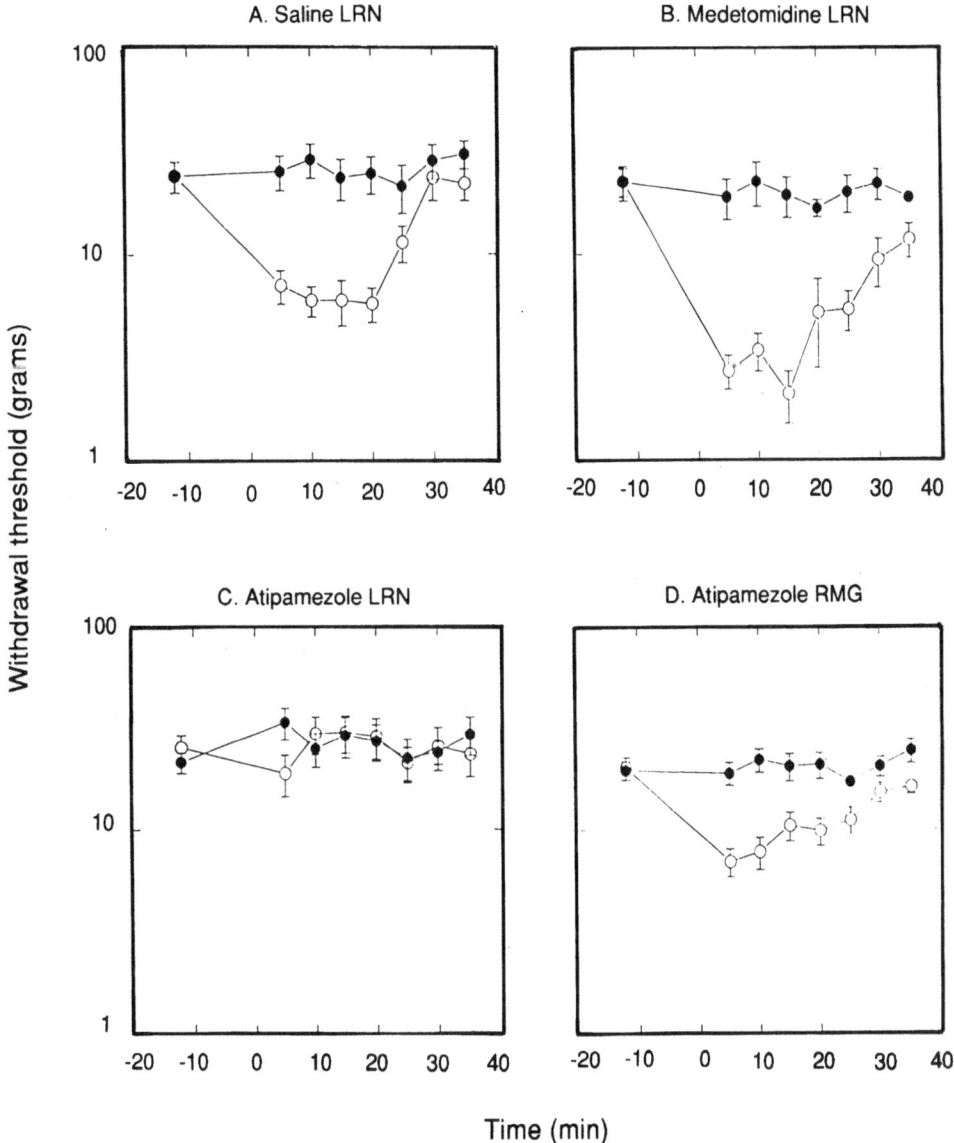

Figure 3. Effect of intramedullary drug administration on the threshold of the hindlimb withdrawal response. A) Saline B) Medetomidine 1 µg C) and D) Atipamezole 2.5 µg. Drugs were administered 12 min prior to mustard oil. LRN = Lateral reticular nucleus and RMG = nucleus raphe magnus.

not attenuate the mustard oil induced mechanical hyperalgesia. However, atipamezole
(2.5 µg) into LRN (Fig.3C) attenuated mustard oil-induced hyperalgesia.

Intrathecal injection of drugs at the lumbar level of spinal cord had a significant effect
on the mustard oil induced hyperalgesia (P=0.0303, 2-w-ANOVA). Intrathecal administration of saline (Fig. 4A) or atipamezole (2.5 µg, Fig.4C) did not attenuate the mustard
oil-induced hyperalgesia, whereas following intrathecal medetomidine (Fig.4B) mustard oil
produced no hyperalgesia.

In all groups, the withdrawal threshold of the untreated limb remained at the same
level following mustard oil and drug administrations as before the administrations.

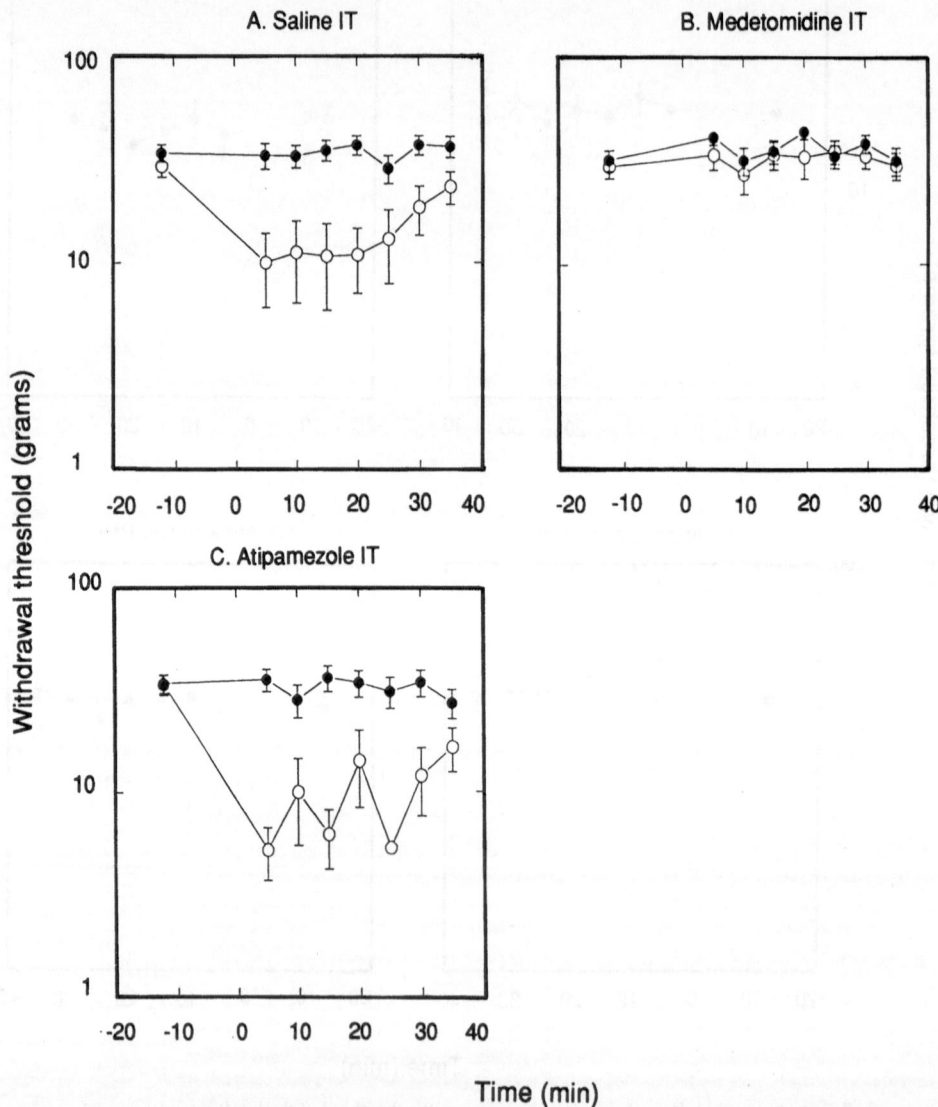

Figure 4. Effect of intrathecal drug administration on the threshold of the hindlimb withdrawal response
following mustard oil treatment. A) Saline B) Medetomidine 1 µg C) Atipamezole 2.5 µg.

4. DISCUSSION

In the current study medetomidine, an α2-adrenoceptor agonist, dose-dependently attenuated the secondary hyperalgesia. Microinjection of medetomidine into the medullary LRN was without an effect on the hyperalgesia, whereas in the lumbar spinal cord level medetomidine was highly effective in reversing the hyperalgesia. Thus, the antihyperalgesic effect of medetomidine was due to a spinal action. Atipamezole, an α2-adrenoceptor antagonist, could also attenuate secondary hyperalgesia at an intermediate dose (100 μg/kg) but was ineffective at low (10 μg/kg) or high (1000 μg/kg) doses. Atipamezole microinjected into the medullary LRN, but not into the RMG nor into the lumbar spinal cord, was effective reversing mustard oil-induced secondary hyperalgesia. This indicates that the paradoxical antinociceptive effect of atipamezole is due to action on the medullary LRN or a structure immediately adjacent to the LRN. The doses of α2-adrenergic drugs that proved effective in reversing the secondary hyperalgesia, were subantinociceptive in non-inflamed limbs.

In the current study the enhanced antinociceptive effect of medetomidine is in line with other studies indicating that α2-adrenoceptor agonists have an increased antinociceptive effect in various pathophysiological models of pain and this is due to action on the spinal cord (Hylden et al., 1991; Idänpään-Heikkilä et al., 1995; Stanfa and Dickenson, 1994).

The antinociceptive effect of atipamezole at an intermediate dose (100 μg/kg) was unexpected, but there is evidence of antinociceptive effects of high doses of yohimbine and idazoxan, other α2-adrenoceptor antagonist, in some pain models (Hayes et al., 1986; Kanui et al., 1993; Kayser et al., 1992). Evidence from our present study indicates that the LRN or a structure immediately adjacent to it is important in α2-adrenoceptor antagonist-induced antihyperalgesia, since atipamezole was ineffective when administered in to the other medullary structure, the RMG, or into the lumbar spinal cord.

Our present results indicate that neurogenic inflammation induces plastic changes in α2-adrenergic pain modulation. These changes involve both spinal and medullary pain modulatory mechanisms. An α2-Adrenergic agonist may provide a useful tool treating patients with inflammatory pain and hyperalgesia. The most effective way of administering an α2-adrenoceptor agonist is direct application to the spinal cord level, since supraspinally α2-adrenoceptor agonist may even counteract the spinal antinociceptive effects (Mansikka & Pertovaara, 1995).

REFERENCES

Basbaum, A.I. and Fields, H.L., Endogenous pain control systems. Brainstem spinal pathways and endorphin circuitry, Annual Review of Neuroscience, 7 (1984) 309

Bennett, G.J., Neuropathic pain. In : Wall, P.D. & Melzack, R. (eds): Textbook of Pain (4th Edition). Churchill Livingstone, London, 1994.

Bousquet, P., Feldman, J., Bloch, R., Schwartz, J., The nucleus reticularis lateralis: A region highly sensitive to clonidine, European Journal of Pharmacology, 69 (1981) 389

Culp, S.R., Ochoa, J., Cline, M., Dotson, R., Heat and mechanical hyperalgesia induced by capsaicin, Brain, 112 (1989) 1317

Gebhart, G.F., Ossipov, M.H., Characterization of inhibition of the spinal nociceptive tail-flick reflex in the rat from the medullary lateral reticular nucleus, Journal of neuroscience, 6 (1986) 701

Hayes, A.G., Skingle, M., Tyers, M.B., Antagonism of alpha-adrenoceptor agonist-induced antinociception in the rat, Neuropharmacology, 25 (1986) 397

Hylden, J.L.K., Thomas, D.A., Iadarola, M.J., Nahin, R.L., Dubner, R., Spinal opioid analgesic effects in a model of unilateral inflammation/hyperalgesia: possible involvement of noradrenergic mechanisms, European Journal of Pharmacology, 194 (1991) 135

Idänpään-Heikkilä, J.J., Kalso, E.A., Seppälä, T., Antinociceptive actions of dexmedetomidine and the kappa-opioid agonist U-50,488H against noxious thermal, mechanical and inflammatory stimuli, Journal of Pharmacology and Experimental Therapeutics, 271 (1995) 1306

Janss, A.J., Gebhart, G.F., Spinal monoaminergic receptors mediate the antinociception produced by glutamate in the medullary lateral reticular nucleus, Journal of neuroscience, 7 (1987) 2862

Kanui, T.I., Tjolsen, A., Lund, A., Mjellem-Joly, N., Hole, K., Antinociceptive effects of intrathecal administration of alpha-adrenoceptor antagonist and clonidine in the formalin test in the mouse, Neuropharmacology, 32 (1993) 367

Kayser, V., Guilbaud, G., Besson, J.M., Potent antinociceptive effects of clonidine systemically administered in an experimental model of clinical pain, the arthritic rat, Brain Research, 593 (1992) 7

LaMotte, R.H., Lundberg, L.E.R., Torebjörk, H.E., Pain, hyperalgesia, and activity in nociceptive C units in human hairy skin, Journal of Physiology, 448 (1992) 749

Mansikka, H. and Pertovaara, A., The role of the medullary lateral reticular nucleus in spinal antinociception in rats, Brain Research Bulletin, 37 (1995) 633

Palacek, J., Paleckova, V., Dougherty, P.M., Carlton, S.M., Willis, W.D., Responses of spinothalamic tract cells to mechanical and thermal stimulation of skin in rats with experimental peripheral neuropathy, Journal of Neurophysiology, 67 (1992) 1562

Paxinos, G. and Watson, C., The Rat Brain in Stereotaxic Coordinates, Academic Press, New York, 1986

Pertovaara, A., Antinociception induced by $\alpha2$-adrenoceptor agonists, with special emphasis on medetomidine studies, Progress in Neurobiology, 40 (1993) 691

Reeh, P.W., Kocher, L., Jung, S., Does neurogenic inflammation alter the sensitivity of unmyelinated nociceptors in the rat ?, Brain Research, 384 (1986) 42

Scheinin, H., Macdonald, E., Scheinin, M., Behavioural and neurochemical effects of atipamezole, a novel $\alpha2$-adrenoceptor antagonist, European Journal of Pharmacology, 157 (1988) 35

Scheinin, M., Lomasney, J.W., Hayden-Hixson, D.M., Schambra, U.B., Caron, M.G., Lefkowitz, R.J., Fremeau, R.T.Jr., Distribution of $\alpha2$-adrenergic receptor subtype gene expression in rat brain, Molecular Brain Research, 21 (1994) 133

Stanfa, L.C., Dickenson, A.H., Enhanced $\alpha2$-adrenergic controls and spinal morphine potency in inflammation, NeuroReport, 5 (1994) 469

Szolcsanyi, J.F., Anton, F., Reeh, P.W., Handwerker, H.O., Selective excitation by capsaicin of mechano-heat sensitive nociceptors in rat skin, Brain research, 446 (1988) 262

Treede, R.D., Meyer, R.A., Raja, S.N., Campbell, J.N., Peripheral and central mechanisms of cutaneous hyperalgesia, Progress In Neurobiology, 38 (1992) 397

Torebjörk, H.E., Lundberg, L.E.R., LaMotte, R.H., Central changes in processing of mechanoreceptive input in capsaicin-induced secondary hyperalgesia in humans, Journal of Physiology, 448 (1992) 765

Willis, W.D., Jr., Central plastic responses to pain. In: Gebhart, G.F., Hammond, D.L., Jensen, T.S. (eds.): Proceedings of the 7th world congress on pain, Progress in pain research and management, Vol 2, IASP Press, Seattle, 1994

Virtanen, R., Savola, J.-M., Saano, V., Nyman, L., Characterization of the selectivity, specificity and potency of medetomidine as an $\alpha2$-adrenoceptor agonist, European Journal of Pharmacology, 150 (1988) 9

Woolf, C.J., Shortland, P., Coggeshall, R.E., Peripheral nerve injury triggers central sprouting of myelinated afferents, Nature, 355 (1992) 75

Woolf, C.J., Wall, P.D., Relative effectiveness of C primary afferent fibres of different origins in evoking a prolonged facilitation of the flexor reflex in the rat, Journal of Neuroscience, 6 (1986) 1433

Yaksh, T.L. and Rudy, T.A., Chronic catheterization of the spinal subarachnoid space, Physiology and Behaviour, 17 (1976) 1031

15

ROLE OF K⁺ CHANNELS IN SEMICIRCULAR CANAL ADAPTATION

Paola Perin and Sergio Masetto

Istituto di Fisiologia Generale
Viale Forlanini 6, 27100 Pavia

1. INTRODUCTION

Vestibular hair cells transduce head movements into receptor potentials. Transduction is based on the modulation of a K current (the receptor current) entering hair cell bodies through mechanosensitive channels and leaving them through voltage- and Ca-activated K channels; voltage-dependent Ca channels modulate the receptor current and trigger transmitter release[10]. The various stages of stimulus conversion in the sensory organ modify the incoming signal, introducing, among the various features, an adaptation to constant or repeated stimulation. In the vestibular system there are many possible causes for adaptation, starting from mechanical relaxation of accessory structures, up to encoder properties in the vestibular nerve fibers and to central mechanisms[6]. The best known form of adaptation in vestibular organs involves a fast relaxation of the mechanoelectrical trasduction apparatus, which is located on the apical membrane stereocilia.[11] This adaptation seems to be due to the sliding of the mechanosensitive complex along the stereocilium lenghth, performed by an actomyosinic complex activated by Ca entering mechanosensitive channels when they open[11]. Channel position would influence the stretch exerted on the gating spring, and thus the open probability of the channel itself.[17] This fast adaptation has been observed in sacculus[11] but not in semicircular canal, since the latter can be stimulated at low frequencies only (up to 10 Hz)[15], while the former responds to frequencies up to several hundred Hz[12]. Accordingly with its lower frequency range, semicircular canal shows a different type of adaptation that takes tens of seconds to develop[8,28]. Semicircular canal sensory adaptation, normally studied by observing the decrease with time of sensory unit discharge, does not appear to be due to the mechanoelectrical transduction apparatus nor to steps ahead of it, since it can be elicited in response to voltage stimulation of the whole organ[25], thus bypassing the mechanoelectrical transduction steps. The site of this adaptation could still be either pre-or postsynaptic. At present, we cannot quantify the importance of postsynaptical factors in slow adaptation; however, we can easily demonstrate an important role for presynaptic factors, since mechanoactivated transepithelial currents (ampullar microphonic currents, AMC) recorded from isolated semicircular canals adapted independently from the whole-nerve discharge. In this work we focus mainly on AMC slow adaptation of semicircular canals. To define better the ion mechanisms involved in this process,

Neurobiology, edited by Torre and Conti
Plenum Press, New York, 1996

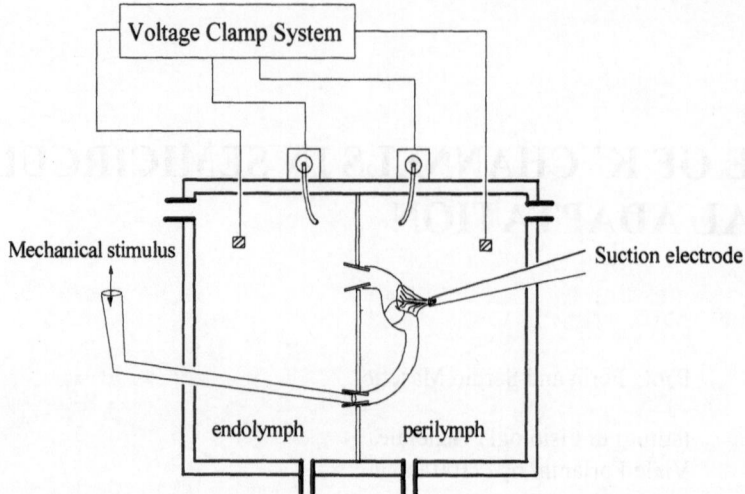

Figure 1. Schematic drawing of the two-compartment chamber used for whole organ registration. Mechanical stimulation device, voltage-clamp system and suction electrode are indicated.

a comparison between the results observed on whole isolated semicircular canal and on single semicircular canal hair cells has been performed.

2. MATERIALS AND METHODS

Experiments were performed, at room temperature (20ºC), on vertical posterior semicircular canals isolated from frogs (*Rana esculenta L.*) previously anesthetized by immersion in 0.1% 3-aminobenzoic acid ethylester methanesulfonate solution (MS-222, Sandoz) and then decapitated.

2.1. Isolated Semicircular Canal Experiments

Isolated canals (Fig.1) were placed in a two-compartment chamber (5 ml each) filled with artificial endolymph and perilymph (see Solutions). The canal was tightly fixed at its ends to two small glass cones. Drugs were perfused in the perilymphatic compartment at a rate of 20 ml/min for 5 min. Ampullar receptors were sinusoidally stimulated (frequency 0.2 Hz) by means of a microsyringe whose plunger (diameter 0.5 mm; displacement ± 5 µm) was operated by a servo-controlled stepper motor[24,26]. The transepithelial receptor current (ampullar microphonic current, AMC), was measured under voltage clamp of the whole sensory epithelium. Ampullar potentials were recorded by means of a pair of agar bridge electrodes, each placed in one compartment, while a second pair of electrodes passed current to clamp transepithelial voltage to 0 mV (i.e. at its normal resting value)[20,26]. Whole nerve slow potentials (Ndc) were recorded using a suction electrode positioned on canal nerve[25]. Ndc was the result of propagation of postsynaptic depolarization up to the electrode: thus, it contained both propagated action potentials and electrotonicaly spread depolarizations.

2.2. Patch Clamp Experiments

Under dissection microscope, canal crista ampullaris was isolated, and sequentially transferred into three dishes filled with: 1) dissociation medium (Solutions) added with 1.2

mg/ml Protease VIII from *Bacillus licheniformis* (Sigma) (1 min); 2) dissociation medium only (3 min); 3) extracellular solution used for patch clamp experiments (See Solutions). Isolated hair cells were obtained by gently twisting and pressing the crista ampullaris against the bottom of the third dish.

Patch clamp experiments were performed in whole-cell configuration, with the perforated patch technique, using a L/M EPC-7 amplifier (List Electronics, Germany) and pClamp software. Membrane perforation was achieved by using amphotericin B (Sigma)[19]. Data were sampled at 2-5 kHz and filtered at 3 kHz. Patch pipettes were pulled from borosilicate glass capillaries (Drummond) coated with Sylgard (Dow Corning, USA) up to about 100 μm from the tip. Electrode resistance was 3-4 MOhm, and series resistance (R_s) 8-15 MOhm. When possible, R_s was actively compensated up to 40%. In voltage clamp conditions, currents were corrected for leakage off-line by subtracting an ohmic current calculated using the cell input resistance measured between -80 and -100 mV, or between -70 and -80 when inward rectifier was present.

2.3. Solutions and Drugs

The used solutions had the following composition (mM): Perilymph: NaCl 135, KCl 2.5, NaHCO$_3$ 1.2, NaH$_2$PO$_4$ 0.17, CaCl$_2$ 1.8, glucose 5.5; pH 7.3; Endolymph: NaCl 19.5, KCl 100, NaHCO$_3$ 1.2, NaH$_2$PO$_4$ 0.17, CaCl$_2$ 1.8, glucose 5.5; pH 7.3; Dissociation: NaCl 135, KCl 2.5, HEPES-NaOH 5, CaCl$_2$ 0.8, MgCl$_2$ 5, EGTA 2, glucose 3; pH 7.25; Extracellular: NaCl 135, KCl 5, NaHCO$_3$ 3.5, NaH$_2$PO$_4$ 0.5, CaCl$_2$ 1.8, glucose 6; pH 7.25; Intracellular: KCl 75, K$_2$SO$_4$ 30, MgCl$_2$ 2, HEPES-KOH 10, glucose 3; pH 7.25. pH was adjusted while adding NaH$_2$PO$_4$, except for dissociation and intracellular solutions where the hydroxide of the major cation was used. The following drugs were tested in perilymphatic or extracellular solutions: Tetraethylammonium (TEA), 4-aminopyridine (4-AP), quinine and cadmium (Cd) from Sigma; apamine and charybdotoxin (CTX) from Alomone Labs (Israel). Osmotic pressure was kept constant by varying the NaCl content of the solutions. For CTX experiments, albumine 50 μg/ml was added to the extracellular solutions[1].

3. RESULTS

3.1. Isolated Semicircular Canal Experiments

Figure 2 shows typical AMC and Ndc responses to a sinusoidal stimulus. Both AMC and nerve depolarization decayed, reaching a constant amplitude after 4-6 cycles of stimulation. As regards AMC, we will call this decay (already observed in previous studies)[27,28] "canal adaptation".

The effects of drug perilymphatic perfusion on AMC are summarized in Fig. 3. All drug effects are measured at the 12th cycle, i.e. at the steady state.

At low concentrations (50 μM), quinine reduced AMC without modifying canal adaptation; at higher concentrations (500 μM; n=6) this drug completely suppressed AMC.

Cd effects were quite different: at saturating concentrations (500 μM), this ion made AMC peak amplitude rigorously constant throughout the whole stimulation period, but only reduced AMC to 44±8% (n=7) of control.

TEA 100 mM reduced AMC amplitude to 46±7% of control (n=4) and reversed canal adaptation (the amplitude of AMC peaks increased during the first stimulation cycles). AMC reduction exerted by the combined perfusion of TEA and Cd was greater (23±5%; n=3) than the effects of each drug alone, suggesting that the two drugs did not act on the same target.

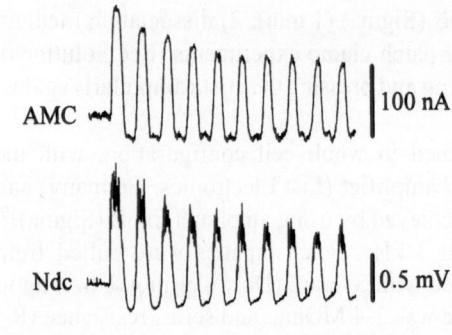

AMC   100 nA

Ndc   0.5 mV

Stimulus ——/\/\/\/\/\/\/\/\/\/\—   10 µm

                                                             10 sec

Figure 2. Ampullar microphonic current (AMC) and whole nerve potentials (Ndc) evoked by 10 cycles of sinusoidal stimulation at 0.2 Hz. Lower trace: stimulation protocol.

CTX produced effects analogous to Cd, although smaller. This toxin, at saturating concentrations (100-150 nM), reduced AMC amplitude to 65±4% (n=3) of control and eliminated canal adaptation.

All drug actions were fully reversible after about 10 min of washing with normal perilymph, except for quinine, whose recovery took about 30 min. Perilymphatic administration of apamine (10 µM; n=6) and 4-AP (40 mM; n=12) had no appreciable effects on AMC (data not shown).

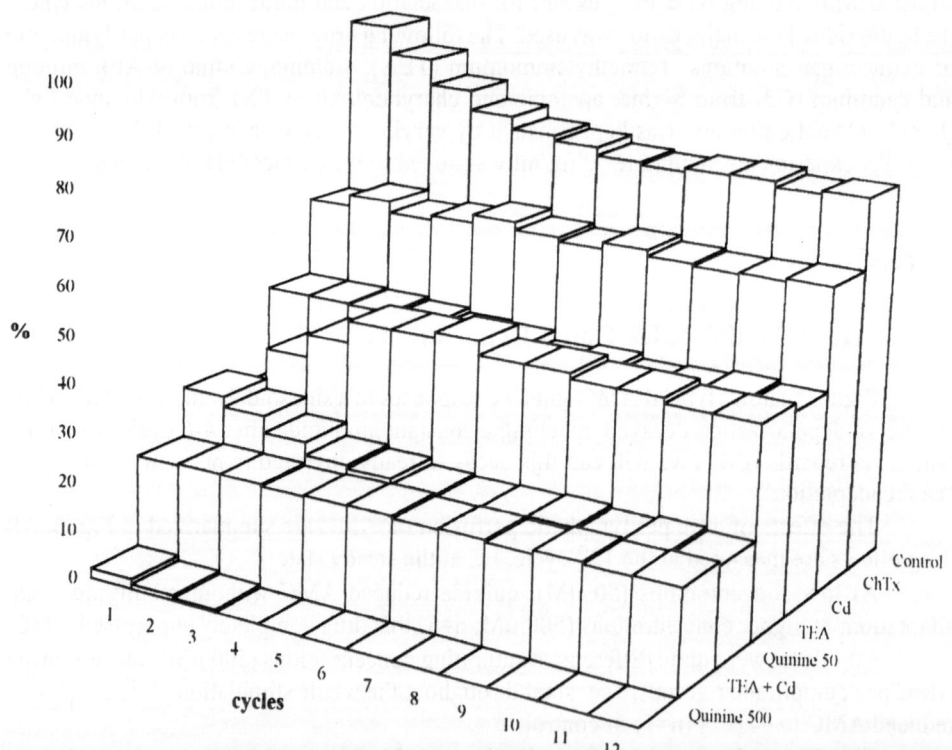

Figure 3. Average peak AMC in control conditions and after 5 minutes of perilymphatic drug perfusion. From back to front: Control; CTX 150 nM; Cd 500 µM (Note the disappearance of adaptation); TEA 100 mM; Quinine 50 µM; TEA 100 mM + Cd 500 µM; Quinine 500 µM.

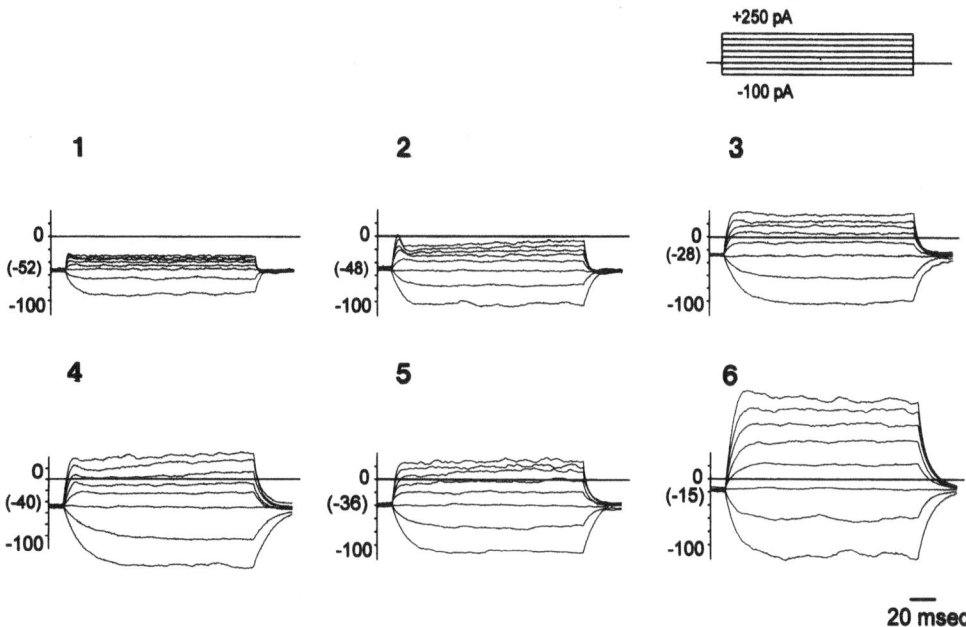

Figure 4. Typical voltage responses from an isolated canal hair cell. Stimulation protocols consisted in 250 msec current steps from -100 to 250 pA in 50 pA increments, delivered from hair cell resting potentials (indicated in parentheses in the figure for each experimental condition) 1. Control. 2. Cd 500 μM. 3. TEA 50 mM. 4. TEA 50 mM + Cd 500 μM. 5. CTX 100 nM (control Vz = -42.1 mV). 6. Quinine 500 μM.

3.2. Patch Clamp Experiments

Traces in Fig. 4 illustrate typical voltage responses of single hair cells to current steps of physiological amplitude[2,5], delivered from zero-current membrane potential (Vz). In control conditions (Fig.4-1), voltage responses showed an evident outward rectification: average steady-state slope conductance calculated between 0 and 100 pA was 7.5 ± 2.7 nS (n=11), and it was 2.6 ± 0.9 nS (n=11) between 0 and -100 pA.

Cd effects on voltage responses are depicted in Fig. 4-2. A reduction of the outward rectification was evident (responses to 100 pA steps changed from -40.7 ± 7.7 mV to -24.1 ± 16.5 mV; n=7); in fact, membrane conductance was reduced both on depolarizatons and on hyperpolarizations. On cell resting membrane potential, Cd did not produce consistent variations (from -52.8 ± 5.8 mV to -52.1 ± 13 mV; n=7). Cd blocked a fast outward current, since voltage responses after Cd showed an initial depolarizing peak.

The effects of TEA (50 mM) are shown in Fig. 4-3. This drug induced a marked cell depolarization and an evident decrease in outward rectification. On average, Vz depolarized from -51.1 ± 8.9 mV to -35.9 ± 22.5 mV (n=9) and voltage-response to 100 pA shifted from -37.4 ± 7.6 mV to -1.2 ± 18.5 mV (n=7). The combined effects of TEA and Cd (Fig.4-4) on Vz and voltage responses to depolarizing current steps were intermediate between those produced by each drug alone, approaching TEA action for potentials near E_{Ca}. On the hyperpolarizing side drug effects were additive, most likely because of the blockade of the inward rectifier current already described in these cells[14].

CTX effects are shown in Fig.4-5. This drug, at a concentration of 100 nM, affected both Vz (from -42.1 mV to -35.8 mV) and voltage responses to current steps (response to a 100 pA step changed from -27 mV to -1 mV).

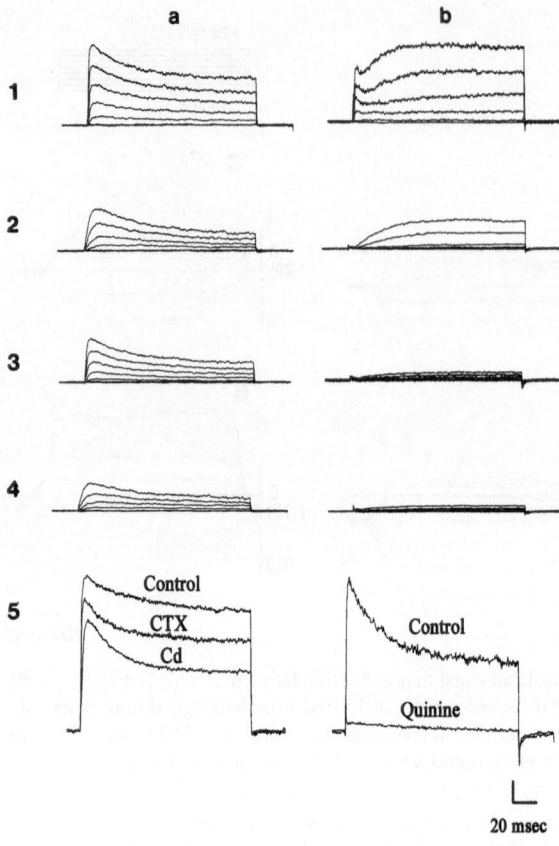

Figure 5. Currents recorded from an I_A-showing (a) and an I_A-lacking (b) cell. Voltage-clamp protocols for 1 to 4 consisted in 250 msec voltage steps from -80 to -10 mV in 10 mV increments, delivered from an holding potential of -80 mV. For 5, a single step to 0 mV was delivered from an holding of -80 mV. 1. Control. 2. Cd 500 μM. 3. TEA 50 mM. 4. TEA 50 mM + Cd 500 μM. 5a. CTX 100 nM and Cd 500 μM. 5b. Quinine 500 μM. Vertical calibration bar: 400 pA (1-4a), 200 pA (1-4b and 5a), 50 pA (5b).

Quinine, at saturating concentration (500 μM), was the most effective (Fig. 4-6) in depolarizing membrane potential (Vz changed from -56.8 ± 13.4 mV to -28.6 ± 11.6 mV; n=5) and decreasing outward rectification (response to a 100 pA step shifted from -45.2 ± 13.2 mV to 10.4 ± 27.6 mV; n=5).

Fig. 5 shows current traces in response to voltage steps in the range of potentials observed in current clamp experiments. Fig. 5-1 shows the typical currents observed in the majority (about 90%) of hair cells (a) and in those cells lacking I_A (b), in control conditions. In both cell types, Cd 500 μM (Fig. 4-2B) affected both transient and steady outward currents. Cd-sensitive currents activated faster than I_A[17] (time-to-peak was 2.8 ± 0.8 msec; n=11). Where I_A was lacking (4-2b), Cd-resistant current showed a slow activation time course (about 90 msec at -10 mV) and no evident inactivation, as the outward delayed rectifier current (I_K) already described[14] in semicircular canal hair cells; on the other side, on the majority of hair cells I_A precluded a kinetic analysis of Cd resistant currents.

TEA did not induce slow down of currents, confirming that currents blocked by TEA and Cd were, to some extent, different. This hypothesis was also supported by the observation that TEA and Cd, when combined, reduced single cell outward current more then each drug alone (Fig. 4-4), analogously to what they did on AMC. The scarce additivity of the two drugs however suggested also that, at the concentrations tested, Cd and TEA effects overlapped, i.e., a consistent fraction of the current blocked by Cd was also blocked by TEA.

The current blocked by 100 nM CTX (Fig. 4-5a) was smaller than that blocked by 500 μM Cd. Moreover, CTX did not appreciably modify its activation time course whereas Cd slowed it, suggesting that Cd blocked, beside CTX-sensitive current, both a steady and a transient component of the whole current. (The possibility that CTX blocked a current different from the Cd-sensitive one may be ruled out since, after Cd perfusion, CTX had no additional effect; data not shown).

The effects of quinine are shown in Fig.4-5b, for a single step to 0 mV. This drug blocks most voltage-dependent currents.

4. DISCUSSION

The present results suggest that, in frog semicircular canal hair cells, $I_{K(Ca)}$ is active in the range of receptor potentials. This is not surprising, since Ca must enter hair cells at rest, in order to sustain the appreciable resting discharge which characterizes most vestibular nerve unit, giving them bidirectional sensitivity.[18] Accordingly, in our experiments both Cd and CTX modified hair cell voltage response to current steps of intensity comparable to those of mechanically evoked transducer currents, and reduced AMC in the whole organ. $I_{K(Ca)}$ appears to be the fastest outward current generated by canal hair cells: therefore, its blockade induced the appearance of a rapid depolarizing peak in current-clamp conditions. However, the fastest $I_{K(Ca)}$ component does not appear to be sensitive to CTX, suggesting the presence on hair cells of two populations of Ca-activated K channels (On the basis of the present experiments the nature of the different $K_{(Ca)}$ channels cannot be positively identified; however, we can exclude S_K presence, given the lack of apamin effects both in the whole sensory organ and on single cells)[4]. From the present study, the steady, CTX sensitive fraction of $I_{K(Ca)}$ appears also important in canal adaptation, since its blockade eliminates this process. It has to be reminded that in semicircular canals $I_{K(Ca)}$ is quite small, if compared to other vestibular organs: in the sacculus, $I_{K(Ca)}$ represents most of the potassium current[13] involved in receptor current, whereas in the canal it only sums up to 25-30% of outward current[14]. The remaining potassium current is represented by a delayed rectifier and, in most cells, also an I_A which is active at rest[21]; moreover, several cells show an inward rectifier (I_{ir}) which activates around -90 mV[14]. Concerning the delayed rectifier I_K, its activation close to -50 mV[14] and the lack of an evident inactivation indicate that this current is always available for hair cells repolarization. Therefore, I_K might play a prominent role during low-frequency mechanical stimulation, avoiding the need for a maintained high Ca in the cytoplasm. Moreover, as observed for the toadfish sacculus, I_K may be important for reducing hair cell resonance[22]. Unfortunately, the only drugs that resulted able to block this current, namely TEA and quinine, appeared nonselective, since they affected I_K, $I_{K(Ca)}$ and I_{ir}. It seems anyway worth to note that nonsaturating concentrations of quinine, that noticeably depolarized hair cells and reduced AMC amplitude, did not cancel canal adaptation. This indicates that canal adaptation is not directly due to hair cell membrane depolarization, an hypothesis supported also by the lack of effects of Cd on hair cell resting potential. An apparent contradiction in our experiments was that, even if the majority of hair cells showed a large 4-AP sensitive I_A, in isolated semicircular canal saturating doses of 4-AP were without effects on microphonic current adaptation and amplitude. This result might be explained taking into account I_A fast inactivation (about 40 msec)[14], which, at the low frequency of mechanical stimulation (0.2 Hz) adopted in the present study, would completely inactivate the channels before a significant fraction of them opens. As proposed by Norris[16], I_A physiological role might be that of an electrical low-pass filter, which counteracts rapid membrane depolarizations. This would improve signal-to-noise ratio, especially during inhibitory efferent system stimulation that, by hyperpolarizing hair cells, would remove I_A inactivation[3]. Cells lacking I_A express an inward rectifier which may hyperpolarize them steadily at rest[14,7]. An involvement of the inward rectifier in canal adaptation appears very unlikely; in fact, on one side quinine, which drastically reduced membrane conductance in isolated hair cells at hyperpolarizing potentials, did not affect canal adaptation; on the other side, CTX, which has never been reported to block inward rectifiers, and consistently did not affect hair cells hyperpolarizing responses, eliminated adaptation in the whole organ.

Semicircular canal hair cells may be divided in two populations,[14,8] one of which, that expresses I_{ir} but not I_A[14], is most likely synapting to the higher-gain phasic afferents[9]. An interesting hypothesis suggests that these cells, which are located in the central part of the crista ampullaris, would show adaptation in their synaptic activity, while perypheral hair cells, innervating tonic fibres, would not. Thus, canal response may encode several different features of incoming stimuli by adopting parallel processing of the same signal; in this view, canal adaptation does not necessarily involve all hair cells. Unfortunately, it was not possible to observe adaptation in isolated cells, most likely because of factors that are missing in the single cell approach, like the electrochemical gradients between the endolymph and perilymph, ion accumulation in subcupular or basolateral spaces, ampullar geometry effects, etc. In fact, canal adaptation was greatly reduced also by perilymphatic ouabain[28], most likely because of a disruption of the K gradient between endolympyh and perilymph which is maintained by dark cells Na-K-ATPase[23].

Although the experimental procedures adopted in the present study had allowed an identification of several K channels involved in AMC modulation, the cellular processes that determinate AMC adaptation still need to be investigated. In fact, the time course of this process appears slower than usual membrane-limited current modulation, and more in agreement with the involvement of several biochemical steps.

ACKNOWLEDGMENTS

The Authors are grateful to Professors Valli and Zucca and to Dr. Botta for technical help and valuable discussion.

REFERENCES

1. Anderson CS, MacKinnon R, Smith C, Miller C (1988) Charybdotoxin block of single Ca^{2+}-activated K^+ channels. J Gen Physiol 91, 317-333
2. Art SS, Fettiplace R (1987) Variation of membrane properties in hair cells isolated from turtle cochlea. J Physiol (Lond) 385, 207-242
3. Art SS, Fettiplace R, Fuchs PA (1984) Synaptic hyperpolarization and inhibition of turtle cochlear hair cells. J Physiol (Lond) 356, 525-550
4. Blatz AL, Magleby KL (1986) Single apamin-blocked Ca-activated K^+ channels of small conductance in cultured rat skeletal muscle. Nature 323, 718-720
5. Corey DP, Hudspeth AJ (1979) Ionic basis of the receptor potential in a vertebrate hair cell. Nature 281, 676-677
6. Eatock RA, Corey DP, Hudspeth AJ (1987) Adaptation of mechanoelectrical transduction in hair cells of the bullfrog's sacculus. J Neurosci 7, 2821-36
7. Griguer C, Sans A, Valmier J, Lehouelleur J (1993) Inward potassium rectifier currents in type I vestibular hair cells from guinea pig. Neurosci Lett 149, 51-55
8. Guth PS, Fermin CD, Pantoja M, Edwards R, Norris C (1994) Hair cells of different shapes and their placements along the frog crista ampullaris. Hearing Res 73, 109-115
9. Honrubia V, Hoffman LF, Sitko S, Schwartz IR (1989) Anatomic and physiological correlates in bullfrog vestibular nerve. J Neurophysiol 61(4), 688-701
10. Hudspeth AJ (1985) The cellular basis of hearing: the biophysics of hair cells. Science 230, 745-752
11. Hudspeth AJ, Gillespie PG (1994) Pulling springs to tune transduction: adaptation by hair cells. Neuron 12, 1-9
12. Hudspeth AJ, Lewis RS (1988) A model for electrical resonance and frequency tuning in saccular hair cells of the bull-frog, Rana catesbeiana. J Physiol 400, 275-297

13. Hudspeth AJ, Lewis RS (1988) Kinetic analysis of voltage- and ion-dependent conductances in saccular hair cells of the bull-frog, Rana catesbeiana. J Physiol 400, 237-274
14. Masetto S, Russo G, Prigioni I (1994) Differential expression of potassium currents by hair cells in thin slices of frog crista ampullaris. J Neurophysiol 72, 443-455
15. Melvill Jones G, Milsum JH (1971) Frequency-response analysis of central vestibular unit activity resulting from rotational stimulation of the semicircular canals. J Physiol 219, 191-215
16. Norris CH, Ricci AJ, Housley GD, Guth PS (1992) The inactivating potassium currents of hair cells isolated from the crista ampullaris of the frog. J Neurophysiol 68, 1642-1653
17. Pickles JO, Corey DP (1992) Mechanoelectrical transduction by hair cells. TINS 15(7), 254-259
18. Precht W (1976) Physiology of the perypheral and central vestibular systems. In: R.Llinàs and W.Precht (eds) Frog Neurobiology, Springer-Verlag, pp 452-512
19. Rae J, Cooper K, Gates P, Watsky M (1991) Low access resistance perforated patch recordings using amphotericin B. J Neurosci Meth 37, 15-26
20. Rüsch A, Thurm U (1989) Cupula displacement, hair bundle deflection, and physiological responses in the transparent semicircular canal of young eel. Pflügers Arch 413, 533-545
21. Russo G, Masetto S, Prigioni I (1995) Isolation of A-type K$^+$ current in hair cells of the frog *crista ampullaris*. NeuroReport 6, 425-428
22. Steinacker A, Romero A (1992) Voltage-gated potassium current and resonance in the toadfish saccular hair cell. Brain Res 574, 229-236
23. Sterkers O, Ferrary E, Amiel C (1988) Production of inner ear fluids. Physiol Rev 68, 1083-1127
24. Valli P, Zucca G (1977) The importance of potassium in the function of frog semicircular canals. Acta Otolaryngol. 84, 344-351
25. Valli P, Zucca G, Botta L (1990) Perilymphatic potassium changes and potassium homeostasis in isolated semicircular canals of the frog. J Physiol (Lond) 430, 585-594
26. Valli P, Zucca G (1976) The origin of slow potentials in semicircular canals of the frog. Acta Otolaryngol 81, 395-405
27. Valli P, Zucca G, Prigioni I, Botta L, Casella C, Guth P (1985) The effect of glutamate on the frog semicircular canal. Brain Res 330, 1-9
28. Zucca G, Botta L, Milesi V, Valli P (1993) Sensory adaptation in frog vestibular organs. Hearing Res 68, 238-242

16

POTASSIUM CURRENTS OF HAIR CELLS IN THIN SLICES OF VESTIBULAR EPITHELIUM

Ivo Prigioni, Giancarlo Russo, and Walter Marcotti

Institute of General Physiology
University of Pavia
via Forlanini 6, I-27100 Pavia, Italy

1. INTRODUCTION

  · Hair cells are the typical mechanoreceptors of sensory organs of the acoustico-lateralis system. They operate the conversion of natural stimuli, such as water flow, sound waves and linear and angular accelerations into a bioelectrical signal. Each hair cell is a multipurpose device whose apex acts as a transducer and whose basal pole is specialized for the secretion of the synaptic transmitter. These functions depend on the complement of ionic channels located in discrete areas of the cell.

  Evidence has accumulated that in hair cells, the natural stimuli act directly on the mechanotransduction channels presumably located at the top of the stereocilia [7, 12]. A fraction of these channels is open even under resting conditions and therefore a small cationic current, mainly sustained by K^+ ions, flows from the endolymphatic compartment into the cell [12]. This current can be modulated positively or negatively depending on the direction of stereocilia deflection. A deflection of stereocilia toward the kinocilium increases the transduction currrent by increasing the opening of transduction channels, whereas deflection of stereocilia away from the kinocilium has an opposite effect [12].

  The depolarization of the hair cell resulting from the transduction current determines the activation of different voltage-dependent channels located at the basolateral membrane. Among these, Ca^{2+} channels allow the inflow of calcium and related transmitter release from the basal pole of the hair cell, while several K^+ conductances are involved in cell repolarization [1, 4].

  Current knowledge of the complement of basolateral ionic channels in hair cells derives from patch-clamp studies on sensory cells isolated enzymatically. Evidence has accumulated that hair cells express probably a single type of Ca^{2+} channel, which is similar to L channel described in neuronal membranes [4, 11], and different types of K^+ channels. The most common K^+ channels which have been described include: a delayed rectifier, a transient current of A-type, a calcium-activated K^+ current and an inward rectifier [2, 4]. The basolateral channels in hair cells determine the receptor potential initiated by the transduction current and functional differences among sensory cells of the acoustico-lateralis system are related at least in part to the distribution of these channels.

Neurobiology, edited by Torre and Conti
Plenum Press, New York, 1996

In this article, we describe the distribution of K^+ conductances in vestibular hair cells based on recordings of electrical activity from single cells in situ. This was achieved by applying the whole-cell patch-clamp technique to cells located in thin slices of sensory epithelium of the frog crista ampullaris [8]. Evidence is provided that K^+ conductances are differentially expressed in hair cells located in the central and in the peripheral regions of the crista respectively.

2. CRISTA SLICE PREPARATION

Slices of vestibular epithelium were obtained from the posterior semicircular canal of adult frogs (*Rana esculanta Limneo*). After decapitation, the head of the animal was pinned to the bottom of a perspex chamber filled with normal Ringer solution having the following composition (in mM): 135 NaCl, 3 KCl, 1,8 $CaCl_2$, 3 D-glucose, 10 N-2-hy-droxyethylpiperazine-N''-2-ethasulfonic acid (HEPES)-NaOH (7.25 pH, 275 mOsm). The otic capsule was opened to expose the posterior semicircular canal according to a procedure described elsewhere [10]. After cutting the posterior ampullar nerve distally, both ends of the semicircular canal were dissected near the utricule. The preparation was removed and the longer end of the canal (4-5 mm) was included into a small block of agar that was fixed at the bottom of a chamber containing normal Ringer solution in which Mg^{2+} was substituted for Ca^{2+}. The solution was maintained at 4°C.

Thin slices of whole crista, about 100 μm thick, were obtained by dissecting the epithelium through planes parallel to the long axis of the crista using a vibroslicer (Campden-Instrument, UK). The slices were transferred to a dish containing normal Ringer solution and immobilized by using a nylon mesh fixed to a silver ring. The dish containing the preparation was mounted on the stage of an upright microscope (Zeiss-Axioskop, Germany) equipped with Nomarski differential interference contrast optics.

The morphological features of a fresh thin slice of a crista ampullaris of the posterior canal (× 200 magnification) are shown in Fig. 1A. The slices were about 800 μm long and showed the typical central isthmus and the planar expansions at the periphery. Three regions have been identified in these slices [9]: 1) the peripheral region, about 240 μm long, which corresponds to the planar expansions of the crista; 2) the central region, about 200 μm long, which corresponds to the central part of the isthmus; and 3) the intermediate region, which is about 100 μm long and is located between the two other regions. The intermediate region was not investigated in our experiments. Despite a pronounced variability in shape and size, it was possible to distinguish two main types of hair cells: club-like cells which exhibited an elongated body projecting to the cuticular surface and an expanded basal portion containing the nucleus, and cylindrical cells, which showed a more uniform diameter from the apex to the base.

3. ELECTRICAL RECORDINGS

Whole-cell current measurements were obtained at room temperature (20-22°C) using a L/M EPC7 amplifer (List-Electronics, Germany). The command protocol and data acquisition were carried out by using a Labmaster computer interface and a pClamp system (Version 5.5, Axon-Instruments) running on an Olivetti M300 computer. Recordings were filtered at 3 kHz and sampled at 2.5 kHz. Data were stored for off-line analysis on hard disk. Patch pipettes were pulled from borosilicate glass capillaries (Drummond), coated with Sylgard (Dow-Corning, USA), and heat polished. The pipette filling solution contained (in mM): 122 KCl, 0.1 $CaCl_2$, 2 $MgCl_2$, 10 ethyleneglycol-bis (β–aminoethylether)-N,N,N',N''-

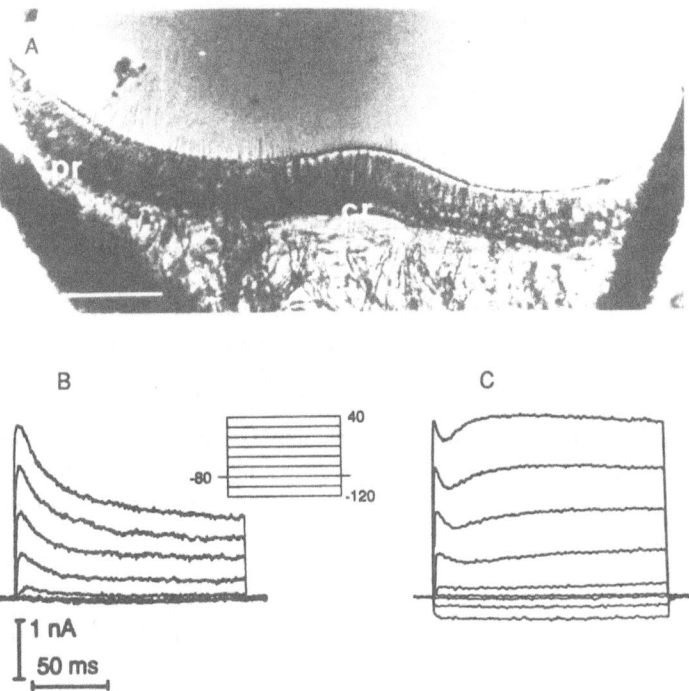

Figure 1. Typical outward K^+ currents. A: photomicrograph of a fresh slice obtained by cutting the crista ampullaris parallel to its longitudinal axis; cr and pr indicate the central isthmus and peripheral regions of the crista respectively. Scale bar, 100 μm. B: K^+ currents elicited in a hair cell from the peripheral regions and showing a large transient component. C, K^+ currents recorded in a hair cell from the central region, showing an initial small and rapidly decaying component. Note the presence of inward currents during hyperpolarizing pulses. Voltage protocol is shown in the inset.

tetraacetic acid (EGTA), 1 adenosine 5'-triphosphate (ATP), 3 D-glucose, 10 HEPES-KOH (7.25 pH, 275 mOsm). After gigaseal formation and cancellation of the residual capacitance of the pipette, the whole-cell mode was obtained by gentle suction. Series resistance ranged from 10 to 20 MΩ and usually was compensated electronically by 50-70%. Recordings were corrected on line for linear leakage currents and residual capacitive transients.

Application of test solution was performed through a multibarreled pipette positioned over the slice close to the hair cell investigated. In Na^+ and K^+- free experiments, cations in the normal Ringer solution were substituted with equimolar amounts of choline choride and NaCl, respectively. In Ca^{2+} -free experiments, Ca^{2+} was substituted with 5.4 mM Mg^{2+} to obtain a similar screening effect on superficial membrane charges [3]. In experiments with the K^+-channel blockers 4-AP (Sigma) and Cs^+, these compounds were simply added to the normal Ringer solution. To block the complex outward K^+ currents [11] generated by hair cells, the following pipette filling solution was used (in mM): 102 CsCl, 2 $MgCl_2$, 20 tetraethylammonium (TEA)-Cl, 10 EGTA, 1ATP, 3D-glucose, 10 HEPES-CsOH (7.25 pH, 275 mOsm).

4. OUTWARD K^+ CURRENTS

The resting membrane potentials of hair cells in crista slices, studied by using the 0-current clamp method, ranged from -35 to -82 mV. A significant difference in average resting potential was found between the two epithelial regions. Cells in the peripheral regions had an average resting potential of about -46mV, whereas cells in the central region exhibited a more negative potential (about -57mV).

When hair cells were held at -80 mV under voltage clamp conditions, 150-ms de- and hyperpolarizing pulses elicited different responses in cells from the two epithelial regions. In sensory cells in the peripheral regions, depolarization invariably produced rapidly activating outward currents, which showed a marked inactivation reaching a steady-state level at the end of the pulses (Fig. 1B). In these regions, no evidence of inward currents was observed during hyperpolarizing pulses. By contrast, in cells from the central region depolarization elicited rapid outward currents, which exhibited little inactivation, whereas hyperpolarizing pulses produced consistently the appearance of inward currents (Fig. 1C). Outward currents usually showed an initial small and rapidly activating transient component. In hair cells from both regions, currents activated at potentials close to -60 mV and their amplitudes, measured at -40 mV, ranged from 1.5 to 4 nA.

The nature of the ions involved in the complex outward currents generated by hair cells from the two regions was assessed by investigating the effects of substituting K^+ in the pipette filling solution with Cs^+ and TEA (see methods). Cs^+ and TEA completely blocked the outward currents, indicating that K^+ is the major ionic species involved in these currents.

5. VOLTAGE AND Ca^{2+} SENSITIVITY OF OUTWARD CURRENTS

The voltage sensitivity of outward K^+ currents in hair cells from both regions was investigated by using the same conditioning voltage protocol. After holding hair cells near their resting potential, the voltage was first conditioned for 5 s between -80 and -40 mV and then increased to a test pulse of 30 mV for 200 ms. As shown in Fig. 2A1 and 2A2, both the transient component and the sustained component of the outward currents were dependent on the conditioning potential: the more negative the conditioning potential, the larger was the current elicited by the depolarizing test pulse.

To investigate the dependence of outward currents on extracellular Ca^{2+}, hair cells were exposed to calcium-free saline. In cells from the peripheral regions, removal of Ca^{2+} reduced significantly both the transient and the sustained component of the outward current (Fig. 2B1). The time course of the residual current was similar to that recorded in normal saline, except that the time to peak was slightly longer. By contrast, in hair cells from the central region, the removal of Ca^{2+} produced not only a reduction of the sustained component, but also a considerable modification of the onset of outward currents with disappearance of the initial rapidly inactivating component (Fig. 2B2). These results indicate that the complex outward K^+ current in hair cells from both the central and peripheral regions consists partially of calcium-activated K^+ currents and shows voltage-dependent steady-state inactivation.

6. INACTIVATING K^+ CURRENT, I_A

The kinetic properties of the outward current in hair cells from the peripheral regions suggest the presence of an I_A. In fact, this current was isolated by using 4-AP, an A-channel blocker in several cells [13]. Use of 4-AP at concentrations < 12 mM failed to provide a satisfactory isolation of I_A, because at these concentrations 4-AP produced a voltage-dependent blockade. Conversely, use of 4-AP at 15-20 mM concentrations allowed a good separation of an uncontaminated I_A [14]. Fig. 3C shows the I_A isolated by subtracting, from the complex outward inactivating current (Fig. 3A), the current remaining after perfusion with

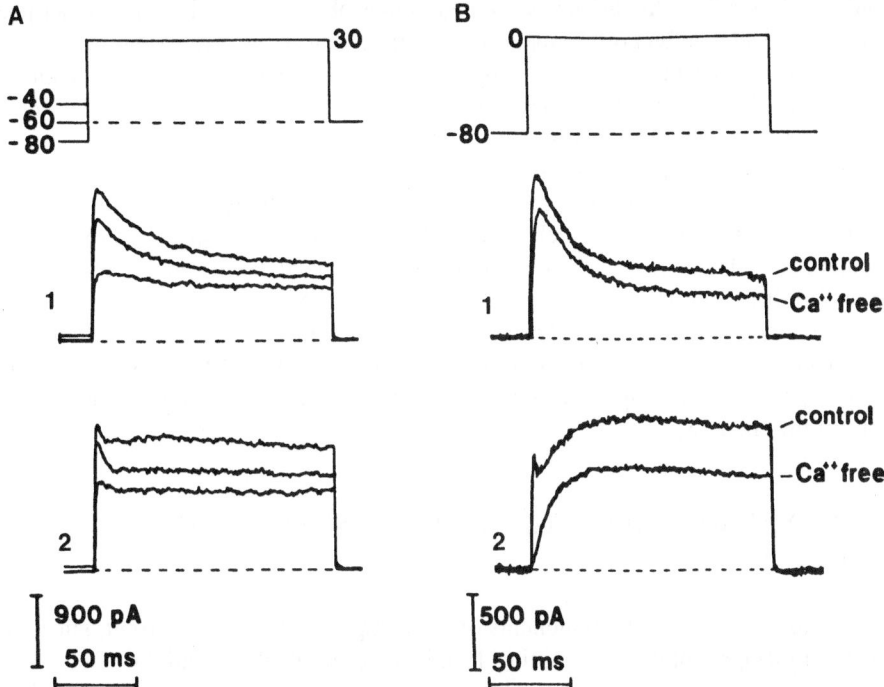

Figure 2. Voltage and calcium sensitivity of K⁺ currents. A: voltage dependence of K⁺ currents in hair cells from the peripheral (1) and central (2) regions. In the voltage protocol shown at the top the conditioning pulses lasted 5 s. B: effects of Ca^{2+} free-saline on K⁺ currents evoked in hair cells from the peripheral (1) and central (2) regions.

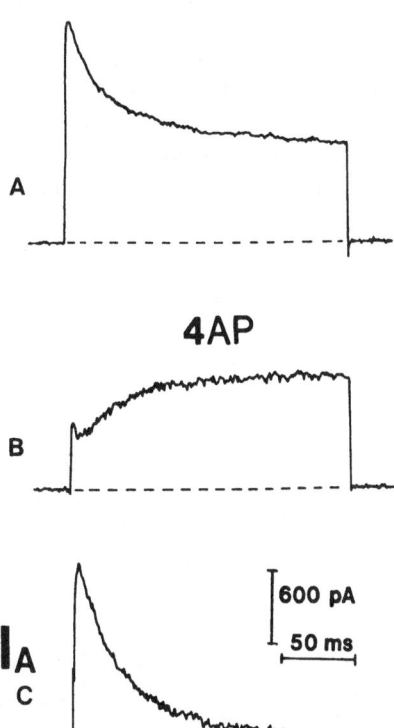

Figure 3. Isolation of an A-type K⁺ current. A: control current. B: current after perfusion with 15 mM 4-AP. C: A-current obtained by subtracting the 4-AP resistant current from the control current. Currents were recorded in response to a single step from -80 mV to 0 mV.

15 mM 4-AP (Fig 3B). At variance with the findings obtained in cells from the peripheral regions, 4-AP had no effect on the majority of hair cells from the central region.

I_A activated near the resting potential of hair cells (-50mV), a potential at which about 40% of A-channels are in a non-inactivated state [14]. The time to peak of I_A was clearly voltage-dependent and decreased from 5 ms at -20 mV to 1.7 ms at 20 mV. I_A inactivation could be fitted by a single exponential and the decay time constant, which showed little voltage dependence, averaged about 33 ms at 0 mV. These features resemble those reported for A-type K^+ currents in other cell types [13]. Since the main suggested role of I_A in vestibular hair cells is to act as a transient buffer of the depolarization induced by the natural stimulus [6], it is clear from the present results that this action is restricted to hair cells in the peripheral regions of the crista. The observation that I_A inactivation may parallel the time course of adaptation of the transduction current [4] suggests that I_A may also be responsible, at least in part, for non-adaptive responses of afferent neurons innervating the peripheral regions of the crista [5].

7. DELAYED RECTIFIER (I_K) AND Ca^{2+}-ACTIVATED (I_{KCa}) CURRENTS

To identify the other components of the complex outward K^+ current, slices were incubated in Ringer solution containing 15 mM 4-AP to block I_A, and the effect of Ca^{2+} removal was tested in hair cells from both regions (Fig. 4). Under these conditions, outward currents activated slowly and showed no significant sign of inactivation during 150 ms test pulses (Figs. 4A2 and 4B2). The time to peak was clearly voltage-dependent and decreased from 140 ms at -40 mV to 65 ms at 40 mV. Outward currents were recruited close to the resting potential (-50/-60 mV), and their amplitude was about twice as large in hair cells from the central region compared to cells from the peripheral regions. The currents were not blocked completely by TEA at concentrations as high as 50 mM. On the basis of these features, these currents could be reasonably classified as delayed rectifiers, or I_K [13]. I_K was the main K^+ current in canal hair cells and it should play a major role in cell repolarization. Being particularly prominent in hair cells from the central region, I_K could contribute to maintain the more negative resting potential found in cells from this region.

When I_K was subtracted from the control current, a Ca^{2+}-sensitive K^+ current could be isolated. On average, the magnitude and the time course of this current were similar in

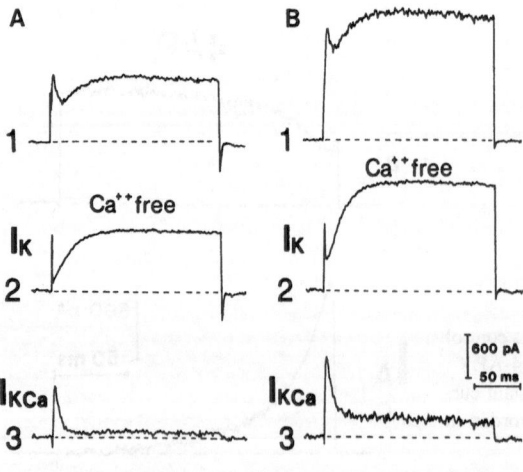

Figure 4. Isolation of delayed rectifier and Ca^{2+}-activated K^+ currents. Hair cells from the peripheral (A) and central regions (B) were exposed to 4-AP to block I_A. Control currents are shown in A1 and B1. The I_K isolated after treatment with Ca^{2+}-free saline is shown in A2 and B2. I_{KCa} obtained by subtracting I_K from the 4-AP resistant current is shown in A3 and B3. Currents were recorded in response to a single step from -80 to 0 mV.

hair cells from both epithelial regions (Figs. 4A3 and 4B3). The Ca^{2+}-sensitive K^+ current consisted invariably of a rapidly activating transient component followed by a sustained component. It was recruited close to -40 mV and the I/V relationship for both the transient and the sustained component showed the N-shaped pattern typical of Ca^{2+}-activated K^+ currents. The time to peak of the transient component was about 3 ms at 0 mV and the inactivation was very fast and almost independent of the membrane potential. The decay time constant was about 4 ms at 0 mV. The possibility that inactivation depends on Ca^{2+} current kinetics can be excluded, because Ca^{2+} currents in canal hair cells have found not to show any evidence of inactivation [11]. Based on these data, the presence of an I_{KCa} showing a transient and a sustained component suggests the presence of two distinct populations of channels. A sustained I_{KCa} has been described in several hair cells and its function, together with the Ca^{2+} current, is to generate the electrical resonance in hair cells of lower vetebrates [4]. A transient I_{KCa} has been described in neurons [13], and it is of interest that, as in our preparation, this current overlaps with I_A and can be differentiated on the basis of its higher threshold, faster kinetics and insensivity to 4-AP. A transient I_{KCa} in canal hair cells may possibly represent a transient buffer of the depolarization induced by Ca^{2+} inflow into the cell.

8. INWARD RECTIFIER CURRENT

As previously reported an inward rectification following hyperpolarizing pulses was observed only in hair cells from the central region of the crista. Inward currents (Fig. 5) were recruited at potentials close to -90 mV and activated very rapidly, reaching a plateau which was maintained throughout at potentials up to -130 mV. At more negative potentials, however, these currents showed an evident outward relaxation, the extent of which increased with increasing hyperpolarization. The outward relaxation disappeared in Na^+-free medium . The inward currents were blocked by exposure to 6 mM Cs^+ and disappeared in K^+-free Ringer solution. These features resemble those described for an inward rectifier current of the I_{IR} type [13]. A general role of the inward rectifier in most cells is to contribute to the membrane resting potential [13]. However, this current is unlikely to play this role in canal hair cells because it was recruited at potentials much more negative than the cell resting potential. A possible role of the inward rectifier is to decrease the membrane time constant to allow hair cells to follow inhibitory stimuli more efficiently. Finally, it is important to consider that the presence of I_{IR} is restricted to sensory cells in the central region of the crista.

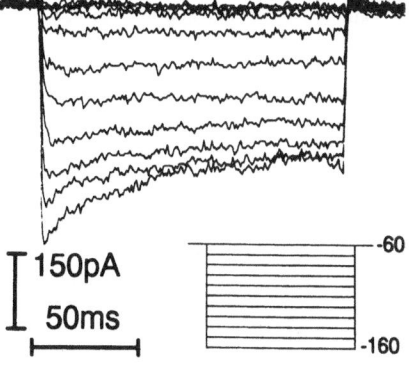

Figure 5. Inward rectifying currents recorded in a hair cell from the central region. Note the marked decay of the current at potentials more negative than -130 mV. Voltage protocol is shown in the inset.

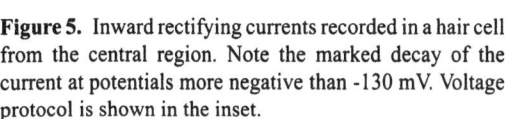

9. DISCUSSION

To our knowledge, this is the first study which provides information on the complement of hair cell ionic currents in thin slices of a sensory organ of the inner ear. We demonstrated that slices of frog crista ampullaris represent a suitable preparation to study the biophysical properties of hair cells in situ using the patch-clamp method. The main finding of the present study was that in semicircular canals K^+ conductances differ in their expression between the central and the peripheral regions of the crista. This may have functional implications, because afferent neurons innervating the central and peripheral regions of the crista also differ in their resting and dynamic responses. In particular, thick afferents innervating the central region show more irregular discharges and higher sensitivity compared with thin afferents innervating the peripheral regions. [5]. It has been proposed that differences in postsynaptic properties are involved in determining the characteristics of the afferent response [5]. Our findings suggest that presynaptic mechanisms may also be involved in differentiating the afferent responses in vestibular receptors and that the complement of K^+ conductances may play an important role.

REFERENCES

1. Ashmore JF (1988) Hair cells. Sci Prog 72: 139-153
2. Ashmore JF (1991) The electrophysiology of hair cells. Ann Rev Physiol 53: 465-476
3. Blaustein G, Goldman DE (1968) The action of certain polyvalent cations on the voltage- clamped lobster axon. J Gen Physiol 51: 279-291
4. Fuchs PA (1992) Ionic currents in cochlear hair cells. Prog Neurobiol 39: 493-505
5. Honrubia V, Hoffman LF, Sitko S, Schwartz IR (1989) Anatomic and physiological correlates in bullfrog vestibular nerve. J Neurophysiol 61: 688-701
6. Housley GD, Norris CH, Guth PS (1989) Electrophysiologocal properties and morphology of hair cells isolated from the semicircular canal of the frog. Hear Res 38: 259-276
7. Jaramillo F, Hudspeth AJ (1991) Localization of the hair cell's transduction channels at the hair bundle's top by iontophoretic application of a channel blocker. Neuron 7: 409-420
8. Masetto S, Russo G, Prigioni I (1994) Differential expression of potassium currents by hair cells in thin slices of frog crista ampullaris. J Neurophysiol 72: 443-455
9. Myers SF, Lewis ER (1990) Hair cell tufts and afferent innervation of the bullfrog crista ampullaris. Brain Res 534: 15-24
10. Prigioni I, Valli P, Casella C (1983) Peripheral organization of the vestibular efferent system in the frog: an electrophysiological study. Brain Res 269: 83-90
11. Prigioni I, Masetto S, Russo G, Taglietti V (1992) Calcium currents in solitary hair cells isolated from frog crista ampullaris. J Vestibular Res 2: 31-39
12. Roberts WM., Howard J, Hudspeth AJ (1988) Hair cells: trasduction, tuning, and transmission in the inner ear. Annu Rev Cell Biol 4: 63-92
13. Rudy B (1988) Diversity and ubiquity of K channels. Neuroscience 25: 729-749
14. Russo G, Masetto S, Prigioni I (1995) Isolation of A-type K^+ current in hair cells of the frog crista ampullaris. NeuroReport 6: 425-428

17

A COMPARTMENT MODEL FOR VERTEBRATE PHOTOTRANSDUCTION PREDICTS SENSITIVITY AND ADAPTATION

Jean-Pierre Raynauld

Centre de recherches en sciences neurologiques
Département de Physiologie
Université de Montréal
B.P. 6128 Succ. Centre-Ville
Montréal, Quebec, H3C 3J7

1. INTRODUCTION

Most vertebrate retinae are duplex in the sense that they contain both rods and cones. Cones have a high threshold, contain different visual pigments, support color vision, and can work under high ambient luminosities. On the other hand, rods have a low threshold, contain a single visual pigment, and operate only at low ambient luminosities. The biochemical machinery which supports visual phototransduction has been the focus of intensive studies over the past ten years and is still an active field of research (Molday and Molday 1987; Stryer 1991; Chabre, Antonny, Bruckert, Vuong 1993; Yau 1994; Koch 1992; Heck and Hofmann 1993; Gillespie 1990; Kaupp and Koch 1992). Although minor differences have been found between the proteins involved in the phototransduction cascade of rods and cones, none of these differences offers a satisfactory explanation of the above fundamental differential behavior of rods and cones. Furthermore, the rods are believed to have evolved from the cones (Walls,1963), and possibly kept the same biochemistry. However, following the lead of Forti, Menini, Rispoli & Torre 1989 in simulation studies, Ichikawa (1994a, 1994b) has been able to mimic important aspects of the differences in gain and kinetics between rods and cones by manipulating the rate constants of the reactions in the cascade.

This paper presents a compartment model of vertebrate phototransduction and proposes that the major difference between rods and cones is the number of compartments present in their respective outer segments. This single difference not only explains the single photon response and the intensity response curve, but also, together with the decay characteristic of the response, the adaptation behavior, that is the reduced sensitivity under non-bleaching background conditions. The only other requirements are that a single isomerization closes all the cationic channels associated with a compartment and that the current decay represents an underlying limiting process. It must, however, be understood that this is a first approximation model in the sense that a number of experimental results are ignored in order to make the model mathematically tractable. An initial account of this model has been presented elsewhere (Raynauld and Gagne 1987).

2. THE MODEL

2.1. The Nature of the Compartment

For the purpose of this model, a compartment does not necessarily need to be a physical compartment in the sense that physical barriers such as membranes would be required to define its boundaries, it can be an operational compartment. An example of such an operational compartment could be the length of tube where a bolus of radioactive nuclei, moving at a given speed, looses 90% of its radioactivity. In this sense, any molecule or protein with a finite lifetime operates within a compartment. Thus excited rhodopsin, excited transducin, and excited phosphodiesterase all operate within compartments whose size depends on the diffusion properties of these molecules and on their respective lifetimes. It is therefore conceivable that the outer segment of both rods and cones is made up of compartments which are biochemically isolated from each other but electrically connected in a linear fashion such that current changes in different compartments can add algebraically.

2.2. The Size of the Response

The model makes the additional hypothesis that the action of a given photon which produces an isomerization is maximal. This means that two or more simultaneous isomerizations within the same compartment will not produce an effect greater than a single isomerization. The kinetics of the rise time of the response may change with multiple simultaneous isomerizations but the maximal amplitude will be the same as that of a single isomerization. The consequence of these two hypotheses is that if an outer segment, made up of N compartments, carries a dark current of magnitude J, then the single photon response will be J/N and the intensity response of that outer segment will be of the exponential saturation type regardless of the shape of the single photon response. It can be any shape, even a square pulse. This model was first proposed by Lamb, McNaughton & Yau (1981) and follows from the following argumentation. Suppose that a flash of intensity I photons/square microns affects n different compartments in an outer segment made of N compartments, the fractional response to this flash will be n/N. This number n/N also represents the probability that a compartment is the locus of one or more isomerizations and $1 - n/N$ represent the probability of no isomerization in a compartment or zero event . For a flash of light illuminating the whole outer segment, the absorption of photons can be considered a Poisson process because individual absorptions are independent of each other and because the incremental probability of absorption is proportional to the incremental size of the photon flux density . This last requirement is met if the optical density is constant along the outer segment and the geometry of the outer segment is considered cylindrical. The probability of zero event in this Poisson process is $Exp[-kI]$. In this expression, k represents the cross-section in square microns for isomerization of the compartment and I the intensity of the flash expressed in photons per square micron. By simple substitution one finds that the fractional response r/r_{max} is given by the following formula which is illustrated in Fig.1:

$$r/r_{max} = n/N = 1 - Exp[-kI]$$

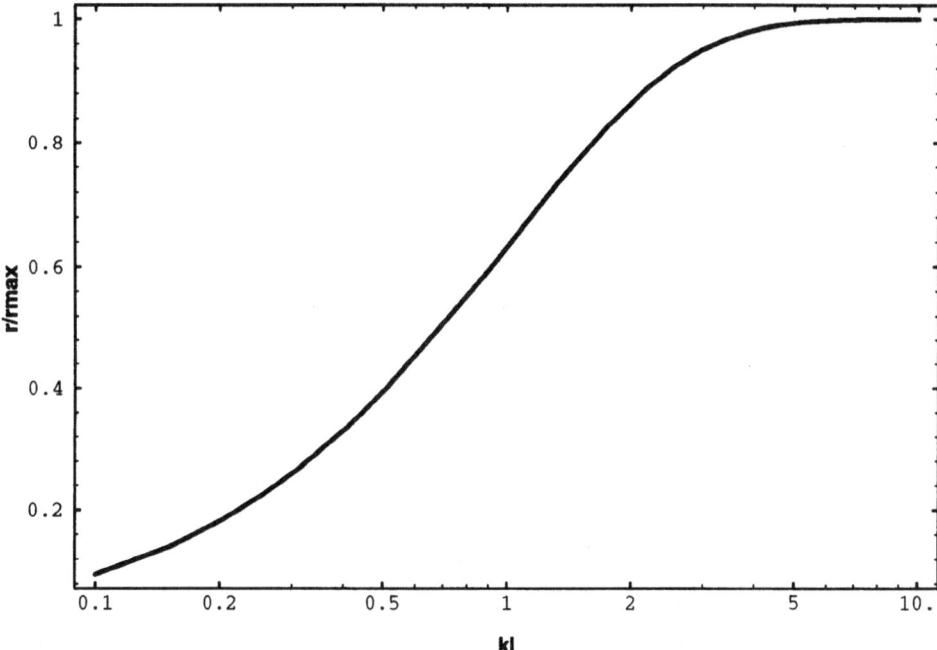

Figure 1. Normalized amplitude response of the photoreceptor as a function of the factor kI, where k is the cross-section for isomerization of the compartment in μm^2 per photon and I, the flash intensity in photons per μm^2.

The above results, that is the size of the single photon response and the shape of the intensity response curve are entirely independent of the shape of the single photon response. The presence of the exponential in this formula comes from the equation describing the Poisson statistics and not from the shape of the rising phase of the response as described by the kinetics of the biochemical cascade. This point is made because Pugh and Lamb(1993) have shown that for $t << t_{peak}$, $r(t)/r_{max}(t)$ also shows exponential saturation under certain assumptions regarding the cascade kinetics. However this latest analysis does not yet give a solution when $t = t_{peak}$.

2.3. The Action of Background Light on the Sensitivity

In order to predict the sensitivity change in this system as a function of background light, it is important to know the time course which characterizes the recovery of the compartment after it has been hit by a photon, that is select a shape for the single photon response. I have chosen the simple exponential $Exp[-t/T]$, where T is the time constant of the exponential current decay. The implication of this choice is that the current response represents a limiting process. No other process having a slower time course should affect the recovery under this hypothesis. One such process is pigment regeneration which takes place with a time course of the order of minutes. Since I am considering a non-bleaching situation, it will not affect the model. I am also ignoring that the decay time changes to a certain extent as the background intensity increases (Schnapf, Nunn, Meister, Baylor 1990).

The approach used to analyze the effect of background light is novel and uses a theorem in statistical mechanics called the *"ergodic theorem or hypothesis"* (McQuarrie 1976) which states that *"for a stationary random process, a large number of observations made on a single system at N arbitrary instants of time have the same statistical properties*

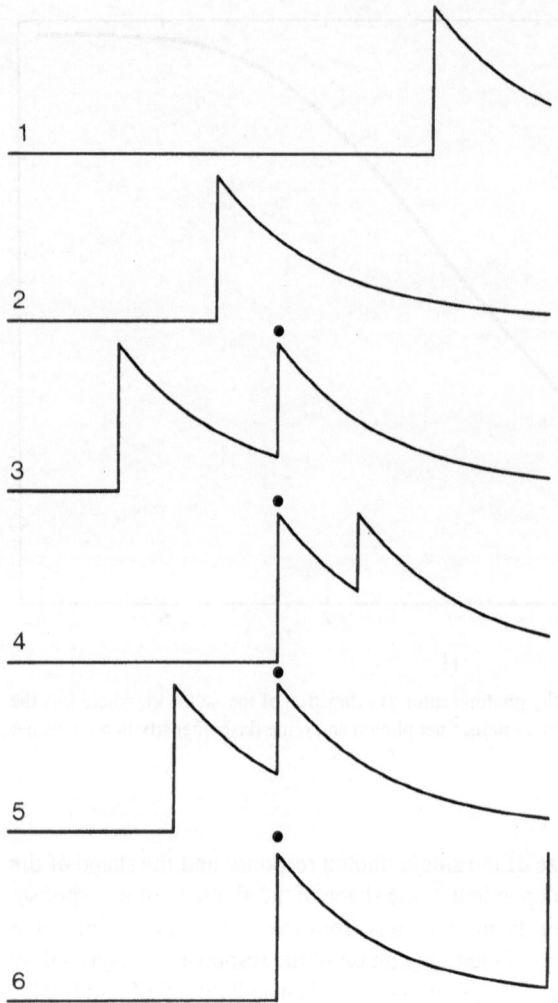

Figure 2. Cartoon representing the current response of six compartments subjected to background photons and to a test flash occurring at mid time and producing an isomerization in four compartments.

as observing N arbitrary chosen systems at the same time from an ensemble of similar systems." If the action of a steady background light on an outer segment can be considered stochastic and stationary, I can apply the ergodic hypothesis by equating compartments to systems.

Due to the absorption of photons which have led to isomerization, a certain number of compartments are not in the dark-adapted state but rather in a process of exponential recovery during which their sensitivity is reduced. The contribution of these compartments to a response of a test flash will be different for each compartment and depend on how long ago each compartment has been the site of an isomerization due to a background photon. This is illustrated in Fig.2.

This figure illustrates the effects of a test flash producing four isomerizations (identified by dots at the peak of the response) in four out of six different compartments. Background photons hit compartments #1 and #2, while the other compartments are hit by both background photons and test photons which occur in the middle of the trace. In compartments #4 and #6, which are completely dark-adapted, the response is maximal, in the other two (#3, #5), the response amplitude depends on how long ago that compartment has been hit by a background photon. The sum of these responses divided by four gives the mean response per isomerization for that very specific background situation. This represents *"observing N arbitrary chosen systems at the same time from an ensemble of similar systems"* in the ergodic hypothesis. If N is large, I would have a very good estimate of the single photon response of my system under a given background. The *"ensemble average,"* as I have done above, can be substituted by a *"time average"* of the observations done on one compartment as described below.

In an outer segment made of N identical compartments and under a total background generating I_{bk} isomerizations per second, each compartment is subject to random isomerizations at a rate equal to I_{bk}/N. To measure the sensitivity of a given compartment, I only need to observe the response to N isomerizations occurring randomly in time and to average the response since all compartments are considered identical. Actually these isomerizations need not come from "test photons"; the photoreceptor cannot tell whether the photon comes from the background or the test flash. I only need to observe the response to N random isomerizations due to photons belonging to the background to test the sensitivity of the system. This part represents *"a large number of observations made on a single system at N arbitrary instants of time"* in the ergodic hypothesis. Because of the nature of light, the intervals between isomerizations in a given compartment are arbitrary and Poisson distributed, and the mean interval is N/I_{bk} second. After a given isomerization, a compartment will start its recovery process with a time constant T and will be the locus of another isomerization after, on the average, a time equal to N/I_{bk}. To find out the mean response to such a process, using Mathematica®, I ran a series of computer simulations. Using the Exponential Distribution, I generated 5000 poisson distributed random intervals having a mean rate of 5 per sec., each event generating a step of amplitude 1 decaying exponentially with a time constant of 200 mscc. According to the model, the amplitude of the response to a given event is:

$$1 - \text{Exp}[-t_i/T]$$

where t_i represents the time since the previous event. The mean amplitude to the 5000 poisson distributed random events was then calculated, it was 0.4997. Keeping the same time constant of 200 msec., the simulation was repeated for mean frequencies of 0.5 to 500 isomerizations per sec. and the mean response to each frequency calculated. The points representing these mean responses were then plotted on a loglog graph as a function of frequency and are presented in Fig. 3, the continuous line is the Weber-Fechner $I_o/(I_o + I)$ function with I_o set to $1/T$. The superposition is perfect.

One can therefore conclude that model adapts according to the Weber-Fechner law. The obvious advantage of this model is that from the knowledge of the number of compartments and of the decay time constant of the photocurrent, one can localize the adaptation curve on the background intensity axis, that is determine I_o as being equal to N/T in a system made of N compartments. Up to now the Weber-Fechner equation was simply a mathematical fit to a series of experimental observations. In phototransduction, if the model is true, it is now linked intimately to structure of the outer segment and the kinetics of the biochemistry responsible for the current recovery.

2.4. Does This Model Fit Reality?

Up to now I have tried to establish that a system constructed of identical compartments which are chemically isolated but electrically connected, carrying the same fraction of the total current, and responding maximally to a single event will saturate exponentially regardless of the shape of the single event in time. Furthermore, if the shape of the single event can be approximated to a sharp rise followed by an exponential decay, then the sensitivity of that system as a function of background will follow the Weber-Fechner function. However, one can legitimately ask to which extent does this apply to vertebrate

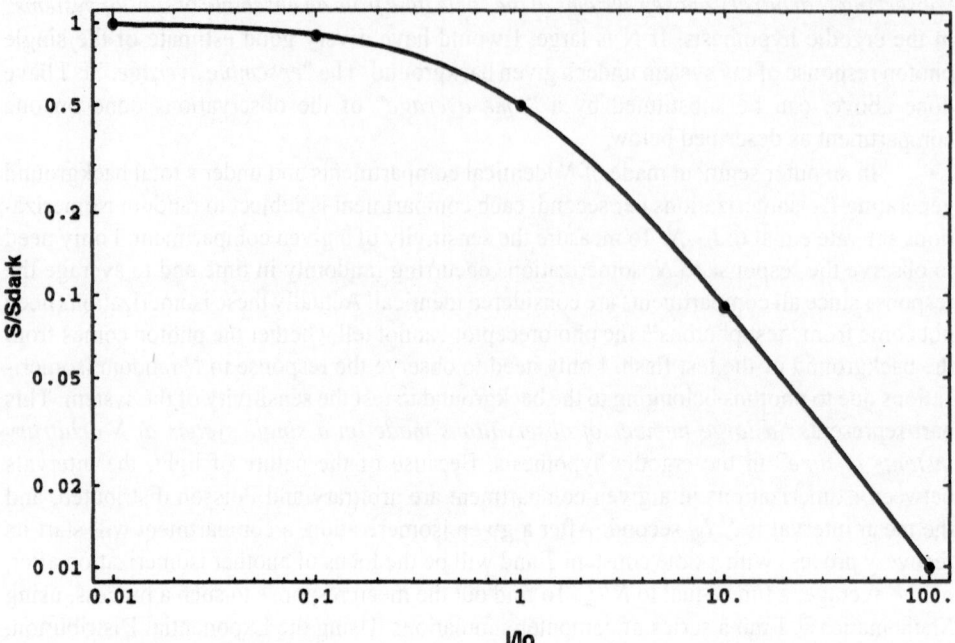

Figure 3. Light adaptation behavior of the model: the solid line is the Weber-Fechner function and the points are the results of the computer simulation.

phototransduction. In other words, are the vertebrate rod and cone outer segments constructed of biochemically isolated compartments which are maximally excited by a single isomerization?

2.5. The Case of the Vertebrate Cone

As it is well known, the major anatomical difference between rods and cones is that, in the cones, the outer segment is made of infoldings of the plasma membrane while in the rods this is true only for the first micron or so. Can I define a credible compartment in the cone outer segment? One of the most recent anatomical studies of the cone outer segment is that of Eckmiller(1987). It clearly shows that each fold hangs out in extra cellular space and its interior is in communication with the rest of the outer segment through only a small part of the circumference of the fold adjacent to the ciliary structure. From the article of Eckmiller(1987), I estimate this arc to be of the order of 20 degrees. This small part is called here a "neck." Can the cone fold therefore constitute an isolated biochemical compartment bound by the membranes which define its geometry? Electrically, this small neck has a small electrical resistance because its length is minute. The cone fold is thus connected to the ciliary structure of the outer segment through a low resistance path.

The biochemical isolation of the cone fold becomes more evident when one considers that any molecule in order to exit or enter the fold has to pass through this neck opening. For example, an activated opsin molecule in a given fold would first have to find the exit, move up or down in the ciliary section, and enter into an adjacent fold in order to active a transducin molecule located there. These motions would have to take place before the opsin became inactive. Random walk in two dimensions is described by the following formula:

$$<r^2> = 4Dt$$

where $<r>^2$ is the mean square distance traveled, D is the diffusion coefficient taken here to be similar to that of rhodopsin (circa 0.5 μm^2-sec^{-1})(Poo & Cone 1974; Liebman & Entine 1974), and t is the time in second. Given that the rise time of the small flash cone response is in the order of 35 msec. in mammals (Schneeweis & Schnapf 1995) and assuming that the opsin is inactivated at the time of the peak, one obtains a value of r equal to 0.26 micron. Opsin molecules which are further than this distance away from the opening have little chance of leaving the fold during the rise time of the cone response. Clearly, these numbers indicate that the excited visual pigment molecule is restricted in its action to the fold it belongs to. Similar calculations can be done for transducins and phosphodiesterases attached to the membrane. The diffusion coefficients of these proteins are not believed to be more than twice as large as that of the visual pigment(Lamb and Pugh 1992). Their action is also estimated to be restricted to the fold that they were belonging to when activation took place.

Molecules which are diffusing in solution such as cGMP or Ca^{2+} have diffusion coefficients which are much greater than membrane bound proteins and one can ask to which extent, cGMP molecules present in neighbouring folds diffuse into the fold which is the site of photo activation. It is true that reduction of cGMP concentration in a given fold will generate a concentration gradient and a number of cGMP molecules present in adjacent folds will diffuse out of theses folds and enter the active fold and be hydrolyzed there. However, the rod disk arrangement constitutes a system of baffles which reduces the effective diffusion coefficient of cGMP in the range of (1.4- 5.5 μm^2-sec^{-1}) (Olson and Pugh 1993). More recent results give a value in the range of (60-70 μm^2-sec^{-1}), but these values are still six to seven times lower than in aqueous solutions (Koutalous et al 1995). After 35 msec. the mean distance traveled is only 3 μm for a molecule diffusing at 70 μm^2-sec^{-1}. Although a similar analysis has not been done for the cone outer segment, given the cone outer segment geometry, the effect should be even stronger. The conclusion is thus that, over the time course of phototransduction, considering that folds of the outer segment of cones are chemically isolated compartments, is probably a good first approximation.

The next question is whether or not a single isomerization closes all the open cationic channels associated with a given fold? The inside volume of a mammalian cone fold is estimated to be 1.6 x 10^{-2} femtoliter (assuming an inside height of 5 nm and a radius of 1 μm); if the resting concentration of free cGMP is approximately 5 μM, then the number of cGMP molecules in the fold is circa 50. Knowing that the PDE surface density is approximately 1000/μm^2 (Dumke, Arshavsky, Calvert, Bownds, Pugh, Jr. 1994), the number of PDE molecules in the fold is circa 6000. This means that, if fully activated, there are 120 molecules of PDE for each molecule of free cGMP to hydrolyze. Actually it is interesting to note that there are two bound molecules of cGMP to each PDE (Gillespie & Beavo 1989) for a total of 12, 000 cGMP bound to PDE non-catalytic sites. If the total dark current of 40 picoamp. is equally divided among the 800 folds of a typical cone outer segment and the channel conductance is 0.1 ps, at the resting potential of 40 mv., then the number of open channels is 13. Since 3 cGMP molecules are required to keep a channel open (Fesenko, Lelesnikov, Lyubarsky 1985; Haynes, Yau 1985) the total number of cGMP molecules bound to the channel proteins is 39, neglecting the channels to which only one or two cGMPs are bound. At 50, the molecules of free cGMP are nearly equal in number to the molecules bound to the channel proteins. Given that PDE has an effective K_m of 600 micromolar, a concentration of 22 μM (Dumke et al. 1994), and a turnover number of 1000 per sec.(Gillespie 1990), the initial number of 50 cGMP would be reduced to 1/e^2 of its value or to 7 in less than 54 msec. This leaves little time for cGMP from neighboring folds to come in, or for the cGMP cyclase to react in order to prevent the closing of all the open channels. Therefore, I am confident that the hypothesis of total closure of all cationic channels associated with a given fold is a possible one. The consequences of accepting the above hypotheses are as follows:

1. the single photon fractional current response of cones should be $1/N$, where N is the number of folds or disks in the outer segment. It also indicates that, although increasing the length of the outer segment (that is N) increases the probability of photon capture by increasing the overall optical density, it decreases the size of the single photon response. Actual length of cone outer segments therefore represents an engineering compromise between the size of the single photon response and the probability of photon capture.

2. from the knowledge of the thickness of the fold (33 folds/μm, the optical density per unit thickness (0.016 OD/μm, Harosi 1975) , the quantum efficiency for isomerization of the visual pigment (0.67, Dartnall 1967), and using equation 3 of Baylor et al. (1984) k for cones in the total occlusion model becomes $0.00059d^2$, where d is the cone diameter in microns.

3. I_o in the Weber-Fechner law is given by N/T where N is the number of folds in the outer segment and T the time constant of the exponential which can be fitted to the decay of the small flash response.

2.6. The Case of the Vertebrate Rod

Considering the geometry of the rod outer segment, it is not evident that we are facing a system made up of compartments. Although it is true that a given rhodpsin molecule, being a transmembrane protein, cannot leave the disk to which it belongs, and that PDE is in the same situation being a permanently membrane attached protein, the case of transducin is more ambiguous since there have been reports of two types of transducin, membrane attached and soluble (Chabre, Antonny, Brucket, Vuong 1993). The soluble kind could leave its native disk and possibly excite a PDE molecule located on an adjacent disks or on the cylindrical plasma membrane. However, to a modeler, the most perplexing fact in rod phototransduction is that a single isomerization produces the same 3- 5% reduction in dark current in the Bufo rod (Baylor et al. 1979) which is one of the largest rods (6 μm x 50 μm) and in the smallest rod (2 μm x 25 μm) in monkey (Baylor et al.1984). This small size rod is also found in bovine, rat and human. The ratio of the volumes of these two rods is 27. Following an isomerization in Bufo rod, the biochemistry activated has therefore to hydrolyze 27 times more cGMP than in the monkey rod in order to reduce the cGMP concentration to the same level in the two rods. Since the proteins involved in the cascades are all initially membrane bound, their number only increases as the square of the dimensions, thus there is, per disk, only 9 times more rhodopsin, transducin, and phosphodiestrerase in a bufo rod as compared to a monkey rod assuming a constant surface density across species for these molecules. Soluble molecules such as Ca^{2+} and cGMP increase in number as the cube of the dimensions. Larger rods would be expected to be less sensitive than smaller ones, but this is not the case as found experimentally.

The approach taken had therefore to be empirical. Analyzing in many rods the amount of plasma membrane involved in the total occlusion model (Lamb, McNaughton, Yau 1981), it was found that this area was equal to the area of membrane present in one disk. That is the occlusion length l could be predicted by equating the cylindrical area of plasma membrane (πdl) to the amount of membrane present in one disk ($2\pi d^2/4$) where d is the diameter of the rod. The solution being $l = d/2$. It was as if in terms of membrane area affected by an isomerization, the effects were the same in rods and in cones of the same diameter, only the site of action was different. In the cones, it is the membrane of the disk shaped fold while in the rods it is the plasma membrane whose area is equal to that of the disk. The consequences of accepting the above are as follows:

1. The fractional current response of a rod to a single isomerization becomes equal to $d/(2l)$, where d and l are respectively the diameter and the length of the outer segment.

2. Since $d/2$ defines the occlusion length, again using equation 3 of Baylor et al. (1984) the value of k in the total occlusion model becomes $0.01\ d^3$, where d is the rod diameter in μm. The dimensions of the constant 0.01 are μm^{-1}.

3. I_o in the Weber-Fechner law is given by N/T where N is the number compartments the outer segment and T the time constant of the exponential which can be fitted to the decay part of the small flash response. N is calculated by dividing l by $d/2$ which should be the same as dividing the total dark current by the amplitude of the rod response to a single isomerization.

2.7. Comparison with Experimental Results

Predictions obtained from the model can be tested against values obtained from experiments for the following three parameters: the single photon response, the value of k in the total occlusion formula, and the value of I_o in the Weber-Fechner formula. Although the single photon response and the value of k are not independent of each other, these two parameters will be used in the comparison tables since experimental results have been reported both ways. Data relating to both rods and cones will be combined.

Table 1 summarizes the data concerning the single photon response of various rods and cones in a number of species. The theoretical response is $1/N$ for the cones where N is the number of folds in the outer segment; for the rods the theoretical response is $d/2l$ where d is the diameter, and l the length of the outer segment.

Table 2 compares the observed k in the intensity-response plot obtained by various experimenters to the k calculated using the theoretical "occlusion length," the quantum efficiency, the optical density of the visual pigment, and the dimensions of the outer segment.

It can be observed that the predicted sensitivity, expressed either as percent of total current or voltage reduction due to a single isomerization (Table I) or as the value of k in the total occlusion formula (Table 2) is in general agreement with the experimentally obtained

Table 1. Single photon response: model vs experiments

Species	Receptor type	Observed %	Predicted %	Ref.
Turtle	cone	0.16	0.125	1
idem	idem	0.05	0.125	2
Walleye	idem	0.14	0.2	3
idem	twin cone	0.14	0.2	3
Salamander	red cone	0.76	0.33	4
idem	blue cone	18	0.39	4
idem	rod	1	23	5
Squirrel	cone	0.02	0.5	6
Toad	rod	5	6	7
Monkey	rod	3-5	4	8
Rabbit	rod	5	5	9

References: 1- Baylor et al. (1973), 2- Schnapf & McBurney (1980), 3-Burkhardt et al. (1986), 4- Perry & McNaughton (1991), 5- Hodgkin & Nunn (1988), 6- Kraft(1988),7- Baylor et al. (1979), 8- Baylor et al. (1984), 9- Nakatami et al. (1991)

Table 2. Intensity reponse: model vs experiments

Species	Receptor type	"k" observed x 1000	"k" predicted x 1000	Ref.
Human	cone	1	1	1
Macaque	red cone	0.38	1	2
idem	green cone	0.42	1	2
idem	blue cone	0.40	1	2
idem	rod	6-20	80	3
Human	rod	18	80	3
Monkey	rod	68	80	4
Bass	rod	24	40	5
idem	single cone	3	9.4	5
idem	twin cone	0.5	15	5
idem	fast twin	0.07	15	5

References: 1- Schnapf et al. (1987), 2-Schnapf et al. (1990), 3- Kraft et al. (1993), 4-Baylor et al. (1984), 5- Miller & Korenbrot (1993)

values. There are exceptions notably in the sensitivity of the Salamander rod and blue cone, and that of the Bass twin and fast twin cones. For the Salamander rod, the presence of deep incisures (see Fig. 5 in Olson, Pugh 1993) in the rod disk represents diffusion barriers to the cascade proteins moving in or on the membrane and this could be an effective mean of reducing sensitivity. For the Salamander blue cone and the Bass twin and fast twin cones, I have no explanation. However, in the Salamander I have difficulty in seeing how a blue cone, which has a sensitivity 24 X greater than that of a red cone, can operate in a color opponent system to generate color vision. The same argument can be used in the case of the Bass cones where the ratio of the sensitivities of the single and fast twin cones is 43. From an engineering point of view, the best color system would be one in which all cones would have the same dark-adapted sensitivity and the same behavior under light adapting conditions. These criteria seem to have been achieved in the Macaque retina (Schnapf et al. 1990). The sensitivity of the cones in the ground squirrel retina is also much lower than predicted. Using the value of 50 pa for the dark current, a channel conductance of 0.1 ps, a resting potential of -40 mv, and 200 folds in the other segment, a single photon response of 0.02% change in dark current (Kraft 1988) corresponds to the closing of only 2.5 channels in a fold. However, in a psychological study, Jacobs and Yolton, 1971 have found that ground squirrel dichromatic color vision has the same threshold as trichomatic human vision. I take this to indicate a possible problem with the electrophysiological data reporting the low sensitivity of ground squirrel cones.

Table 3 compares the experimentally obtained value for I_0 in various Weber- Fechner plots to our predictions based on the number of compartments and the decay time constant of the small flash response. Predicted I_o is N/T, for the cones, the number of compartments is equal to the length of the outer segment in μm. multiplied by 33 which represents the number of folds per μm.. For the rods, the number of compartments is equal to the length of the outer segment divided by the occlusion length $d/2$ except for the Salamander rod as described below. The decay time of the single photon or small flash response was measured from the figures in the referred publications with the help of a micrometer and represents the time from the peak to amplitude equal to 37% of the peak.

The formulation of the adaptation behavior where I_o is related to the time constant T offers an explanation of a phenomena which was considered bizarre. J.L. Schnapf (1981) studying the sensitivity, kinetics and adaptation along the length of toad rods found that rods have a lower sensitivity and slower kinetics at the tip than at the base, but were bizarrely adapting at lower background intensities at the tip than at the base, seemingly at odds with

Table 3. Background adaptation: model vs experiments

Species	Receptor type	I_0 Observed Rh*/sec.	I_0 Predicted Rh*/sec.	Ref.
Turtle	cone	2,000	7,273	1
Toad	rod	4	11	2
Cat	rod	35	83	3
Bass	single cone	8,500	3,800	4
idem	twin cone	33,700	13,200	4
idem	rod	3	75	4
Human	rod	120	110	5
Monkey	rod	100	133	6
Rabbit	rod	42	83	7
Salamander	rod	80	65	8

References: 1- Baylor et al. (1974), 2- Fain (1976); Baylor et al. (1979), Tamura et al. (1989), 4- Miller & Korenbrot (1993), 5-Kraft et al. (1993), 6-Baylor et al. (1984), 7- Nakatani et al. (1991), 8- Torre et al. (1992).

the lower sensitivity of that region. Similar findings had been described (Hemilia and Reuter 1981) from electroretinogram recordings. However, from the above formula, slower kinetics means an increased value of T and therefore a reduced value for N/T which represents the background isomerization rate at which the sensitivity is reduced by 50%. Using the ratio of the time constants at the tip and at the base to predict the ratio of I_0's, one obtains 1.35. The actual value from J.L. Schnapf (1981) is 1.66 in good agreement with the prediction. This approach offers a way of testing the model. Kinetics of the photoresponse are known to vary with temperature (Lamb 1984), adaptation properties can be tested as a function of temperature to see if the variation of the decay time has the effect on the adaptation parameter I_0 as predicted by N/T formula.

3. DISCUSSION

In this paper I have presented a model which addresses the differential behavior of vertebrate rods and cones with respect to sensitivity and adaptation. This differential behavior has been the focus of many investigations both experimental and theoretical. In 1981, Lamb, McNaughton and Yau proposed the total occlusion model in order to explain the shape of the intensity-response curve of photoreceptors; if the light intensity was expressed in photons per square microns, then the single parameter k was representing the isomerization cross-section in microns square sustained by the compartment where the total occlusion was taking place. This expression was providing an adequate fit to experimental data and could be used in place of the Michaelis-Menten type of formula. Here I have extended this approach in identifying the compartments by an analysis of the confinement of the biochemistry by the ultrastructure of the outer segment of cones and rods and I have obtained a value for the size of these compartments. This analysis leads to a numerical value for the parameter k in the total occlusion model (Lamb et al. 1981) for both types of receptors.

The hypothesis that a single isomerization closes all the ionic channels associated with a single cone fold receives support from human cone threshold experiments. Psycho-physical experiments indicate that a test spot of 1 min of arc must deliver 600 photons (550 nm) at the cornea in order to be perceived reliably (Hood and Finkelstein, 1986). According to Schnapf et al. 1990, this corresponds to 32 isomerizations per cone. At this intensity, the probability of two or more isomerizations in the same fold is negligible, this means that 32 folds have been the site of an isomerization. Since there are approximately 825 folds in a 25 μm cone, and if each fold carries the same fraction of dark current, the total occlusion of 32 folds represents a 4% (32/825) change in the dark current. This would mean that, at perception threshold, both rods and cones must see their dark current reduced by the same fraction (circa 4%). In a sense, this is not surprising since the anatomy at the synaptic junction located at the cone pedicule or at the rod spherule is very similar as seen in electron microscopy. In order to carry the message to the next level reliably, the same current change at the level of the outer segment would be required in the two types of receptors. When tested against published experimental data the model gets good marks even if it is not perfect.

The finding that, in rods, the area of plasma membrane occluded is related to the disk area may be more than fortuitous. It has been proposed a few years ago (Liebman et al. 1987) that the disk size is a major factor controlling rod sensitivity. Furthermore, it is surprising that the surface density of PDE is smaller than that of transducin by a factor of 9 in mammalian rods (Reviewed in Pugh and Lamb, 1992). Since transducin remains bound to PDE when this one is activated, there would be an overproduction of transducin assuming that all the disk transducins are activated. One could argue that we are in the presence of a gain loss in the sense that more PDE's could be activated if they were present. Another problem with the current rod model, in which the PDE is located on rod disk membrane, is that the minimum cGMP concentration is located at the center of the disk (Dumke et al. 1994) when the optimum location for this minimum would be at the plasma membrane where the cGMP activated channels are located. One solution to that problem would be to locate rod PDE on the plasma membrane close to the channels. This new location would explain why Roof et al. (1982) failed to find PDE on rod disk membranes with the electron microscope when, according to surface density numbers and its larger size than transducin, it should have been clearly visible. This was not due to elution since PDE was present in SDS PAGE columns of the same preparation. This new location would also equate the surface density of PDE on the plasma membrane in amphibian and mammals to 16,500 PDE per μm^2 (value calculated with the data given below and based on 33 disks per μm of outer segment) since the diameter of toad rod is about three times that of bovine or human rod. Current values based on disk location give a surface density of PDE three times lower in amphibian (167 per μm^2) when compared to mammals (500 per μm^2) (Sitaramayya et al, 1985; Hamm and Bownds, 1986). A lower surface density of PDE in toad rod should normally result in a lower transduction gain but it does not since a single isomerization produces the same percent current reduction in these two rods.

An hypothesis which would accommodate the model would be that phototransduction is a surface phenomenon. A single isomerization would result in the closure of channels located over an area determined by the area of the disk. For the cone, the area is the cone disk itself. For the rod, activated transducin, in number proportional to the size of the disk, would migrate from the disk to plasma membrane where PDE would be located. Activated PDE molecules would then sweep the plasma membrane hydrolyzing cGMP close to the membrane or as it becomes unbound from the channel. Because of the finite lifetime of activated PDE, the area affected would be limited and would define the rod compartment.

The most striking feature of the model resides in its ability to predict the isomerization rate required to reduce the sensitivity by 50% of both rods and cones in warm as well as in cold blooded species as shown in Table 3. It does that by making use of the "ergodicity theorem." The result of this analysis is that the single parameter I_o in the Weber-Fechner adaptation formula can now be calculated. It is simply N/T where N is the number of compartments in the outer segment used to predict the intensity-response curve and T is the time constant of the decaying exponential which can be fitted over the decay part of the small flash response. When comparing rods and cones, the model offers an explanation for the different sensitivity and adaptation by proposing that the number of compartments supporting the dark current is much greater in cones than in rods. The effect of this difference is much stronger when comparing the adaptation properties because of the way the number of compartments appears in the formulae.

When comparing rods of warm blooded species to cold blooded ones, the difference is not in the sensitivity since a toad rod and a monkey rod show the same 3-5% change in dark current following an isomerization, but in the adaptation properties these are shown here to be related to the time constant T of the small flash response which is much longer in cold blooded vertebrates, thus offering an explanation for the much lower background intensities required to reduce the sensitivity by 50% in these species.

Regarding the intensity-response curve, the data for the Salamander indicates that the size of the compartment is much smaller than predicted. This could be due to the presence of lobules in the outer segment disks which restrict the diffusion of membrane bound proteins leading to a smaller number of excited transducins and/or phosphodiesterases molecules in a given amount of time. This is equivalent to raising the number of compartments. A 1% current reduction (Hodgkin and Nunn 1988) corresponds in this model to the presence of 100 compartments. When this number is used in the N/T formula using 1.54 sec. as the decay time constant of the Salamander rod small flash response, one predicts a value for I_0 close to that found experimentally (Torre, Straforini, Campani 1992), validating in a sense the approach.

The model does not preclude a role for calcium or any other molecules in adaptation (Tamura et al 1991; Kawamura, 1993; Kaupp and Koch, 1992) though it suggests that their action must result in changing the value of the decay time constant T and possibly changing the value of N for the rods where the size of the compartment could be affected by factors such as diffusion coefficients, or lifetimes of various excited proteins. Perturbations of the biochemical cascade, which renders the decay of the photocurrent no longer limiting, kill this model since this hypothesis is, with "total occlusion" (Lamb et al 1981), one of the two pillars on which the model rests.

ACKNOWLEDGMENTS

The author wishes to express his gratitude to S. Gagné who took part in the initial development of the model, to R. Roy who helped with statistics leading to the computer simulation, and to K.-W. Yau and D.W. Corson, A.T. Ishida who read previous versions of this manuscript and made many suggestions.

REFERENCES

Baylor DA, Lamb TD, Yau K-W (1979) The membrane current of single rod outer segments. J Physiol 288: 613-634

Baylor DA, Hodgkin AL (1973) Detection and resolution of visual stimuli by turtle photoreceptors. J. Physiol. 234:163-168

Baylor DA, Hodgkin AL (1974) Change in time scale and sensitivity in turtle photoreceptors. J. Physiol. 242:729-758

Baylor DA., Nunn BJ, Schnapf JL (1984) The photocurrent noise and spectral sensitivities of rods of the monkey Macaca fascicularis. J Physiol 357: 575-607

Brucket F, Chabre M, Vuong TM (1992) Kinetics analysis of the activation of transducin by photoexcited rhodopsin. Biophys J 63: 616-619

Burkhardt DA, Kraft TW, Gottesman J. (1986) Functional properties off twin and single cones. Neurosci Res supp : S45-S58

Chabre M, Antonny B, Bruckert F, Vuong TM (1993) The G protein cascade of visual transduction: kinetics and regulation. In Ciba Foundation Symposium 176, pp. 112-124. Wiley, Chischester

Dartnall HJA (1972) Photosensitivity. In Handbook of Sensory Physiology, vol. VII/1, Photochemistry of Vision, ed Dartnall, HJA, pp. 191-264. Plenum, New York

Dumke CL, Arshavsky VY, Calvert PD, Bownds MD, Pugh,Jr. EN (1994) Rod Outer Segment Structure Influences The Apparent Kinetics Parameters of Cyclic CMP Phophodiesterase. J Gen Physiol 103: 1071-1098

Eckmiller MS (1987). Cone Outer Segment Morphogenesis: Taper Change and Distal Invaginations. J Cell Biol 105: 2267-2277

Fain G (1976). Sensitivity of toad rods: dependence on wavelength and background illumination. J Physiol 261: 71-101

Fesenko EE, Lelesnikov SS, Lyubarsky AL (1985) Induction by cyclic GMP of cationic conductance in plasma membrane of retinal rod outer segment. Nature 313 310-313

Forti S, Menini A, Rispoli G, Torre, V (1989) Kinetics of phototransduction in retinal rods of the newt Triturus cristatus. J Physiol 55: 563-573

Gillespie PG (1990). In Cyclic Nucleotide Phosphodiesterase: Structure, Regulation and Drug action., ed Beavo, J. & Houslay, MD, John Wiley and Sons, New York.

Gillespie PG, Beavo JA (1989) cGMP is tightly bound to bovine retinal rod phosphodiesterase. Proc Natal Acad Sci USA 86: 4311-4315

Hamm HE, Bownds MD (1986) Protein complement of rod outer segments of frog retina. Biochemistry 25: 4512-4523

Haynes L, Yau K-W (1985) Cyclic GMP-sensitive conductance in outer segment membrane of catfish cones. Nature 317: 61-64

Harosi F.I. (1975) Absorption spectra and linear dichroism of some amphibian photoreceptors. J Gen Physiol 66: 357-382

Hemila S, Reuter T (1981) Longitudinal spread of adaptation in the rods of the frog's retina. J Physiol 310: 501-528

Hood DC, Finkelstein MA (1986) In Handbook of perception and human performance, ed. Bolf, K.R., Kaufman, L. & Thomas, J.P., John Wiley and Sons, New York

Ichikawa K (1994a) Critical processes which characterize the photocurrent of retinal rod outer segment to flash stimuli. Neurosci Res 19: 201-212

Ichikawa K (1994b) Transduction steps which characterize retinal cone current response to flash stimuli. Neurosci Res 20: 337-343

Heck M, Hofmann KP (1993) G-protein-effector coupling: a real-time light-scattering assay of transducin-phosphodiesterase interaction. Biochemistry 32(32): 8220-8227

Hodgkin AL, Nunn BJ (1988) Control of light sensitive current in salamander rods. J Physiol 403: 439-471

Jacobs GH, Yolton, RL (1971) Visual sensitivity and color vision in ground squirrels. Vision Res 11: 511-537

Kaupp B, Koch KW (1992) Role of cGMP and Ca2+ in vertebrate photoreceptor excitation and adaptation. Ann Rev Physiol 54:153-175

Kawamura S (1993) Molecular aspects of photoreceptor adaptation in vertebrate retina. International Rev. Neurobiology 35: 43-86

Koch K.W. (1992) Biochemical mechanism of light adaptation in vertebrate photoreceptors. TIBS 17: 307-311

Kraft TW (1988) Photocurrents of cone photoreceptors of the golden-mantled ground squirrel. J Physiol 404: 199-213

Kraft TW, Schneeweis DM, Schnapf JL (1993) Visual Transduction in human photoreceptors. J Physiol 464: 747-765

Koutalous Y, Nakatani K, Yau KW (1995) Cyclic GMP diffusion coefficient in rod photoreceptor outer segments. Biophysical J. 88(1): 373-382

Lamb TD (1983) Effects of temperature changes on toad rod photocurrents. J Phsysiol 346: 557-578

Lamb TD, McNaughton PA, Yau KW (1981) Spatial spread of activation and background desensitization in toad rod outer segments. J Physiol 319: 463-496

Lamb TD, Pugh EN (1992) G-protein cascades: gain and kinetics. Trends Neurosci 15: 291-298.

Liebman PA, Entine,G (1974) Lateral diffusion of visual pigment in photoreceptor disk membranes. Science 185: 457-459

McQuarrie DA (1986) In Statistical mechanics, ed Woods, JA, Harper & Row, Inc. New York.

Miller JL, Korenbrot JI (1993) Phototransduction and adaptation in rods, single cones, and twin cones of the striped bass retina: A comparative study. Visual Neurosci 10: 653-667

Molday RS, Molday LL (1987) Differences in protein composition of bovine retinal rod outer segment disk and plasma membranes isolated by a ricin-gold-dextran density perturbation method. J. Cell Biol. 105: 2589-2601

Nakatani K, Tamura T, Yau KW (1991) Light Adaptation in Retinal Rods of the Rabbit and Two Other Nonprimate Mammals. J Gen Physiol 97: 413-435

Olson A, Pugh EN (1993) Diffusion coefficient of Cyclic GMP in salamander rod outer segments estimated with two fluorescent probes. Biophys J 65: 1335-1352

Perry RJ, McNaughton PA (1991) Response properties of cones from the retina of the tiger salamander. J Physiol 433: 561-587

Poo MM, Cone RA (1974) Lateral diffusion of rhodopsin in visual receptor membrane. Nature 247: 438-441

Pugh Jr. EN, Lamb TD (1993) Amplification and kinetics of the activation steps in phototransduction. BBA 1141: 111-149

Raynauld JP, Gagne S (1987). Geometrical considerations predict the response of both vertebrate rods and cones following a single isomerization. Society for Neuroscience. Abstracts 13: 1398

Schnapf JL (1983) Dependence of the single photon response on longitudinal position of absorption in toad rod outer segments. J Physiol 343: 147-159

Schnapf JL, Nunn BJ, Meister M, Baylor DA (1990) Visual transduction in cones of the monkey Macaca fascicularis. J Physiol 427: 681-713

Schnapf JL, McBurney RN (1980) Light-induced changes in membrane current in cone outer segments of tiger salamander and turtle. Nature 287: 239-241

Schneeweis DM, Schnapf JL (1995) Photovoltage of Rods and Cones in the Macaque Retina. Science 268: 1053-1056

Sitaramayya A, Harkness J, Parkes JH, Gonzalez-Oliva C, Liebman, PA (1985) Kinetic studies suggest that light-activated cyclic GMP phosphodiesterase is a complex with G-protein subunits. Biochemistry 25: 651-656

Stryer L (1991) Visual Excitation and Recovery. J Biol Chem 266: 10711-10714

Tamura T, Nakatani K, Yau KW (1989) Light Adaptation in Cat Retinal Rods. Science 245: 755-758

Tamura t, Nakatani K, Yau KW (1991) Calcium feedback and sensitivity regulation in primate rods. J. Gen. Physiol. 98: 95-130

Torre V, Straforini M, Campani M (1992) A quantitative model of phototransduction and light adaptation in amphibian rod photoreceptors. The Neurosciences 4: 5-13

Walls GL (1994) in Vertebrate eye and its adaptative radiation. Hafner Publishing Co. New York, London.

Yau KW (1994) Phototransduction mechanism in retinal rods and cones: The Friedenwald Lecture. Inv Ophthal Vis Sci 35: 9-32

18

MODELLING ODOR INTENSITY AND ODOR QUALITY CODING IN OLFACTORY SYSTEMS

Jean-Pierre Rospars,[1] Petr Lánskỳ,[2] and Jean-Claude Fort[3]

[1]Laboratoire de Biométrie
 INRA, Versailles, France
[2]Institute of Physiology and Center for Theoretical Study
 Academy of Sciences, Prague, Czech Republic
[3]Département de Mathématiques
 Université de Nancy I
 Vandœuvre lès Nancy, France

1. INTRODUCTION

Various pieces of information can be extracted from an odor source. The main ones are the quality of the odor which is related to its chemical composition, the intensity of the odor which is related to the concentrations of its components, and the temporal fluctuations in intensity and quality. The aim of this paper is to present models of information processing in the olfactory systems that we have recently developed for odor intensity and quality. We will consider mainly insects as examples (Fig. 1), although the models presented are sufficiently general to be applicable to both insects and vertebrates.[2,11]

In the first part of the paper models of intensity coding in single receptor neurons are presented, which are based mainly on mathematical analyses. In the second part a model of quality coding in the first two neuronal layers is studied based mainly on computer simulations.

2. INTENSITY CODING IN SINGLE RECEPTOR ANTENNAS

In insects odors are collected by antennae and detected by receptor neurons housed in specialized olfactory organs called sensilla.[16,18,27,32] The antenna of male moths, for example, bears two types of olfactory sensilla. The most abundant are the long sensilla trichodea specialized in the detection of the sex pheromone emitted by conspecific females; they are part of a very sensitive system which allows males to locate conspecific females. The second type consists of the small sensilla basiconica that detect complex odors such as plant odors. Both types have basically the same structure (Fig. 2). A cuticular hair with many nanometer-size holes set in it houses two or more receptor neurons which are

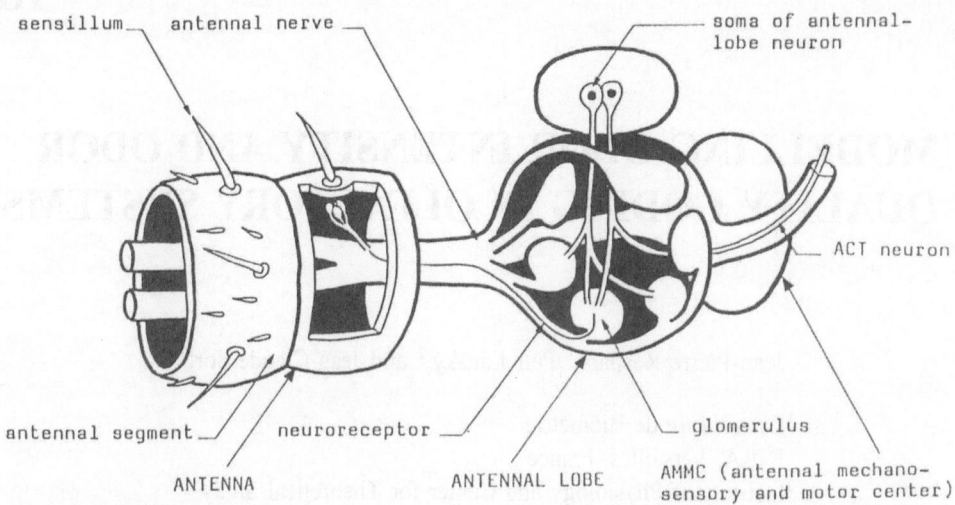

Figure 1. Organization of the first two neuronal layers in the insect olfactory system. Olfactory receptors neurons (RNs) housed into sensilla at the surface of the antenna project to the ipsilateral antennal lobe. They synapse with local (LNs) and principal neurons (PNs) within spheroidal glomeruli. PNs project to higher brain centres.

surrounded by auxiliary cells. The receptor neuron presents three main parts,[17] the outer segment with the transduction system,[43] the inner segment with cell body and the axon.

What we want to model first of all is the final response of the neuron, the firing frequency, as a function of stimulus intensity. It is well known that the firing frequency increases according to a monotonous sigmoid curve with the stimulus intensity, which is usually measured for a pure chemical as the logarithm of the odorant concentration (e.g., Refs. 16184147; in vertebrates see Ref. 5). The purpose of the models presented is to describe the neural mechanisms that finally give rise to this sigmoid curve. We are also interested in other features such as the spontaneous activity in the absence of stimulation and the variability of the responses to stimulation which both reveal stochastic mechanisms. A deterministic model that describes various neuron responses using the moth sex-pheromone receptor neuron as an example is presented first, then, briefly, a stochastic model that extends the former.

2.1. Deterministic Model of the Sex-Pheromone Receptor Neuron

The model of the receptor neuron is based on the following simplifications. The neuron is considered in isolation, the accessory cells of the sensillum are not taken into account. The sensory dendrite bears only one type of receptor sites, which is believed to be the case for sex-pheromone receptor cells. The neuron is considered as a long cylinder cable; this is a reasonable assumption in the case of a sex-pheromone neuron.[21] The neuron is assumed to be in a steady state under constant stimulation, so time can be ignored. With these simplifications analytical solutions can be proposed for the three main signal conversions that occur in the neuron.[25,39] (1) The first conversion occurs in the sensory dendrite. The interaction of pheromone molecules with the receptor sites triggers the opening of ion channels. This is modelled as a conductance change g for ions with equilibrium potential E, so that each element of membrane is described by a circuit consisting in a variable conductance in series with a battery (Fig. 2c). (2) The second

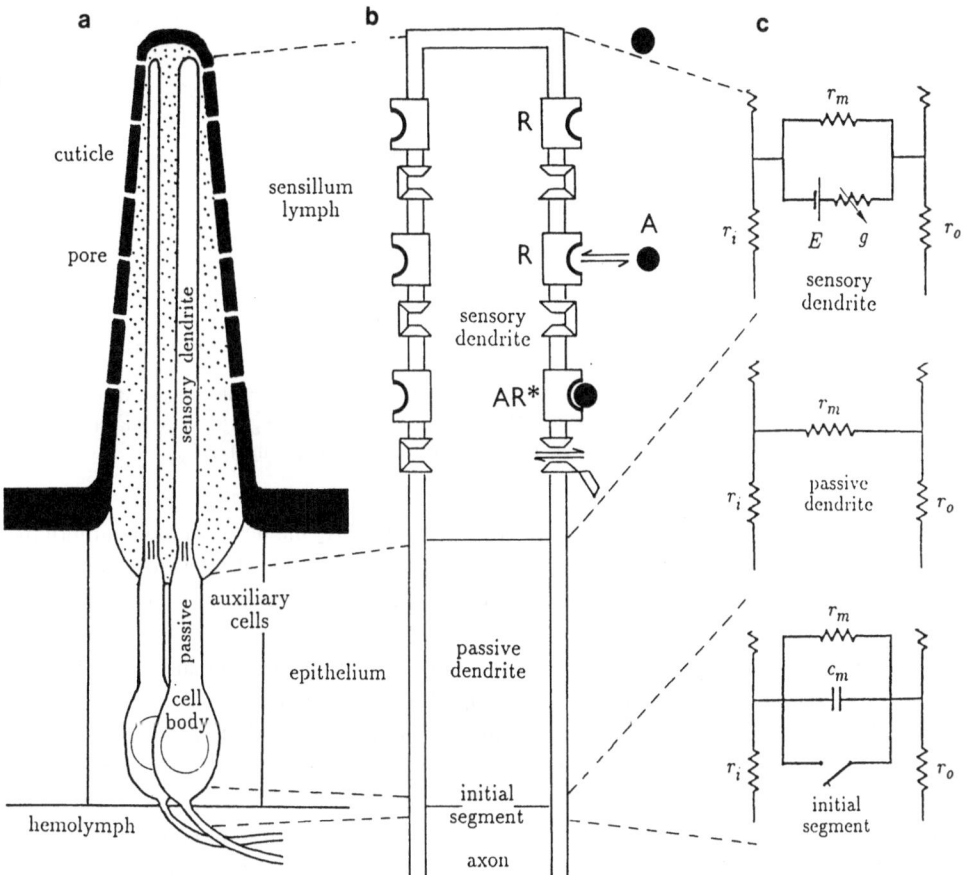

Figure 2. Organization of a sensillum (**a**, adapted from Kaissling, 1987) and model of the moth sex-pheromone receptor neuron (**b**) with equivalent elementary circuit (**c**). The odorant-sensitive dendrite bears receptor sites R. When activated by odorant molecules A they trigger a second-messenger system that ultimately opens ion channels (variable conductance g). Ions flowing along the electrochemical gradient (battery E) generates a receptor potential that spreads along the sensory and passive (including cell body) dendrites and triggers action potentials at the axon initial segment (RC circuit).

step is a conductance-to-voltage conversion which results from the flow of ions through the circuit; a receptor potential is generated that spreads along the neuron. (3) Finally, the receptor potential is converted to a train of spikes at the axon initial segment.

Transduction module. Let's consider the first module with its concentration-to-conductance conversion which can be modelled as a two-step process with first receptor activation followed by conductance change.[16,18]

For receptor activation the classical sequence of binding and activation can be used which involves two forward and two backward reactions,

$$A + R \underset{k_{-1}}{\overset{k_1}{\rightleftharpoons}} AR \underset{k_{-2}}{\overset{k_2}{\rightleftharpoons}} AR^*$$

At equilibrium the forward and backward reactions proceed at the same velocity. The number $[AR^*]$ of activated receptor sites per unit area of membrane at equilibrium can be derived from the total number of receptor sites $[R_T]$, the concentration $[A]$ of the odorant and

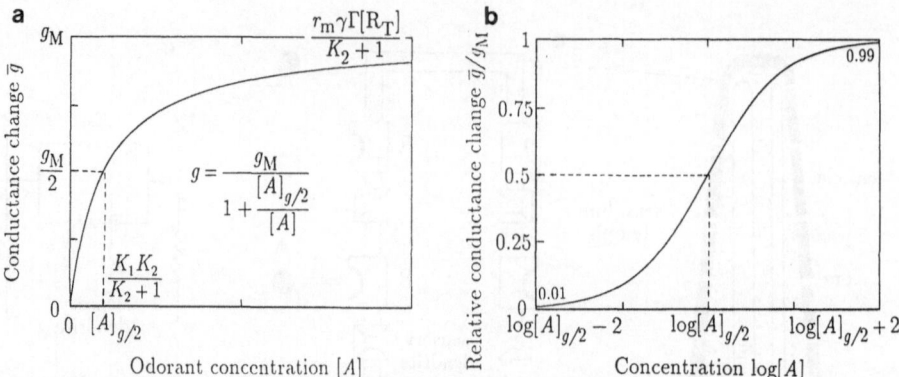

Figure 3. Concentration-to-conductance conversion. The conductance change g at the sensory dendrite as a function of odorant concentration $[A]$ is a branch of hyperbola, whereas g vs. $\log[A]$ is a logistic curve. For g/g_M rising from $\epsilon = 0.01$ to $1 - \epsilon = 0.99$, the rise in concentration (coding range) is $\Delta_g = 4$.

two constants, K_1 the dissociation equilibrium constant and K_2 the deactivation equilibrium constant which are respectively the ratios of the velocity constants (backward-to-forward) for the first (binding, k_{-1}/k_1) and second (activation, k_{-2}/k_2) reactions.[6,8] The curve of $[AR^*]$ vs. $[A]$ is a branch of hyperbola.

For the second step of the transduction process the simplest possible model is to assume that the conductance change g is proportional to the number of activated sites (in the following g is expressed as a multiple of the conductance at rest, which is the inverse of the membrane resistance r_m) (for more realistic models see)[22,19]. The constant of proportionality depends on the mean number of ion channels opened per activated receptor sites (Γ) and the unit conductance of the ion channels (γ). The curve of conductance as a function of $[A]$ is a branch of hyperbola (Fig. 3) with a horizontal asymptote (maximum response g_M). A more natural representation of the same function consists in using the logarithm of concentration because the concentration range is usually quite large. Then the hyperbolic function is transformed in a logistic function (sigmoid curve) with the same asymptote. Although the experimentally measured conductance curves slightly differ from a logistic (Ref. 28 and this volume), suggesting that cooperative reactions take place in the transduction cascade, the minimal model considered remains a useful idealization of the actual process.

Three aspects of the concentration-response curves characterize their coding properties. (1) The *magnitude* of the response is measured by the maximum response and is related to the amplification of the response. The main function of the activation-to-conductance conversion is to amplify the initial reaction and thus to transform a weak signal (activation) into a strong one (ionic current). (2) The concentration at half-maximum response measures the *sensitivity* of the response: the smallest this concentration the more sensitive the response is. The activation and conductance curves have the same concentration at half-maximum response $[A]_{g/2} = K_1 K_2/(K_2 + 1)$. So, the activation-to-conductance conversion entails no gain in sensitivity. (3) The *coding (or dynamic) range* (Torre, this volume) characterizes the range of concentrations over which the system responds. In the case of conductance this is the range of concentration Δ_g in log units for which the ratio g/g_M varies between a value ϵ close to zero, the threshold level $[A]_{g_t}$, and a value $1 - \epsilon$ close to 1, the saturation level $[A]_{g_s}$, so $\Delta_g = \log[A]_{g_s} - \log[A]_{g_t}$. With $\epsilon = 0.01$ it can be shown that the coding range is 4 log units. When changing K_1 the position of the curve along the concentration axis can be changed but not the saturation level, whereas when changing

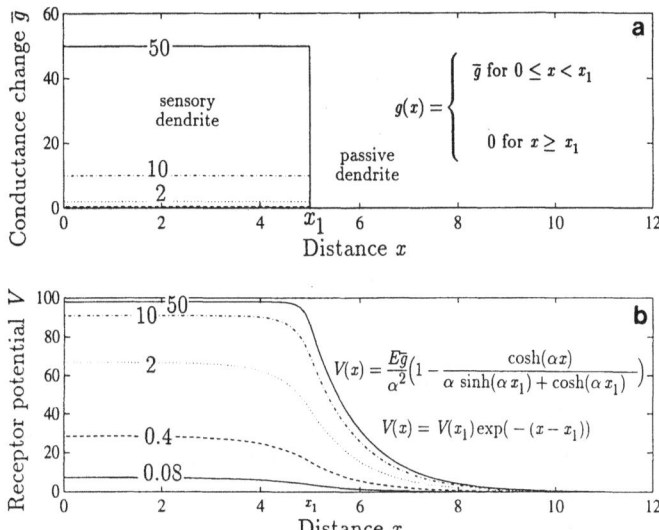

Figure 4. Receptor potential $V(x)$ along the neuron for different conductance change $\bar{g}$ at the sensory dendrite. In b the equations of $V(x)$ for the sensory dendrite (first line) and passive dendrite (second line) are given with $\alpha = \sqrt{1+\bar{g}}$.

K_2 both the saturation level and the position of the curve can be changed. However in all cases the coding range remains exactly the same.[39]

One of the questions we want to answer is how the last two coding properties, the position of the response curve along the concentration axis described by the concentration at half-saturation and the coding range, are changed by the other conversion steps, conductance-to-voltage and voltage-to-frequency.

Receptor potential module. The second module describes the conductance-to-voltage conversion. The dendrite is considered to be a cable, so the module is based on the cable equation.[29,45] This second-order differential equation gives at each point X along the cable the potential $V(X)$ knowing the conductance change $g(X)$ at the same point. The constants are the internal r_i, external r_o and membrane r_m resistances, and the equilibrium potential E of the permeating ion. A simpler equivalent expression of the cable equation can be obtained when a dimensionless variable is used for abscissa, $x = X/\lambda$ where $\lambda = \sqrt{r_m/(r_i + r_o)}$ is the membrane space constant. We have solved this differential equation in the case of a uniform stimulation of the sensory dendrite,[39] i.e., in the case where the odorant and receptor molecules are uniformly spread along the sensory dendrite so that the conductance change is the same everywhere on this dendrite $g(x) = \bar{g}$, and of course $g(x) = 0$ elsewhere because there are no receptor sites on the passive dendrite, cell body and axon. The analytical solutions are given by a couple of equations which apply to the sensory dendrite and the rest of the neuron respectively (Fig. 4). The voltage change depends on the conductance change $\bar{g}$; it is almost uniform on the sensory dendrite then falls exponentially beyond it. The voltage $V(0)$ at the distal end of the dendrite gives the potential over much of the sensory dendrite. The $V(x_1)$ voltage at the proximal end of the sensory dendrite is the most important to know because the potential at the initial segment $V(x_2)$ is merely proportional to that at x_1. For this reason we will consider only $V \equiv V(x_1)$ in the following.

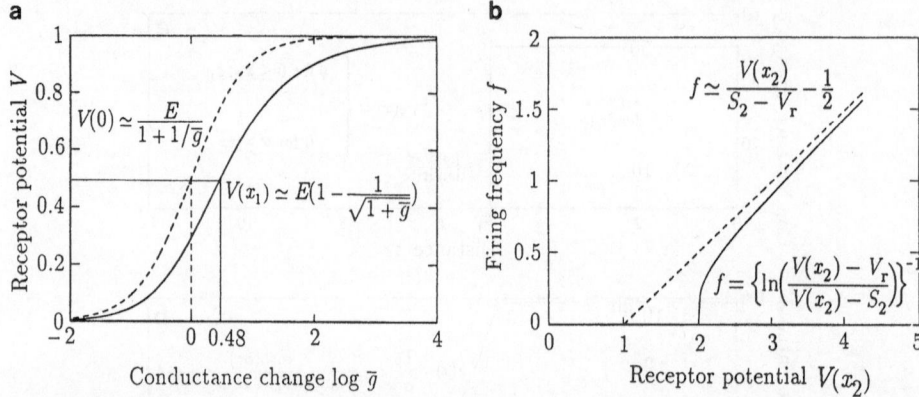

Figure 5. Conductance-to-voltage (**a**) and voltage-to-frequency (**b**) conversions. In **a** for a length of sensory dendrite $x_1 > \lambda/3$, potentials $V(0)$ and $V(x_1)$ at the distal and proximal ends of the sensory dendrite and $V(x_2)$ at the axon initial segment are independent of x_1 (see equations). $V(x_1)$ is compared to the logistic curve $V(0)$ at $x = 0$. In **b**, the firing frequency as a function of $V(x_2)$ starts abruptly from $S_2 = 2$ (threshold) then tends rapidly to the straight line shown. $V_r = 0$, resting potential.

The equations shown in Fig. 4 can be simplified, at least in the case of the sex-pheromone receptor cell, because the receptor potential at any point along the neuron does not depend on the length x_1 of the sensory dendrite when this length is greater than about 0.3 space constant. Then the asymptotic version of V for large x_1 can be used as a very good approximation. This independence on x_1 is important from a biological point of view because it suggests that increasing the length of the sensory dendrite of the sex-pheromone receptor cell beyond 100 μm ($\lambda \simeq 300\mu m$[18]) does not increase the magnitude of the receptor potential. It suggests that the very long length of the dendrite in sex-pheromone neuroreceptor contributes mainly to an increase of the surface area for gathering molecules. This might also explain the small length of the dendrites of neurons within sensilla basiconica that respond to general odorants with lower sensitivities.

The conductance-to-voltage conversion $V(\bar{g})$ at the proximal end of the sensory dendrite (x_1) is a non-hyperbolic function with conductance at half-saturation equal to 3 (whereas at point 0 this is an hyperbolic function with conductance at half-saturation equal to 1). In log scale the curve is a sigmoid that is less steep that a logistic curve (Fig. 5a). Putting together the modules of transduction and of the receptor potential the concentration-to-voltage conversion $V([A])$ can be studied directly (Fig. 6). This conversion presents three main properties. (1) The magnitude of the voltage response for any $[A]$ depends on the saturation level V_M that depends in turn on the maximum conductance g_M. This shows that the magnitude of the receptor potential can be increased by increasing g_M i.e., by decreasing the deactivation constant K_2 and increasing the unit conductance of channels and the amplification factor of transduction. (2) The sensitivity of the response, as characterized by the concentration $[A]_{V/2}$ at half-maximum response, is shifted towards low concentrations when g_M increases. Thus, the maximum conductance also controls the sensitivity of the receptor-potential response. (3) The coding range Δ_V increases with g_M and tends to 5.7 for large g_M. So the second conversion tends to increase the coding range yielded by the first one.

Action potential module. The third module is based on the "leaky integrator" model (e.g., Ref. 45). It describes the voltage-to-frequency conversion in a simplified way which retains only the basic lineaments of the real phenomena. The shape of the spike is not

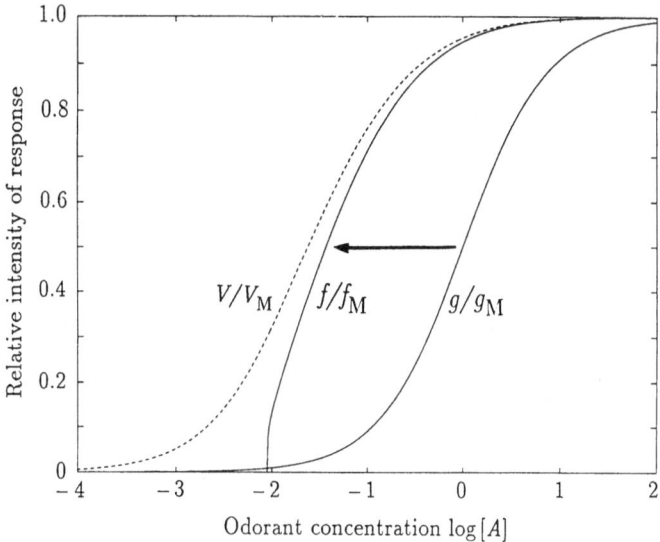

Figure 6. Sensitivity of the responses (conductance g, voltage V at x_1, firing frequency f at $x_2 = x_1 + 1$) for $g_M = 100$ times the resting conductance. The overall gain in sensitivity (arrow) is entirely due to the conductance-to-voltage conversion. In this deterministic model the f curve starts abruptly at $[A]_{f_i}$.

described but only the time of firing. When the receptor potential at the axon initial segment x_2 crosses the firing threshold S_2 the switch is closed and the condenser discharges (see circuit in Fig. 2c). Then the switch is opened again and the voltage returns exponentially to the receptor potential. One of the main simplification here is to assume that the dendritic receptor potential remains constant and is not affected by the time variation of the potential at the initial segment. This is the price to pay to get an analytical solution (see Fig. 5).

It can be seen that the neuron begins to fire when the voltage reaches the threshold. The firing frequency f increases very quickly at first then depends linearly on the receptor potential. The modules 2 and 3 together gives a nearly hyperbolic curve $f(g)$. The most interesting curves are for the three modules together in normal $f([A])$ and log scale $f(\log[A])$ as shown in Fig. 6. The curve $f(\log[A])$ is not sigmoid, which is a major deviation with respect to experimentally observed curves (see below).

The coding properties of the concentration-to-frequency conversion can be summarized as follows. (1) The saturation level f_M depends on the maximum conductance g_M. (2) The frequency curves are shifted to the right of the voltage curve. This shows that the voltage-to-frequency conversion corresponds to a slight loss in sensitivity, however much of the gain in sensitivity due to the second conversion is conserved, so that globally the concentration-to-frequency conversion corresponds to a gain in sensitivity. This gain (arrow in Fig. 6) increases with the maximum conductance. (3) The coding range Δ_f also depends on the maximum conductance. Although the curve is truncated to the left with respect to the conductance and voltage curves, this is compensated for g_M greater than 10 by a shift to the right of the concentration at saturation $[A_{f_s}]$. Δ_f tends to 5.7 for large g_M and small S/E. This is consistent with experimental observations although Δ_f larger than 6 have been reported (e.g., Kaissling, 1987).

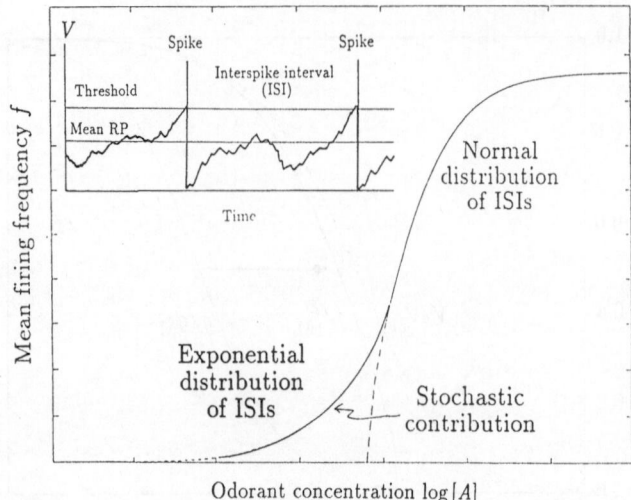

Figure 7. Stochastic model of firing frequency. In the subthreshold range $[A] < [A]_{f_t}$ (see Fig. 6) spikes are fired because the receptor potential $V(x_2) < S$ fluctuates randomly around its mean and sometimes crosses the threshold S (inset). For $V(x_2) << S$, the spikes obey a Poisson process (exponential ISIs).

2.2. Stochastic Model and Application to Spontaneous Activity

The previous model presents many simplifications. For example, the precise geometry of the neuron has not been considered, nor the effect of accessory cells in the sensillum,[16,20] nor the passive backpropagation of spikes along the dendrite,[49,50] nor the time-dependency of stimulation,[46] nor the presence of noise and other stochastic features. The last aspect is one of the most important. A source of randomness is for example the fluctuation in the number of activated receptors especially when the odorant concentration is very low. We have proposed a model of these fluctuations based on a birth and death process which describes the Poissonian binding and release at receptor sites.[23,37] In this case it can be shown that the receptor potential follows an Ornstein–Uhlenbeck process, which means that the potential has a normal distribution with mean $\overline{V}$ and standard deviation σ, the values of both can be derived from the characteristics of the birth and death process. Such a model is significantly different of the previous deterministic model because even when $\overline{V}$ is less than the threshold spikes are fired due to random fluctuations of V at the initial segment. It follows that the frequency transfer function does not start abruptly at the concentration $[A]_{f_t}$ given by the deterministic model but at a much lower concentration. The corresponding part of the response curve is the stochastic contribution that transforms the deterministic curve into a sigmoid curve (Fig. 7).

Furthermore the model predicts that at low concentration the interspike intervals (ISIs) follow an exponential distribution, which corresponds to a Poissonian firing (spikes appear completely at random), whereas at high concentration the ISIs are normally distributed (the random component appears as a mere fluctuation around a fixed mean ISI). At medium concentrations the distribution is intermediate, close to a gamma distribution which is a two-parameter law.[24]

We have begun to test this model using the spontaneous activity in the frog olfactory system.[38] It has been found that the firing of mitral cells is indeed Poissonian but surprisingly not that of receptor cells. The deviation from exponential distributions in these cells resulted from the smallest intervals only, the longer ones being exponentially distributed.

This means that receptor neurons fire bursts of spikes and the bursts follow a Poisson process. In the framework of the stochastic model presented these properties mean that, in mitral cells, the mean potential is always much below the firing threshold. In receptor cells, the same condition usually holds, except that sometimes the potential remains close to S for some time. The underlying mechanisms and specific ion conductances that give rise to this behavior remains to be specified.

3. QUALITY CODING IN THE ANTENNAL-LOBE NEURAL NETWORK

The problem of coding odor quality is very different from that of coding odor intensity. One basic difference between them is that intensity is coded more or less by individual cells, whereas quality is coded by a population of cells. By recording from one cell one may be able to estimate the intensity of an odor but not its quality. Some exceptions to this rule are known, such as the sex pheromones of insects, but for most odors (e.g., from food and plant) the rule holds.

The information coded by receptor neurons is sent along the antennal nerve to a well delimited brain area, called antennal lobe (Fig. 1). There, the receptor neurons terminate within specific structures called glomeruli.[1,11,13,32] Within glomeruli they make synaptic contacts[26] with several types of cerebral neurons, including uniglomerular projection (or principal) neurons whose axons project to higher brain centres and multiglomerular local neurons. The local neurons have been found to be mostly or only inhibitory.[3] All these neural components are also present in the olfactory bulbs of vertebrates.[9,42]

It has been shown in several species that the glomeruli are identifiable units (e.g., Ref. 31333644. This means that a given glomerulus can be found in the same place, with the same size and the same shape, in all individuals of the same sex and stage of development in a given species. The macroglomerular complex receives the projections from the sex-pheromone receptor cells, whereas the other glomeruli receive the projections from other types of receptor cells.[4,10]

This anatomical identifiability of glomeruli is believed to be important from a physiological point of view. An attractive hypothesis linking quality coding and identified glomeruli distinguishes specialized (labeled) and relatively unspecialized (cooperative) glomeruli as follows. Some glomeruli, like the macroglomerular complex, might process information related to a given odorant or a small family of related odorants. The problem with such a specific coding is that the number of odors that can be coded is strictly limited by the number of glomeruli, which is typically about 60 in moths and a hundred in cockroaches.[32] A more economical system consists in using an array of glomeruli and to code quality in a pattern of active and inactive glomeruli.[31] With an array of 16 glomeruli which can only be in two states each, it would be possible to code 2^{16} different odors, that is more than 65,000 odors. There are experimental results that support this spatial coding scheme, in both insects[30] and vertebrates.[40] The model summarized now develops this idea further.[35]

3.1. The Model

Model of odor quality. The notion of odor quality must be first defined in a way that is usable in a quantitative model. For this we consider that the odority of a molecule is related to several independent properties P_1, P_2, P_3, that describe for example the size, the shape, the polarity of the molecule and that the contribution of each property is measured by coefficients x_1, x_2, x_3, with $\sum_i x_i = 1$. In the case of a 3D odor space for example, all odors would be located on a plane and a specific odor could be made of 1/5 of P_1, 1/5 of P_2 and 3/5 of P_3. In our simulations we have chosen an odor space with 9 dimensions and we have used a systematic sample in which different odor qualities are separated by a Δx of 0.33. This sample contains 165 odors. In this definition the intensity of the odor is not taken into account, which means that, for the sake of simplicity, each odor quality is considered at a fixed concentration.

Model of receptor neurons. The existence of several types of receptor neurons has been assumed. Each type r responds with a specific firing frequency f_{rk} to a given property k. For example, type R_1 responds with low frequency to P_1 and P_3 and high frequency to P_2. This is different for types R_2 and R_3. The resulting matrix of frequencies f_{11}, f_{21} etc. characterizes the receptor types. In simulations we have taken 5 types. Now, for an odor which is a mixture of properties P_k in proportions x_k the response frequency is the weighted sum $f_r = \sum_k f_{rk} x_k$. In this way one can specify the response frequency of any receptor neuron when stimulated with any odor quality. Such a model can be easily interpreted in terms of molecule-receptor interactions and electrical properties of neurons. Of course a much more realistic model could be used instead, based for example on the model presented above for intensity coding, but the present one retains the main features we want to have and is easier and faster to simulate.

Model of glomeruli and principal neurons. The global organization of the neural network used is shown in Fig. 8. Five hundred receptor neurons (RNs) converge onto one principal neuron (PN) in each glomerulus. They belong to different types in a specific proportion for each glomerulus, so that all glomeruli are different. An array of 4 glomeruli can generate 16 black and white patterns and so code for 16 different odors.

When stimulated by an odor the RNs fire action potentials. Each spike contributes to an EPSP in the principal neuron i. All the EPSPs at time t are summated to give the input $E_i(t)$. This input potential varies with time in a different way for different neurons and different odors (Fig. 9). All the information about the odorant stimulus is assumed to be included in these signals.

Model of local neurons. Processing of the input signals involves local neurons. The 4 PNs are completely interconnected by 12 local neurons (LNs). The output S_i of any PN i results from the input E_i of the RNs and the action L_i of the LNs, which is the sum of the actions of the 3 LNs connected to i. Each LN sends the output of a PN multiplied by a coefficient c_{ij} (less than 1), usually called "synaptic weight" although in the present case this name is misleading. For example, the action exerted by the PNs number 2, 3 and 4 on PN number 1 is the weighted sum of their outputs (see Fig. 9).

The role of the LNs has been studied by comparing 3 different models. (A) In model A we have put no LNs at all, the synaptic weights are all zero. (B) In model B the LNs have a fixed inhibitory effect, all coefficients are constant in time and equal, for example to 0.25. (C) In model C the coefficients change with time according to an algorithm proposed by Hérault and Jutten.[7,14,15] Each time a new odor is presented the coefficients are reset to zero and evolve in time as the covariance of two different functions of the outputs S_i and S_j of the connected PNs (Fig. 10). This means that the coefficients tend to make the output of PNs as different as possible. Hérault and Jutten have shown that their algorithm presents

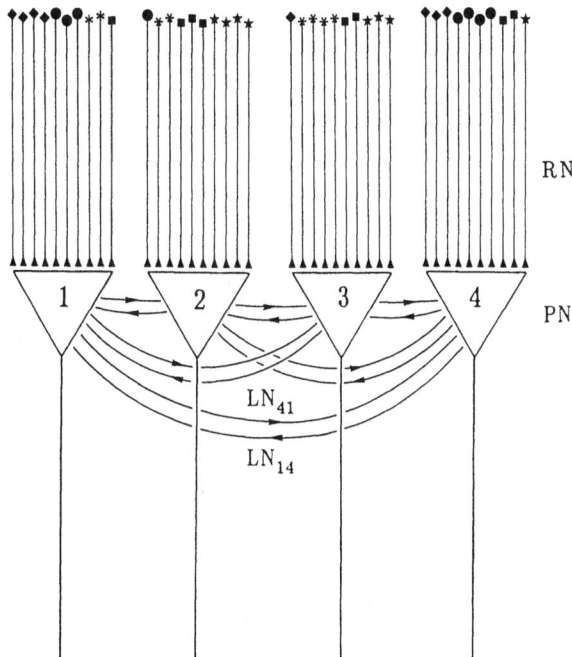

Figure 8. Model of the antennal-lobe neural network (see Fig. 1). 500 RNs terminate onto each PN with 1 PN per glomerulus. All PNs (glomeruli) are interconnected with 12 LNs. In model B and C (see text) the 4 PNs (glomeruli) can generate 2^4 two-state patterns of activity (black and white patterns).

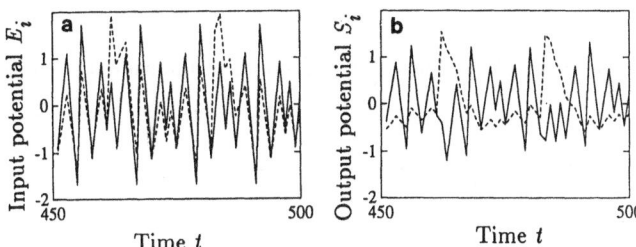

Figure 9. Examples of input signal (**a**) and output signals (**b**) for a principal neuron when stimulated by a given odor. Input $E_i(t)$ is the summated EPSPs received by neuron i; it depends on the number of RNs terminating on i that fire at time t. Output $S_i(t) = E_i(t) - L_i(t)$ involves the action L_i of LNs; e.g $L_1(t) = c_{12}S_2(t) + c_{13}S_3(t) + c_{14}S_4(t)$ with $c_{ij} < 1$ characterizing the LN connecting PN_i to PN_j. In the case shown all $c_{ij} = 0.25$ (model B; $c_{ij} = 0$ in model A and are time-dependent in model C). Due to inhibition outputs are much less similar than inputs (decorrelation). Time is measured in ms.

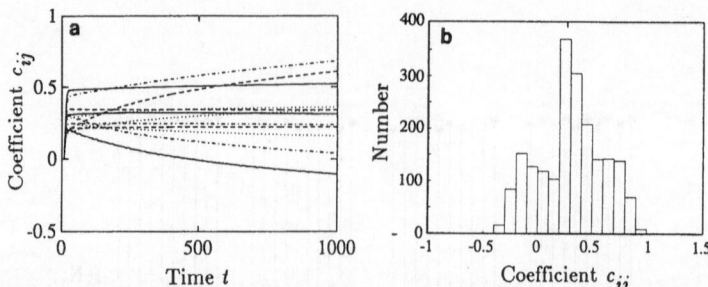

Figure 10. Coefficients c_{ij} giving the activity of local neurons in model C. **a** Change in time for the 12 LNs when stimulated by a given odor according to $\frac{dc_{ij}(t)}{dt} = \text{covariance}(\tan^{-1}(S_i^3(t), S_j(t))$; the coefficients are found to evolve towards a steady state. **b** Distribution at steady-state ($t = 1000$) of 1980 coefficients (12 LNs $\times$165 odors); L being subtracted from E (see Fig. 9), a positive coefficient means an inhibitory action and a negative one an excitatory action. LNs in model C are inhibitory or exert no action in 80% of cases.

useful features and a wide range of applications. It has also been used in an olfactory context by Hopfield.[12] Figure 10 shows the evolution in time of the 12 coefficients for different odors. It can be noted that they begin to evolve quickly then stabilize to constant values after 500 msec.

As a result the fluctuations in time of the membrane potential of the PNs are more and more different from one another. To convert these membrane potentials in spike trains we merely put a firing threshold. Each time the potential crosses the threshold a spike is fired. The activity of the neuron is measured by the number of spikes fired during every 100 milliseconds. Figure 11 shows the activity of a PN as a function of time during the first 500 msec of stimulation for an odor. It is not easy to analyze these activities at a given time for all odors. To simplify the picture we considered only two states of activity, either below or above the median activity observed for the whole sample of odors in the different models A–C. Then, the neuron activities are converted into black-and-white activity maps (Fig. 11). Of course more states might have been considered. However, this more refined analysis is not needed at the present time.

3.2. The Results

Spatial coding. To analyze how the 165 odors of the sample were coded into the

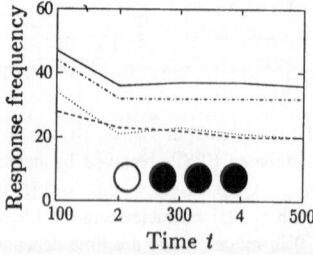

Figure 11. Firing frequency as a function of time of the 4 principal neurons when stimulated by a given odor (same as in Fig. 9) and corresponding activity pattern. In this case (model B) the median activity of all PNs for all 165 odors is 36. For this odor PN$_2$–PN$_4$ with $f < 36$ at $t = 500$ are considered inactive (in black); only PN$_1$ is active (in white).

16 available black and white patterns considered, the number of odors that generate each pattern were counted for each model. (1) It was observed that all patterns are generated: thus a systematic scanning of the odor space gives a systematic scanning of all patterns. So, the spatial pattern appears as a good code. (2) All patterns were also generated in the model A, in the absence of local neurons. This shows that the differential convergence of different types of receptor neurons in glomeruli is the main factor that explains the generation of spatial patterns. However, the number of odors per pattern is the most uniform in model B and the least uniform in model C.

Inhibition. Why are there inhibitory local neurons in the natural systems? (1) A first piece of response is given by the remarkable fact that in 80% of cases in model C the coefficients c_{ij} were positive which indicates an inhibitory action (Fig. 10b). This shows that inhibition is more efficient than excitation to decorrelate the outputs of principal neurons and suggests that inhibition may be interpreted as a decorrelation mechanism. (2) But why to decorrelate? One of the most obvious benefit of decorrelation is to increase the contrast of the activity maps. Defining contrast as the difference of activity between the least active and the most active PN for any odor, we have shown that the average contrast depends on the models; it is 20% of the median activity in model A (no decorrelation), 30% in model B (a decorrelation is present as shown in Fig. 9) and 60% in model C. So, it appears that the patterns obtained with the Hérault–Jutten algorithm are much more contrasted than the other two. This increased contrast is useful for increasing the robustness of the system and also, in case where more than two states are used, to increase the number of odors that can be discriminated.

4. DISCUSSION

Models of the receptor neuron and of the first two neuronal layers of the olfactory system, were studied. These models were developed with the aim of investigating to what extent a set of assumptions, kept as small and simple as possible, could account for known properties of the intensity and quality coding of odorants. This approach presents obvious advantages and limitations. Among advantages is the possibility of studying in details, either analytically or with numerical experiments, the effect of all selected variables on the properties under investigation and thus to determine the most important ones. The limitations are that such models, ignoring many aspects present in the real systems, cannot duplicate exactly experimental results and when applied at the proper level of complexity (e.g., receptor potential for the receptor neuron) can only account for the orders of magnitude of the actual measurements.

However, these models offer a sound framework for developing more realistic models of increasing complexity. In the single neuron approach we are adding such factors as, for example, the size and shape of the dendrites and other neuron parts, the presence of accessory cells and comparing the effect on the coding properties of various locations of the voltage sources. Similarly, for the neural network approach other problems and variants, such as the projection of only one type of receptor neuron per glomerulus (see e.g., Ref. 48), are being considered. Models of increasing complexity and specificity must be developed in close interaction with experimental investigations. Yet, a complementary effort directed at simplification and generalization is also increasingly needed. It is believed that simple and global models, retaining only the most important factors, are irreplaceable tools to explore the principles of operation of the actual systems in the same way as, for example, the notion of ideal gas in thermal physics is irreplaceable to understand the behavior of an actual gas.

REFERENCES

1. Boeckh J, Tolbert LP (1993) Synaptic organization and development of the antennal lobe in insects. Microsc Res Tech 24:260–280
2. Boeckh J, Distler P, Ernst KD, Hösl M, Malun D (1990) Olfactory bulb and antennal lobe. In: D Schild (ed) Chemosensory information processing. Springer, Berlin, pp 201–227
3. Christensen TA, Waldrop B, Harrow ID, Hildebrand JG (1993) Local interneurons and information processing in the olfactory glomeruli of the moth Manduca sexta. J Comp Physiol A 173:385–399
4. Christensen TA, Harrow ID, Cuzzocrea C, Randolph PW, Hildebrand JG (1995) Distinct projections of two populations of olfactory receptor axons in the antennal lobe of the sphinx moth *Manduca sexta*. Chem Senses 20:313–323
5. Duchamp-Viret P, Duchamp A, Vigouroux M (1989) Amplifying role of convergence in olfactory system: a comparative study of receptor cell and second order neuron sensitivities. J Neurophysiol 61:1085–1094
6. Ennis DM (1991) Molecular mixture models based on competitive and non-competitive agonism. Chem Senses 16:1–17
7. Fort J-C (1991) Stabilité de l'algorithme de séparation de sources de Jutten et Hérault. Traitement du Signal 7:407–418
8. Getz WM, Akers RP (1995) Partitioning non-linearities in the response of honey bee olfactory receptor neurons to binary odors. BioSystems 34:27–40
9. Halász N (1990) The Vertebrate Olfactory System. Akadémiai Kiadó, Budapest
10. Hansson BS, Anton S, Christensen TA (1994) Structure and function of antennal lobe neurons in the male turnip moth, *Agrotis segetum* (Lepidoptera: Noctuidae). J Comp Physiol A 175:547–562
11. Hildebrand J (1995) Analysis of chemical signals by nervous systems. Proc Natl Acad Sci USA 92:67–74
12. Hopfield JJ (1991) Olfactory computation and object perception. PNAS USA 88:6462–6466
13. Homberg U, Christensen T, Hildebrand JG (1989) Structure and function of the deutocerebrum in insects. Annu Rev Entomol 34:477–501
14. Jutten C, Hérault J (1988) Une solution neuromimétique au problème de la séparation de sources. Traitement du Signal 5:389–403
15. Jutten C, Hérault J (1991) Blind separation of sources. Part I: An adaptive algorithm based on a neuromimetic architecture. Signal Processing 24:1–10
16. Kaissling K-E (1971) Insect olfaction. In: LM Beidler (ed) Handbook in Sensory Physiology 4(1):351–431. Springer, Berlin
17. Kaissling K-E (1986) Chemo-electrical transduction in insect olfactory receptors. Annu Rev Neurosci 9:121–145
18. Kaissling K-E (1987) RH Wright Lectures on Insect Olfaction. K Colbow (ed). Simon Fraser University, Vancouver
19. Kaissling K-E (1994) Elementary receptor potentials in insect olfactory cells. In: Kurihara K, Suzuki N, Ogawa H (eds) Olfaction and Taste XI, pp 812–815
20. Kaissling K-E, Thorson J (1980) Insect olfactory sensilla: structural, chemical and electrical aspects of the functional organization. In: Sattelle DB, Hall LM, Hildebrand JG (eds) Receptors for neurotransmitters, hormones and pheromones in insects. Elsevier/North-Holland, Amsterdam, pp 261–282
21. Keil T (1984) Reconstruction and morphometry of silkmoth olfactory hairs: A comparative study of sensilla trichodea on the antennae of male *Antheraea polyphemus* and *Antheraea pernyi* (Insecta, Lepidoptera). Zoomorphology 104:147–156
22. Lamb TD, Pugh EN (1992) G-protein cascades: gain and kinetics. Trends in Neurosci 15:291–298
23. Lánský P, Rospars J-P (1993) Coding of odor intensity. BioSystems 31:15–38
24. Lánský P, Rospars J-P (1995) Ornstein–Uhlenbeck neuron revisited. Biol Cybern 72:397–406
25. Lánský P, Rospars J-P, Vermeulen A (1994) Basic mechanisms of coding stimulus intensity in the olfactory sensory neuron. Neural Proc Letters 1:9–12
26. Malun D (1991) Inventory and distribution of synapses of identified uniglomerular projection neurons in the antennal lobe of *Periplaneta americana*. J Comp Neurol 305:348–360
27. Masson C, Mustaparta H (1990) Chemical information processing in the olfactory system of insects. Physiol Rev 70:199–243

28. Menini A, Picco C, Firestein S (1995) Quantal-like current fluctuations induced by odorants in olfactory receptor cells. Nature 373:435–437

29. Rall W (1977) Core conductor theory and cable properties of neurons. In: Kandel ER, Brookhardt JM, Mountcastle VB (eds) Handbook of Physiology: The Nervous System, 1:39–98 Williams and Wilkins, Baltimore

30. Rodrigues V, Pinto L (1989) The antennal glomerulus as a functional unit of odor coding in *Drosophila melanogaster*. In: Singh RN and Strausfeld NJ (eds) Neurobiology of Sensory Systems. Plenum, New York, pp 387–396

31. Rospars J-P (1983) Invariance and sex-specific variations of the glomerular organization in the antennal lobes of a moth, *Mamestra brassicae*, and a butterfly, *Pieris brassicae*. J Comp Neurol 220:80–96

32. Rospars J-P (1988) Structure and development of the insect antennodeutocerebral system. Int J Insect Morphol Embryol 17:243–294

33. Rospars J-P, Chambille I (1981) The deutocerebrum of the cockroach *Blaberus craniifer* Burm. Quantitative study and automated identification of the glomeruli. J Neurobiol 12:221–247

34. Rospars J-P, Chambille I (1989) Identified glomeruli in the antennal lobes of insects: invariance, sexual variation and postembryonic development. In: Singh RN, Strausfeld NJ (eds) Neurobiology of Sensory Systems. Plenum, New York, pp 355–3375

35. Rospars J-P, Fort J-C (1994) Coding of odor quality: roles of convergence and inhibition. Network 5:121–145

36. Rospars J-P, Hildebrand JG (1992) Anatomical identification of glomeruli in the antennal lobe of the male sphinx moth *Manduca sexta*. Cell Tiss Res 270:205–227

37. Rospars J-P, Lánský P (1993) Stochastic model neuron without resetting of dendritic potential: application to the olfactory system. Biol Cybern 69:283–294

38. Rospars J-P, Lánský P, Vaillant J, Duchamp-Viret P, Duchamp A (1994) Spontaneous activity of first- and second-order neurons in the frog olfactory system. Brain Res 662:31–44

39. Rospars J-P, Lánský P, Tuckwell HC, Vermeulen A (1996) Coding of odor intensity in a steady-state deterministic model of an olfactory receptor neuron. J Comput Neurosci 3:51 72

40. Royet J-P, Sicard G, Souchier C, Jourdan F (1987) Specificity of spatial patterns of glomerular activation in the mouse olfactory bulb: computer-assisted image analysis of 2-deoxyglucose autoradiograms. Brain Res 417:1–11

41. Selzer R (1984) On the specificities of antennal olfactory receptor cells of *Periplaneta americana*. Chem Senses 8:375–395

42. Shepherd GM, Greer CA (1990) Olfactory bulb. In Shepherd GM (ed) The Synaptic Organization of the Brain. Oxford Univ Press, Oxford, pp 133–169

43. Stengl M, Hatt H, Breer H (1992) Peripheral processes in insect olfaction. Annu Rev Physiol 54:665–68

44. Stocker RF, Lienhard MC, Borst A, Fischbach KF (1990) Neuronal architecture of the antennal lobe in *Drosophila melanogaster*. Cell Tissue Res 262:9–34

45. Tuckwell HC (1988) Introduction to theoretical Neurobiology. Cambridge Univ Press, New York

46. Tuckwell HC, Rospars J-P, Vermeulen A, Lánský P (1996) Time-dependent solutions for a cable model of an olfactory receptor neuron. J. Theor. Biol. 181:25–31

47. Vareschi E (1971) Duftunterscheidung bei der Honigbiene. Einzelzell-Ableitungen Verhaltensreaktionen. Z Vergl Physiol 75:143–173

48. Vassar R, Chao SK, Sitcheran R, Nunez JM, Vosshall LB and Axel R (1994) Topographic organization of sensory projections to the olfactory bulb. Cell 79:981–991

49. Vermeulen A, Rospars J-P, Lánský P, Tuckwell HC (1995a) Some new results on the coding of pheromone intensity in an olfactory sensory neuron. In: Verleysen M (ed) Third Europ Symp on Artif Neural Networks. D facto, Brussels, pp 105–110

50. Vermeulen A, Rospars J-P, Lánský P, Tuckwell HC (1995b) Coding of stimulus intensity in an olfactory receptor neuron: Role of neuron spatial extension and dendritic passive backpropagation of action potentials. Bull Math Biol 58:493–512

19

THE ELECTRONIC EAR

Towards a Blueprint

André van Schaik[1][*] and Ray Meddis[2][†]

[1] Mantra Centre for Neuromimetic Systems
Swiss Federal Institute of Technology
CH-1015 Lausanne, Switzerland
[2] Speech and Hearing Laboratory
Department of Human Sciences
University of Technology
Loughborough LE11 3TU, United Kingdom

1. INTRODUCTION

In this chapter we propose to review recent developments and to establish an achievable research agenda for developing VLSI models of auditory signal processing in the mammalian brainstem. Considerable progress can be reported for models of the transduction of acoustic signals into the electrical response of auditory nerve fibres. A useful start has also been made in developing circuits to represent individual nerve cells similar to those found in the auditory brainstem. Only tentative beginnings have been made in the construction of massively parallel circuits to represent the activity of large ensembles of auditory nerve cells but these are enough to establish the feasibility of the enterprise. This progress in the field of electronic modelling capability has been accompanied by a substantial increase in our knowledge of the detailed functioning of individual nerve cells in the early stages of the auditory signal processing system. The time is therefore right for a substantial investment of effort in the development of these new modelling technologies.

In recent years, several computer models of low-level processing in the brain have been developed (Arle and Kim, 1990, 1991a, 1991b; Banks and Sachs, 1991; Cooke, 1986; Ghoshal et al, 1992, 1991; Giguere and Woodland, 1992, 1994; Hewitt and Meddis, 1993, 1991, 1994; Hewitt et al, 1992; Meddis, 1986, 1988; Meddis and Hewitt, 1993; Meddis et al, 1990a, 1990b; Pont and Damper, 1991; Young et al, 1993). As our understanding of the actual processes in the brain increases, our models will become more and more detailed. Furthermore, while early attempts at modelling focused on the peripheral (cochlear) mecha-

[*] eMail: Andre. van_Schaik@di. epfl. ch
[†] eMail: R. Meddis@lut. ac. uk

Neurobiology, edited by Torre and Conti
Plenum Press, New York, 1996

nisms of hearing, more recent modelling efforts are attempting to characterise higher levels of auditory processing taking place in the auditory brainstem. A serious problem with this development is that the computer models are becoming more and more computationally intensive and memory demanding. Current trends threaten to take the simulation of these models beyond the range of even the most powerful digital computers. Alternative methodologies are therefore required.

One alternative is to build auditory models using analogue electronic circuits. These are more difficult to construct and modify than computer programs but they do have some important advantages. For example, analogue VLSI architectures can be made very similar to neural wetware architectures. In general neurons in the auditory system are organised in two-dimensional sheets, each performing a certain signal processing function. The topology is also often conserved in the communication from one sub-nucleus to another. It is fairly straightforward to map these kind of structures onto silicon using analogue VLSI, since the silicon surface provides a corresponding two-dimensional support for the neuron circuits.

Communication of analogue information in real time between subsystems on a single VLSI chip, or between multiple VLSI chips has been a major problem for analogue chips. However, in recent years several communication techniques have been proposed, which solve this problem. Most of the proposed communication protocols use pulse frequency to code the analogue activity and conserve topology from sender to receiver (Lazzaro et al, 1993; Lazzaro and Wawrzynek, 1995; Mortara and Vittoz, 1994; Mortara, 1995; Murray et al, 1991). These protocols therefore add another similarity between neural wetware and hardware architectures.

Analogue VLSI enables one to create small, but imprecise building blocks (Vittoz, 1994), corresponding to, for example, individual neurons. These can be replicated many thousands of times and put on a single chip. The imprecision of the building blocks is not an issue in neural architectures, since they obtain accuracy over large numbers of units rather than from the accuracy of individual units. In addition, neural wetware and analogue hardware are similar in the fact that saturating non-linearities are more a rule than an exception for both.

One advantage of conserving the architecture of neural wetware in VLSI implementations, is that all processing happens in parallel and in continuous time. Unlike digital circuits, there is no time-multiplexing in order to simulate multiple neurons. This means that, with care, the circuits can function in real time. An important further advantage over computational approaches and digital circuits arises from the fact that expanding the number of neurons does not introduce a time penalty, but only increases the surface area needed to implement the model. Since multiple chips can be used in parallel, there is no obvious limit to the number of neurons in an analogue VLSI model. This is an important consideration when modelling neural activity.

Neural modelling may also be better suited to analogue Very Large Scale Implementations than other applications. All silicon chips are subject to a certain failure density, caused by contamination of the silicon wafer. A larger chip therefore implies a higher probability of having a failure on this chip. With conventional electronic architectures, this leads to a higher number of rejected chips and seriously constrains the cost-effectiveness of the process. In the case of neural models it may be possible to fabricate larger chips without paying this penalty. This is because the actual functionality in a neural architecture is distributed in the sense that every neuron contributes only a small amount to the collective function. Considerable redundancy is therefore built in to the system. The circuits based on these kind of architectures may, therefore, be more tolerant to failures in individual building blocks. In the case of a neural architecture, a number of defective neurons will hardly influence the collective function and these failures need not lead to the rejection of the chip.

2. MODELLING THE AUDITORY PERIPHERY

Hearing like vision is a largely parallel system. A broadband acoustic stimulus arriving at the eardrum is subjected to mechanical frequency selectivity in the cochlea even before the sound is transduced into auditory nerve action potentials. As a consequence, the inner hair cells which are responsible for the conversion from a mechanical to an electrical signal see only the output from a narrow bandpass filter. It follows that, the one dimensional audio input signal is converted into a parallel representation in which every channel only responds to a narrow frequency range in the input signal. This filtering process has been characterised in terms of a travelling wave passing along a membrane in the cochlear duct subjected to a low pass filter whose cut off frequency is lower the further the wave travels (see Pickles (1982) for general review).

Lyon and Mead (1988) realised this effect electronically in terms of a cascade of low pass filters with taps taken from between each filter. They successfully implemented this idea as an analogue VLSI chip and have used this circuit as the input stage of a number of other more complex models. Others have produced different versions of the same idea (Watts et al, 1992; Schaik et al, 1995) incorporating a range of improvements. A sample of the filter functions obtained at different positions along our silicon is shown in figure 1. A disadvantage of the cascade approach is that the propagation delay of the travelling wave from the input to a certain output tap is controlled by the same variable that controls the cut-off frequency of the filters. This leads to the problem that the accumulated delay increases as a function of the number of filters per given frequency range, i.e., as the frequency resolution, increases. This problem can be avoided by having separate variables controlling the propagation delay and the cut-off frequency (Bhadkamkar, 1993). All these models work in real time and can be adapted to produce any number of outputs and could be useful, for example, as a front end to an automatic speech recognition device since these typically use computationally intensive spectral analysis not unlike the frequency selectivity observed in the mammalian auditory periphery.

Current analogue VLSI cochlear models are basically linear models. Physiological measurements made in the cochlea however show that the mechanical filters are in fact non-linear in the sense that the narrow bandpass function observed at low acoustical signal

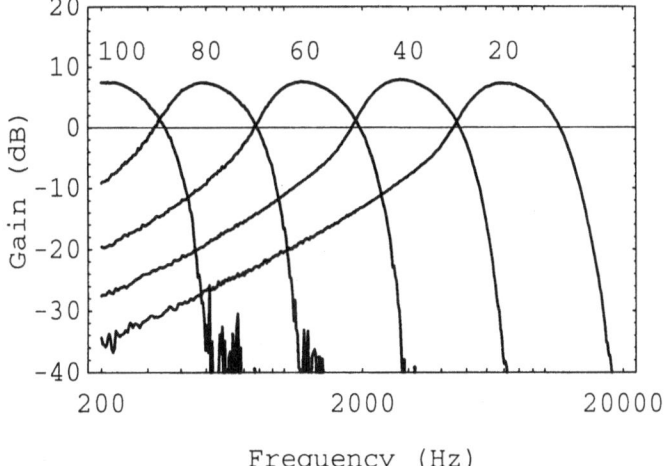

Figure 1. Frequency response of the silicon cochlea by Schaik et al. (1995), measured at equal distances along the cochlea. The filter number which output is measured is indicated above the maximum of each curve.

levels are gradually replaced by wider bandpass functions at higher signal levels. It remains a research issue to design a non-linear filter which can function appropriately in a VLSI implementation but a start has been made here too. (Bhadkamkar, 1993; Watts, 1993). The specification for a non-linear filter able to simulate the response of the basilar membrane in the cochlea are quite demanding. Firstly, the input-output function for frequencies close to the filter's best frequency (BF) is linear up to about 30 dB SPL but is strongly compressed above this frequency. However, frequencies below the BF show no compression and at high signal levels can produce a stronger output than an equally intense tone at BF. As a result the centre frequency of the 'filter' appears to move towards a lower frequency when signals are more intense. The filter function also becomes wider, i.e. less selective as signal levels increase (Johnstone et al, 1986; Sellick et al., 1982). Secondly, the biological system generates two-tone distortion products but only within a narrow range (±10%) of BF (Ruggero, 1992). Thirdly, the system manifests 'two tone suppression,' a phenomenon where the output of a filter to tones at BF can be *reduced* substantially by the introduction of a second tone to the input (Pfeiffer, 1970). These curious features may have important repercussions for the behaviour of the system at later stage and it is important that they become a feature of peripheral models at the earliest possible date. Secker-Walker and Searle (1990), for example, have argued that identification of vowel formant frequencies is enhanced by the non-linearity of the filters, especially when the identification is carried out in the time domain.

A start has also been made in using silicon to simulate the next stage in auditory signal processing. This is the conversion of the mechanical vibration of the membrane to an electrical signal in the auditory nerve. Physiologically this process has a number of interest-

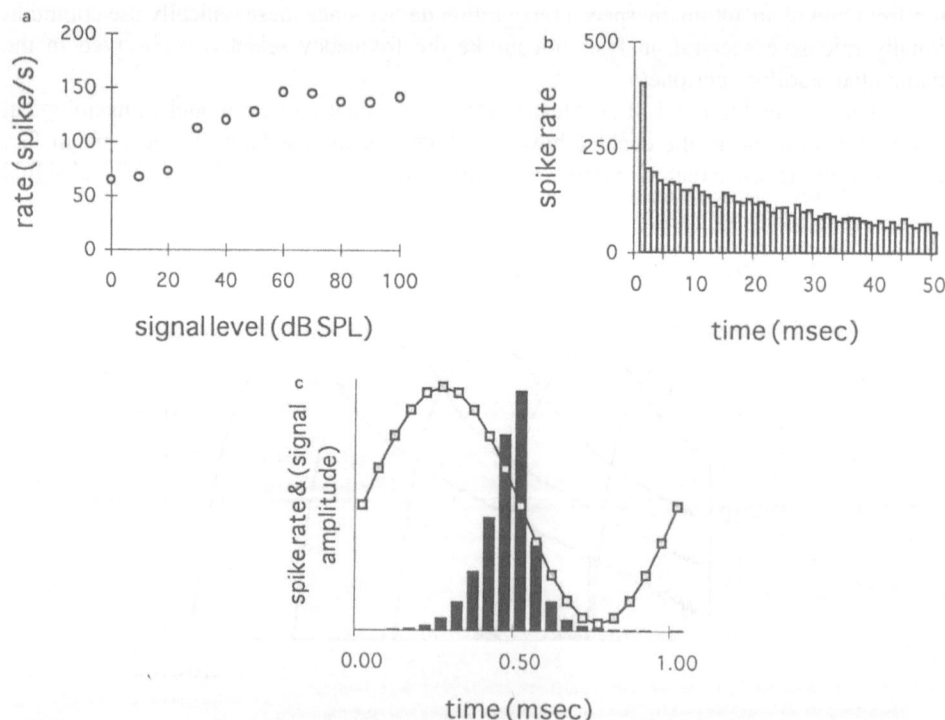

Figure 2. Inner hair cell response. a) IHC rate /intensity function, b) Post Stimulus Time Histogram, c) Period Histogram shown with stimulating sinusoid.

ing features. Firstly, the spike rate of most AN fibres increases with sound stimulus intensity but only over a narrow range of signal levels (say from 20-60 dB SPL). Above this level the output saturates, showing no further rate increase (figure 2a). Secondly, the rate of firing also adapts to a new stimulus; the high rate of firing at onset is quickly replaced by a much lower, adapted rate within about 3 msec of the onset of the stimulus. With a continued stimulus, the rate then reduces even further, but at a slower pace, during the following 100 ms (figure 2b). Thirdly, at low frequencies (<3 kHz) the spike rate in an individual AN fibre can be shown to be modulated by one direction of the basilar membrane motion only (figure 2c). Finally, in the absence of any signal at all, an AN fibre shows spontaneous action potentials which can be as frequent as half of the maximum rate of firing. Hewitt and Meddis (1991) review these required features in the context of existing computer models of this process. Ideally, they should all be present in an analogue circuit that simulates these processes.

It is fairly easy to design an IHC circuit that quantitatively shows the above properties. Using a simple transconductance amplifier made of only five MOS transistors a saturating non linearity can be obtained of a form which resembles the saturation of the biological IHC output (figure 3). Passing the output current of the transconductance amplifier through a current mirror yields a half wave rectification, allowing only the positive half of the waveform to influence the spike probability. Temporal adaptation can then be obtained by subtracting a weighted, low-pass filtered copy of this rectified current from the rectified current itself. Furthermore, a spontaneous rate can be obtained by adding a constant current to the output current of this circuit which represents the instantaneous spike probability.

An alternative way of designing an analogue electronic model that matches physiological data is to transfer a well tested computer model in a straightforward way onto silicon. We are currently exploring a circuit that is closely modelled on Meddis' (1986, 1988) computational model of inner hair cell functioning that shows all of the above properties. This circuit captures more of the biological detail of the operation of an IHC, although at greater cost and it remains to be decided whether it is really necessary to capture all this detail in order to make the models of higher level auditory processing work. If the circuit described in the previous paragraph, which only performs a crude approximation of the IHC's function, is sufficient, this will be our preferred model, since it is about five times smaller than the circuit based on the computer model. It is a research issue to decide on the minimum amount of detail required to satisfy the functional requirements of later processing stages.

Circuits of varying levels of complexity have already been manufactured that combine the frequency selectivity stage with the hair cell transduction stage (Lazzaro, 1992;

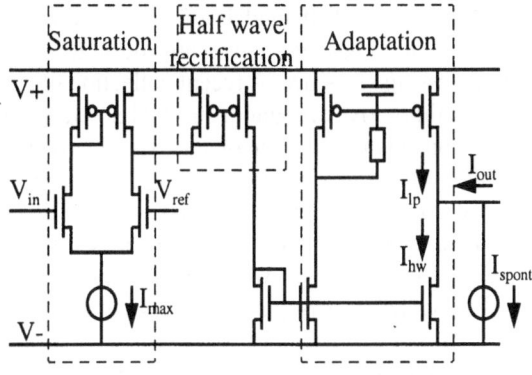

Figure 3. An electronic IHC circuit, implementing a saturating rate intensity function, half wave rectification, and temporal adaptation.

Liu et al, 1992). While we might want to improve the biological verisimilitude further, these studies have already established the feasibility of combining these two stages on the same chip and encourage plans to use such a circuit as the input to the next stage of auditory processing, the cochlear nucleus in the brainstem.

One question remains to be answered however. This concerns the form of the output of the combined model. The AN transmits information in the form of a series of spikes whose timing is influenced by the IHC activity but also contains a large random element. Using VLSI architectures it is relatively easy to generate an analogue output representing the instantaneous probability of occurrence of an action potential. Converting this probability function to a genuinely stochastic series of spikes requires substantial extra circuitry. There is no theory, however, concerning the value of stochastic nature to the system. It might be that the unpredictable response of the individual nerve fibre reflects the fact that the fibre can only generate a maximum of 1000 spikes per second because of the refractory nature of neurons. The spikes are therefore distributed randomly across a large number of fibres so that detailed information concerning the fine structure of the stimulus can be preserved. If that is the case we might proceed more economically with electronic fibres that can fire more frequently. Alternatively, the noisiness of the transmission channel may serve some other function, like establishing a noise floor to eliminate responses to insignificantly quiet stimuli. If this is the case then fully stochastic models may be essential if the system aims to reproduce psychophysical performance.

Identifying the function, if any, of the stochastic nature of the AN response is an important research question which has been highlighted by the modelling approach. Such problems arise whenever we seek to simplify a simulation in order to reduce manufacturing costs or even make its construction possible. In this respect they pose important research questions concerning the functional significance of individual features of the system.

3. SINGLE UNITS IN THE COCHLEAR NUCLEUS

While further design improvements are certainly possible, we can say that the problems of simulating the response of auditory nerve fibres using electronic circuitry have been solved, to a first approximation at least. It is, therefore, only a matter of time before there is wide availability of chips generating real-time output that is almost indistinguishable from recordings of large numbers of auditory nerve fibres. This will provide the necessary basis for extending the scope of VLSI circuits into the first auditory relay station in the mammalian brain stem, the cochlear nucleus (CN).

While many questions remain unanswered concerning the CN, we now know enough to begin the process of developing VLSI models of CN functioning. A surge of recent anatomical and physiological studies has revealed an unexpectedly complex structure at this early stage in the signal processing sequence. A recent collection of papers on the cochlear nucleus gives an unusually authoritative account of current knowledge in this area (Merchan et al, 1993). It is clear that a considerable amount of analysis takes place as soon as the signal enters the brain. This is achieved using a parallel processing strategy. When the AN enters the CN it immediately divides into two branches and later subdivides further, so the single AN fibre makes contact with a large number of morphologically different types of cell, each of which processes the signal in a different way. For example each AN fibre innervates spherical bushy, globular bushy, stellate, octopus, vertical, giant and fusiform cells (figure 4). Each type of cell is now known to react differently to the same acoustic stimulus, emphasising (or filtering out) particular features of the input. The various cells also project out of the nucleus to a range of different destinations. Even within a given morphological cell type, there are further indications of variation in the signal processing operations carried out. This

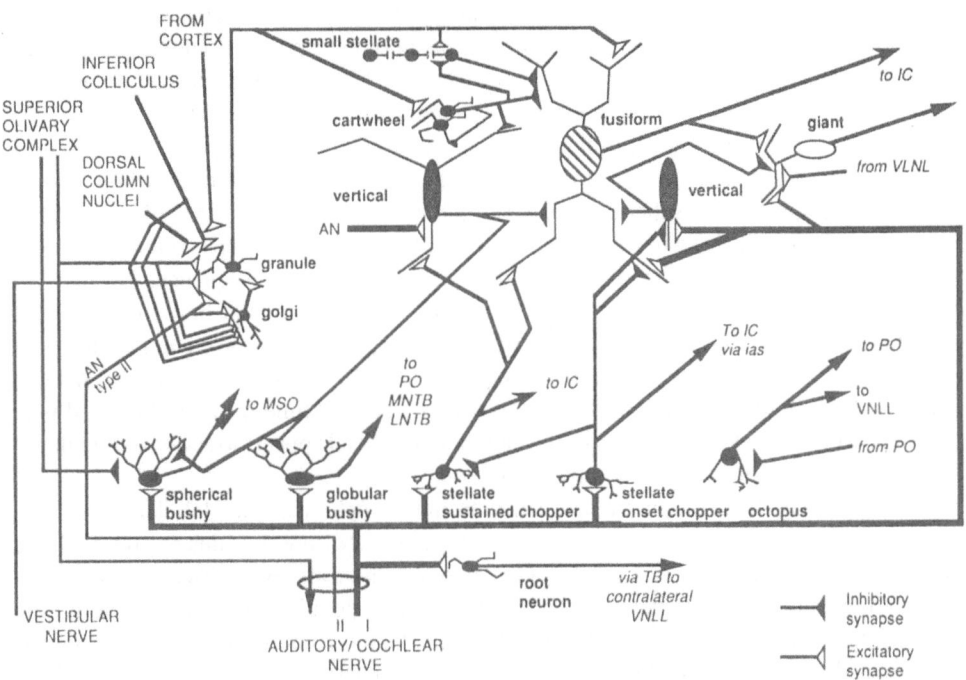

Figure 4. Different cell types in the cochlear nucleus and their innervation. Abbreviations: ias: intermediate acoustic stria; AN: auditory nerve; IC: inferior colliculus; MSO /LSO: medial /lateral superior olive; MNTB /LNTB: medial/lateral nucleus of the trapezoid body; PO: periolivary nuclei; VNLL: ventral nucleus of the lateral lemniscus.

is particularly true of the stellate cell classification that appears to encompass a number of response types some of which are excitatory and some inhibitory.

Almost all of the simulation work so far has used conventional computer programming techniques. This has proved satisfactory particularly when modelling the response of a single cell to a short burst of acoustical energy (Arle and Kim, 1990, 1991a, 1991b; Banks and Sachs, 1991; Ghoshal et al, 1992, 1991; Hewitt and Meddis, 1993, 1991, 1994; Hewitt et al, 1992; Meddis and Hewitt, 1993; Pont and Damper, 1991; Young et al, 1991). However, the computer runs are typically very time-consuming and it is increasingly difficult to envisage the simulation of large assemblies of such neurons without recourse to the real-time parallel processing capabilities of VLSI implementations.

Progress in modelling auditory signal processing has been greatly assisted by the realisation that the variation in the responses of individual types of neurons can be understood in terms of two critical features of the cells; firstly, the input connectivity of the cell and secondly, the membrane characteristics of the cell. By input connectivity we mean the number of AN fibres contacting the cell, their origin along the cochlear partition and whether these inputs contact the cell soma itself or the cell dendrites. Synaptic events impinging on the soma give rise to rapid and strong membrane voltage changes at the axon hillock where the output spikes are generated while contacts on the dendrites give rise to only slow and weak changes. By membrane characteristics we refer loosely to the response of the cell to depolarising and hyper polarising (sub-threshold) currents. For example, Oertel (1983) has shown that stellate cells have linear voltage-current relationships (ohmic) while bushy cells show a saturating voltage-current relationship, i.e., the membrane voltage increases in response to an increase of current up to a certain voltage, after which the voltage tends to stay constant. This saturation is caused by an increase in the membrane's conductance resulting from the opening of fast potassium ion channels, in response to depolarisation. This also results in a rapid membrane time constant when the cell is excited. Differences in these key respects can result in substantial differences in the response of the cell to stimuli.

Figure 5 illustrates some of the variation across cell types in response to a pure tone burst. For such a stimulus the AN input takes the form of a high rate of firing which rapidly adapts. The output from a spherical bushy cell looks very similar to an AN fibre response. This comes as no surprise when it is appreciated that spherical bushy cells typically receive contact from only one AN fibre directly onto the cell soma. As a consequence, the spherical bushy cell generates one output spike for every input spike. Globular bushy cells, on the other hand, receive many inputs from AN fibres and many of these make contact only with the dendrites. It is likely, therefore, that a single input spike is not enough to drive the cell to respond. More than one spike must arrive within a narrow time window if the cell is to generate an output spike. Functionally, the globular bushy cell appears to act as a coincidence detector among a large number of input fibres.

At first sight, the output train of spikes from both spherical and globular bushy cells looks very similar to a train of spikes recorded from an AN fibre. For this reason, both are said to have a 'primary-like' response pattern. However, there are some differences and these will need to be reflected in analogue models. Firstly, the globular bushy cell has very little spontaneous activity. This is probably because the spontaneous firing of the AN input is uncorrelated across fibres and gives rise to few coincidences. The spherical bushy cell on the other hand has an approximately one to one relationship between input and output events and therefore fires spontaneously at the same rate as an AN fibre. Secondly, the globular bushy cell has a much shorter latency to firing following the onset of a stimulus. This is because it is receiving input from many fibres all with varying latencies. Statistically, some of these will give an early indication of the signal onset so that the globular bushy cell is more likely to respond sooner than the spherical bushy cell (Young et al, 1993). In terms of their function, the globular bushy cell appears to be a low-noise unit that emphasises changes in signal level. The spherical bushy cell, on the other hand, appears to relay without distortion information concerning individual AN fibres to a second level of analysis. It is possible that this unit amplifies weak AN fibre action potentials. Both spherical and globular bushy cells are also subject to inhibitory influences from other parts of the auditory brainstem although relatively little is known about these as yet.

Stellate cells respond quite differently (Smith and Rhode, 1989; Rhode and Smith, 1986). Like the globular bushy cells they receive many AN fibre contacts, mainly on the dendrites. Consequently, they respond only when a number of inputs are coincidentally active. However, stellate cells integrate their many inputs over a much longer time period. The net result of this is that many stellate cells respond to a continuous acoustic stimulus with a steady depolarisation of the membrane potential during which period the cell fires regularly. This regular firing is called 'chopping' and the rate of chopping appears to be determined by the internal physical characteristics of the cell. Cells that respond in this way are known as sustained choppers.

There is considerable variation in the patterns of response attributed to stellate cells, however (Smith and Rhode, 1989). Some chop briefly only at the onset of the stimulus (onset choppers). The reason for this is likely to be that the depolarisation of the membrane potential is only great enough to provoke firing at the stimulus onset, when many AN fibres fire simultaneously. The difference between sustained and onset chopping may be accounted for by any of a number of factors which depress the membrane voltage below threshold. This could be either a high firing threshold, or a smaller number of inputs, or weaker inputs (e.g.

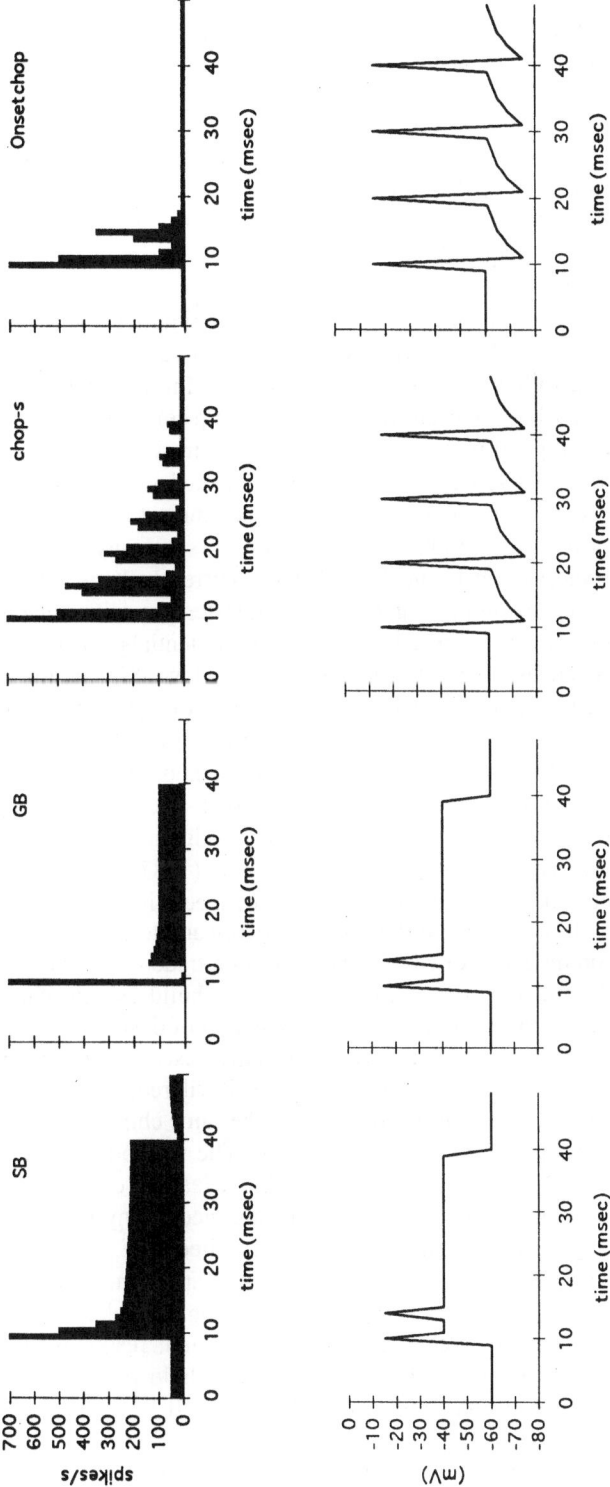

Figure 5. Response types of different cells in the cochlear nucleus: spherical bushy cells (SB), globular bushy cells (GB), stellate sustained chopper (chop-S), stellate onset chopper (onset-chop). Left: PSTH for pure tone stimulation; Right: schematised evolution of the membrane potential, when stimulated with a depolarising DC current.

dendritic contacts distant from the soma), or inhibition from other cells, or a combination of these.

Some cells (possibly a stellate type or maybe octopus cells) fire once only at the onset of the acoustic stimulus (Palmer and Winter, 1993). This may represent an extreme case of the chopping response where only the first spike in the sequence occurs. This could occur because the cell has a very high firing threshold which is only exceeded at the onset of the tone. Careful physiological studies suggest that these cells are contacted by hundreds of AN fibres and respond to a very wide range of input frequencies. This contrasts with most other cells in the cochlear nucleus that receive input from only a narrow region of the cochlea, representing a narrow range of frequencies.

Variation in membrane characteristics among cells has also been shown to be partly responsible for the observed variation in responses to the same stimuli. Chopper cells, as we have seen, respond to an injected steady current with a steady stream of spikes. Each spike consists of a depolarisation followed by a hyper-polarisation. Intracellular investigations indicate that these cells have a linear current voltage relationship controlled mainly by a fast sodium and a single slower type of potassium channel. By contrast, primary-like response cells such as bushy cells have been found to respond to an injected steady current with only one or possibly two spikes occurring only at the onset of the current injection (Feng et al, 1994). The spikes have a depolarised phase only and do not show any hyper-polarising undershoot. Intracellular measurements indicate that the current/voltage relationship is non-linear (saturating) in a manner indicating that a second, faster, potassium channel is also functional. This results in a rapid response to post-synaptic potentials consistent with the proposed function as a coincidence detector with a narrow window of integration.

These examples are merely illustrative of the complexity of signal processing taking place in individual cells in the CN. Studies using computer models have already shown that the known physiological observations of single cells in the auditory brainstem can be successfully replicated using relatively simple models. This is true whether we use very detailed Hodgkin/Huxley type representations of the cells (Hodgkin and Huxley, 1952) or more schematic accounts such as those proposed by MacGregor (1987). It appears that good analogue models can be built if, as a minimum, the known connectivity is replicated and the membrane characteristics of the cells are reflected in the hardware design.

An analogue neuron model is, of course, the result of a trade-off between the detail incorporated into the model and the actual size of the circuit. To build a single neuron model is in most cases not a goal, but only a means to allow simulation of the collective behaviour of a large group of neurons. It is thus important to determine the least amount of detail needed to ensure the collective behaviour one is looking for. Less detail will reduce the neuron circuit size and will allow a larger group of neurons to be put on the same chip.

A very simple neuron model is shown in figure 6. The membrane of a neuron is modelled by a membrane capacitance, C_{mem}, and the membrane leakage current is controlled by the gate voltage, V_{leak}, of an NMOS transistor. In the absence of any input ($I_{ex}=0$), the membrane voltage will be drawn to its resting potential (controlled by V_{rest}), by this leakage current. Excitatory inputs simply add charge to the membrane capacitance, whereas inhibitory inputs discharge C_{mem}. If an excitatory current I_{ex}, larger than the leakage current of the membrane, is injected, the membrane potential will increase from its resting potential. This membrane potential, V_{mem}, is compared with a controllable threshold voltage V_{thres}, using a basic transconductance amplifier driving a high impedance load. If V_{mem} exceeds V_{thres}, an action potential will be generated.

The generation of the action potential happens as follows. The output voltage of the comparator will be close to the positive power supply if V_{mem} is larger than V_{thres}, and it will be somewhat lower than V_{mem} if V_{mem} is smaller than V_{thres}. This output can be converted in to a binary signal using an inverter. The output of the inverter will be close to the positive

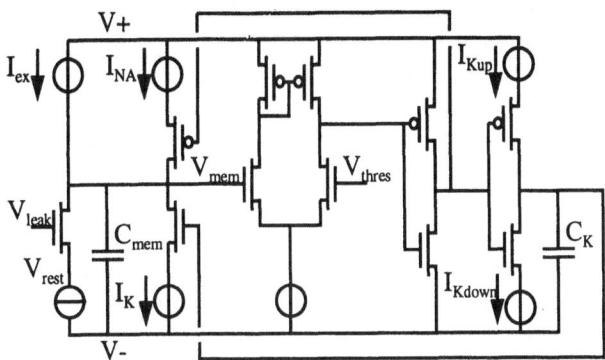

Figure 6. An electronic neuron circuit. See text for details.

power supply if V_{mem} is smaller than V_{thres} and close to the negative power supply if V_{mem} is larger than V_{thres}. In the second case, the PMOS transistor, that is controlled by this voltage and is used as a switch, will start conducting and will allow the "sodium current" I_{Na} to pull up the membrane potential, i.e. the upswing of the 'spike'. At the same time however, a second inverter will allow the capacitance C_K to be charged at a speed which can be controlled by the current I_{Kup}. As soon as the voltage on C_K is high enough to allow conduction of the NMOS to which it is connected, the "potassium current" I_K will be able to decrease the membrane potential and restore it to its resting value, with or without an undershoot.

Two different potassium channel time constants govern the opening and closing of the potassium channels. The current I_{Kup} which charges C_K will control the spike width, since the delay between the opening of the sodium channels and the opening of the potassium channels is proportional to I_{Kup}. If V_{mem} now drops below V_{thres}, the output of the first inverter will become high, cutting off the current I_{Na}. Furthermore, the second inverter will then allow C_K to be discharged by the current I_{Kdown}. If I_{Kdown} is small, the voltage on C_K will decrease only slowly, and, as long as this voltage stays high enough to allow I_K to discharge the membrane, it will be impossible to stimulate the neuron if I_{ex} is smaller than I_K. Therefore I_{Kdown} can be said to control the refractory period of the neuron.

Despite its simplicity, this neuron model already allows us to simulate different spiking behaviour, characteristic of different neuron types by changing its biases. If the amount of charge, which is transferred onto C_{mem} when an input pulse arrives, is very high, the neuron will spike after every input pulse, as long as the inputs do not arrive during its refractory period. If the transferred charge per pulse is reduced, more input pulses need to arrive within a certain time window in order to provoke an output spike. If the membrane leakage current is increased, inputs have to arrive within a narrower window to generate a spike.

This simple model can be expanded to more detailed models of the different neuron types. For instance, a circuit can be added which causes the threshold voltage to adapt, and more detailed models of the output synapses can be added, including dendritic filtering. Alternatively, a neuron model might be created that sticks as close as possible to the operation of the biological neuron, yielding a more accurate but also much larger silicon neuron (Mahowald and Douglas (1991)).

The research agenda, here, is to develop a range of silicon neurons that robustly reproduce the key physiological behaviours of different types of cells. Existing circuits have established the feasibility of the operation but none have been very thoroughly tested. Key design decisions will also need to be made in the light of the requirement that a useful model may require thousands of these neurons operating in parallel. A general purpose neuron, representing complex dendritic effects and large numbers of different types of ion channels would obviously be useful. Modelling a particular cell type would then require the designer merely to delete those sub-circuits not needed. It would, however, occupy a large area on the chip. Designing simpler circuits from scratch for each cell type would lead to more economical designs but the development effort could be much greater.

Testing protocols need to be established and routinely deployed. For example, the engineer needs to apply the same tests to the neuron models as the physiologist applied originally when studying the cells properties. These include current and voltage clamping techniques under normal conditions and conditions where the individual ion channels have been rendered inactive by manipulating the intra and extracellular ion concentrations. Ideally the cell should show similar behaviour to the arrival of exitatory pre-synaptic potentials and inhibitory pre-synaptic potentials, i.e. similar membrane voltage excursions and time constants. The more properties reproduced, the more acceptable the model.

Every model is, however, necessarily a compromise and it is essential that the designer list all of the properties of the model. These include the good and the bad features, the successful and the unsuccessful aspects of the model's functioning. Only in this way can potential users of the circuit make informed decisions about whether to deploy it in a future system. Current practice typically restricts reports to descriptions of one or two successful features of the circuit's operations leaving the reader to discover the hard way what the circuit does in more general terms. It is understood that reductions in size are desirable and that small size inevitably compromises the functionality. Only a full account of the properties of the model can allow the wisdom of a particular compromise to be fully appreciated.

4. MODELLING COLLECTIVE BEHAVIOUR

While the simulation of the activity of individual neural units is of scientific interest in its own right, the goal of our research is to understand the signal processing properties of large ensembles of such units. In fact, VLSI architectures are probably not the best way of modelling single unit activity. Too many compromises are required to make the circuit design problem tractable. The flexibility of computer programming is much better suited to this purpose. However the massive parallel processing capability of VLSI circuits will make them the preferred medium for studying large numbers of interacting circuits.

Anatomy can supply us with some information concerning the connectivity between nuclei in the auditory brainstem. To a more limited extent it can even identify which specific cell types within a nucleus project to which other specific cell type within another nucleus, although this information is sometimes speculative. Anatomical techniques also exist for identifying whether connections are excitatory or inhibitory. The job of the modeller is to propose and evaluate anatomically feasible circuits which could give rise to the observed physiological response of the cell to acoustic stimulation. This necessarily involves models which begin with the acoustic input and follow the process through all of the peripheral stages as well as the neuronal responses. The anatomical and physiological picture is far from complete but we have enough information to begin the process of generating some simple models which mimic some known circuits in the auditory brainstem.

Processing schemes implicated in sound source localisation give the clearest examples of circuits that generate an output of obvious functional significance to an animal and can be modelled. Sounds coming from the left of an animal deliver a more intense signal to the left ear than the right ear and vice versa. Cells in the lateral superior olive (LSO) have been shown to respond only when sounds delivered to the ipsilateral ear are the more intense (Glendenning et al, 1985). The greater the disparity in sound levels between the two ears, the greater the firing rate of the cell. This appears to be achieved using an arrangement similar

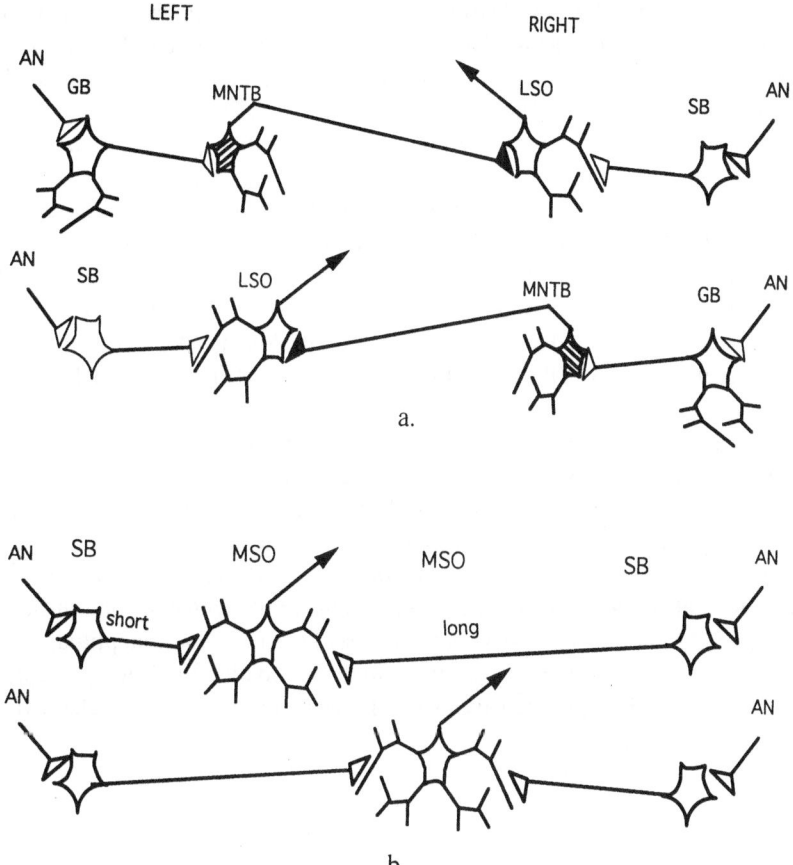

Figure 7. a) Neural circuit for binaural intensity difference sensitivity in the Lateral Superior Olive; b) Neural circuit for binaural time difference sensitivity in the Medial Superior Olive.

to that illustrated in figure 7a. Ipsilaterally the LSO is excited by activity in the SB cells that relay activity from the ipsilateral auditory nerve fibres. The same LSO cell is inhibited by activity in the medial nucleus of the trapezoid body (MNTB), that is, in turn, excited by activity in the contralateral GB cells. In effect the MNTB converts contralateral activity into inhibition (Aitkin, 1986). The LSO responds to the combined signal (excitatory from the same side and inhibitory from the opposite side). As a result it responds only when the signal is more intense on the ipsilateral side of the head.

In the medial superior olive (MSO) there are cells which also respond selectively to the azimuthal location of a sound source but use a completely different principle (Goldberg and Brown, 1969). The operation of these cells relies upon the fact that a sound to the left will arrive at the left ear a few hundred microseconds earlier than it will at the right ear. Cells in the MSO appear to act as coincidence detectors, driven to fire only when action potentials arrive simultaneously from spherical bushy (SB) cells in both the left and right CN. Activity transmitted from the contralateral CN has a longer journey and that introduces a delay (figure 7b). When the difference in transmission time is the same as the delay in the time of arrival of the sound at the left and right ear, the MSO cell will be persuaded to fire. This is because the action potentials arriving from the left and the right will arrive at the same time. By varying the distance travelled by the contralateral activity it is possible to tune individual MSO cells to respond to specific left/right sound arrival times.

The MSO and LSO are clearly processing the input signals in order to extract information relevant to the azimuthal angle of the sound source. This information is represented spatially in the sense that different sound source locations (azimuthal angles) will generate different patterns of activity across the nucleus. The actual location of the sound can not be deduced from the activity of a single cell. Rather the source location is represented in the distribution of unit activity across the ensemble of thousands of neurons. We can study these representations using conventional computers only with difficulty because of the sheer scale of the problem. VLSI implementations however make the problem more tractable. A real time VLSI implementation of interaural time difference detection for source localisation, using an architecture closely resembling the architecture sketched in the previous paragraph, has already been implemented and tested (Lazzaro and Mead, 1989). In fact the LSO and the MSO are more complex in their structure and connections than this model would suggest but it represents an encouraging start.

A less obvious form of signal processing concerns the response of brainstem units to amplitude modulated (AM) sounds (figure 8). These are of particular interest because this class of signals includes music and voiced speech where the pitch of the sound is related to the rate of modulation. It has recently been discovered that stellate cells can be entrained to AM sounds, that is they fire in synchrony with the peaks of the modulation envelope (Frisina et al, 1990; Kim et al, 1990). Sustained chopper cells, in particular, are entrained by only a limited range of AM frequencies. In other words they have a bandpass amplitude modulation transfer function and this suggests that they might be involved in pitch perception. This theory could be tested through extensive use of animal recordings but a good hardware model

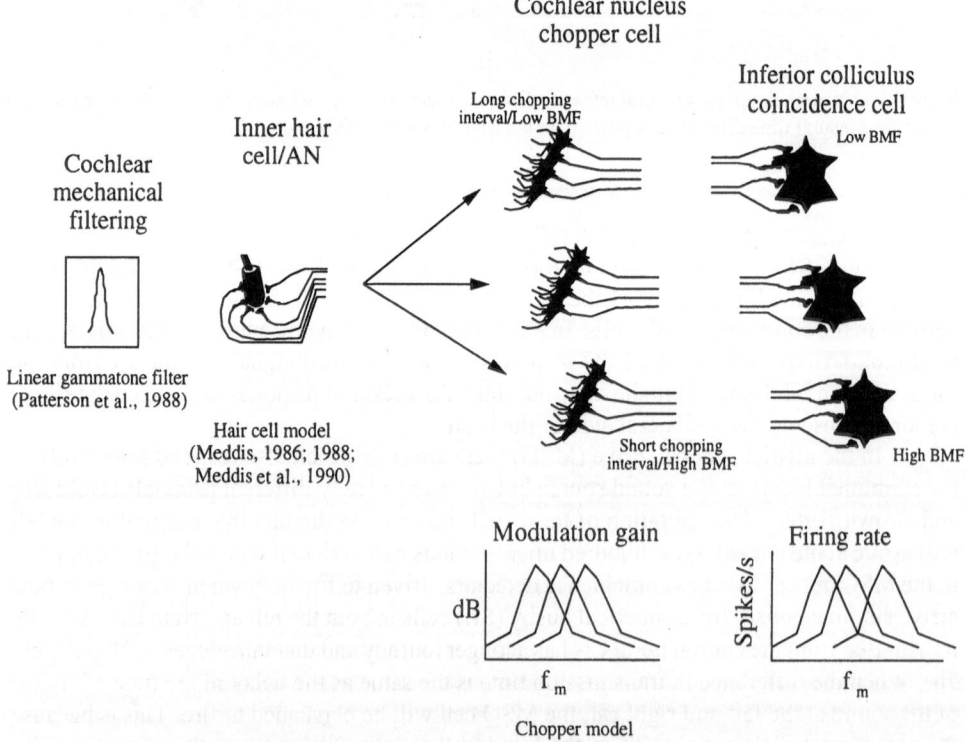

Figure 8. Proposed neural circuit for the Amplitude Modulation sensitivity in the Inferior Colliculus. Abreviations: BMF best modulation frequency.

of this process would allow us to explore the complex implications of the theory more conveniently and humanely.

It is important to note that the firing rate of the chopper cell is unaffected by changes in the frequency of modulation of the signal, only the coherence of the cell activity with respect to the acoustic stimulus is influenced. Because different cells vary in the particular frequency of AM that they respond to, we infer that pitch is represented spatially in the CN. Computer studies that simulate sustained chopping activity in response to AM signals suggest that the preferred rate of modulation is limited by the natural rate of chopping of the cell (Kim et al, 1990). This is, in turn, related to some intrinsic property of the cell such as membrane capacitance or the time constant of recovery of potassium channels. Computer simulations of sustained chopper cells give an almost perfect match to the physiological data measured on single stellate cells (Hewitt et al, 1992).

We can however go further than this because it is known that some cells in the inferior colliculus (IC) also respond selectively to AM rates but this time by increasing the *firing rate* of the cell as the rate of modulation approaches the cell's preferred frequency. We have shown that this effect can be replicated by directing the output of a number of simulated CN chopper cells (with the same amplitude modulation best frequency) into a coincidence detector type of neuron (Hewitt and Meddis, 1994). This latter cell responds with a high firing rate only when the inputs from the CN are synchronised with one another. In the case of sustained chopper cells, this condition is met only when they are all driven by an AM acoustic stimulus at their preferred AM frequency. This idea is supported (but not proved) by anatomical studies showing that some sustained chopper cells in the CN have excitatory projections to the IC (Smith and Rhode, 1989).

The neuron model described in section 3 could, for instance, be used to implement a pitch extraction model. This neuron circuit can be set to simulate a sustained chopper cell by using a relatively high threshold voltage and low membrane leakage current, yielding long integration times. The equivalent of the neural time constant of recovery of the potassium channels can be controlled using the current I_{Kdown} (figure 6). Chopper cells with different time constants will yield different best frequencies in the amplitude modulation transfer functions. A population of chopper cells having similar time constants, receiving input from the same group of auditory nerve fibres, will then synchronise if the modulation frequency of their input signal is close to their preferred frequency. A VLSI implementation will allow the system to run in real time, irrespective of how many hundreds or thousands of neurons are used. At this level of complexity the time taken to create and test the circuit would not necessarily compare unfavourably with the development of a corresponding computer progam.

A population of synchronised chopper cells, when driving a coincidence detecting neuron, will yield a rate/place coding of the amplitude modulation frequency, with the position of the coincidence detector having maximum activity corresponding to the dominant modulation frequency. The same neuron circuit used to create the sustained chopper cells can also be used to create a coincidence detecting neuron. If the threshold voltage is set so that no single input spike can create an output spike, several spikes have to arrive at the neuron, within a time period controlled by the membrane leakage current, in order for the neuron to spike.

Using the simple neuron model described above, it will be possible to put about thousand of these neurons on 10 mm^2 of silicon, all operating in parallel. This will allow simulation of the operation of one or several channels of the model in real time on a single chip, where one channel can be defined as a population of choppers and coincidence detectors driven by the same group of inner hair cells. More channels can of course be simulated using several identical chips in parallel, but connected to different groups of inner hair cells.

Fortunately, the anatomy of the auditory brainstem suggests very little communication between the channels and this would be consistent with a bank of chips working in parallel.

This model is an example of a large group of interconnected neurons, creating a representation in which the output is activated by a certain feature (in this case the amplitude modulation frequency) within a restricted band of input frequencies. The auditory brainstem appears to contain several similar systems containing representations of different features versus signal frequency. It should be possible to connect several tens (or even hundreds) of chips together, creating several of these representations in real time. This will enable us to study models of even higher levels of auditory processing, which require these representations to create very sophisticated representations of the auditory environment.

5. CONCLUSION

Computer programming methods have proven useful in developing models of the auditory periphery and the behaviour of individual units in the auditory brainstem. However, it is clear that we shall need faster and more flexible methods for studying ensembles of cells on the kind of scale routinely reported for even small brainstem nuclei. VLSI technology offers us this possibility in the form of massively parallel analogue circuits that function in real time. Enough progress has been made in developing models of the auditory periphery, individual neurons and large scale integration of neuron-like units, to reassure us that the proposal is realistic. Enough progress has also been made in studies of the anatomy and physiology of the auditory brainstem to inspire models of plausible signal processing systems that can be implemented using the new technology.

The way forward appears to require more detailed work at the level of modelling individual neurons because these will be the essential building bricks of the final systems. It is quite likely that we shall need to design different circuits for each type of cell to be used if we are to keep circuit size to a minimum. We already know enough about the properties of the relevant cells in the auditory brainstem to make this possible. The testing of the cells will require the development of evaluation protocols closely modelled on existing techniques used by physiologists. The need to restrict circuit size will focus the scientific debate on which aspects of a cell's functioning are critical to the signal processing carried out by that cell. For example, we need to know whether the stochastic nature of AN fibre action potentials are an essential feature of the input to the CN or whether they can be replaced by a more readily computable signal.

Other questions arise when we integrate many thousands of model neurons into biologically plausible systems. Will the need to distribute the model across many separate chips create insuperable problems? Alternatively, will the redundancy of the system allow us to develop very large circuits on the same wafer and tolerate a substantial number of individual component failures, as the brain itself obviously does? Can the modelling be extended beyond simulating the results of scientific studies to genuinely useful technologies? Existing models of auditory localisation and pitch perception are indicative of future progress in this respect but, in reality, acceptable proof can only be based on working examples.

REFERENCES

Aitkin, L. (1986) "The Auditory Midbrain," Humana Press.
Arle J.E., Kim D.O. (1990) "A Modelling study of Single neurons and neural circuits of the ventral and dorsal cochlear nucleus," in 'Analysis and Modelling of Neural Systems,' Edited by F.Eeckman, Kluwer, Boston.

Arle, J.E., and Kim, D.O. (1991a) "Neural modeling of intrinsic and spike-discharge properties of cochlear nucleus neurons," Biological Cybernetics. 64, 273-283.

Arle, J.E., and Kim, D.O. (1991b) "Simulations of Cochlear nucleus neural circuitry: Excitatory- inhibitory response-area types I-IV," Journal Acoustical Society America 90 (6), 3106-3121.

Banks, M. I., Sachs M. B. (1991) "Regularity Analysis in a Compartmental Model of Chopper Units in the Anteroventral Cochlear Nucleus," Journal of Neurophysiology 65 (3).

Bhadkamkar, N. (1993) "A Variable Resolution, Nonlinear Silicon Cochlea," Stanfort University Technical Report CSL-TR-93-58.

Cooke, M.P. (1986). "A computer model of peripheral auditory processing incorporating phase-locking, suppression and adaptation effects," Speech Communication. 5, 261-281.

Feng, J J., Kuwada, S., Ostapoff, E.M., Batra, R., Morest, K. (1994) "A Physiological and Structural Study of Neuron Types in the Cochlear Nucleus. I. Intracellular Responses to Acoustic Stimulation and Current Injection," Journal of Comparative Neurology 346, 1-18.

Frisina, Robert .D., Smith, R .L., and Chamberlain S.C. (1990) "Encoding of amplitude modulation in the gerbil cochlear nucleus: I. A hierarchy of enhancement," Hearing Research 44, 99-122.

Ghoshal, S., Kim, D.O., and Northrop R.B. (1991) "Modeling amplitude-modulated (AM) tone encoding behavior of cochlear nucleus neurons," Proc. of the 17th Annual IEEE Northeast Bioengineering Conference, Hartford CT.

Ghoshal, S., Kim, D.O., and Northrop R.B. (1992) "Ampitude-modulated tone encoding behaviour of cochlear nucleus neurons: Modeling study," Hearing Research 58, 153-165.

Giguere, C., and Woodland, P.C. (1992) "A Composite Model of the Auditory Periphery with Feedback Regulation," Cambridge University, Engineering Dept.

Giguere, C., and Woodland, P C. (1994) "A computational model of the auditory periphery for speech and hearing research. I. Ascending path," Journal Acoustical Society America 95, 331-342.

Glendenning, K.K., Hutson, K.A., Nudo, R.J., and Masterton, R.B. (1985) "Acoustic Chiasm II: Anatomical Basis of Binaurality in Lateral Superior Olive Cat," The Journal of Comparative Neurology 232, 261 285.

Goldberg, J. M., and Brown, P. B. (1969) "Response of Binaural Neurons of Dog Superior Olivary Complex to Dichotic Tonal Stimuli: Some Physiological Mechanisms of Sound Localization," Journal of Neurophysiology 32, 613-636.

Hewitt, M J., and Meddis R. (1991) "An evaluation of eight computer models of mammalian inner hair-cell function," Journal Acoustical Society America 90, 904-917.

Hewitt, M. J., Meddis R., and Shackleton, T. M.(1992) "A computer model of a cochlear-nucleus stellate cell: Responses to amplitude-modulated and pure-tone stimuli," Journal Acoustical Society America 91, 2096-2109.

Hewitt, J.M., and Meddis R. (1993) "Regularity of cochlear nucleus stellate cells: A computational modeling study," Journal Acoustical Society America 93, 3390-3399.

Hewitt, M. J., and Meddis, R. (1994) "A computer model of amplitude-modulation sensitivity of single units in the inferior colliculus," Journal Acoustical Society America 95, 1-15.

Hodgkin, A.L., and Huxley, A.F. (1952) "A Quantitative description of membrane current and its application to conduction and excitation in nerve" Journal of Physiology 117, 500-544.

Johnstone, B.M., Patuzzi, R., and Yates, G.K. (1986) "Basilar membrane measurements and the travelling wave" Hearing Research 22, 147-153.

Kim, D.O., Sirianni, J.G., and Chang, S.O.(1990) "Responses of DCN-PVCN neurons and auditory nerve fibers in unanesthetized decerebrate cats to AM and pure tones: Analysis with autocorrelation /power-spectrum," Hearing Research 45, 95-113.

Lazzaro J., and Mead C.M. (1989) "A Silicon Model of Auditory Localisation," Neural Computation (1), 47-57.

Lazzaro, J. (1992) "Temporal Adaption in a Silicon Auditory Nerve," in 'Neural Information Processing Systems 4,' edited by J. E. Moody, S. J. Hanson, R. P. Lippmann. Morgan Kaufmann, California, 813-820.

Lazzaro, J.,Wawrzynek, J.., Mahowald, M., Sivilotti, M, Gillespie D. (1993) "Silicon Auditory Processors as Computer Peripherals," in 'Advances in Neural Information Processing Systems,' edited by Hanson, S., Cowan J., and Giles, C., Morgan Kaufmann, San Mateo: CA, and IEEE Transactions on Neural Networks, 4 (3), 523-528.

Lazzaro J.P., and Wawrzynek, J. (1995) "A Multi-Sender Asynchronous Extension to the Address-Event Protocol," in '16th Conference on Advanced Research in VLSI,' edited by Dally, W.J., Poulton, J.W., and Ishii, A.T., 158-169.

Liu, W., Andreou, A G., and Goldstein, M H. Jr. (1992) "Voiced-Speech Representation by an Analog Silicon Model of the Auditory Periphery," IEEE Transactions on Neural Networks 3 (3), 477-487.

Lyon, R. .F., and Mead, C. A. (1988) "An Analog Electronic Cochlea," IEEE Transactions on Acoustics, speech and Signal Processing 36, 1119-1134.

MacGregor, R.J. (1987) "Neural and Brain Modelling," Academic Press, San Diego.

Mahowald, M., and Douglas., R. (1991) "A Silicon Neuron," Nature 354, 515-518.

Meddis, R. (1986). "Simulation of mechanical to neural transduction in the auditory receptor," Journal Acoustical Society America 79, 702-711.

Meddis, R. (1988) "Simulation of auditory-neural transduction: Further studies," Journal Acoustical Society America 83, 1056-1062.

Meddis, R., Hewitt, M. J., and Shackleton, T M.(1990a) "Non-linearity in a computational model of the response of the basilar membrane" in 'Lecture Notes in Biomathematics,' edited by P. Dallos., C.D. Geisler., J.W. Matthews., M.A. Ruggero., C.R. Steele. Springer-Verlag.

Meddis, R., Hewitt, M.J. and Shackleton, T. (1990b) "Implementation details of a computational model of the inner hair-cell/auditory-nerve synapse," Journal of the Acoustical Society of America, 87, 1813-1818.

Meddis, R. and Hewitt, M.J. (1993) "Computational modeling of cochlear nucleus functioning. The Mammalian cochlear nuclei: organisation and function," in 'The Mammalian Cochlear Nuclei: Organization and Function,' edited by Merchan, M.A., Godfrey, D.A. and Mugnaini, E., Plenum Press, New York.

Merchan, M.A., Juiz, J.M., Godfrey, D.A. and Mugnaini (1993) "The Mammalian cochlear nuclei: organisation and function," (Eds) Plenum Press, New York.

Mortara, A., and Vittoz, E.A. (1994) "A Communication Architecture Tailored for Analog VLSI Artificial Neural Networks: Intrinsic Performance and Limitations," IEEE Transactions on Neural Networks 5 (3), 459-466.

Mortara, A. (1995) "Communication Techniques for Analog VLSI Perceptive Systems," PhD Thesis 1329, Ecole Polytechnique Fédérale de Lausanne, Lausanne, Switzerland.

Oertel, D. (1983) "Synaptic responses and electrical properties of cells in brain slices of the mouse anteroventral cochlear nucleus," Journal of Neuroscience 3, 2043-2053.

Palmer, A R., and Winter, I M. (1993) "Coding of the fundamental frequency of voiced speech sounds and harmonic complexes in the cochlear nerve and ventral cochlear nucleus," in 'The Mammalian Cochlear Nuclei: Organization and Function,' edited by Merchan, M.A., Godfrey, D.A. and Mugnaini, E., Plenum Press, New York.

Patterson,R.D., Nimmo-Smith,I., Holdsworth,J. and Rice, P. (1988) "Spiral vos final report. Part A: The auditory filter bank," Cambridge Electric Design. Contract Rep. (Apu 2341).

Pfeiffer, R.P. (1970). "A model for two-tone inhibition of single cochlear-nerve fibres," Journal Acoustical Society America 48, 1373-1378.

Pickles, J O. (1982) "An Introduction to the Physiology of Hearing," Academic Press.

Pont, M.J., and Damper, R.I. (1991) "A computational model of afferent neural activity from the cochlea to the dorsal acoustic stria," Journal Acoustical Society America 89, 1213-1228.

Rhode, W. S., and Smith, P H. (1986) "Encoding Timing and Intensity in the Ventral Cochlear Nucleus of the Cat," Journal of Neurophysiology 56, 261-285.

Ruggero, M A., Robles L, and Rich N C. (1992) "Two-Tone suppression in the Basilar Membrane of the Cochlea: Mechanical Basis of Auditory-Nerve Rate Suppression," Journal of Neurophysiology 68 (4).

Schaik, F.A. van, Fragnière, E., and Vittoz, E.A. (1995) "Improved Silicon Cochlea using Compatible Lateral Bipolar Transistors," to be published in 'Advances in Neural Information Processing Systems 8,' edited by Touretzky, D., Mozer, M., and Hasselmo, M., MIT press, Cambridge.

Secker-Walker, H.E., and Searle, C.L. (1990) " Time-domain analysis of auditory-nerve fiber firing rates," J. Acoust. Soc.Am. 88(3), 1427-1436

Sellick, P.M., Patuzzi, R., and Johnstone, B.M. (1982) "Measurement of basilar membrane motion in the guinea pig using the Mossbauer technique," Journal Acoustical Society America 72, 131-141.

Smith, P.H., and Rhode, W.S. (1989) "Structural and functional properties distinguish two types of multipolar cells in the ventral cochlear nucleus," Journal of Comparative Neurology, 282, 595-616.

Vittoz, E.A. (1994) "Analog VLSI Signal Processing: Why, Where and How?" Journal of VLSI Signal Processing, 8 & Analog Integrated Circuits and Signal Processing, 27-44.

Watts, L., Kerns, D.A., Lyon, R.F., and Mead, C.A.(1992)"Improved Implementation of the Silicon Cochlea," IEEE Journal of Solid-state circuits 27, 692-700.

Watts, L. (1993) "Cochlear Mechanics: Analysis and Analog VLSI," PhD thesis, California Institute of Technology, Pasadena.

Young, E. D., Rothman, Jason, S., and Manis, P. B. (1993) "Regularity of discharge constrains models of ventral cochlear nucleus bushy cells," in 'The Mammalian Cochlear Nuclei: Organization and Function,' edited by Merchan, M.A., Godfrey, D.A. and Mugnaini, E., Plenum Press, New York.

20

COUPLING OF NETWORKS OF NEURONS TO SUBSTRATE PLANAR MICROTRANSDUCERS

A Review

Marco Bove, Massimo Grattarola, and Sergio Martinoia

Bioelectronics Laboratory and Bioelectronic Technologies Laboratory
(c/o Advanced Biotechnology Center)
Department of Biophysical and Electronic Engineering
University of Genoa
Via Opera Pia 11a, 16145, Genoa, Italy

1. INTRODUCTION

Cultures of dissociated neurons (Barinaga, 1990; Bulloch & Syed, 1992) constitute a promising method for characterizing the auto-organization properties of populations of neurons under controlled physico-chemical conditions. Neurons can survive for weeks in culture, where they reorganize into two-dimensional networks. Especially in the case of populations obtained from vertebrate embryos, these networks cannot be regarded as faithful reproductions of in vivo situations, but rather as new rudimentary neurobiological systems whose activity can change over time spontaneously or as a consequence of chemical/physical stimuli (Gross, Rhoades, Jordan, 1992). Quite a recent technique, appropriate for recording the electrical activity of networks of cultured neurons, lies in using substrate transducers, i.e., arrays of planar microtransducers forming the adhesion surface for the reorganizing networks.

This non-conventional electrophysiological method has several advantages over standard intracellular recording that are related to the possibility of monitoring/stimulating noninvasively the electrochemical activities of several cells, independently and simultaneously for a long time (Jimbo & Kawana, 1992; Jimbo, Robinson & Kawana, 1993; Gross & Kowalsky, 1991; Gross et al., 1992).

In consideration of the relative novelty of this technique, its key features will be briefly reviewed in comparison with other more traditional electrophysiological methods. Intracellular and patch-clamp recordings are single-cell electrophysiological techniques requiring that a thin glass capillary be brought near a cell membrane. Intracellular recording involves a localized rupture of the cell membrane. Patch-clamp methods can imply the rupture and (possible) isolation of a small membrane patch or, in the case of the so called "whole-cell-loose-patch configuration" (Stühmer, Roberts & Almers, 1983), the sealing between the microelectrode tip and the membrane surface.

Neurobiology, edited by Torre and Conti
Plenum Press, New York, 1996

The novel use of planar substrate microtransducers presents some similarities to the latter technique, in the sense that it involves the sealing between the neuron membrane and the underlying planar microtransducer surface the neuron is growing on.

Peculiar to this technique are at least two features: A) several (i.e., tens-hundreds) neurons are brought into simultaneous contact with several underlying microtransducers, with a neuron-to-microtransducer correspondence which can be supported by mechanical and/or chemical means (Jimbo et al. 1993; Curtis, Breckenrigde, Connolly, Dow, Wilkinson, Wilson, 1992). Simultaneous multisite recording from units at well-localized positions is a unique feature of this technique, though an exact one-to-one coupling between neurons and electrodes is not always feasible. B) Recording/stimulation can be protracted for days (Jimbo & Kawana, 1992). During this period, which is very long as compared with the typical time intervals allowed by intracellular techniques, the neuronal population in culture is continuously developing and synaptic contacts change in the presence of different physiological conditions producing changes in the network functions and dynamics.

The neuron-to-transducer junction can be appropriately characterized by using an equivalent-circuit approach, as previously done for other, more traditional, electrophysiological methods (Robinson, 1968; Stühmer et al., 1983). Such a characterization is especially effective, in consideration of the above mentioned feature B), which implies that, during the very long period (days) of an experiment, neurons are alive (i.e., possibly showing continuous changes in shape, adhesion and arborizations) on top of the recording transducer; therefore, the sealing might change over time, with possible variations in the recorded signals.

Most of the data reported in the literature refer to noble metal microelectrodes. i.e., passive transducers. An interesting exception lies in the use of insulated field effect transistors (Fromherz, Offenhäuser, Vetter & Weis, 1991).

These two distinct categories of microtransducers will then be considered in the following.

2. METHODS

2.1. Planar Microelectrode Arrays

Several technological alternatives have been described in the literature (Novak & Wheeler, 1986; Regehr, Pine, Cohan, Mischke, Tank, 1989; Boppart, Wheeler, Wallace, 1992; Gross, Rhoades, Reust & Schwalm, 1993; Wilson, Breckenridge, Blackshaw, Connoly, Dow, Curtis, Wilkinson, 1994). In the following, two examples are given that refer to arrays presently utilized in our Lab.

1. One kind of microelectrode array, kindly supplied by the Center for Integrated Systems - Stanford University (USA), is made of 32 gold electrodes, 10μm x 10μm and 50μm apart; the substrate is silicon and the passivation layer is silicon nitride (Martinoia, Bove, Carlini, Ciccarelli, Grattarola, Storment & Kovacs, 1993);

2. A second kind of microelectrode array, kindly supplied by the Material Science Laboratory of NTT Basic Research Laboratories, Atsugi (Japan), consists of 64 ITO (Indium Tin Oxide) tracks covered with biocompatible insulators (Al_2O_3 and polyamide) on top of a silica glass substrate (Jimbo & Kawana, 1992).

To reduce the impedance of each electrode, the electrochemical platinization of the electrodes is usually performed. In our Lab, when platinization is performed, a solution of

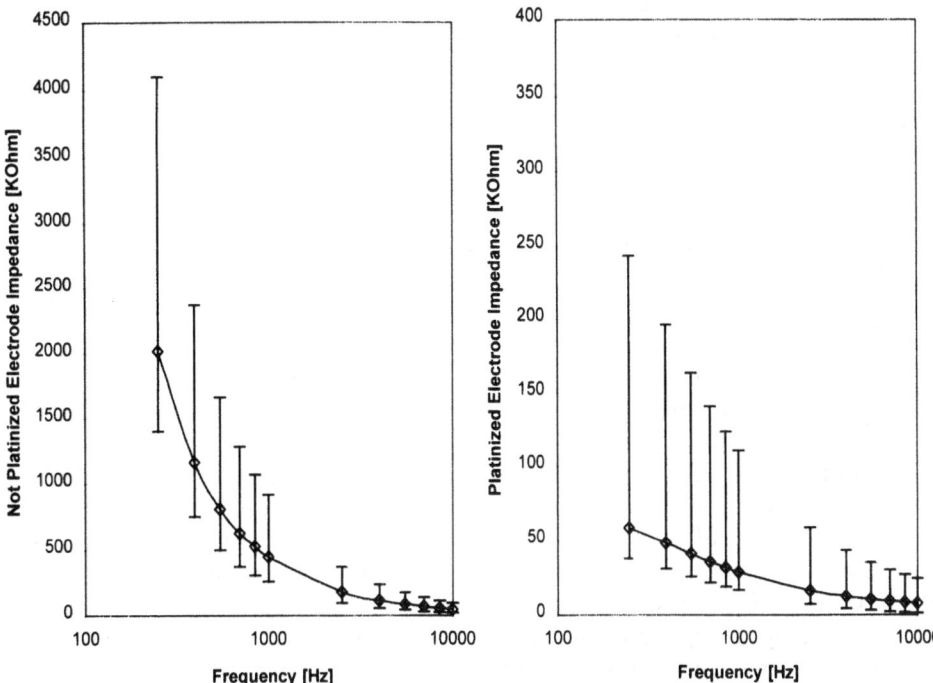

Figure 1. Impedance of six microelectrodes of an array, before and after the platinization process, as a function of frequency. For each data point, minimum, mean, and maximum are shown.

1g of H_2PtCl_6 in 28.1 ml of 0.025% lead acetate is used, and a DC current of 300 mA/cm^2 is applied for about 1 min (Bove, Grattarola, Martinoia, Verreschi, 1995). As an example, a comparison of the impedance values of 6 microelectrodes of an array (designed and fabricated in the NTT Basic Research Laboratories) before and after the platinization process is shown in Figure 1 (Figure 1 reproduced from (Bove et al., 1995)).

2.2. Insulated Field Effect Transistors

"Active" microtransducers are insulated-gate FETs showing ideal capacitive coupling between the patch of a neuron membrane and the transducer gate. This device is usually described as a pH-sensor (Bousse, de Rooij & Bergeveld, 1983; Grattarola, Massobrio & Martinoia, 1992), whereas, in this case, the technology and the fabrication process are not devoted to optimizing the sensitive gate-layer but rather the geometry of the device. An example of design is reported in (Fromhertz et al., 1991). The device is constituted by a 4 x 4 array of transistors (distance 300μm) on a 10 mm x 30 mm chip of n-type silicon. The distance between source and drain is 6μm. The openings in the field oxide (1μm thickness) have a size of 20μm x 34μm. They are covered with a gate oxide of 12 nm thickness.

It should be noted that typical H$^+$-sensitive FETs with a Si_3N_4 or Al_2O_3 gate layer could be used for this kind of application.

2.3. Cell Types

For the sake of simplicity, we can divide the kinds of neuronal cultures utilized connected to substrate planar microtransducers into two broad categories:

a) Neurons from ganglia of adult invertebrates.

b) Neurons from nervous tissue of vertebrate embryos.

Cultures from the first category are characterized by the use of a small (e.g., 10^2) number of large (e.g., 50μm in diameter) identified neurons.

Most of the results described in the literature refer to the leech and are concerned with basic biophysical questions, such as electrophysiological signal propagation on a single, arborized neuron (Wilson et al., 1994). Attempts at chemically guiding the arborizations of a single neuron on patterned substrata are also reported (Fromherz & Schaden, 1994).

Cultures from vertebrate embryos are characterized by a large (e.g., 10^6) number of small (e.g., 10μm in diameter) "similar" neurons. These cultures are mostly considered as random networks of (supposedly) identical units which form a dense, highly connected, layer on the top of the microelectrodes. Attempts to mechanically (Jimbo et al., 1993) or chemically (Clark, Connolly & Moores, 1990) position the network over the array are described in the literature.

Most of these cell cultures are based on neurons obtained from the chick or rat embryo. In both cases the networks of neurons developing in vitro cannot be regarded as faithful reproductions of in vivo situations, but rather as new rudimentary networks whose activity can change over time spontaneously or as a consequence of chemical/physical stimuli (Gross *et al.*, 1992; Gross *et al.*, 1993; Jimbo *et al.*, 1993).

As already anticipated in the Introduction, the systematic study of these networks can therefore become a new powerful tool for addressing questions concerning the coherent behaviour and the computational properties of neurobiological networks. Original suggestions for the field of formal networks can also be foreseen (Gardner, 1993). It should be stressed that a key feature of the substrate microelectrode technique, as compared for example with the use of potentiometric fluorescent dyes (Fromherz & Vetter, 1992), is the capability of stimulating, in a virtually noninvasive way, a selected subset of the neuronal population, thus allowing the experimenter to simulate a kind of rudimentary learning process.

For the sake of completeness, it should be mentioned that substrate microelectrode arrays have also been used for recording the synchronized activity of layers of embryo heart cells (Israel, Barry, Edell, Mark, 1984; Martinoia *et al.*, 1993).

2.4. Workstation for Signal Recording and Processing

A sketch of a measurement and recording system is shown in Figure 2. It refers to a workstation which has been developed in our Lab over the years in co-operation with G.T.A. Kovac's Group (CIS, Stanford University, USA). It consists of a Personal Computer 486 equipped with an A/D conversion board (National Instruments AT-MIO16F5; maximum sampling rate: 200kHz), with 2 DMA channels and 16 input channels (Martinoia *et al.*, 1993). Thus, extracellular signals from multiple sites of a microelectrode array can be simultaneously recorded.

Signals collected from a microtransducer array have typical amplitudes in the range 20-600μV and are embedded in biological and thermal noise ranging between 10μV and 20μV from peak to peak. In addition, the biological material has to be isolated from the electronics to avoid electrically induced artifacts. To meet these specifications, a high-impedance amplification custom stage (with a gain equal to 190) and a filtering custom stage (with a bandwidth ranging from 400Hz to 8kHz) are introduced before signal digitization. Each experiment is performed in a Faraday cage to avoid electromagnetic interference. During each experiment, a stereomicroscope with a TV camera is used to check, from time to time, the positions of the cells on the microtransducer array.

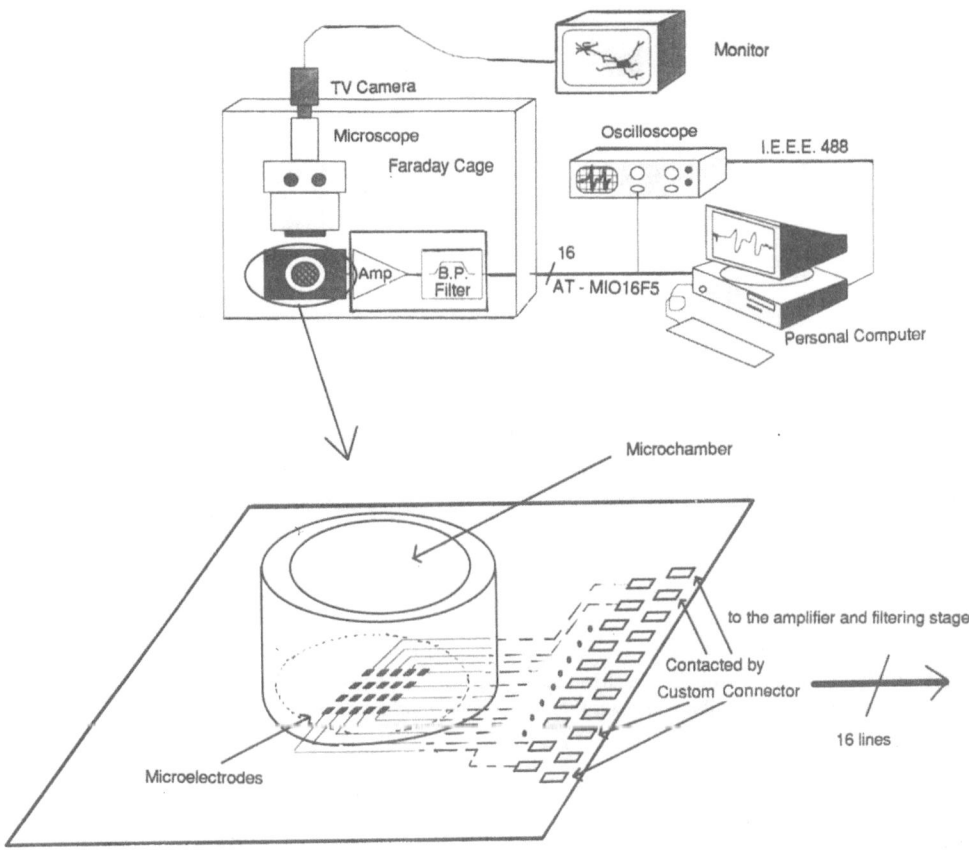

Figure 2. Sketch of the workstation for signal recording from a planar microelectrode array.

Figure 3 is representative of a typical recording session. More precisely, it refers to four simultaneously recorded signals from chick embryo Dorsal Root Ganglia (DRG) neurons.

The recorded signals were processed off line by using a software package that included frequency-filtering, peak-detection and feature extraction programs (Martinoia *et al.*, 1993). We utilized peak-detection and feature-extraction programs to detect extracellularly recorded action potentials with different shapes, amplitudes, and durations in order to compare them with the signals resulting from computer simulations.

3. SIMULATIONS TOOLS

3.1. SPICE Implementation and Compartmental Approach

In the following, we will mostly refer to research activity developed over the years in the Bioelectronics Lab and Bioelectronic Technologies Lab (c/o Advanced Biotechnology Center) of the Department of Biophysical and Electronic Engineering (DIBE).

An ad hoc modified version of the electric circuit analysis program SPICE, which has been optimized for detailed simulations of the electrical behaviour of neurons, has been developed.

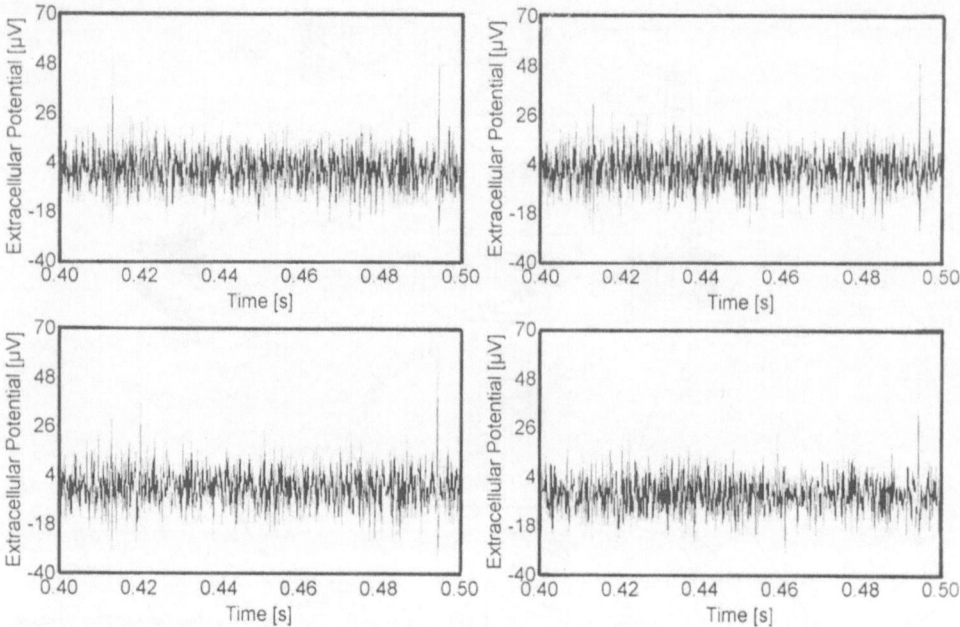

Figure 3. Multichannel recording of electrical activity by using a planar microelectrode array. Four simultaneously recorded signals of cultured Dorsal Root Ganglia neurons are shown.

The implementation of the neuronal model in the circuit simulation program SPICE was performed as described in (Bove, Massobrio, Martinoia & Grattarola, 1994). The modified version of SPICE made it possible to model the neuron by the compartmental approach according to the Hodgkin-Huxley kinetics (1952), (Segev, Fleshman, Miller & Bunow, 1985; Bunow, Segev & Fleshman, 1985). This approach allows one to divide a nerve cell into compartments that represent different, isopotential nerve membrane patches described by the Hodgkin & Huxley kinetics. Thus, it is possible to obtain a neuronal model that takes into account the anatomical and functional properties of a neuron in terms of the electrical properties of each compartment.

Of specific relevance for this review, the efficiency of SPICE in simulating electronic devices makes our modified version an appropriate tool for the characterization of the neuron-to-microtransducer junction.

3.2. The Neuron-To-Microelectrode Junction Equivalent Circuit

The coupling strength between a neuron and a planar metallic substrate microelectrode is a very critical parameter in determining the 'quality' (shape and amplitude) of a recorded signal (Grattarola & Martinoia, 1993). This issue, which has been fragmentarily considered in the literature (Regehr, Pine, & Rutledge, 1988), has been systematically addressed in our Lab. More specifically, a circuit model has been developed to simulate a nerve cell membrane patch coupled to a noble metal planar microelectrode. The circuit model, described in (Bove *et al.*, 1995) and representing the neuron-to-microelectrode junction used for the simulations, is shown in Figure 4. The meanings of the symbols are the following: Ri denotes the cytoplasmic resistance connecting two adjacent compartments; RCl and VCl are the chloride resistance and the chloride equilibrium potential, respectively; Iact represents the sum of the sodium, potassium, and calcium currents (Epstein & Marder,

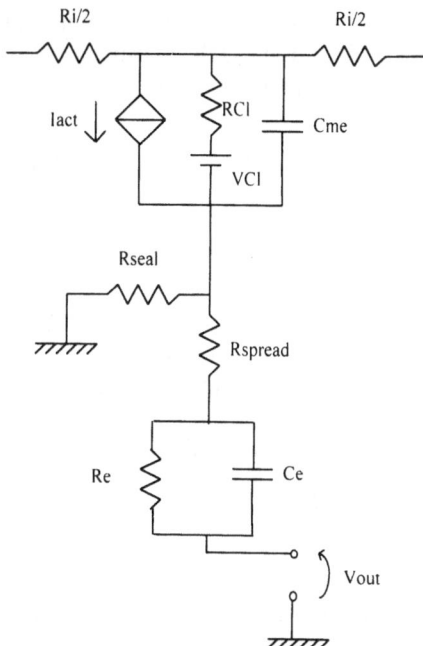

Figure 4. Equivalent circuit of the junction between a neuron membrane patch (described by the compartmental model) and a planar noble metal microtransducer. More details are given in the text.

1990); Cme is the cell membrane-electrolyte capacitance; Rseal and Rspread denote the sealing resistance between a cell and a microelectrode and the spreading resistance, respectively; Re and Ce are the equivalent elements of the electrode-electrolyte interface. The sealing impedance is represented by the RC coupling circuit consisting of Cme, Rseal, and Rspread.

Cme models the capacitive component of the membrane-electrolyte interface (Figures 5a, 5b). Cme can be represented as a series of at least three equivalent capacitances: the membrane capacitance (Cm), the Helmoltz capacitance (C_h), (i.e., the capacitance modeling the IHP (Inner Helmholtz Plane) and the OHP (Outer Helmholtz Plane) layers just outside the cell membrane) and the diffuse layer capacitance (C_d), (i.e., the capacitance modeling

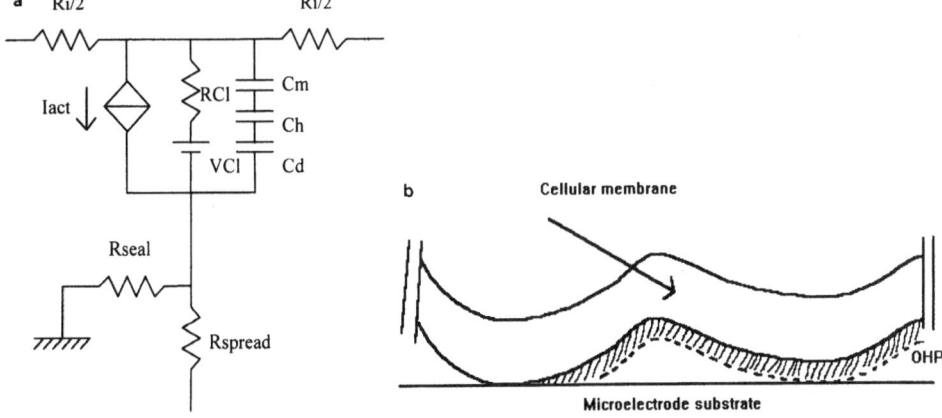

Figure 5. (a) RC coupling circuit representing the sealing impedance. (b) Cell adhesion to a microelectronic substrate. The drawing shows the membrane glycocalix and a local area of molecular adhesion. More details are given in the text.

the diffuse layer of ions in the solution) (Bockris & Reddy, 1970; Grattarola & Martinoia, 1993). Rseal models the resistive component of the thin layer of solution between the cell membrane and the microelectrode.

3.3. Simulation of the Microelectrode Transduction

In the following, we will refer to simulation results already described elsewhere (Bove *et al.*, 1995). The microelectrode area considered in the simulations was $10 \times 10 \mu m^2$, and a constraint $\omega ReCe=1$ at a frequency of 1kHz was imposed, according to the convention adopted in the literature (Robinson, 1968). Moreover, the compartmental approach was used to define the neuron area coupled to the planar microelectrode.

In a set of simulations, a patch of neuron membrane 10μm in diameter was considered that covered completely the planar microelectrode. Figures 6a, 6b, and 6c show the results of three simulations of neuronal electrical activity recording which differed only in the sealing impedance (Cme=0.7pF and Rseal=0.2MΩ; Cme=1pF and Rseal=1 MΩ; Cme=1.4pF and Rseal=4MΩ, respectively).

Figure 6a represents a simulation result for Cme=0.7pF and Rseal=0.2MΩ, which corresponds to a weak coupling between a membrane patch of the neuron and a microelectrode. For this coupling, we can hypothesize that the small Cd dominates, and that the total membrane capacitance is lower than the capacitance of the membrane itself. The amplitudes and durations of signals increase as the values of the parameters Cme and Rseal increase, too. Simulation results corresponding to increasingly stronger contact conditions between a neuronal membrane patch and a microelectrode for Cme=1pF and Rseal=1MΩ and for Cme=1.4pF and Rseal=4MΩ are shown in Figures 6b, and 6c, respectively.

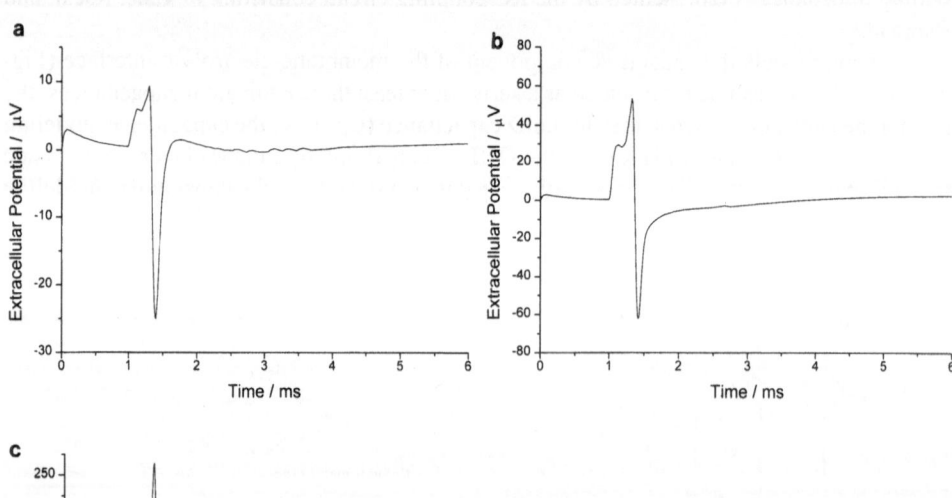

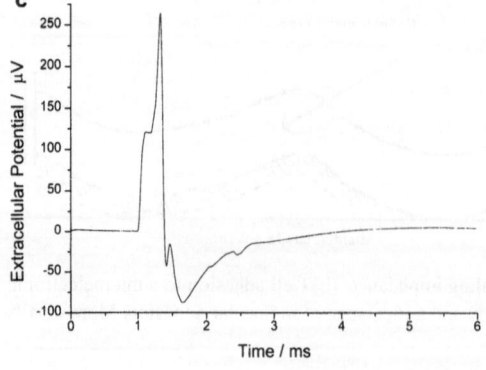

Figure 6. Simulation results for different values of the sealing impedance: (a) Cme=0.7pF and Rseal=0.2MΩ; (b) Cme=1pF and Rseal=1MΩ; (c) Cme=1.4pF and Rseal=4MΩ.

3.4. The Neuron-To-Transistor Junction Equivalent Circuit

Whenever an insulated FET is utilized as a transducer, the sealing region is bounded by the patch of membrane on one side and by the transducer surface on the other side, i.e., by the surface of an insulator (typically Si_3N_4 or SiO_2). Consequently, in this configuration, the patch-to-transducer coupling is a capacitive one. Due to lack of direct experimental data on the thickness and physico-chemical properties of the volume surrounded by the leaky capacitor (the membrane) and by the other capacitor (the FET insulator), we have made the following assumptions (Bove, Martinoia, Grattarola, Ricci, 1996), which will be utilized to obtain the equivalent circuit description of the junction. Previous attempts at such a description can be found in Fromherz *et al.* (1991); Grattarola, Martinoia, Massobrio, Rosichini, Tetti (1991); Fromherz, Müller, Weis (1993); Grattarola & Martinoia (1993). A pioneering description of the electrophysiological use of insulated (or ion-sensitive) FETs was given by Bergveld (1972).

3.4.1. Thickness d. The thickness d is the (average) patch-to-insulator distance. It affects the sealing resistance, Rseal, through the relation:

$$Rseal = \frac{\rho_{seal}}{d} \frac{l}{w} \tag{1}$$

where ρ_{seal} is the sealing resistivity, and l and w are the length and width of the portion of the insulated gate FET coupled to the patch of neuronal membrane, respectively.

3.4.2. Resistivity of the Sealing Region ρ_{seal}. For a weak sealing, we assume a ρ_{seal} value typical of an electrolyte solution (i.e., $\cong 0.7$ Ωm). For a tight sealing, we assume that a considerable portion of space is occupied by the cell glycocalix and its associated fixed charge; therefore a larger value of ρ_{seal} (e.g., $1 \div 5$ Ωm) can be assumed.

3.4.3. Water Permittivity ε_w. Water permittivity affects the Inner Helmholtz Plane (IHP), the Outer Helmholtz Plane (OHP) and diffuse layer capacitances, which model the polarization of the electrolyte solution in front of the membrane patch and in front of the insulator through the relation:

$$C_h = \frac{\varepsilon_{IHP}\varepsilon_{OHP}\varepsilon_0}{\varepsilon_{OHP}d_{IHP} + \varepsilon_{IHP}d_{OHP}} \tag{2}$$

where eIHP and eOHP are the dielectric permittivities of the Inner and the Outer Helmholtz Planes, respectively (the values 6 and 32 were assumed (Bockris et al., 1970), respectively); dIHP =0.2 nm, and dOHP =0.7 nm are the insulator-nonhydrated ion and the insulator-hydrated ion distances, respectively. A value ew =78.5 is assumed for the permittivity of the bulk solution.

Figure 7 shows the equivalent circuit of the patch-to-FET sealing. The meanings of the symbols used in Figure 7 are the following (Bove *et al.*, 1996): Cm is the cell membrane-electrolyte capacitance; INaK represents the sum of the sodium and potassium currents that flow through the membrane; Rl and Vl represent the leakage resistance and the equilibrium potential, respectively; Ri is the resistance of the axoplasm; Rseal is the sealing resistance between the neuron membrane and the insulated FET; Rspread is the spreading resistance; Vgs is the polarization voltage of the insulated FET (I-FET).

As mentioned earlier, the neuron membrane is described according to the Hodgkin-Huxley model and is derived by the compartmental approach (Bove *et al.*, 1994). As a result,

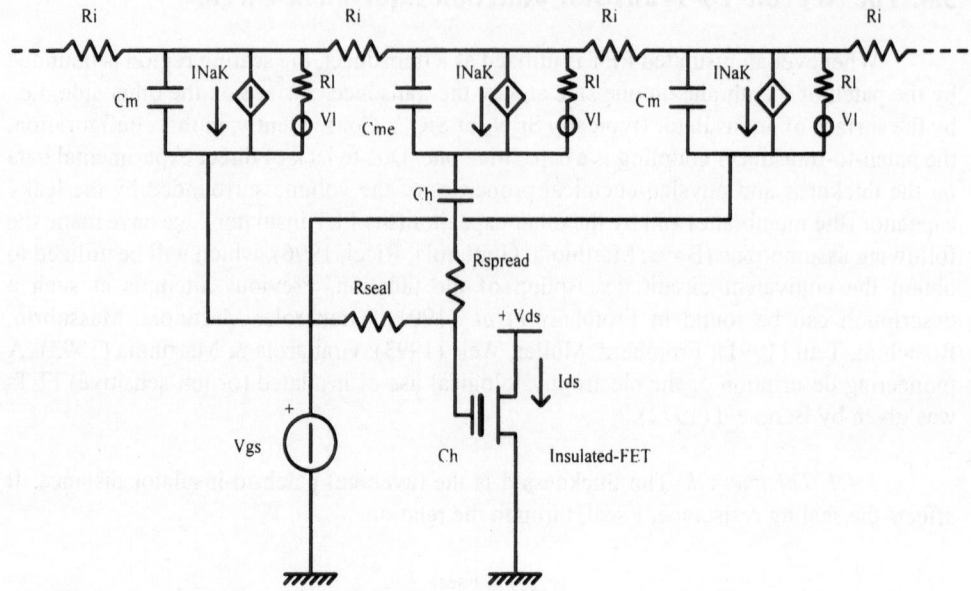

Figure 7. Equivalent circuit of the junction between a patch of neuronal membrane (described by the compartment model) and an insulated FET microtransducer.

it is possible to consider a space extended model of the neuron and to define the dimensions of the patch of membrane coupled to the FET.

The insulated FET (I-FET) equivalent circuit is based on descriptions previously given in the literature (Fromherz *et al.*, 1991; Grattarola & Martinoia 1993).

3.5. Simulation of the I-FET Transduction

In the following, we will refer to simulation results already described elesewhere (Bove *et al.*, 1996). A portion of excitable membrane 10μm in diameter, divided into compartments, was considered in the simulations. The patch of membrane interfacing the insulated FET was represented by a specific Hodgkin-Huxley compartment 10μm long.

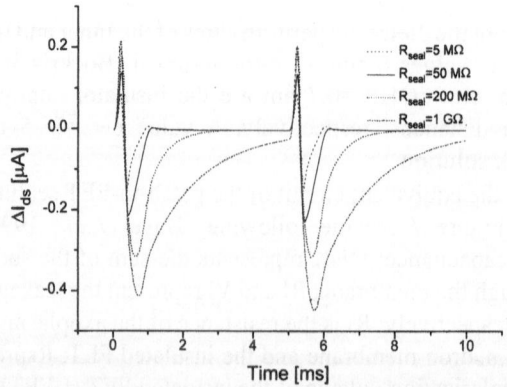

Figure 8. Simulation of action potentials transduced by the insulated FET for different sealing conditions (Rseal=5 MΩ, 50MΩ, 200MΩ, and 1GΩ).

To simulate the insulated FET, the parameters reported in (Fromherz *et al.*, 1991) were utilized. The value of Ch, which models the polarization layers of the electrolyte solution in front of the patch and in front of the insulator, was assumed to be 15pF.

Figure 8 gives the results of simulations of action potentials obtained by considering different sealing conditions. By taking into account experimental data reported in the literature (Fromherz *et al.*, 1991; Fromherz *et al.*, 1993) which describe two kinds of recorded signals and suggest a relationship between the shape of the signal and the strength of the junction, a loose sealing (Rseal: 5MΩ-50MΩ) and a tight sealing (Rseal: 200MΩ-1GΩ) were considered. A loose sealing implies a distance d greater than 30nm; a tight sealing implies a distance d smaller than 20 nm.

Figures 9a and 9b show the effects induced on the recorded signals by variations in the densities of the ionic channels of the membrane patch coupled to the transistor. The simulations were based on the following assumptions:

a) It was assumed that a very tight sealing could preclude channel opening (Figure 9a).

b) It was assumed that, under looser sealing conditions, channel migration phenomena (Fromherz, 1988) could be induced by the presence of a component of the local electric field parallel to the membrane patch, thus causing a local increase in the densities of the ionic channels (Figure 9b).

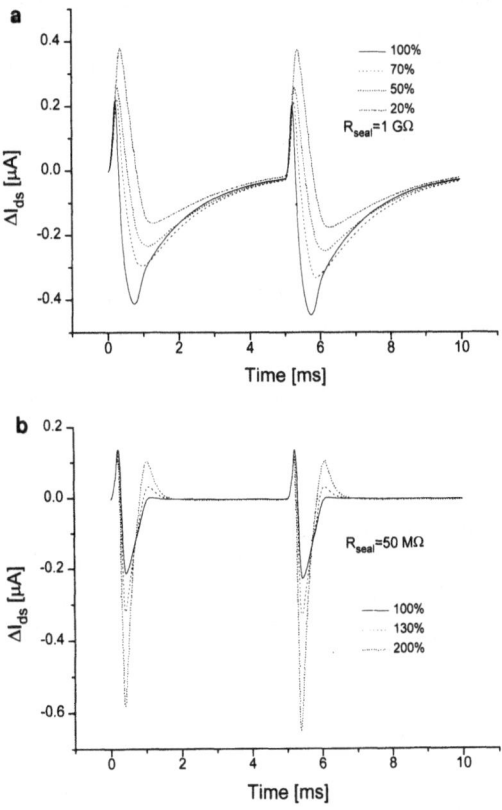

Figure 9. Simulation of action potentials transduced by the insulated FET (a) for a constant tight sealing (Rseal=1GΩ) and decreasing densities of the ionic channels. (b) For a constant loose sealing (Rseal=50MΩ) and increasing densitites of the ionic channel.

4. INDICATIONS ON THE COLLECTIVE BEHAVIOR OF NEURONAL NETWORKS

Especially in the case of neurons from the nervous tissue of vertebrate embryos, the experimenter's interest is focused on the characterization of the collective behavior displayed by the network. The growth of neuronal networks on planar microtransducer arrays permits one to monitor in a simultaneous way and for a period ranging from hours to weeks, the internal network dynamics (i.e., competitive and cooperative neuronal interactions), which was not measurable with standard electrophysiological techniques.

Data in the literature related to this research field refer to the induction and the maintenance of epileptic activity in cultured monolayer networks (Rhoades & Gross, 1991) and to the study of the mechanisms of generation and propagation of synchronized bursting in developing networks of cortical neurons (Maeda, Robinson & Kawana, 1995). In addition, in this last experiment, the use of optical techniques for measuring ionic fluxes in living neuronal networks is also described (Robinson, Kawahara, Jimbo, Torimitsu, Kuroda & Kawana, 1993; Jimbo et al., 1993).

Another application deals with the use of a planar microelectrode array coupled to the flat mosaic of retinal ganglion cells for recording the extracellular action potentials elicited by a randomly flickering display in order to study how a population of optic nerve fibers encode a visual scene (Meister, Pine, Baylor, 1994).

The characterization of the internal network dynamics can be used for quantifying the response of the network to specific chemical changes (Gross et al., 1992). Neuronal networks in culture show remarkable sensitivities to small chemical changes and mimic some of the properties of sensory tissue. These sensitivities could be enhanced by receptor upregulation and altered by the expression of unique receptors (Gross, Rhoades, Azzazy & Wu, 1995).

5. DISCUSSION

Starting from the seventies, intracellular recording from single neurons by means of a tiny glass microelectrode has become a standard tool for the detailed characterization of action potentials.

During the eighties, the development of patch clamp techniques has allowed the experimenter to record also single channel currents from micron-sized patches of excitable membranes.

One of the neurobiological challenges of the nineties is the study of the collective behavior of networks of neurons. Science and technology influence each other: in the near future arrays of planar substrate microelectrodes may happen to become a new powerful neurobiological tool perfectly suited for characterizing this collective behavior, thus opening new vistas for systematic experiments dealing with neurocomputation.

ACKNOWLEDGMENTS

The authors are very grateful to Dr. A. Kawana (Material Science Research Laboratory of NTT Basic Research Laboratories, Atsugi, Japan) and to Dr. G.T.A. Kovacs (Center for Integrated Systems, Stanford University, USA) for kindly supplying the microelectrode arrays.

This work was supported by the European Joint Research Project CELLENG (Contract No. ERB-CIPA-CT93-0235).

REFERENCES

Barinaga, M. (1990). The high culture of neuroscience. *Science* **250**, 206-207.

Bergveld, P. (1972). Development, operation and application of the ISFET as a tool for electrophysiology. *IEEE Trans. on Biom. Eng.* **19**, 342-351.

Bockris, J. O'M., Reddy, A. K. N. (1977). *Modern Electrochemistry 2*. New York: Plenum Press.

Boppart, S., Wheeler, B., Wallace, C. (1992). A flexible perforated microelectrode array for extended neural recording. *IEEE Trans. on Biom. Eng.* **39**, 37-42.

Bove, M., Massobrio, G, Martinoia, S, Grattarola, M. (1994). Realistic simulations of neurons by means of an ad hoc modified version of SPICE. *Biol. Cybern.* **71**, 137-145.

Bove, M., Grattarola, M., Martinoia, S, Verreschi G. (1995). Interfacing cultured neurons to planar substrate microelectrodes: characterization of the neuron-to-microelectrode junction. *Bioelectrochemistry and Bioenergetics* **38**, 255-265.

Bove, M., Martinoia, S., Grattarola, M., Ricci, D. (1996). The neuron-transistor junction: linking equivalent circuit models to molecular descriptions. *Thin Solid Films* **in press**.

Bousse, L., de Rooij, N. F. and Bergveld, P. (1983). Operation of chemically sensitive field effect sensors as a function of the insulator-electrolyte-interface. *IEEE Trans. Electron Devices* **ED-30**, 1263-1270.

Bulloch, A. G. M., Syed, N.I. (1992). Reconstruction of neuronal networks in culture. *TINS* **15**, 422-427.

Bunow, B., Segev, I., Fleshman, J. W. (1985). Modeling the electrical behavior of anatomically complex neurons using a network analysis program: Excitable membrane. *Biol. Cybern.* 53, 41-56.

Clark P., Connolly, P. and Moores, G. R. (1990). Cell guidance by micropatterned adhesiveness in vitro. *J. Cell. Sci.* **103**, 287-292.

Curtis, A. S. G., Breckenridge, L., Connoly, P., Dow, J. A. T., Wilkinson, C. D. W., Wilison, R. (1992). Making real neural nets: design criteria. *Medical and Biological Engineering & Computing* CE33-C36.

Epstein, I. R. and Marder, E. (1990). Multiples modes of a conditional neural oscillator. *Biol. Cybern.* **63**, 25-34.

Fromhertz, P. (1988). Self-organization of the fluid mosaic of charged channel proteins in membranes. *Proc. Natl. Acad. Sci. USA* 85, 6353-6357.

Fromhertz, P., Offenhauser, A., Vetter, T., Weis, J. (1991). A neuron-silicon junction: a retzius cell of the leech on an insulated-gate-field effect transistor. *Science* 252, 1290-1293.

Fromhertz, P. & Vetter, T. (1992). Cable properties of arborized Retzius cells of the leech in culture as probed by a voltage-sensitive dye. *Proc. Natl. Acad. Sci. USA* 89, 2041-2045.

Fromhertz, P., Muller, C. O. & Weis R. (1993). Neuron transistor: electrical transfer function measured by the patch-clamp technique. *Physical Review Letters* **71**, 4079-4082.

Fromhertz, P. and Schaden, H. (1994). Defined neuronal arborizations by guided outgrowth of leech neurons in culture. European. *J. of Neurosci.* **6**, 1500-1504.

Grattarola, M., Martinoia, S., Massobrio, G., Cambiaso, A., Rosichini, R., Tetti, M. (1991). Computer simulations of the responses of passive and active integrated microbiosensors to cell activity. *Sensors and Actuators* **B4**, 261-265.

Grattarola, M., Massobrio, G. and Martinoia, S. (1992). Modeling H^+-sensitive FET's with SPICE. *IEEE Electron Dev.* **39**, 813-819.

Grattarola, M. and Martinoia, S. (1993). Modeling the neuron transducer junction: from extracellular to patch recording. *IEEE Trans. on Biom. Eng.* **40**, 35-41.

Gross, G. W., and Kowalsky, J. M., (1991). Experimental and theoretical analysis of random nerve cell networks dynamics. *In Neural Networks Concepts, Applications and Implementations*, ed. Antognetti, P. and Milutinovic V. vol. **IV**, pp. 47-110 New Jersey: Prentice Hall.

Gross, G. W., Rhoades, B. K., Jordan, R. J (1992). Neuronal networks for biochemical sensing. *Sensor and Actuators* **B6**, 1-8.

Gross, G. W., Rhoades, B. K., Reust, D. L., and Schwalm, F. U. (1993). Stimulation of monolayer networks in culture through thin film indium-tin oxide recording electrodes. *J. Neurosci. Methods* **50**, 131-143.

Gross, G. W., Rhoades, B. K., Azzazy, H. M. E. and Wu, M. (1995). The use of neuronal networks on multielectrode arrays as biosensors. *Biosensors & Bioelectronics* **10**, 553-567.

Hodgkin, L., and Huxley, A. F. (1952). A quantitative description of membrane current and its applications to conduction and excitation in nerve. *J. Physiol.* **117**, 500-544.

Israel, D.A., Barry, W.H., Edell, D., Mark, R.G. (1984). An array of microelectrodes to stimulate and record from cardiac cells in culture. *Am. J. Physiol.* **247**, 669-674.

Jimbo, Y. & Kawana, A. (1992). Electrical stimulation and recording from cultured neurons using a planar electrode array. *Bioelectrochemistry and Bioenergetics* **29**, 193-204.

Jimbo, Y., Robinson, H. P. C. & Kawana, A. (1993). Simultaneous measurements of intracellular calcium and electrical activity from patterned neural networks in culture. *IEEE Trans. Biom. Eng.* **40**, 804-810.

Maeda, E., Robinson, H. P. C. and Kawana, A. (1995). The mechanisms of generation and propagation of synchronized bursting in developing networks of cortical neurons. *J. of Neurosci.* **15**, 6834-6845.

Martinoia, S., Bove, M., Carlini, G., Ciccarelli, C., Grattarola, M., Storment, C., Kovacs, G. T. (1993). A general purpose system for long-term recording from a microelectrode array coupled to excitable cells. *J. Neurosci. Methods* **48**, 115-121.

Meister, M., Pine, J., Baylor, D. A. (1994). Multi-neuronal signals from the retina: acquisition and analysis. *J. Neurosci. Methods* **51**, 95-106.

Novak, J. L. and Wheeler, B. C. (1986). Recording from the Aplysia abdominal ganglion with a planar microelectrode array. *IEEE Trans. Biomed. Eng.* **33**, 196-202.

Regehr, W. G., Pine, J., and Rutledge, D. B. (1988). Long-term in vitro silicon-based microelectrode-neuron connection. *IEEE Trans. Biomed. Eng.* **35**, 1023-1032.

Regehr, W. G., Pine, J., Cohan, C. S., Mischke, M. D., and Tank, D.W. (1989). Sealing cultured invertebrate neurons to embedded dish electrodes facilitates long-term stimulation and recording. *J. Neurosci. Methods* **30**, 91-106.

Robinson, D. A. (1968). The electrical properties of metal microelectrodes. *Proceedings of IEEE* **56**, 1065-1071.

Robinson, H. P. C., Kawahara, M., Jimbo, E., Torimitsu, K., Kuroda, E. and Kawana, A. (1993). Periodic synchronized bursting and intracellular calcium transient elicited by low magnesium in cultured cortical neurons. *J. Neurophys.* **70**, 1606-1616.

Rhoades, B. K. and Gross, G. W. (1991). The effects of extracellular potassium on epilitic-form burst dynamics in cultured monolayer networks. *Soc. Neurosci. (Abstr.)* **17**, 571.5.

Segev, I., Fleshman, J. W., Miller, J. P., Bunow, B. (1985). Modeling the electrical behavior of anatomically complex neurons using a network analysis program: Passive membrane. *Biol. Cybern.* **53**, 27-40.

Stühmer, W., Roberts, W. M. and Almers, W. (1983). The loose patch clamp. In *Single-Channel Recording*, ed. Sackmann, B. Neher, E., pp. 123-132. New York and London: Plenum Press.

Wilson, R. J. A., Breckenridge, L., Blackshow, S. E., Connollt, P., Dow, J. A. T., Curtis, A. S. G., Wilkinson, C. D. W. (1994). Simultaneous multisite recordings and stimulation of single isolated leech neurons using planar extracellular electrode arrays. *J. Neurosci. Meth.* **53**, 101-110.

21

ELECTRICAL ACTIVITY IN THE LEECH NERVOUS SYSTEM CAN BE STUDIED USING A CCD IMAGING TECHNIQUE

Marco Canepari and Marco Campani

INFM, Dipartimento di Fisica
Università di Genova
Via Dodecaneso 33, 16146 Genoa, Italy

1. INTRODUCTION

The analysis of the integrative properties of a small nervous system, such as that of many invertebrate animals, requires the study of electrical events originating in different locations of the neuronal network and occurring on different time scales. Even though a detection of electrical signals at a very precise spatial and temporal resolution is at present beyond available experimental techniques, useful information can be obtained from a coarse analysis of the activity in the nervous system.

Optical recording techniques [Cohen & Salzberg, 1975] can be used for this purpose. Optical signals produced by voltage sensitive dyes were used to study the electrical activity in the Aplisia nervous system [Zecevic, Wu, Cohen, London & Hopp, 1989; Nakashima, Yamada, Shioro, Maeda, & Satoh, 1992]. These signals were detected with photodiodes, arranged in a suitable array with a maximal space resolution of 20X20, but with a temporal resolution in the kHz range. CCD cameras were used to measure fluorescence changes of voltage sensitive dyes caused by slow electrical waves [Kauer, 1988; Delanay, Gelperin, Fee, Flore, Gervais, Tank & Kleinfeld, 1994]. In all these experiments the neurons originating the optical signals could not be identified.

In order to detect electrical signals from well identified neurons and to provide new and useful information for the understanding of the dynamics of small neuronal networks, a CCD imaging technique has been developed. Experiments have been performed in the nervous system of the leech *Hirudo Medicinalis* [Muller, Nicholls & Stent, 1981]. Isolated ganglia of the leech were incubated with the voltage sensitive dye Di-4-Anepps whose fluorescence depends on the membrane voltage of stained neurons. Optical measurements were coupled with extracellular electrical recordings from the roots or from the connective fibers using suction electrodes. The same suction electrodes were also used for stimulation.

In order to detect evoked action potentials, images were taken at fixed times. Due to the high stability in the occurrency of single spikes, images taken in this way could be averaged, thus leading to a significant increase of the S/N ratio. Signals produced by firing of action potentials from well identified neurons corresponded to changes in light intensity between .2% and .5% and could be detected at an effective time resolution of 50Hz. By taking images at different times and focal planes, a coarse three dimensional description of the electrical events was obtained.

Afterhyperpolarizations and afterdischarges could be evoked by using much stronger and more prolonged stimulations and most of the times produced optical signals corresponding to changes in light intensity greater than 1%. These signals could be imaged in a "real time mode" acquiring sequences of about 20 images with a space resolution of 59 × 82 and a time resolution of about 4 Hz.

These preliminary results indicate that our CCD technique is able to detect both fast signals produced by action potentials and long lasting signals such as afterhyperpolarization and the high space resolution of the system allows in many cases to identify the neurons originating the optical signals.

2. METHODS

2.1. Preparation

A single ganglion from the leech *Hirudo Medicinalis* was mechanically isolated and placed in a chamber mounted on the stage of an inverted microscope. The isolated ganglion was one between 7th and 18th [Muller et al., 1981]. Two roots from the same side or from different sides of the ganglion and a bundle of the connective fibers were sucked into suction electrodes used either to record the extracellular activity or to provide electrical stimulations. Fig. 1A shows the ventral side of an isolated ganglion in which the two roots of same side were sucked.

Figs. 1B and 1C show recordings from the posterior root of the ganglion when a positive polarity (Fig.1B) or a negative polarity (1C) stimulation of .2V amplitude were used. Using stimulation of this intensity, evoked action potentials could be observed only in the case of a negative polarity stimulus. Fig. 1D shows the average of 10 electrical recordings in the case of negative polarity stimulations. As the occurrency of single spikes was highly stable, the S/N ratio increased when several recordings taken at a fixed time after the onset of stimulation were averaged together.

Ganglia were incubated in a saline medium (110 mM NaCl, 2.5 mM KCl, 8 mM CaCl, 1.5 mM MgCl, buffered to pH 7.6 with 10 mM HEPES and NaOH) containing the fluorescent dye. The Dye Di-4-Anepps (Molecular Probes) was used. The final concentration of the dye used to stain the ganglia ranged between .2 and 1 mg/ml and the incubation time varied between 20 and 60 minutes. The experiments were performed at room temperature.

2.2. Optical System

The exciting light was provided by a tungsten lamp or a mercury lamp. The wavelength of the exciting light was selected at 498 nm and the fluorescence was measured at 610 nm. As shown by Fromherz and Lambacher [Fromherz & Lambacher, 1991], in leech neurons a decrease of fluorescence of the dye Di-4-Anepps at the measured wavelength indicates membrane depolarization and therefore excitation and an increase of fluorescence indicates membrane hyperpolarization and therefore inhibition.

The fluorescence light was monitored with a nidrogen cooled CCD camera (Astromed CCD 3200 Imaging System). The imaging system was able to digitize each image with a resolution of 16 bits.

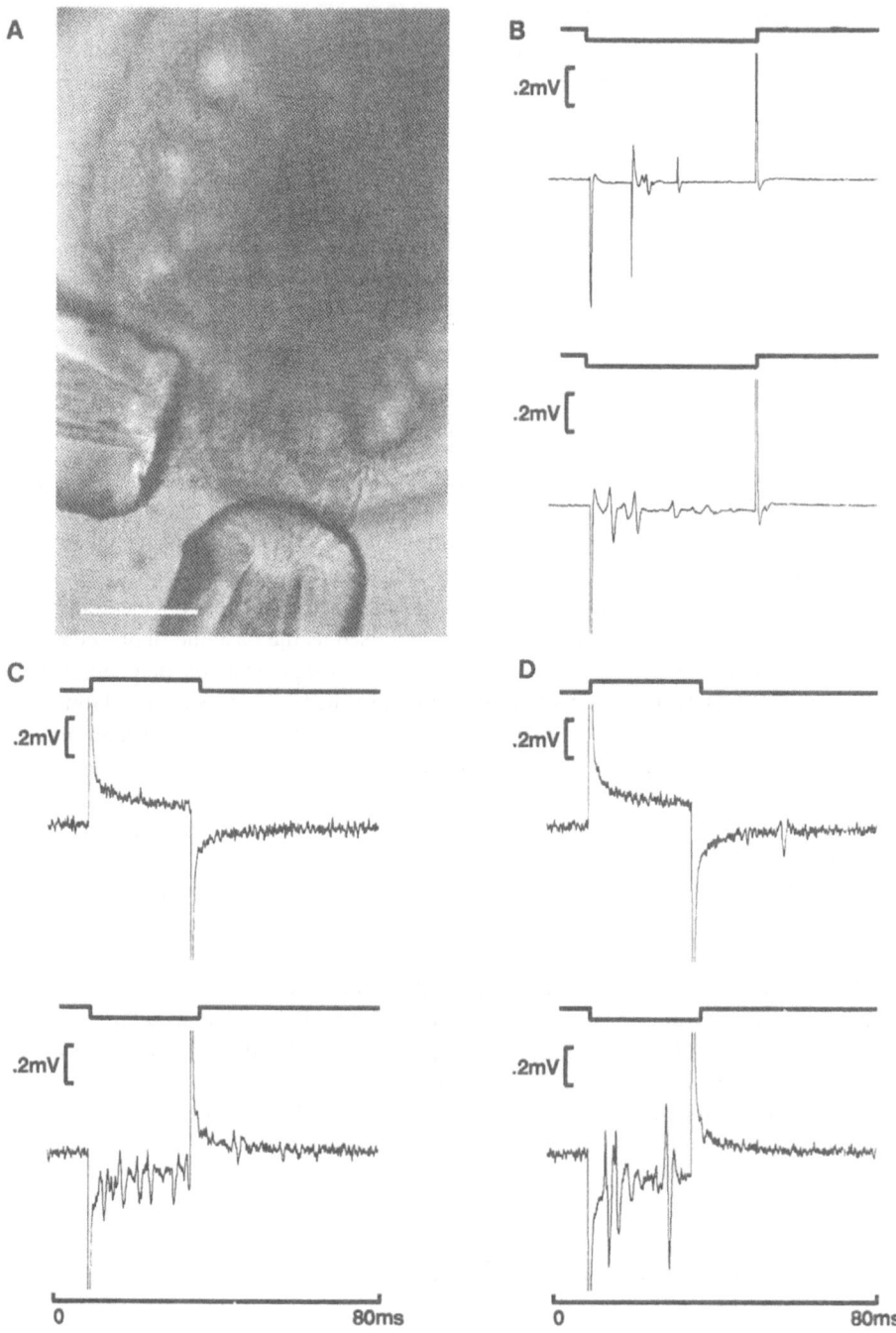

Figure 1. The isolated leech ganglion used in the experiment. A: A CCD image in trasmitted light at full resolution (416X578 pixels) of the ventral side of an isolated leech ganglion, with the anterior (left) and posterior (right) roots sucked into two suction electrodes. 16X objective. The calibration bar indicates 150 μm. B: Averages of 10 extracellular recordings from the anterior root when the posterior root was stimulated (up) and from the posterior root when the anterior root was stimulated (down). B,C: Extracellular recordings from the posterior root when the anterior root was stimulated (B) and from the anterior root when the posterior root was stimulated (C). The upper case is the one in which the polarity of the stimulus was positive. In B,C and C the solide line shows the artifact of the stimulation and its polarity. The intensity of the voltage pulse was .2V.

The maximal space resolution of the system was 416×578 and the time required to digitize and transfer a full resolution image was more than one second. In order to increase the temporal resolution, the images were binned by a factor between 5 and 15. With a binning factor 7, leading to a space resolution of 59×82, the system was able to acquire about 4 frames/second.

The major source of noise of our system was photon noise. When N photons were absorbed by the sensor, the amplitude of fluctuations was in the order of $\sqrt{N}$. As a consequence, in order to have a S/N ratio of at least 500, it was necessary to have signals corresponding to at least 250,000 photons. The procedure of binning the images was also used to collect photons from many pixels in order to increase the S/N ratio. However, the system was not able to measure more than 1,300,000 photons at any binning. The shortest exposure time of the system was 20 ms.

2.3. Recordings

Two different modes of acquiring images were developed. In the "Real time mode," image sequences at a reduced resolution were taken; each image sequence contained from 3 to 7 images taken before the voltage stimulation and from 15 to 25 images taken during and following the stimulation. In this mode the exact timing of image acquisition was not controlled, but could be determined by measuring the artifact of the T.V. camera shutter. Images taken before stimulation were averaged and used as the baseline fluorescence. The fluorescence signal related to the electrical activity was computed as the percentual change in the fluorescence intensity of each image from the baseline fluorescence. In real time mode, it was possible to detect signals larger than .5%. Assuming that the dye Di-4 Anepps provides a signal of some % for a change in the membrane potential produced by an action potential and the time duration of this signal is less than 2 ms, the optical signal detectable by our system will be some ‰. Therefore by using the real time imaging mode we could not measure signals produced by firing of action potentials.

In order to detect signals corresponding to changes in light intensity smaller than .5%, a "fixed time mode" of acquiring images was developed. In this mode 10 sequences of two frames were acquired. The first image was taken before the stimulation and the second image was taken at a fixed time after the onset of stimulation. The 10 images taken before stimulation and the 10 images taken after the onset of stimulation were averaged. Since the occurrency of single spikes was highly stable, fixed time mode allowed to increase the S/N ratio by a factor $\sqrt{10}$. Signals produced by action potentials could be detected using this mode. By changing the fixed time of acquiring the second image of each sequence, an effective time resolution of 50 frames/second was achieved.

2.4. Analysis

During each experiment, full resolution images were acquired in order to localize the signals.

Fig. 2A illustrates a fluorescence image of the ganglion taken by the CCD camera. The image was processed by suitable contrast enhancement procedures, as shown in Fig. 2B. By a semiautomatic procedure from this image, the neuronal map shown in Fig. 3C was obtained. Some neurons could be identified in the atlas of the leech nervous system [Muller et al., 1981]. Fractional fluorescence changes were visualized in a false colour scale. Green and blue signals corresponded to increases of light intensity and therefore to hyperpolarizations, while yellow and red signals corresponded to decreases of light intensity and therefore to depolarizations. Signal images were often smoothed by the convolution with an appro-

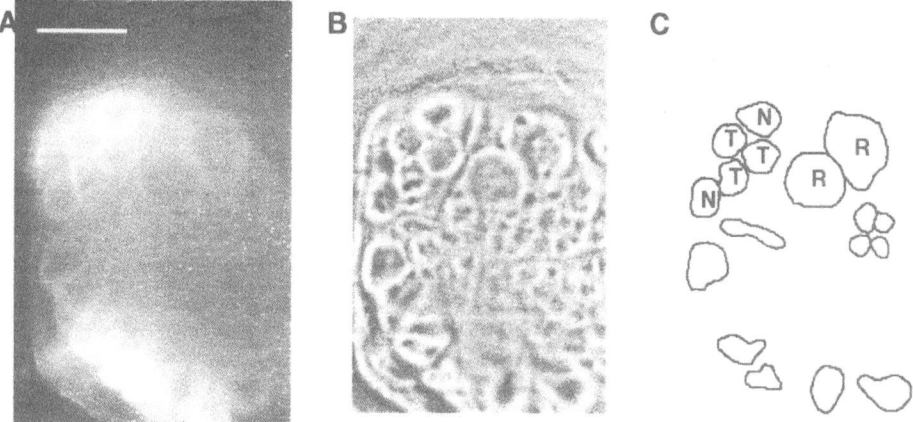

Figure 2. The localization of optical signals. A: A CCD image of the fluorescence from the stained ganglion at full resolution (416 × 578 pixels). The ventral side of the ganglion was focussed and the posterior connective was sucked into a suction pipette. The calibration bar indicates 150 μm. B: the image of A after the processing with suitable contrast enhancement procedures. C: some neuronal profiles extracted from B. Retzius cells are indicated by R touch neurons by T and noxius neurons by N. Each image was obtained with an exposure of 20 ms, with a binning 5, leading to images of 82 × 115. 25 × objective, N.A.1.

priate filter. Images showing the signals were superimposed to full resolution images of the ganglion, to extracted neuronal maps or to both.

3. RESULTS

3.1. Imaging the Action Potentials

As shown in Fig. 1, when one root was stimulated with a negative polarity stimulation and with an amplitude smaller than .3V, action potentials could be recorded from the other root on the same side of the ganglion. Firing of action potentials occurring in this case could be imaged using fixed time mode as shown in Fig. 3B. The signal image was superimposed to a neuronal map extracted with the procedure described in methods.

The posterior root of the ganglion was stimulated with a voltage step of .2V. The high space resolution of the signal image taken using a binning factor 8 allows to correlate the optical signal to an identified Retzius cell.

When a positive polarity stimulation of the amplitude was applied, no evoked activity was recorded from the other electrode and therefore no optical activity was detected in the ganglion as shown in Fig. 3B. The decrease of light intensity that can be observed on the stimulated root is due to the effect of direct polarization produced by the stimulation.

3.2. Imaging the Dynamics of the Network

The time of the second image taken in the fixed time mode could be varied in order to reach an effective time resolution of 50Hz. Signal images taken at two different times after the onset of the stimulation are shown in Figs. 4B and 4F, in the case of stimulation of the posterior root, Figs. 4C and 4G in the case of stimulation of the anterior root and Figs. 4D and 4H in the case of simultaneous stimulation of the anterior and posterior roots. The

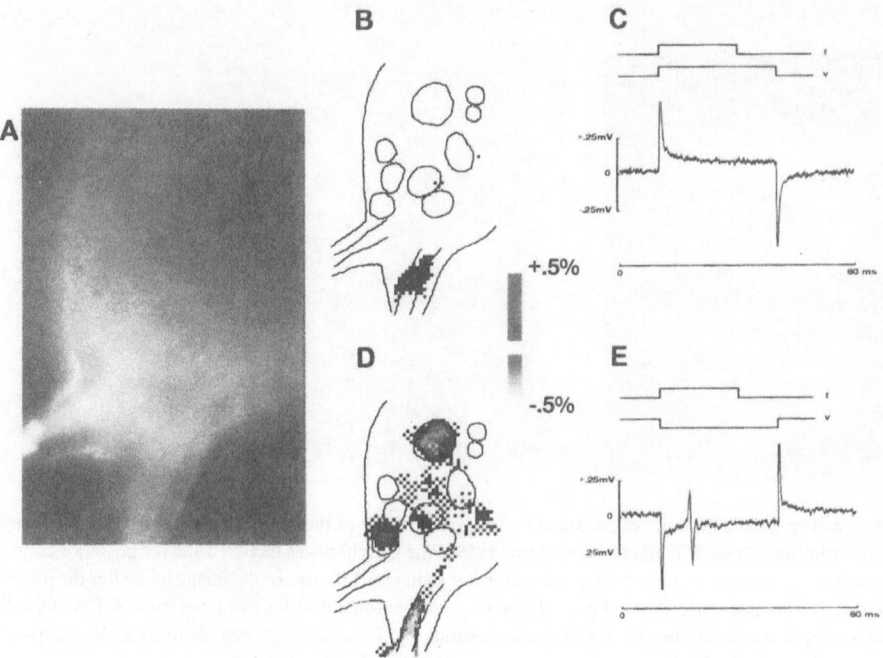

Figure 3. Imaging the action potentials. A is a full image of the viewed ganglion. B (C) are optical signals obtained at the times indicated in D (E). Ten runs were averaged in fixed time mode to obtain a signal image and the signals were superimposed to a map of the ganglion. Signal image B (C) corresponds to the electrical activity recorded from the anterior root shown in D (E) when the posterior root was stimulated with a positive (negative) pulse of .2V. The binning factor of the signal images was 8.The line indicated by v is the stimulus artifact that indicated by f is the shutter artifact. (A color reproduction of this figure appears following page 270).

signal images were superimposed to the neuronal map extracted from Fig. 4A. The timing of image acquisition is indicated in the upper trace in Fig. 4E.

In this experiment the 5 sensory neurons in the upper left side of the ganglion were visible and could be easily recognized.

At the onset of the stimulation, in the case of stimulation of the posterior root, the excitation was restricted to one neuron, but rapidly propagated to other neuronal structures. At the cessation of the stimulation the electrical activity extinguished within 50 ms as evident from the electrical response in Fig. 4E. A different dynamics was observed when the anterior root was stimulated. In both cases the stimulation produced the activation of neuronal structures located in the middle of the ganglion and an optical signal was recorded in the stimulated root. This signal can be originated in the soma or in the axons of several neurons that project nerve fibers in the stimulated root. Possible candidates are the sensory neurons, the Retzius cells, the controlateral Anterior Pagoda cell and the motoneurons which are located in the dorsal side of the ganglion. In this case this optical signal could not be correlated with any identified neuron.

Our CCD imaging technique is also particularly suitable to study interactions and linear and non-linear integrative properties of nervous systems. Optical signals elicited by the simultaneous stimulation of both roots are shown in Figs. 4D and 4H. By computational analysis, it is possible to evaluate the degree of linear summation between the two different inputs. This kind of analysis can provide new and relevant information in order to better understand the processing of information in the nervous system [Torre & Poggio, 1978].

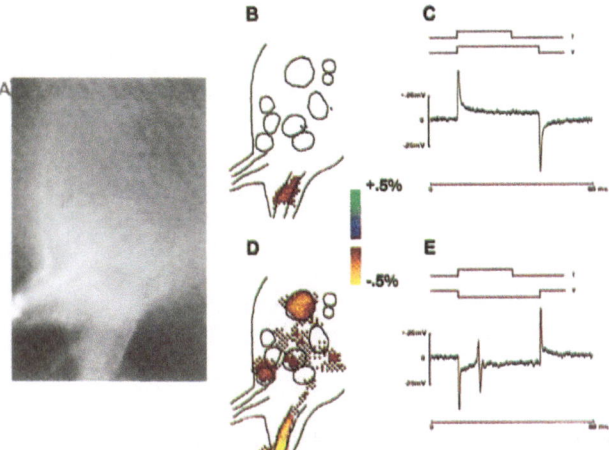

Figure 21.3. Imaging the action potentials. A is a full image of the viewed ganglion. B (C) are optical signals obtained at the times indicated in D (E). Ten runs were averaged in fixed time mode to obtain a signal image and the signals were superimposed to a map of the ganglion. Signal image B (C) corresponds to the electrical activity recorded from the anterior root shown in D (E) when the posterior root was stimulated with a positive (negative) pulse of .2V. The binning factor of the signal images was 8. The line indicated by v is the stimulus artifact that indicated by f is the shutter artifact.

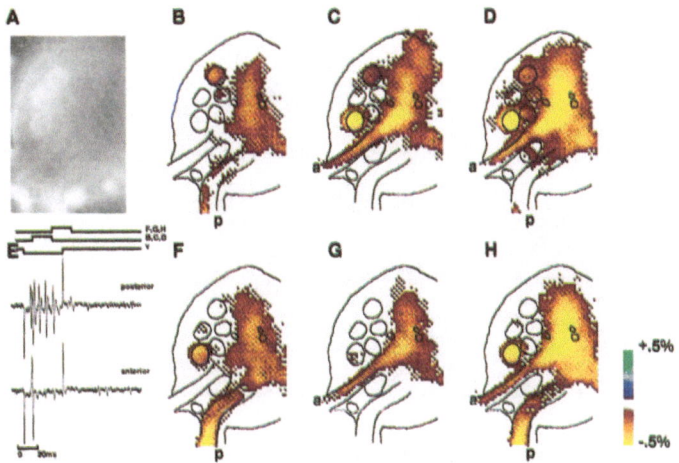

Figure 21.4. Imaging the dynamics of the network. A is a full image of the ventral ganglion. Signal images B, C, and D were taken 10 ms after the onset of a .1V stimulation of the anterior root (B), posterior root (C), and anterior and posterior roots simultaneously (D). Signal images F, G, and H were taken 30ms after the onset of the stimulation of the anterior root (F), posterior root (G), and anterior and posterior roots simultaneously (H). Each optical signal was obtained as the average of 10 runs. The electrical recordings from the posterior and anterior root are shown in E. The line indicated by v is the stimulus artifact, the lines indicated by B,C,D and F,G,H refer to the shutter artifacts.

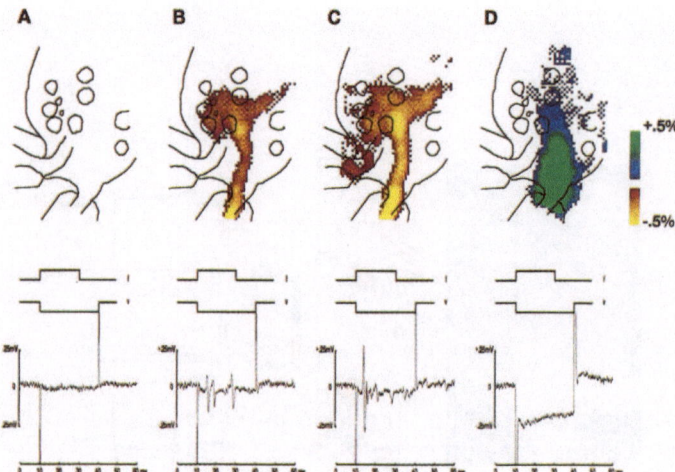

Figure 21.5. Imaging the effect of increasing the stimulation. Lower row reproduces extracellular recordings from the anterior root following the stimulation of the posterior root with a voltage step of increasing intensities: .075V (A and E), .125V (B and F), .175V (C and G), and .75V (D and H). The line indicated by v is the stimulus artifact. The upper column reproduces the optical signals obtained during the corresponding stimulation at the time indicated by the shutter artifact (f). Each optical signal was obtained as the average of 10 runs with a binning factor 8.

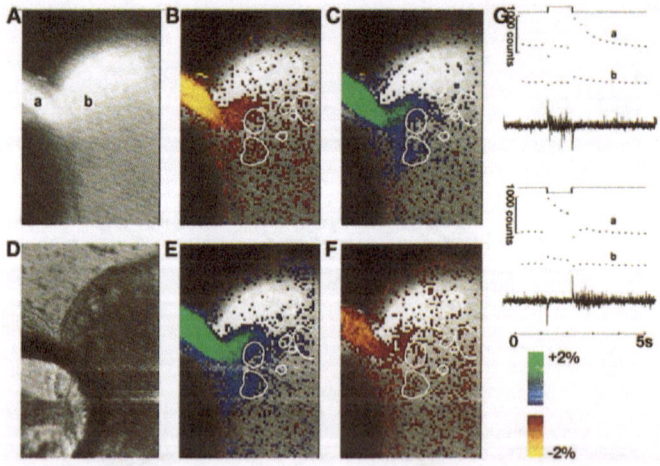

Figure 21.6. Imaging the afterpolarization and the afterdischarge. CCD images in real time during the stimulation of a root with a depolarizing or hyperpolarizing voltage pulse lasting 1 second. The amplitude of the stimulation was 1V. A is a full image of the ventral ganglion and B is a full image of the same ganglion of figure A, but taken in trasmitted light. B, C, F, and G are signal images superimposed to the image shown in A and the extracted neuronal map. B, C correspond to the 5th and 9th frames of the series in the case of depolarizing stimulation, while F and G correspond to the same frames, but in the case of hyperpolarizing stimulation. Each image was obtained with an exposure of 20 ms and a binning factor 7. In D and H the upper series of points indicate the fluorescence measured at the two locations indicated by a and b in A. These dots also indicate the timing of image acquisition by the video camera. The lower trace is the electrical signal measured from the posterior connective fibers.

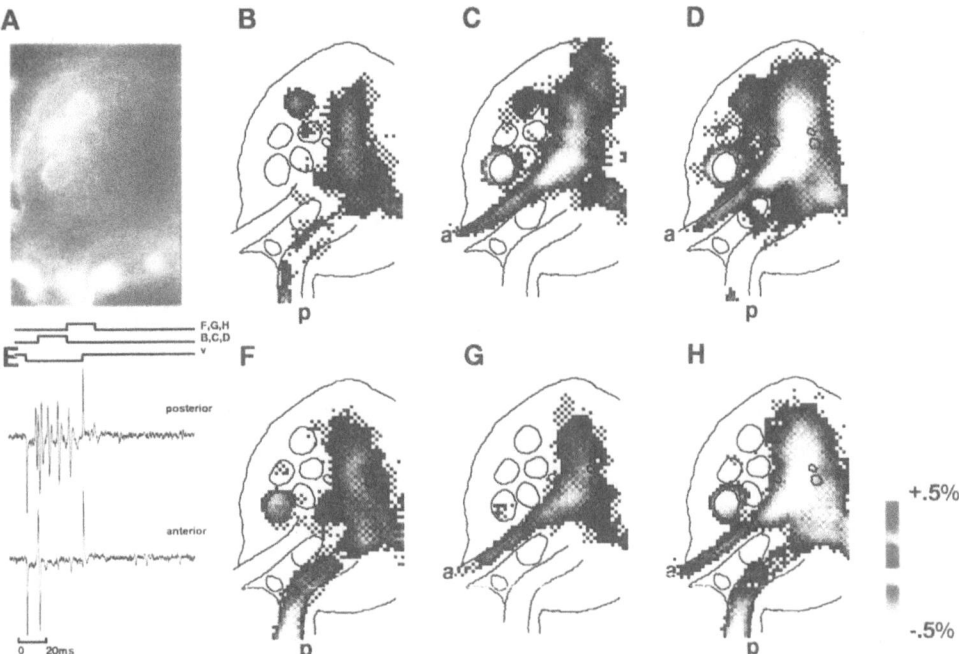

Figure 4. Imaging the dynamics of the network. A is a full image of the ventral ganglion. Signal images B, C, and D were taken 10 ms after the onset of a .1V stimulation of the anterior root (B), posterior root (C) and anterior and posterior roots simultaneously (D). Signal images F, G, and H were taken 30ms after the onset of the stimulation of the anterior root (F), posterior root (G) and anterior and posterior roots simultaneusly (H). Each optical signal was obtained as the average of 10 runs. The electrical recordings from the posterior and anterior root are shown in E. The line indicated by v is the stimulus artifact, the lines indicated by B, C, D and F, G, H refer to the shutter artifacts. (A color reproduction of this figure appears following page 270).

By analysing optical signals at different focal depths, it was also possible to extract information about the three dimensional dynamics of the electrical activity.

3.3. Imaging the Effect of Increasing the Stimulation

When very weak electrical stimulations with a negative polarity were applied to one root, usually with an amplitude lower than .1V, no evoked activity was observed from the other root. Different patterns of electrical activity were recorded increasing the amplitude of the stimulus. When stimulations higher than .5V were applied, no action potential could be observed. This effect is shown in the second row of Fig. 5.

The posterior root was stimulated with a voltage pulse of 75 mV, (see Fig. 5E), 125 mV, (see Fig. 5F), 175 mV, (see Fig. 5G) and 750 mV, (see Fig. 5H), and the electrical activity was recorded from the anterior root of the same ganglion. The signal images corresponding to the electrical recordings are shown in the first row of Fig. 5. In the case of 750 mV stimulation, the pulse propagated through the nerve fibers and reached the middle of the ganglion. Therefore, neurons that innervate the posterior root of the ganglion were directly hyperpolarized by the stimulation and could not fire. A diffuse signal of hyperpolarization could be observed in this case (see Fig. 5D).

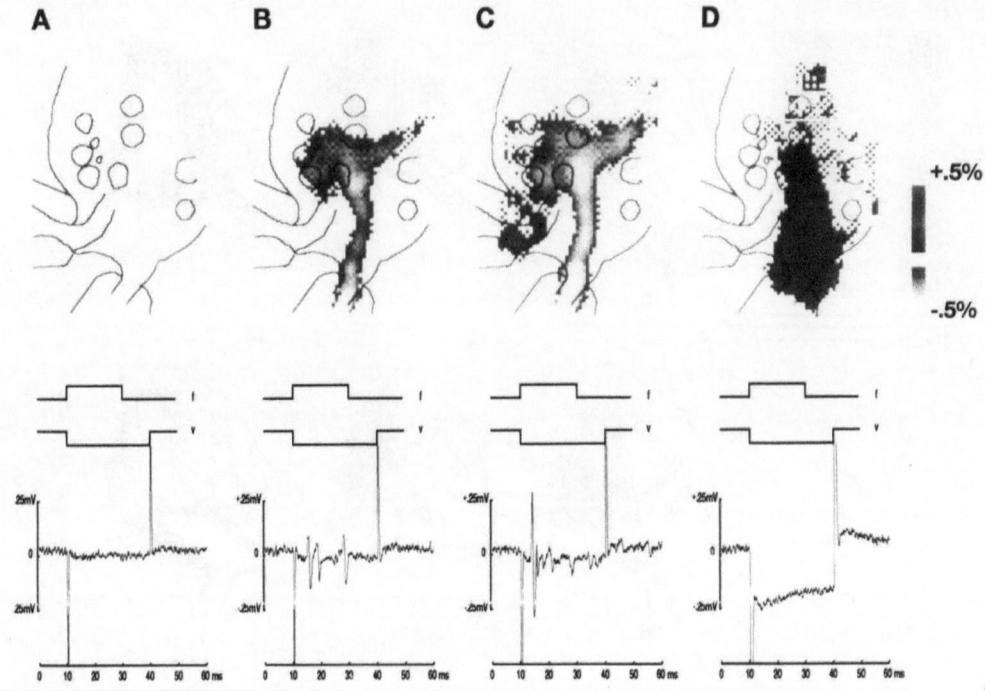

Figure 5. Imaging the effect of increasing the stimulation. Lower row reproduces extracellular recordings from the anterior root following the stimulation of the posterior root with a voltage step of increasing intensities: .075V (A and E), .125V (B and F), .175V (C and G) and .75V (D and H). The line indicated by v is the stimulus artifact. The upper column reproduces the optical signals obtained during the corresponding stimulation at the time indicated by the shutter artifact (f). Each optical signal was obtained as the average of 10 runs with a binning factor 8. (A color reproduction of this figure appears following page 270).

3.4. Imaging the Afterhyperpolarization and the Afterdischarge

A prolonged membrane depolarization induces a strong long-lasting afterhyperpo-larization in some leech neurons [Baylor & Nicholls, 1969; Jansen & Nicholls, 1973]. Signals produced by the afterhyperpolarization could be imaged in real time mode.

Fig. 6A illustrates a fluorescence image of a leech ganglion, which is imaged in trasmitted light in Fig. 6E. In the experiments shown in Fig. 6, the posterior root, which is hardly visible in the fluorescence image, was stimulated with a 1V pulse in order to have a massive direct excitation of the ganglion.

A sequence of 20 images was taken before, during and after the stimulation and the percentual changes of light intensity were computed for each image using the procedure described in methods. Figs. 6B and 6C show signal images obtained at the 5th and 9th frame of the sequence and the time course of the measured fluorescence at the two locations indicated by a and b in Fig. 6A is reproduced in Fig. 6D. A clear, but transient excitation of the root which was not stimulated could be observed during the stimulation. At the cessation of the stimulus, a large and long lasting afterhyperpolarization was imaged.

Similar data, but for a hyperpolarizing voltage pulse are shown in Figs. 6F and 6G. During the stimulation, a sustained hyperpolarization was observed, as shown in Fig. 6F. At the cessation of the voltage step, a brief depolarization (see Fig. 6G) could be imaged followed by a small, but long lasting afterhyperpolarization (data not shown).

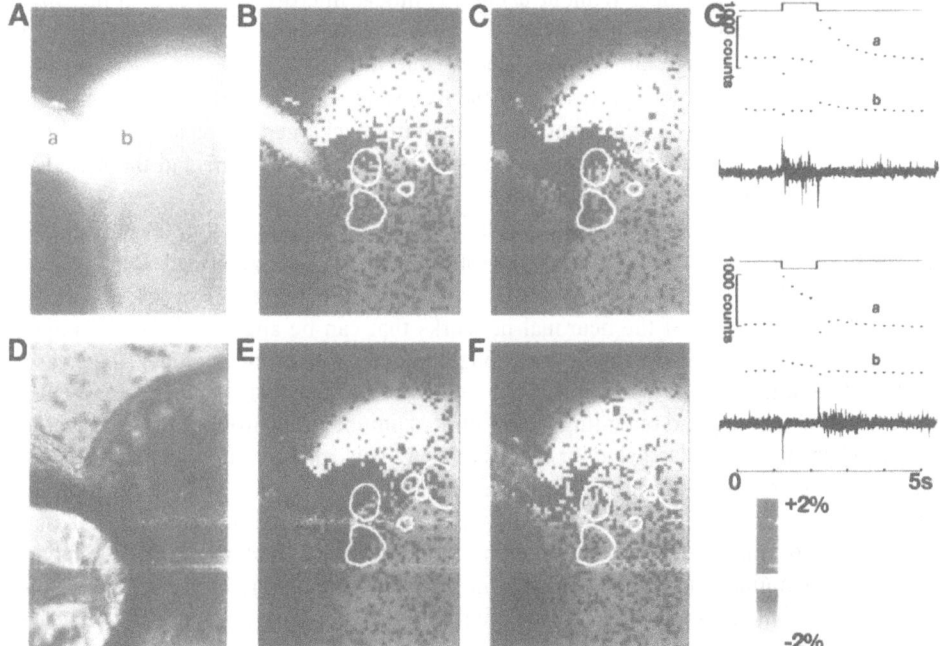

Figure 6. Imaging the afterpolarization and the afterdischarge. CCD images in real time during the stimulation of a root with a depolarizing or hyperpolarizing voltage pulse lasting 1 second. The amplitude of the stimulation was 1V. A is a full image of the ventral ganglion and B is a full image of the same ganglion of figure A, but taken in trasmitted light. B,C,F and G are signal images superimposed to the image shown in A and the extracted neuronal map. B,C correspond to the 5th and 9th frames of the series in the case of depolarizing stimulation, while F and G correspond to the same frames, but in the case of hyperpolarizing stimulation. Each image was obtained with an exposure of 20 ms and a binning factor 7. In D and H the upper series of points indicate the fluorescence measured at the two locations indicated by a and b in A. These dots also indicate the timing of image acquisition by the video camera. The lower trace is the electrical signal measured from the posterior connective fibers. (A color reproduction of this figure appears following page 270).

In both sets of experiments the optical signal was correlated with the gross electrical activity recorded by the posterior connective fibers. In the case of depolarizing stimulation, (see Fig. 6D), a strong electrical activity was recorded during the stimulus. The spontaneous activity, which was always observed without stimulating the preparation, was reduced for about one second after the end of the stimulus. A complementary behavior of the electrical activity recorded from the posterior connective was observed when the anterior root was stimulated with a hyperpolarizing pulse. The spontaneous activity was almost completely abolished during the stimulation while an afterdischarge was recorded at the end of the stimulus.

4. DISCUSSION

Fluorescence images originating from the leech ganglion can be detected by our CCD imaging technique and their time course and spatial resolution can be resolved with a moderate accuracy.

Even though our CCD camera is quite slow and the exposure time of each frame can not be less that 20 ms, the present study shows that it is possible to image spiking events at a temporal resolution of 50 Hz and to obtain a good localization of the signal as originating from well identified neurons. Without using a confocal microscope, it is also possible to record the optical signals at different focal planes and therefore to obtain a study of the electrical activity at a three dimensional space resolution.

Several physiological properties of the electrical activity of the nervous system of the leech, such as the afterhypolarization and the afterdischarge, can be analysed with this technique. When the time course of these signals is sufficiently slow and the signals are larger than .5%, it is possible to follow them even in real time.

Signal images of the electrical activity can also be manipulated in several ways and interesting properties of the occurring neuronal events can be described. Our system is particularly suitable to study the integrative properties of the network.

Many properties of the neuronal networks that can be analysed with our imaging technique cannot be studied with the traditional methods of neurophysiology. In the case of the leech, neural basis of specific behaviors of the animal were studied mainly using intracellular electrodes and recording from a limited number of neurons of the leech nervous system when the particular behavior was activated. This method allowed to identify motoneurons and interneurons involved in specific rhytmic behaviors such as "swimming" [Stent, Kristan, Friesen, Ort, Poon & Calabrese, 1978] or "crawling" [Baader & Kristan, 1994] and to formulate theories about possible neural networks underlying these behaviors. However, a simultaneous monitoring of the activity from many neurons during these behaviors could not be achieved in this way and therefore, most of the models that have been proposed could not be directly tested. This lack of information can be filled in the future by using our imaging system and it will be possible to obtain a description of the neuronal dynamics during specific behaviors of the animal.

By developing this imaging technique using new special imaging devices to improve the space and time resolution of the system, new and useful information can be obtained.

ACKNOWLEDGMENTS

We are indebted to Drs. Franco Conti, Larry Cohen, Peter Fromherz, Massimo Grattarola, Marco Bove, Maria Teresa Tedesco, John Nicholls for many helpful discussions and valuable suggestions. This work was supported by the EC grant B.R.A. SSS 6961.

REFERENCES

Baader, A.P. & Kristan Jr, W.B. (1995) Parallel pathways coordinate crawling in the medicinal leech, Hirudo Medicinalis. J. Comp. Physiol. 173: 715-726

Baylor, D.A. & Nicholls, J. (1969) After-effect of nerve implulses on signalling in the central nervous system of the leech. J. Physiol. (London) 203: 571-589

Cohen, L. & Salzberg, M.B. (1975) Optical measurment of Membrane potential. Rev. Physiol. Biochem. Pharmacol. 83: 287-310

Delaney, K.R., Gelperin, A., Fee, M.S., Flore, J.A., Gervais, R., Tank, D.W. & Kleinfeld D. (1994) Waves and stimulus-modulated dynamics in an oscillating olfactory network. Proc. Natl. Acad. Sci. USA 91: 669 673.

Fromerz, P. & Lambacher, A. (1991) Spectra of voltage sensitive fluorescence of styryl-dye in neuron membrane. Biochim. Biophys. Acta. 1068: 149-156

Jansen, J.K.S. & Nicholls, J.G. (1973) Conductance changes, an electrogenic pump and the hyperpolarization of leech neurones following impulses. J. Physiol. (London) 229: 635-665

Kauer, J.S. (1988) Real-time imaging of evoked activity in local circuits of salamander olfactory bulb. Nature (London) 331: 166-168

Muller, K.J., Nicholls, J.G. & Stent, G.S. (1981) Neurobiology of the leech. Cold Spring Harbor Laboratory, New York.

Nakashima, M., Yamada, S., Shioro, S., Maeda, M. & Satoh, F. (1992) 448-detector optical recording system: development and applications to Aplysisa Gill-Withdrawal reflex. IEEE trans. biomed. engen. 39: 26-36

Stent, G.S., Kristan Jr., W.B., Friesen, W.O., Ort, C.A., Poon, M. & Calabrese, R.L. (1978) Neuronal Generation of the Leech Swimming Movement. Science 200: 1348-1356.

Torre, V. & Poggio, T. (1978) A synaptic mechanism possibly underlying directional selectivity to motion. Proc. R. Soc. Lond. B 202: 409-416.

Zecevic, J., Wu, J.Y., Cohen, L.B., London, J.A. & Hopp, H.P. (1989) Hundreds of neurons in the Aplysia abdominal ganglion are active during the gill-withdrawal reflex. J. Neurosci. 9: 3681-3689.

22

HIGHER COGNITIVE FUNCTION OF THE HIPPOCAMPUS AS THE INTEGRATION OF A HIERARCHICAL MODEL DERIVED FROM AN n-LEVEL FIELD THEORY

G. A. Chauvet[1,2]* and T. W. Berger[2,3]†

[1] Institut de Biologie Théorique
Université d'Angers
49100 Angers, France
[2] Department of Biomedical Engineering and
[3] Program in Neuroscience
University of Southern California
Los Angeles, California 90089

1. INTRODUCTION

The present paper gives an introduction to the neurobiological basis of the cognitive functions of learning and memory performed by the hippocampus. We show that to do meet this objective requires describing the variation in time and space of activity in the hippocampus in terms of its anatomical structure and of the elementary physiological mechanisms at each of the network, cellular, and molecular levels. In other words, the problem of relating cognitive function to neurobiological mechanism requires the mathematical integration of the physiological mechanisms that occur at each level of organization in a neural system. Four steps are described in the following sections, which identify: (i) specific biological concepts (non-locality of the propagation of activity in the nervous tissue) (Chauvet, 1993e), (ii) a specific mathematical formalism using these concepts (n-level field theory of biological neural networks) (Chauvet, 1993a), (iii) the emergence of specific learning rules in a hierarchical neural network (Chauvet, 1995), (iv) a mathematical description of the activity of granule cells in the dentate gyrus (Chauvet & Berger, 1994).

Because of the biological constraints and the theoretical implications of the hierarchical organization described below, at least four issues have to be solved:

---

* email: chauvetg@ibt.univ-angers.fr
† email: berger@bmsrs.usc.edu

Neurobiology, edited by Torre and Conti
Plenum Press, New York, 1996

- The representation of the hippocampus according to its hierarchical functional organization,
- The determination of the fields that are associated with each function at each level of the hierarchy in order to apply the n-level field theory to the hippocampus.
- The identification of the specific global learning rules in the hippocampus, as it has been proved to exist in the cerebellar cortex,
- The mathematical integration of each level in order to obtain the behaviour at the higher level.

Applying results of the general theory to the specific case of neural networks, then to the hippocampal tissue, the paper will be organized as follows. In the second section, biological constraints are presented with respect to modeling real neural network: learning rules with mechanisms of synaptic efficacy, hierarchical organization, and the relation between network architecture and network function. In the third section, we show the theoretical implications of the hierarchical organization on the time and space evolution of biological systems, i.e. how a theoretical approach of biological organization may lead to new properties, i.e. the emergence of simpler new learning rules at the upper level (the groups of neurons that insure stability), based on the behavior of elements (the neurons) at the lower level. In the fourth section, the particular case of a biological neural network is considered, and finally in the fifth section these results are specifically applied to the link between modeling and extracellular field potential experiments for hippocampal tissue.

2. BIOLOGICAL CONSTRAINTS IN THE STUDY OF A REAL NEURAL NETWORK BASED ON MOLECULAR MECHANISMS

The primary current approach to studying the neurobiological basis of learning and memory has been the use of formal (more commonly termed, "artificial") neural networks with "adaptive" properties that derive from the incorporation of a "learning rule." As in any real neural network, the study of the variation in time and space of activity in hippocampus, in terms of molecular events, needs to solve several problems regarding biological constraints and theoretical implications.

Three classes of constraints raise several fundamental issues with respect to modeling biological neural networks.

2.1. Learning Rules Consistent with Mechanisms of Synaptic Efficacy

A first issue concerns the biological relevance of learning rules incorporated into models of real neural networks. Learning rules, i.e., activity-dependent changes in synaptic efficacy, must be the consequence of biochemical kinetics at the molecular level (Brown *et al.* 1990) as well as the consequence of changes in excitability due to synaptic interactions with other, distant neurons. Hebbian learning rules, expressing synaptic efficacy in direct relation with the product of X, presynaptic activity, and Y, postsynaptic activity (Fig.1), do not incorporate these molecular mechanisms and the corresponding parameters of their kinetics, because a model describing learning rules in terms of the kinetics has not been established. Such a model would describe the effect of some parameters of molecular kinetics on the learning rules, and consequently, the effect of the learning rule on activity in the neural network. So only if such phenomena are *explicitly* (mathematically) incorporated in the learning rules, can the effect of molecular kinetics on learning function be studied. Using the n-level field theory, it has been shown (Chauvet, 1993a) that the physiological dynamics

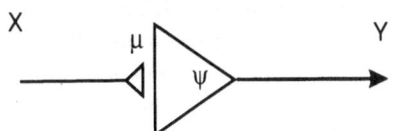

Figure 1. X and Y are the input and output activities. μ is the synaptic efficacy and ψ is the membrane potential. Activity ψ is deduced from membrane potential by the non-linear function f.

Classical Hebbian learning rule:

$$\frac{d\mu}{dt} = \alpha XY \qquad Y = f(\psi)$$

of the network can include biologically relevant learning rules that result from the dynamics at the molecular level.

2.2. Hierarchical Organization of Real Neural Networks

The second issue arises from the hierarchical organization of the nervous system, because the physiological dynamics of the network have to be described in terms of subcellular components, neurons, and populations of neurons. Although it is possible to understand the relation between various molecular mechanisms underlying synaptic plasticity by using a kinetic model, and to understand the relation between various stages of information processing with a model of cognitive processes (such as the Rescorla-Wagner model), the formalisms used for representing the dynamics at each level are sufficiently different that it is not possible, for example, to relate changes in receptor-channel function to behavioral learning.

The issue to be resolved is whether a formalism can be identified which describes and relates the physiological dynamics at each of these levels. Because different formalisms are used for each level, the issue of passing from one level to another cannot be solved. A general theory which leads to the same formalism at each level gives a solution to this issue. The problems introduced by a hierarchical organization also extend to the issue of learning rules discussed above.

In identifying hierarchical levels of organization in the nervous system, it is obvious that one must consider the levels of receptor-channels, synapses, and neurons, but there also is evidence for subpopulations of neurons that function as a "unit" or a "local network," i.e., their activity is highly correlated (Berger & Bassett, 1992). The issue is if there are specific conditions under which the functional synaptic connectivity between two or more such "units" or "local networks" becomes strengthened or weakened? An example of that is given

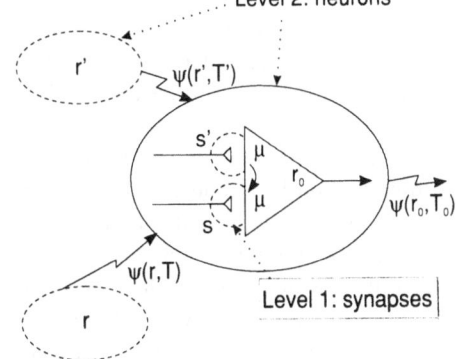

Figure 2. The two levels of organization in the nervous tissue. Neurons are balloons, synapses are balloons in the neurons. Activity is propagated from point r at time T to point r_0 at time T_0. In the same way, synaptic efficacy is propagated from a point s' to a point s.

by the multiple spatial, behavioral, and other properties that is observed in the activity of hippocampal neurons in the behaving animal (Eichenbaum *et al.* 1989). In other words, as a consequence of the hierarchical organization, do there exist specific learning rules at the level of groups of neurons (Finkel & Edelman, 1985)?

2.3. The Relation between Network Architecture and Network Function

A third issue concerns the relation between network architecture and network function.

1. Most formal neural networks include topology but not geometry. It is important to consider the issue of geometry for two reasons: i) it is clear that there are precise geometrical constraints on nervous system organization which must play a functional role; and ii) there are important mathematical consequences, because if geometrical relations are represented, *delays* are introduced in the equations. And the effects of delays are well known in mathematics: they create specific effects like stability, periodic effects, etc. To ignore delays and their effects is too great a simplification of nervous system function: delays exist because distance between neurons is always different. One of the main advantages of a field theory is that it can take into account distances between neurons.

2. The geometry of neural systems is specified by the *continuous densities* of their neurons. Therefore, a continuous formalism is needed to represent the geometry. Including experimentally determined densities in a continuous representation constitutes a major advantage with respect to the coupling between geometry (the location of cells) and topology (the connectivity between cells) of the network. In contrast, models based on automata theory can represent only the topology of the system.

3. THEORETICAL IMPLICATIONS OF THE HIERARCHICAL ORGANIZATION OF BIOLOGICAL SYSTEMS

3.1. A Dynamical Approach of the Hierarchical Functional Organization (O-FBS)

A theory of functional organization has been proposed for biological systems (Formal Biological System, FBS), specifically on the nervous system, based on the concept of *"functional interaction"* between *structural units*, i.e. the action of a physical structure, e.g. a cell, onto another structure, e.g. another cell. From the specific properties of a combination of functional interactions, i.e., *non-symmetry, non-locality*, and *non-instantaneity*, a biological system can be considered as constituted by two hierarchical systems (Chauvet, 1993b): (i) the (O-FBS) that describes the topology of the FBS results from the construction of the set of interactions, i.e., the functional organization; it is represented by a hierarchical *directed* graph due to the non-symmetry of the interactions; (ii) the (D-FBS) that describes the continuous non-local and non-linear dynamics of the FBS, resulting from physiological processes that are represented by non-local fields.

An optimum principle has been established, due to the *non-symmetry* of functional interactions, which could explain the stability of an FBS, and a criterion of evolution for the hierarchical topological organization of a FBS has been deduced from that principle (Chauvet, 1993c). In the case of nervous system, this problem can be expressed as the

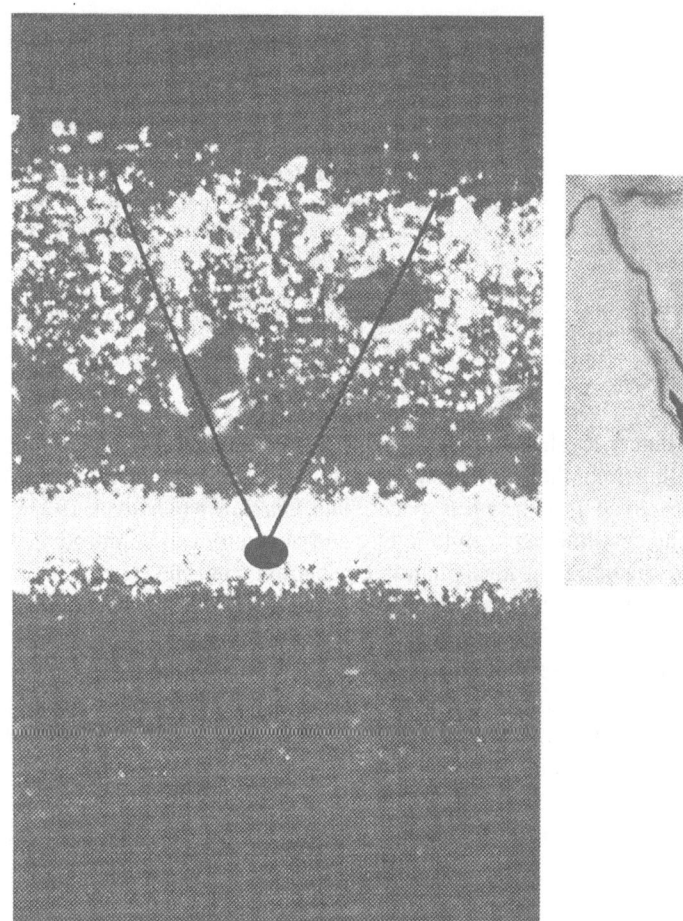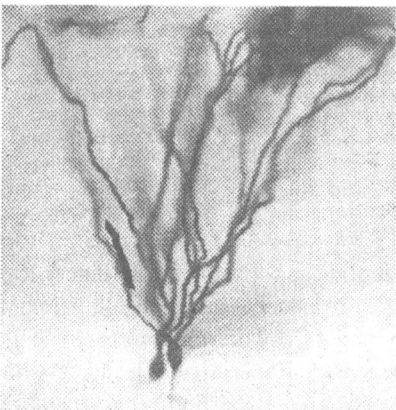

Figure 3. Microscopical view of the granule cell layer in the dentate gyrus of the hippocampus. On the right, one granule cell with its dendritic cone. The related drawing has been reported on the left.

re-distribution of the neurons (sources) and the synapses (sinks), when one of them is suppressed.

Because a physiological process evolves on its own time scale, the levels of functional organization can be classified according to distinct time scales, and, therefore, a "decoupling" of dynamics at each level is obtained (Chauvet, 1993d). It is shown that properties deduced with this formalism give the relationship between topology and geometry in an FBS, and particularly, the geometrical re-distribution of units. In the framework of this field theory, a *statistical distribution function* of the states of the fields shows that the collective behavior of the population of structural units is not a simple summation of the individual elements, and gives a solution to the problem of the passage from one level to another.

3.2. An N-Level Field Theory for the Dynamics of Functional Processes of a (D-FBS)

As presented further in the case of a neural tissue, in a hierarchical continuous system the finite velocity of the functional interaction between structural units at the lower level

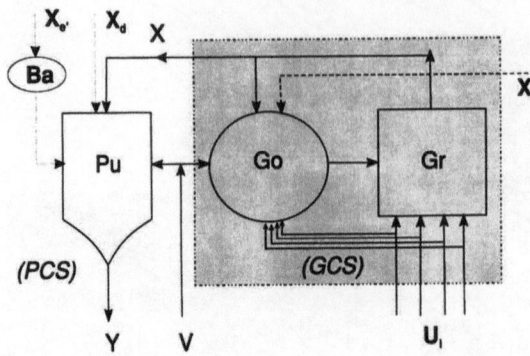

Figure 4. A network of Purkinje units (PU). Each PU is constituted of the two subsystems shown in Figure 6, GCS and PCS. The connectivity of this network determines the output of the GCS at rank u as an input for the PU at rank u-1, i.e., the external context from u onto u-1. New variational learning rules are proved to apply *at the level of the PUs based on the sense of variation of the external context that drives the behaviour of the network of the PUs.*

implies non-locality at the higher level. Two other properties of the functional interaction are introduced in the formulation: the *non-symmetry* between units, e.g. neurons and synapses, and the *non-uniformity* of the extra-units space. Thus, it is shown (Chauvet, 1993e; Chauvet, 1993a) that: (i) The coupling between topology and geometry can be introduced via two functions, the *density* of units at the level of units, e.g. neurons at the level of neurons, and the *density-connectivity* of units at the lower level, e.g. synapses, between two points of the space of units at this lower level. With densities chosen as Dirac functions at regularly spaced points, the dynamics of a discrete, purely connectionnist network becomes a particular case of the n-level field theory. (ii) The dynamics at each of the lower levels are introduced, at the next upper level, both in the source and in the non-local interaction of the field to integrate the dynamics at this upper level.

3.3. The Emergence of New Properties at the Upper Level of a Hierarchical System

The precise role of cerebellum is not yet clearly elucidated, despite other researches on the sensorimotor system (Giszter *et al.* 1989) or on the sensory systems in mammals (Kaas, 1989). Considering an interpretation of the coordination of movement by the cerebellar cortex, from Purkinje unit to network with variational learning rules, has elicited a new global property at the higher level of the hierarchy, resulting from local properties at the lower levels (Chauvet, 1995).

In this recent study on the cerebellum, an interpretation of the coordination of movement that depends on three main features has been proposed:

1. One feature involves the definition of a *unit of function*, the Purkinje unit, a local circuit made of a Purkinje cell and the closely associated cells and connections (Fig.4).
2. Secondly, the most important feature is a new kind of learning rule called a *"Variational Learning Rule"* (VLR) which operates at the level of the Purkinje units by combining the senses of variation of activities (the signs of derivatives) and not their value.
3. The coordination of movements is then interpreted in terms of excitatory and inhibitory interactions between Purkinje local circuit. An emergent property of the model is that Purkinje units occur in groups which act in opposition in the sense that when one has increasing output, the other has decreasing output (Fig.5).

Thus, two questions may be asked for the network of Purkinje units considered as a hierarchical adaptive system: 1) *Is this finding general enough to be applied to various*

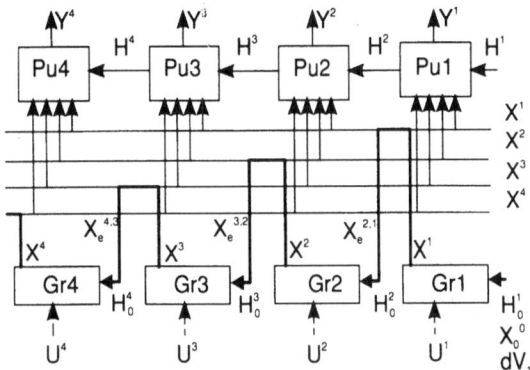

Figure 5. Learning in the hierarchical system. Diagrammatic representation of a Purkinje unit defined by two subsystems, GCS including the network of granule cells (Gr) and the Golgi cell (Go), and PCS including the unique Purkinje cell (Pu) and the basket and star cells (Ba). There are three inputs for the GCS: U_i along the mossy fibers, and X_e, the external context constituted of parallel fibers that originate in other Purkinje units, and V along the climbing fibre. There are four inputs for the PCS: X along the parallel fibres that originate in the GCS, X_d and X_e that originate in other GCSs. There is a unique output Y, from the Purkinje cell. Hebbian learning rules are assumed at the level of the neurons.

network structures, different from the cerebellum, as the hippocampus? 2) *Is there such an adaptive "control" in the hippocampus?*

All these issues will now be considered for the hippocampal neural network.

4. THEORETICAL IMPLICATIONS OF THE HIERARCHICAL ORGANIZATION OF BIOLOGICAL NEURAL NETWORKS

4.1. The Geometry of Neurons and Dendrites Implies a Field Theory

According to the theory shortly recalled in the above sections III.1 and III.2, because a neuron, the source, "emits" activity, i.e., a frequency of action potentials that act on the postsynaptic site of another neuron, the sink, we can use a field theory to describe the continuous variation in time and space of neuronal activity. In a field theory, a field variable, say ψ, is transported from a source, the neuron, represented by Γ, to a sink, the postsynaptic site of another neuron, under the action of the field operator H (Fig.6). Such a theory resolves some of the biological constraints raised previously because the interaction is represented as occurring between geometrically located sources and sinks, and is dynamically described by a *non-local* interaction operator. Also resolving one of the theoretical constraints raised earlier, the equation which represents *the variation in space and time of the field variable has the same form at each hierarchical level of the nervous system: $H\psi = \Gamma$, where H is a* space and time differential operator, e.g.:

$$ H \equiv \frac{\partial}{\partial T} - D \nabla^2 + H_I $$

with the diffusion operator ∇^2 and the interaction non-local operator H_I.

Specifically, a field theory has the following characteristics:

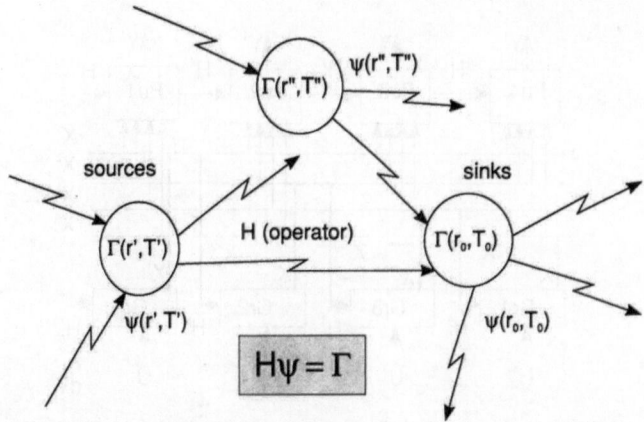

Figure 6. Formal representation of the fields that propagate from (r', T') and (r'', T'') to (r_0, T_0). The field ψ at (r', T') is moved to (r_0, T_0) under the action of an operator H. This is mathematically described by the equation $H\psi = \Gamma$ where Γ represents the source.

    (i) a field theory can represent the dynamics at several levels of organization within the nervous system, subcellular, cellular, and multicellular;

   (ii) as shown above, a field theory can include non-locality by means of a non-local interaction operator.

  (iii) a field theory relates topology and geometry because it describes the space-time propagation of patterns in terms of the connections between neurons (described by the topology) and the space densities (dependent on geometry). This relation is given by a synaptic density- connectivity function.

  (iv) Field variables have a physical meaning, e.g., neuronal soma membrane potential and synaptic efficacy.

   (v) Field equations include the mechanism of learning through the source terms, so learning rules can be deduced from the continuous dynamics at the lower level.

New learning rules are deduced from the structure of the field equations: Hebbian rules result from strictly local activation; non-Hebbian rules result from homosynaptic activation with strict heterosynaptic effects, i.e., when an activated synaptic pathway affects the efficacy of a non-activated one; non-Hebbian rules and/or non-linearities result from the structure of the interaction operator and/or the internal biochemical kinetics.

4.2. The Hierarchical Organization of the System with Using a Field Theory Implies Non-Locality

The second biological constraint has an important consequence in the formulation of the model. If we analyze what happens inside the neuron, we see a set of transformations at different levels of organization, cellular, subcellular, or molecular. Because the space of synapses is included in the space of neurons, and because, functionally, two time scales are attached to these spaces, a hierarchical functional system is defined from the field variables (Chauvet, 1993d). When several levels of organization are taken into account in a neural network, the new concept of *non-locality* must be introduced (Chauvet, 1993e). Figure 7 gives an illustration of this basic concept for a hierarchical network.

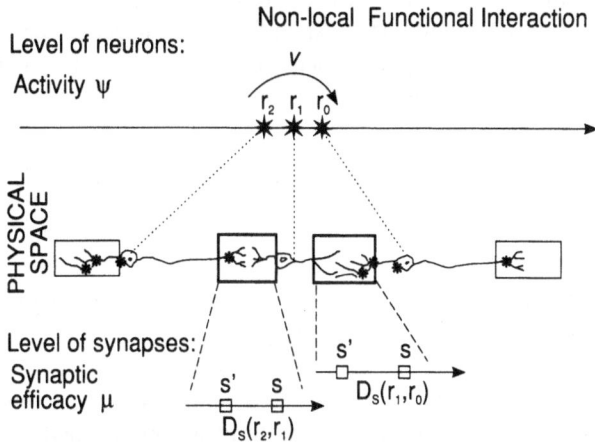

Figure 7. Non-locality of the functional processes. The action from one neuron to another in the space of neurons is, in fact, the action from one volume to another (including the boxes). With a field theory, we must consider *the action from one point to another*. We may solve the problem by considering *spaces of units*. In the figure, this space is the space of axon hillocks, in which the membrane potential is propagated from a point r_2 to a point r_1, then to r_0. These points can be very close (infinitesimally) in this space, although they have a finite extent in the physical space. Boxes show the extent of the synapses which constitute the space of synapses. Thus, the representation of these actions in terms of fields needs the introduction of *non-local* terms in the field equations.

4.3. The Hierarchical Organization of the System Could Imply Global Learning Rules in the Hippocampus

Considering the case of the cerebellum, it is likely that the expression of learning rules is greatly simplified for some structures organized in a specific function, e.g. groups of neurons assembled in a *microzone*, that have to be identified in the hippocampus. This "simplification" consists in replacing classical Hebbian rules (see Fig.1) by new learning rules, called *Variational Learning Rules* (VLR), which operate at the level of the groups of neurons (see Fig. 5). They combine the senses of variation of activities (the signs of derivatives) and not their value. Thus, from this point of view, there are two issues for the hippocampus:

1. The hierarchical system of units is driven by new learning rules that represent the dynamics of one group, each of them being constituted by neurons driven by classical Hebbian learning rules. With these learning rules, the values of activities of the units are replaced by their time derivatives. *May this emergence of learning rules at the level of groups be applied to various network structures, different from the cerebellum, as the hippocampus?*
2. The interpretation of the propagation of activity of a network is based on the mathematical *stability* of its dynamics, specifically on our definition of learning as a monotonic increase in synaptic efficacy. However, we also have to determine the ability of learning and retrieving patterns under some control, which corresponds to the usual definition of learning by a neural network in a closed loop, such that the learned pattern becomes a given "target pattern." *Because the hippocampus is a closed loop, is there such an adaptive "control" in the hippocampus?*

Usual experiments to observe changes in the activity of a nervous tissue as the hippocampus consist in stimulating presynaptic fibers and in recording extracellular field potentials in the population of postsynaptic neurons. We show now the possible link between the above theoretical aspects and the experimental observations.

5. THE HIPPOCAMPAL TISSUE AND THE EXTRACELLULAR FIELD POTENTIAL

5.1. Experimental Implications: Synaptically-Evoked Population Activity of Dentate Granule Cells of the Hippocampus

The higher cognitive function is identified by the space and time propagation of activity in the population of neurons. Activity is derived from the mathematical solution of the field equations. Obviously, the field equations contain physiological and geometrical parameters. The only way to determine these parameters, thus activity, is to simulate experiments with a model using the field theory. Then, we identify parameters using experimental waveforms that have to be as close as possible to the calculated waveforms (Chauvet & Berger, 1994).

The anatomical components of the hippocampus currently included in the field theory representation are perforant path fibers and dentate granule cells. *To maximize the identity between the elements included in the model and the biological system used for its evaluation,* one of us has developed a reduced *in vitro* hippocampal slice preparation which consists almost exclusively of perforant path terminals/fibers and dentate granule cells. He has developed this preparation in stages, with the functional properties at each stage determined experimentally using an application of nonlinear systems analysis (Berger *et al.* 1991). Because it provides the best biological approximation of the neural elements included in the proposed field theory representation, this simplified *in vitro* slice preparation has been used to evaluate the new parameters of the model.

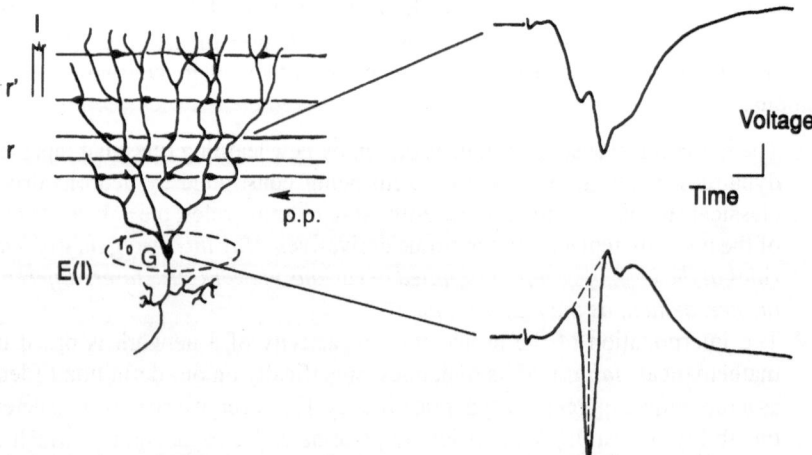

Figure 8. Two waveforms voltage-time recorded at two different locations: on the top, the recording microelectrode is located in the median part of the dendritic tree, on the bottom, at the level of the soma. The perforant path (p.p.) fibers issued from r and r' make synapse with the dendrites of the granule cell G (located at r_0) which belongs to an area E(I) that depends on stimulating intensity I.

Properties in the hippocampus are mainly characterized by synaptically-evoked population activity experiments: presynaptic neurons are stimulated for various intensities and the extracellular field potential (EFP) is then recorded at various locations (Fig.8). These EFPs are the images of the variation in time of the fields at the corresponding points in the space.

The extracellular waveform can be viewed as the image of the number of synapses which create an extracellular potential at a given time, i.e., in relation to the number of micropotentials created by the corresponding synapses at one point of the extracellular space. Specifically, we may interpret an EFP waveform as resulting from two processes: (i) firing of stimulated neurons when the membrane potential ψ (the value of the field variable) is overthreshold; (ii) synaptic activation which results from all the ionic currents variations in the recorded volume of granule cells.

The first process is completely determined by the resolution of the field equations. The second process reflects the time distribution of the summed EPSP in this volume, due to the large number of causes, as the orientation of currents in space, the location of dendrites and synapses, the distance between the synapse and the recording electrode.

Because of these independent influences on the extracellular potential measured at a distance, the Central Limit Theorem in the theory of probability establishes that this sum is a Gaussian variable. Thus, we have a statistical interpretation of the extracellular waveform, confirmed by the observed waveforms (Fig.9). A specific statistical method based on the meaning of the field variables (Chauvet, 1993d), see below, allows the deduction of the behavior of a population of neurons from the fields at all levels.

The state of a synapse is defined by two field variables, the postsynaptic potential Φ and the synaptic efficacy μ. By using a "time distribution function $f(s_0, t; \psi(r_0, t), \mu) \equiv f_0(t)$ of the states of the fields" at a current synapse at s_0, it is possible to describe the time distribution of the extracellular potentials. But, because it is not possible (up to our knowledge) to share the respective influences of the EPSP and of the distance of the recording electrode, we can consider f as the new time distribution function $F(t) = N^*(t)/N$ of the sum of the micropotentials created by the activated synapses. Therefore, $F(\psi, \mu, t) = N^*(t)/N$ will be interpreted as the time distribution of the micropotentials v_e created by the activated

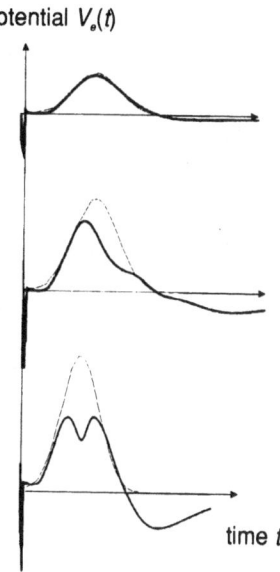

Potential $V_e(t)$

time t

Figure 9. Observed waveforms $V_e(t)$ are bold curves. The calculated Gaussian curves $V_{calc}(T) = A/\sigma \exp[-(T-<T>)/\sigma^2]$ depend on three parameters: standard deviation σ, time average $<T>$ considered as defined by the maximum of the curve, and A, a scale factor. For increasing values of stimulating intensity, the values of these parameters are: $\sigma = 0.23, 0.25, 0.27$; $<T> = 0.67, 0.75, 0.83$; $A = 13, 4.5, 3.2$. As predicted by the theory, discrepancy between the Gaussian curve and the experimental waveform increases with intensity (from the top to the bottom), which means that the number of random micropotentials created by synaptic activation decreases.

synapses. This method provides a means to define the relation between the intracellular and the extracellular potentials *using new parameters having a physiological interpretation.*

Following the above presentation, the time distribution function of micropotentials includes three kinds of variations representing: (i) the fraction $Q(t)$ of synapses that modify their state as a consequence of an external stimulus; (ii) the modification of the internal state of the cell and the corresponding synaptic states, either as a consequence of stimulation that leads to firing (feedback from the action potential to the emitting cell) or a spontaneous "relaxation," i.e., a modification of potential without external stimulation but with internal modification $\Phi(\psi)$ (such as from a voltage-dependent conductance); (iii) any long-term variation in synaptic efficacy. This sum can be written as:

$$dF(\psi, \mu, t) = Q(t)F + (dF)_\psi + (dF)_\mu \qquad\qquad F(t) = \frac{N^*(t)}{N}$$

at time t corresponding to r_0, or, with respect to variables ψ and μ:

$$\frac{dF}{dt} = Q(t)\, F + \frac{\partial F}{\partial \psi}\frac{d\psi}{dt} + \frac{\partial F}{\partial \mu}\frac{d\mu}{dt}$$

where the relationship $\Phi(\psi)$ is included. This equation joined to the field equations gives an expression of ψ, μ and F, i.e., determines the global intracellular potential. Numerical simulations given in Figure 10 show the role of the macroscopical parameters that describe synaptic activation and the propagation of activity for a given intensity.

5.2. Cognitive Function of the Hippocampus

It is clear that developing a model of the learning and memory functions of the hippocampus will allow cognitive operations to be related to the spatio-temporal distribution of activity in the hippocampus, and the molecular mechanisms of synaptic plasticity. In humans and primates, the hippocampal formation and parahippocampal gyrus have been characterized as essential for the formation of 'declarative' or 'data-based' memories, e.g., facts, names, etc. (Cohen & Squire, 1980), whereas the learning and remembering of skilled motor movements, or 'procedural memory', is the domain of other brain systems. In lower animals, an analogous function of the hippocampus has been described in terms of its role in forming and manipulating 'mnemonic labels' to identify 'abstract relations' among environmental events (Hirsh, 1974). For example, the hippocampus is essential for rats to utilize the multiple spatial relationships among environmental cues to reach a goal through multiple routes, and not for cue-directed stereotypic movements (Morris *et al.* 1982). Likewise, hippocampal function is necessary for classical conditioning of higher-order and conditional discriminations; in contrast, nonconditional discriminations are learned and remembered readily (Ross *et al.* 1984).

At the electrophysiological level, these characterizations of hippocampal function are observed as multiple, distributed representations of specific features of environmental cues or components of the behavioral response required to meet task requirements (Berger & Bassett, 1992). For example, when rats are required to navigate to a food source on the basis of their spatial location relative to a configuration of distal visual cues, the electro-physiological activity of individual hippocampal neurons correlates with an animal's position within a particular subspace of the test area, such that the activity of a subpopulation of neurons is sufficient to describe all subspaces (Eichenbaum *et al.* 1989; O'Keefe, 1976). Thus, there is a 'distributed' neural representation of the total space, or the total possible spatial locations. Furthermore, the activity of only a relatively small number of hippocampal

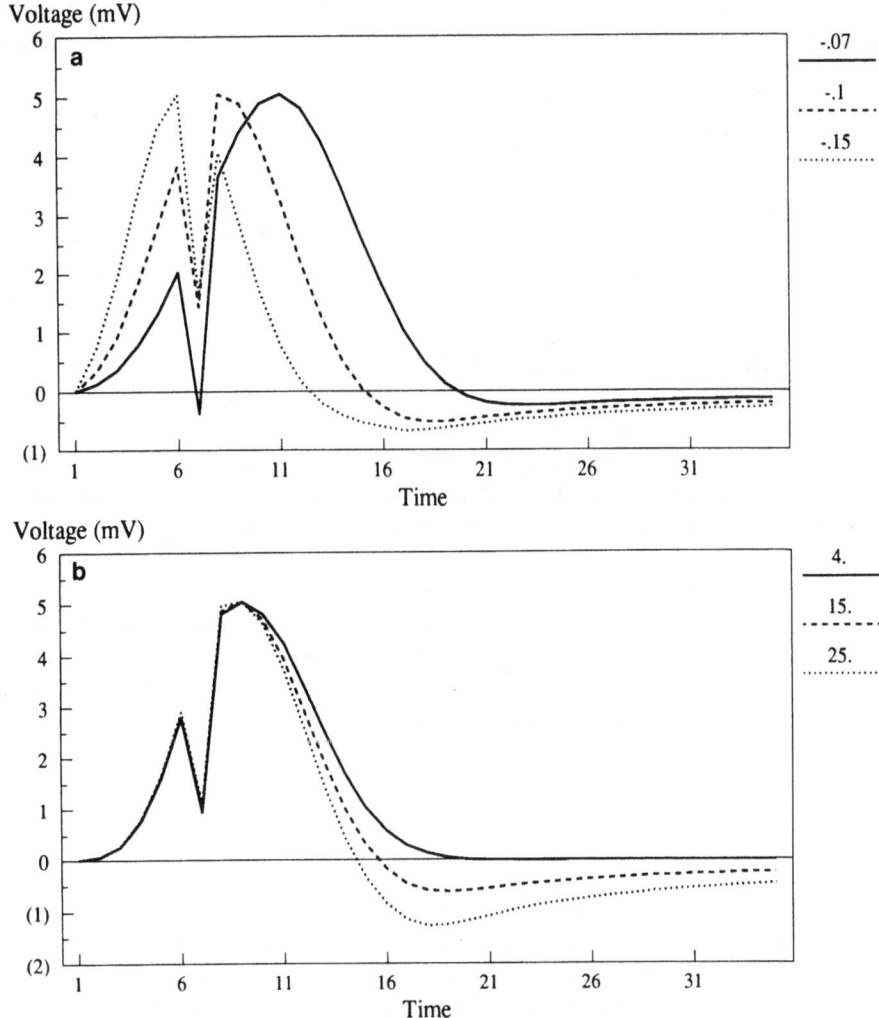

Figure 10. Extracellular field potential waveform obtained by numerical simulation for three values of the parameter included in the model. a) Effect of the parameter $Q(t)$ that represents the fraction of synapses that modify their state as a consequence of an external stimulus; b) Effect of the source of the field equation, represented by a firing coefficient.

neurons is necessary to account for all spatial locations, so that multiple representations of the space can be detected. In addition, neural representations of different elements of the learning task, e.g., spatial location, behavioral response, conditioned stimulus, etc., are generated by different subpopulations of hippocampal neurons (Wiener *et al.* 1989). Finally, the activity of any one cell may be involved in the representation of different features of the same element (e.g., a different subspace), or different elements (e.g., discriminative stimulus vs. spatial location), depending on the environmental context.

At the molecular level, hippocampal neurons can express multiple forms of activity -dependent, long-term synaptic plasticity, the two principal forms being long-term potentiation (LTP) and long-term depression (LTP) (Higashima & Yamamoto, 1985). The form of LRP typically expressed within the dentate and CA1 cell fields is induced only if the postsynaptic neuron is depolarized significantly by a high frequency unput (50-400 Hz for 20-100 ms). The resulting LTP is homosynaptic (i.e., specific to the synapses stimulated), and can be associative (induction of LTP in one input can lead to the induction of LTP in a second pathway that is activated simultaneously) (Bahill & Karnavas, 1989). Lower frequencies of stimulation (e.g., 1-10Hz) result in LTD, particularly if the postsynaptic neuron is hyperpolarized during the period of activation (Xie *et al.* 1992). The form of LTP characteristic of dentate input to CA3 (mossy fibers) is induced by burst of excitatory input that are lower in frequency (e.g., 15-50 Hz) than those which are effective in dentate and CA1, and is heterosynaptic (i.e., all synapses to the same neuron are potentiated, whether or not they were activated) (Bradler & Barrionuevo, 1989). Other inputs to CA3 express the homosynaptic form of LTP observed in dentate and CA1, although when LTP is induced in the non-mossy fiber inputs to CA3, mossy fiber input expresses LTD (Bradler & Barrionuevo, 1989).

In summary, the cognitive function of the hippocampus must be a product of the collective activity of its intrinsic neurons. The information represented in that collective activity can only be understood if one considers the spatio-temporal distribution of active cells, given the complexity of the representations for any one set of conditions and the fact that the information represented in the activity of hippocampal neurons changes dynamically as a function of the environmental constraints. Dynamics of the spatio-temporal distribution of active cells must reflect the induction requirements for synaptic plasticity imposed by mechanisms at the molecular level. Presumably, co-activation of neurons of different subpopulations so as to induce synaptic plasticity leads to a change in the strength of associations among the environmental features represented in the firing of the co-activated neurons. Thus the importance of the system geometry, because the connectivity patterns and delays will be essential for determining whether or not the induction requirements are met for the various forms of hippocampal synaptic plasticity.

6. DISCUSSION

The dynamics of a neural network, and thus the extent to which those dynamics mimic the cognitive processes of learning and memory, are a product of the properties of the elements, the neurons, including: (i) the topology and the geometry of the network (connections and distance between neurons), and (ii) the physiological mechanisms of each cellular type, particularly the influence of molecular kinetics on learning rules. We propose here to deduce the network activity from the anatomy and the physiological mechanisms at the lower levels. That needs at least, as shown above, a non-local n-level field theory of biological neural networks and, in the framework of this theory, a mathematical expression of the extracellular field potential given by experiments in various situations. Can the higher cognitive function of the hippocampus be described, up to the present knowledge, in terms of the propagation of the activity in the network?

Although the field equations give the values of soma membrane potential at each point of the neural network and the values of the synaptic efficacy at each point of the neuron for every neuron, some effects like the modification of internal state after firing, the spontaneous relaxation of potentials due to the internal cytoplasmic machinery, or the statistical nature of the observed micropotentials created in the extracellular space by synaptic activity or firing, cannot be taken into account in the field equations. A joined statistical formalism brings a solution.

From the integrative point of view considered here, properties of the hippocampus could be deduced from specific learning rules for "functional units" determined by the functional hierarchy, as it was done for cerebellum. For this reason, the first step is to determine the hierarchical functional organization of the hippocampus, and, simultaneously, the fields that are associated with each function at each level. The second step is to study the learning rules in the framework of the n-level field theory based on the principles found in the case of cerebellum, i.e., variational learning rules for the "functional units" at the higher level of the functional organization. Finally, these results will have to be included in the expression of the extracellular field potential. *This link with experiments will give an interpretation of the space and time propagation of activity in the hippocampal neural network, in terms of the lowest level, i.e., the molecular level.*

REFERENCES

BAHILL, A. T. & KARNAVAS, W. J. (1989). Determining ideal baseball bat weights using muscle force-velocity relationships. *Biological Cybernetics* **62**, 89-97.

BERGER, T. W., BARRIONUEVO, G., LEVITAN, S. P., KRIEGER, D. N. & SCLABASSI, R. J. (1991). Nonlinear systems analysis of network properties of the hippocampal formation. In *Neurocomputation and Learning:Foundations of Adaptive Networks*, eds. MOORE, J. W. & GABRIEL, M., pp. 283-352. Cambridge, MA: MIT Press.

BERGER, T. W. & BASSETT, J. L. (1992). System properties of the hippocampus. In *Learning and Memory: the Biological Substrates*, eds. GORMEZANO, I. & WASSERMAN, E. A., Hillsdale, NJ: Lawrence Erlbaum.

BRADLER, J. E. & BARRIONUEVO, G. (1989). Long-term potentiation in hippocampal CA3 neurons: tetanized input regulates heterosynaptic efficacy. *Synapse* **4**, 132-142.

BROWN, T. H., KAIRISS, E. W. & KEENAN, C. L. (1990). Hebbian synapses: biophysical mechanisms and algorithms. *Annual Review of Neuroscience* **13**, 475-511.

CHAUVET, G. A. (1993a). An n-level field theory of biological neural network. *Journal of Mathematical Biology* **31**, 771-795.

CHAUVET, G. A. (1993b). Hierarchical functional organization of formal biological systems: a dynamical approach. I. An increase of complexity by self-association increases the domain of stability of a biological system. *Philosophical Transactions of the Royal Society of London Series B-Biological Sciences* **339**, 425-444.

CHAUVET, G. A. (1993c). Hierarchical functional organization of formal biological systems: a dynamical approach. II. The concept of non-symmetry leads to a criterion of evolution deduced from an optimum principle of the (O-FBS) sub-system. *Philosophical Transactions of the Royal Society of London Series B-Biological Sciences* **339**, 445-461.

CHAUVET, G. A. (1993d). Hierarchical functional organization of formal biological systems: a dynamical approach. III. The concept of non-locality leads to a field theory describing the dynamics at each level of organization of the (D-FBS) sub-system. *Philosophical Transactions of the Royal Society of London Series B-Biological Sciences* **339**, 463-481.

CHAUVET, G. A. (1993e). Non-locality in biological systems results from hierarchy: Application to the nervous system. *Journal of Mathematical Biology* **31**, 475-486.

CHAUVET, G. A. (1995). On associative motor learning by the cerebellar cortex: from Purkinje unit to network with variational learning rules. *Mathematical Biosciences* **126**, 41-79.

CHAUVET, G. A. & BERGER, T. W. (1994). A hierarchical model derived from an n-level field theory to study the effects of long-term potentiation on system properties of the hippocampus. In *Long-Term Potentiation, Vol. 2*, eds. BAUDRY, M. & DAVIS, J. L., pp. 337-369. Cambridge: The MIT Press.

COHEN, N. J. & SQUIRE, L. R. (1980). Preserved learning and retention of pattern-analyzing skill in amnesia: dissociation of knowing how and knowing that. *Science* **210**, 207-210.

EICHENBAUM, H., WIENER, S. I., SHAPIRO, M. L. & COHEN, N. J. (1989). The organization of spatial coding in the hippocampus: a study of neural ensemble activity. *Journal of Neuroscience* **9**, 2764-2775.

FINKEL, L. H. & EDELMAN, G. M. (1985). Interaction of synaptic modification rules with populations of neurons.. *Proceedings of the National Academy of Sciences of the United States of America* **82**, 1291-1295.

GISZTER, S. F., MCINTYRE, J. & BIZZI, E. (1989). Kinematic strategies and sensorimotor transformations in the wiping movements of frogs. *Journal of Neurophysiology* **62**, 750-767.

HIGASHIMA, M. & YAMAMOTO, C. (1985). Two components of long-term potentiation in mossy fiber-induced excitation in hippocampus. *Experimental Neurology* **90**, 529-539.

HIRSH, R. (1974). The hippocampus and contextual retrieval of information from memory: A theory..
    Behavioral Biology **12**, 421-444.
KAAS, J. H. (1989). The evolution of complex sensory systems in mammals. *Journal of Experimental Biology*
    146, 165-176.
MORRIS, R. G. M., GARRUD, P. & RAWLINS, J. N. P. (1982). Place navigation impaired in rats with
    hippocampal lesions. *Nature* **297**, 681-683.
O'KEEFE, J. A. (1976). Place units in the hippocampus of the freely moving rat. *Experimental Neurology* **51**,
    78-109.
ROSS, R. T., ORR, W. B., HOLLAND, P. C. & BERGER, T. W. (1984). Hippocampectomy disrupts acquisition
    and retention of learned conditional responding. *Behavioral Neuroscience* **98**, 211-225.
WIENER, S. I., PAUL, C. A. & EICHENBAUM, H. (1989). Spatial and behavioral correlates of hippocampal
    neuronal activity. *Journal of Neuroscience* **9**, 2737-2763.
XIE, X., BERGER, T. W. & BARRIONUEVO, G. (1992). Isolated NMDA receptor-mediated synaptic
    responses express both LTP and LTD. *Journal of Neurophysiology* **67**, 1009-1013.

23

FUNCTIONAL CONNECTIONS BETWEEN THE ARCHITECTURE OF THE DENDRITIC ARBORIZATION AND THE MICROARCHITECTURE OF THE DENDRITIC MEMBRANE

P. Gogan,[1] Suzanne Tyč-Dumont,[1] S. M. Korogod,[2] and L. P. Savtchenko[1]

[1]CNRS UPR 9041, Unité de Neurocybernétique Cellulaire
280 Bd Sainte-Marguerite
13009 Marseille, France
[2]Laboratory of Biophysics and Bioelectronics
Dniepropetrovsk State University, Gagarin Ave 72
320625, GSP10, Dniepropetrovsk, Ukraine

1. INTRODUCTION

The living neuron is the building block of all nervous systems from the simplest one to the human brain. In contrast to other living cells which are of simple and regular geometrical shapes, most neurons display an extraordinary structural complexity of their dendritic arborization (Fig. 1). Together with the complexity, the immense variety of shapes characterize the neuronal elements of the brains. This unique feature of structural complexity has been acquired during a long evolutionary period lasting more than five hundred millions years. The intuition that the structure of a dendritic tree may reflect its functional complexity was early proposed at the end of the last century by Santiago Ramon y Cajal.[16] However, it was only recently that this "shape hypothesis" could be tested by morphometry analysis and simulations.

2. THE MACROSTRUCTURE OF THE NEURON

Modern neuromorphometry analysis based on computer-aided microscopes allowed to digitize neurons which were functionally identified in the brain.[1] It revealed that the dendritic arborization of a single neuron can span up to 1.5 to 2.0 millimeters away from the soma and constitutes more than 95% of its total receptive surface area upon which converge thousands of synaptic inputs. The dendrites display marked differences in diameters and

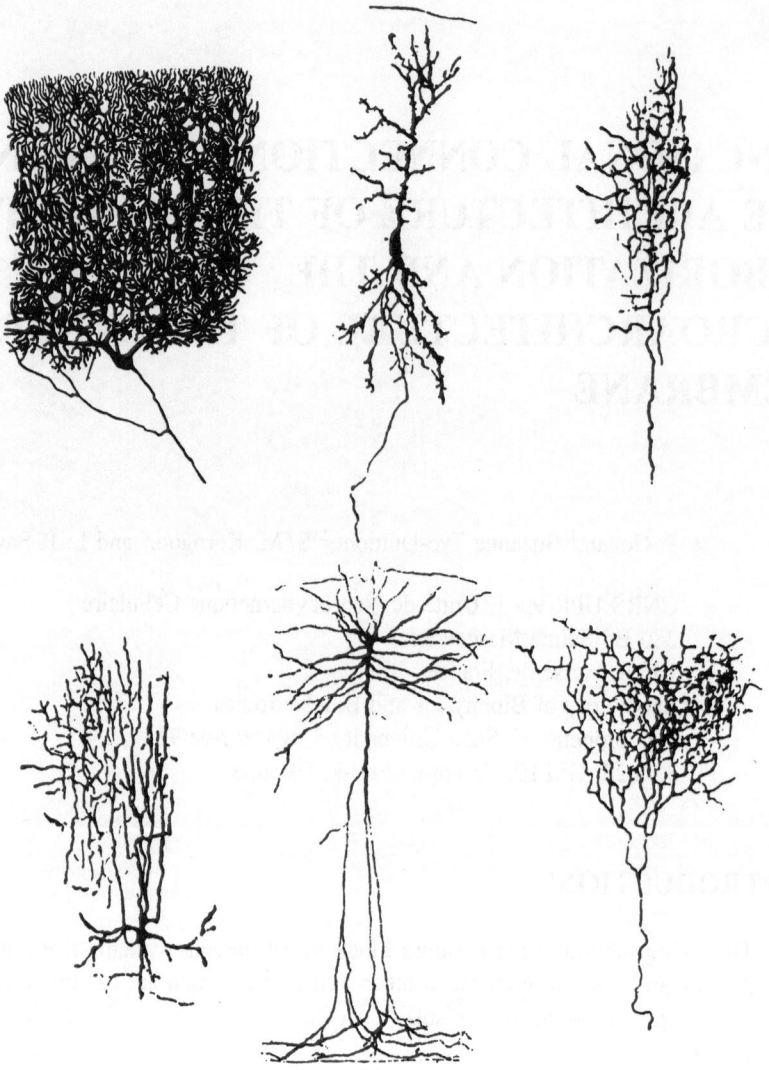

Figure 1. Examples of different shapes of dendritic arborizations of neurons observed in different regions of the brains of various mammals and drawn by Ramon y Cajal[16] about one hundred years ago.

a great diversity of branching patterns. The tapering in the dendritic segments varies greatly without clear relationship to the position of the segment in the dendritic arborization. Whether or not this geometrical heterogeneity is a manifestation of randomness or is the result of an underlying organization remains unknown. Indeed, these structural disparities must have dramatic effects on the channeling and processing of the synaptic inputs.[1,6]

2.1. The Electrotonic Structure of the Dendritic Arborization

The 3D geometry of an arborization directly determines its *electrotonic structure* which is defined by the voltage decay expressed as a function of the somatofugal paths in physical distances, i.e. the voltage gradient (Fig. 2). The passive spread of voltage and charge throughout the neuron depends on the variations in diameter and on the non-uniform branching arrangement.[1,2,13] It was recently demonstrated that the distribution of the charge

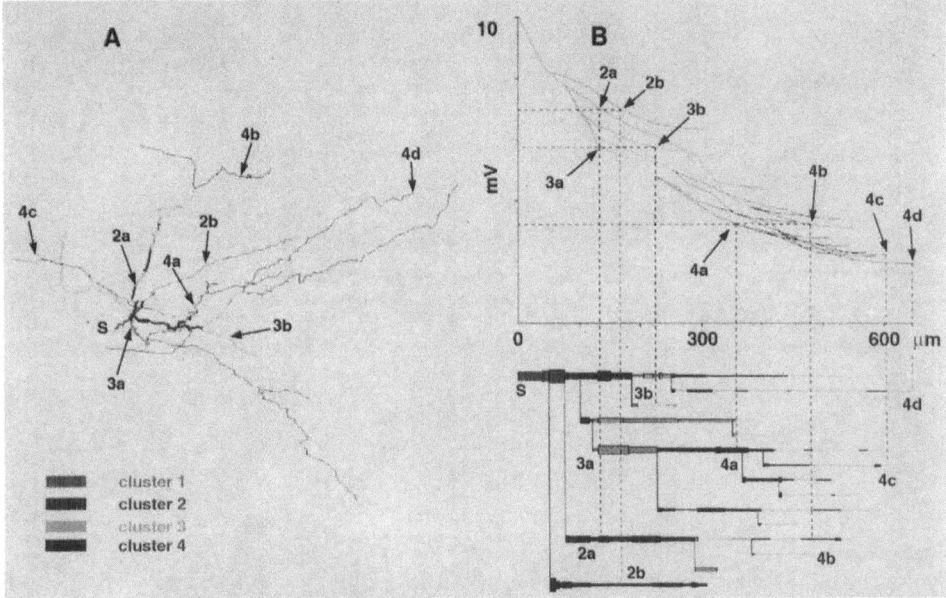

Figure 2. *A*: A single dendrite of a neuron intracellularly recorded, physiologically identified and stained in the brain stem of a rat. The neuron was digitized with a computer-aided microscope and reconstructed in three dimensions. The electrotonic macrostructure of the dendrite is identified by colors indicating the extension and the spatial distribution of the four electrotonic clusters of dendritic branches with similar electrotonic characteristics. *B*: Electrotonic profiles of the dendritic branches of the same dendrite. Following a 10 mV steady-state depolarization applied at the soma (0 μm), the voltage along the branches (ordinates) is expressed as a function of the physical distance from the soma in micrometers (abscissae). The 2D representation of the same dendrite is illustrated below with the path distance from soma in micrometers and the 4 clusters delineated by the same colors. Arrows indicate the simulated sites of recordings (a to d) and the numbers (2 to 4), the corresponding clusters (for details see Ref. 11).

transfer effectiveness factor is identical to the somatofugal distribution of the steady-state electrotonic voltage normalized to the voltage at the soma.[13] Therefore, the representation of the electrotonic structure by the somatofugal distribution of the steady electrotonic voltage over the dendritic tree is of general validity. It allows to evaluate simply and directly the somatopetal synaptic efficacy as well as the somatofugal back invasion.[1,2]

Simulation experiments have revealed that the *electrotonic structure* determines the very features most critical to the information processing functions. It governs the back invasion of the somatic action potentials into the dendritic arborization. Such invasion can shift the dendritic membrane potential over an extended part of the tree (Fig. 3). These membrane potential shifts are unequally distributed over the dendritic paths, thus affecting differentially the electrochemical gradients of postsynaptic currents and the voltage dependent dendritic conductances when they are present. Such heterogeneous changes in membrane conductances can be the source of subtle time dependent dendritic reconfigurations.[2]

Recent biological experiments performed in vitro have shown that such somatofugal dendritic invasion does occur in live neurons.[18,14] This feed-back mechanism confers to the neuron the capability of continuously adjusting and modifying the synaptic weights of its thousands of inputs and to differently modulate the local processing occurring in different regions of the tree.

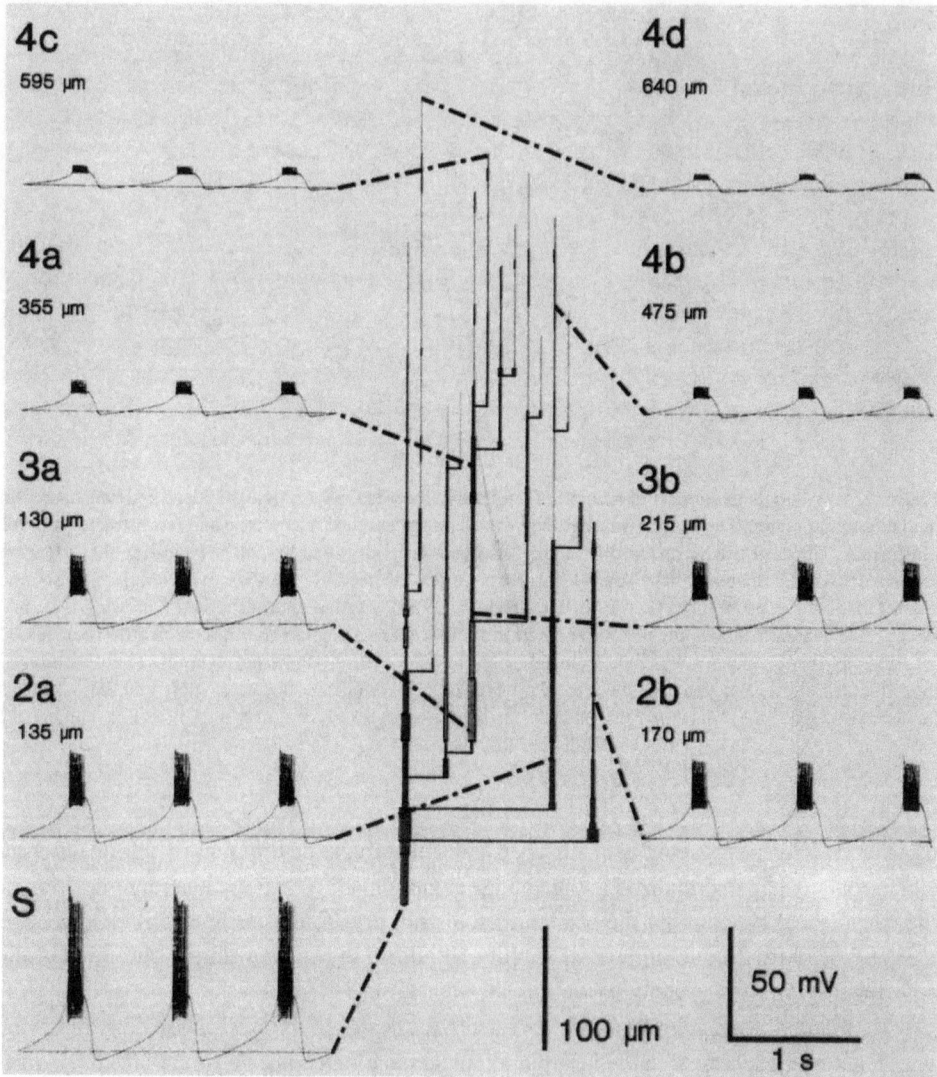

Figure 3. Simulation of the electrotonic invasion of the same dendrite as in Fig. 2 when the neuron discharges in bursts of action potentials produced by ionophoretic application of NMDA at the soma. In the simulation experiments, the typical oscillatory neuronal discharge produced by the soma (S) is shown to invade the whole dendrite in the somatofugal direction. This back-invasion occurs differentially according to the specific geometry of each dendritic branch. The voltage attenuation depends on the clusters. The slow oscillations and fast transients are differentially attenuated, but both contribute to local dendritic depolarization as the fast transients are "carried" on the top of the slow components.

2.2. Electrotonic Clusters

Statistical analysis of both 3D geometry and *electrotonic structure* reveals a functional organization of the dendrites, consisting of clusters of dendritic branches with similar electrotonic properties, expressed in terms of steady electrotonic voltages and gradients.[11] The branches belonging to a cluster may occupy completely different areas of the 3D space of the arborization (Fig. 2). This grouping of dendritic branches into a small number of clusters is preserved when the uniform specific resistance of the dendritic membrane is changed by distributed synaptic bombardment. Thus, the functional properties of the dendrites are stabilized by their 3D geometry whatever the fluctuations of the inputs. However, some dendritic branches (10 to 30 %) may "travel" from one cluster to the next upon the effects of the distributed inputs. This produces a partial reconfiguration of the dendritic *electrotonic structure*, creating conditions for postsynaptic plasticity.[11] The demonstration of the clusters suggests that the processing of synaptic inputs might be more organized than at first implied by the complexity of the electrotonic structure of the neuron. The post-synaptic membranes of all the branches in a given cluster may operate similarly in terms of voltages and gradients. The synapses located in the same space domain (path distance or 3D) may be processed similarly or differently according to the cluster to which the post-synaptic sites are assigned.

3. THE MICROSTRUCTURE OF THE NEURON

The *macrostructure* is made of a highly organized *molecular microstructure* in which the dendritic membrane constitutes an interface where computational elements such as proteins are spatially distributed. Protein molecules are able to perform a variety of computational operations providing a toolkit of components from which the functional circuits are built in the dendritic membrane.

To address the problem of the structural organization of the dendrites at the cellular and molecular levels, we have recently developed a new tool which combines the use of voltage-sensitive dyes and a liquid-nitrogen-cooled astronomical camera.[7] The membranes of live neurons in culture are stained with the fluorescent styril voltage-sensitive dye RH237. The dye acts as a molecular transducer, transforming changes in membrane potential into optical signals. In our conditions, the styril dye, acting as a molecular electric field sensor, explores the spatial distribution of the local electrical events occurring in the membrane of excited neurons on the microscopic scale. An imaging system based on a liquid-nitrogen-cooled CCD astronomical camera, mounted on an inverted microscope, allows to detect and record small changes in the fluorescence of the voltage-sensitive dye. We obtain high spatial resolution ($1~\mu m^2$/pixel) images captured in synchrony with conventional intracellular microelectrode recordings.

Results showed that small fluorescence changes occurred in localized, unevenly distributed membrane areas during an action potential. The images displayed clusters of depolarized sites of different sizes and intensities (Fig. 4). When fast conductances were blocked by tetrodotoxin, the number and intensities of the depolarized sites were reduced both in current and voltage-clamp conditions. The number of depolarized sites varied also according to the degree of activation of the voltage-sensitive channels. The patterned membrane topography changed from trial to trial while similar action potentials were recorded by a conventional intracellular microelectrode.

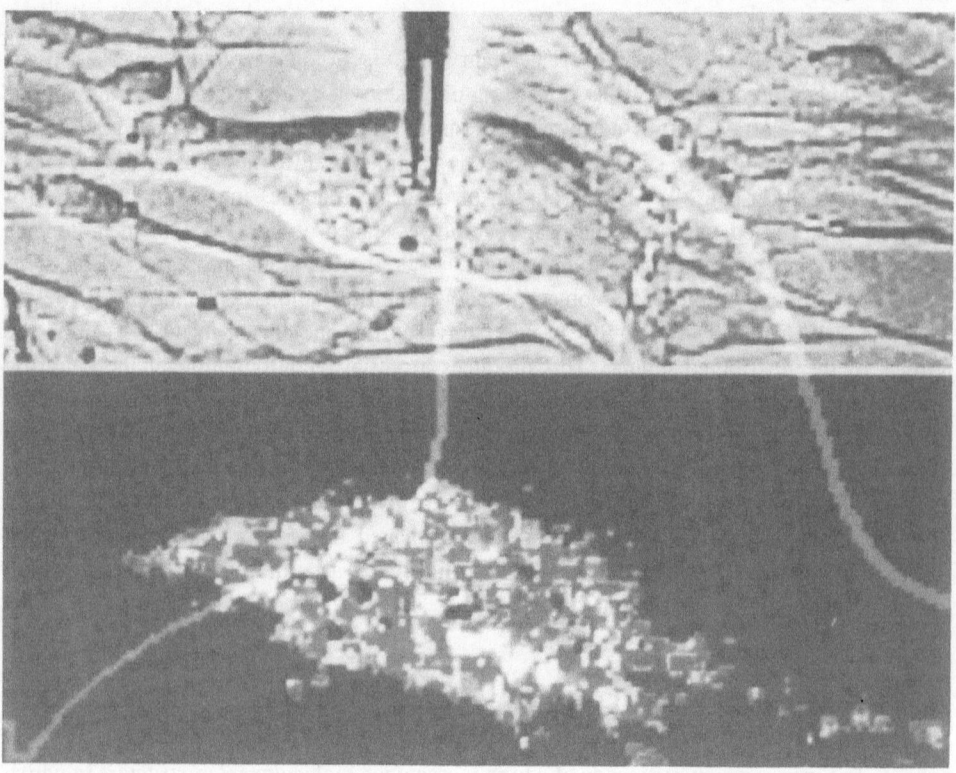

Figure 4. Spatial distribution of membrane depolarization sites in a single neuron during an action potential. *Upper part*: CCD image (DIC optics) of a nodose ganglion neuron in culture, impaled with an intracellular microelectrode. *Lower part*: After staining with the voltage sensitive dye RH237, processed dF/F image of the same neuron during its excitation by an intracellular step depolarizing current (0.5 nA, 20 ms). Every pixel (1 μm^2) in the image integrated the changes in fluorescence which occurred during the action potential shown superimposed. The image processing reveals an heterogeneous spatial distribution of membrane depolarization sites (for details see Ref. 7). Depolarization is increasing from blue to red.

We conclude that the spatial patterns of depolarization observed during excitation on the microscopic scale are related to the operation of ionic channels in the membrane. The variability of the spatial patterns is not in contradiction with the fact that single-site time recorded macroscopic electrical events remain similar. Recent studies have revealed that global coupling across large spatial membrane areas enables clusters of stochastic ion channels to produce stable macroscopic electrical events such as resting potential and action potentials.[3,4] During excitation, a given number of channels must open but, in view of the stochastic behavior of channels, there is no reason to expect the same channels located in the same membrane region to open from trial to trial.

3.1. Theoretical Comments

Changes in fluorescence imaged by averaging over the duration of the action potential revealed an unexpected patchyness. "Hot" patches were intermingled with "cold" patches over the cell membrane and this spatial pattern changed randomly from trial to trial in a kaleidoscopic manner. One possible interpretation can be proposed on the basis of mechanisms discussed by De Felice and Isaac in 1992.[3] "Cold" state of the whole membrane at rest and occurrence of clusters of "hot" and "cold" patches during excitation reflect strong activity-induced coupling of the channels within and between "hot" patches, whereas "cold" patches remain discoupled. Cooperative behavior of the channels when synchronized enough to generate the action potential, is thought to appear when open channels are present in sufficient number and coupled by shared transmembrane voltage. Concentrations of ions carrying transmembrane current could be another coupling factor. If the concentration of a given type of ions is shared by a population of channels, this means also sharing the equilibrium and driving potentials of the ionic current through these channels. Thus, "cold" patches would be uncoupled if: (i) the number of channels was insufficient for eliciting the action potential; (ii) the channels conducting inward and outward currents were present in an improper proportion, distorting the local conditions necessary for generating the action potential; (iii) the inward-outward current relationship is distorted by improper driving potentials of the component currents as a result of change in the local ion concentrations. However, these possible explanations for uncoupling some patches from other coupled patches do not explain how the stochastic variations of the pattern occur during excitation. Mechanisms are needed for maintaining uneven distributions of channels and/or ion concentrations in the nearby intra- and extracellular space.

We have recently studied conditions of generation and maintenance of spatially heterogeneous quasi-steady distribution of the density of ion channels in the membrane of a simulated neuron.[12,17] Our simulations suggest that self-organization phenomena could be the relevant mechanism responsible for heterogeneous spatial patterns of channels in the membrane and ion concentrations in the sub-membrane cytoplasm. These phenomena are similar to those observed in other stochastic non-linear systems far from thermodynamic equilibrium.[8] The neuronal membrane is a non-linear, thermodynamically open system which participates in energy and substance exchanges. Uneven distributions of channels and ions can be maintained at the expense of chemical energy. It is also known that channels may exist in the membrane in both immobile and mobile states.[9] One possible scenario of self-organization is based on the assumption that macromolecules of mobile channels bear electrophoretic charges and are able to diffuse and migrate in the tangential gradients of voltage and channel surface density.[10,5] Electrophoretic charge of the channel protein is composed of uncompensated charges of amino-acid residuals exposed to the intra- and extracellular spaces. If the mobile channels of inward current have negative electrophoretic charges exposed to the intracellular space and/or positive electrophoretic charge exposed to the extracellular space, then the local depolarizing fluctuation of the transmembrane voltage, attracting extracellular cations and repulsing intracellular ones, would also attract the charged channel molecules. This elicits condensation of the channels into a cluster. Thermal motion and coulombian repulsion oppose to this attraction. Inactivation of the channels also reduces inward current thus decreasing the tangential voltage gradients and weakening the attractive forces. A faster mechanism for such lateral pattern formation could be a redistribution of the ions leading to self-maintained heterogeneous concentration profiles, thus defining restricted areas of the submembrane space where cooperative behavior of the channels could take place.

4. DISCUSSION

The "shape hypothesis" assumes that the processing is determined by the local geometry of all dendritic branches in the arborization, the types of proteins embedded in the dendritic membrane, their spatial organization and their dynamic behavior. Both the macrostructure and the microstructure confer to the neuron its functional status as a highly complex information processing element. During the last decades, the discovery of patch-clamp technique[15] has opened a new conceptual field which is focused on the molecular elements, revealing a great variety of ionic channels, pumps and receptors which constitute the basic mechanism for interfacing all living cells with their environment. But only neurons display such an immense variety of dendritic shapes. This unique structural attribute makes the neuron a highly specialized functional unit capable of processing complex information leading to behavior, memory and cognition. The challenge is to discover the rules of the connection between the microstructure which is made of stochastic elements which open and close in a random way and the macrostructure of single neurons which create the condition for collective behavior of the molecular kits to generate the well known electrical events in the neuron.

REFERENCES

1. Bras H, Gogan P, Tyč-Dumont S(1987) The dendrites of single brain-stem motoneurones intracellularly labelled with HRP in the cat. Morphological and electrophysiological differences. Neuroscience 22:947–970
2. Bras H, Korogod SM, Driencourt Y, Gogan P, Tyč-Dumont S(1993) Stochastic geometry and electrotonic architecture of dendritic arborization of a brain stem motoneuron. Eur. J. Neurosci. 5:1485–1493
3. DeFelice LJ, Isaac A(1992) Chaotic states in a random world: relationship between the nonlinear differential equations of excitability and the stochastic properties of ion channels. J. Stat. Phys. 70:339–354
4. Fox RF, Lu YN(1994) Emergent collective behavior in large numbers of globally coupled independently stochastic ion channels. Phys. Rev. 24:5749–5755
5. Fromherz P(1989) Self organization of a membrane in synaptic geometry. Biochem. Biophys. Act. 986:341–345
6. Gogan P, Tyč-Dumont S(1989) How do dendrites process neural information? News in Physiol. Sci. 4:127–130
7. Gogan P, Schmiedel-Jakob I, Chitti Y, Tyč-Dumont S(1995)Fluorescence imaging of local membrane electric fields during the excitation of single neurons in culture. Biophys. J. 69:299–310
8. Gransdorff P, Prigogine I(1971) Thermodynamic Theory of Structure, Stability, and Fluctuations. Wiley-Interscience, London.
9. Hille B(1992) Ionic Channels of Excitable Membranes. 2nd Ed. Sinauer Associates Inc., Sunderland, MA.
10. Jaffe LF(1977) Electrophoresis along cell membranes. Nature 265:600–602
11. Korogod SM, Bras H, Sarana VN, Gogan P, Tyč-Dumont S(1994) Electrotonic clusters in the dendritic arborization of abducens motoneurons of the rat. Eur. J. Neurosci. 6:1517–1527
12. Korogod SM, Savtchenko LP(1994) Formation conditions of the spatially periodical quasi-stationary distribution of density of open ion channels in the membrane. Mathematical Modeling (Moscow) 6(8):33–44
13. Korogod SM(1995) Electro-geometrical coupling in non-uniform branching dendrites. Biol. Cyb. (in press)
14. Larkum ME, Rioult MG, Lüscher HR(1995) Propagation of action potentials in the dendrites of neurons from rat spinal cord slice cultures. J. Physiol. (in press)
15. Neher E, Sackman B(1976) Single-channel currents recorded from membrane of denervated frog muscle fibres. Nature 260:799–802
16. Ramon y Cajal S(1909–1911) Histologie du Système Nerveux de l'Homme et des Vertébrés. Maloine, Paris.
17. Savtchenko LP, Korogod SM(1995) Domains of calcium channels as dissipative structures in the simulated neuron. Neurophysiology (Kiev) 26(2):99–107
18. Stuart GJ, Sackman B(1994) Active propagation of somatic action potentials into neocortical pyramidal cell dendrites. Nature 367:69–72

24

THE FUNCTIONAL SIGNIFICANCE OF CEREBELLAR ANATOMY

Theory and Experiments

Detlef H. Heck[*]

Max-Planck-Institute for Biological Cybernetics
Spemannstr. 38, 72076 Tuebingen, Germany

1. INTRODUCTION

The cortex of the cerebellum is made up by only a few different types of neurons which are arranged in a highly regular and geometrical way. Because of its unusual regularity the cerebellar cortex has attracted many theoreticians and experimentalists, and research on the cerebellum has now been done for more than a hundred years. As a result, the cerebellum is today in many respects one of the best described parts of the vertebrate brain. Its anatomical structure has been well known since the beginning of the century, electrophysiology started in the sixties (Eccles *et al.* 1967) and was continued since then with increasing intensity. In spite of that, however, we are not yet able to relate the anatomical structure and the neuronal activity of the cerebellum to what is widely considered its main function: the control of posture and movement.

Numerous attempts have been made to figure out what the functional meaning of the peculiar cerebellar anatomy might be. Some of the theories on the cerebellum consider anatomical aspects of the cerebellar cortex in great detail (e.g. Gilbert, 1975). However, although sign and direction of the neuronal connections are almost always incorporated in the various theories, questions concerning the geometry of the structure like: "Why are the parallel fibers parallel?" or "Why are the dendritic trees of Purkinje cells, basket cells and stellate cells flat?" remain unanswered.

The aim of the present paper is firstly to discuss a theory of cerebellar cortex which offers an answer to the before mentioned questions (Braitenberg & Atwood, 1958; Braitenberg, 1983) and secondly to discuss experiments which strongly support that theory (Heck, 1993a; Heck, 1993b; Heck & Braitenberg, 1993; Heck, 1995a; Heck, 1995c). It will be suggested that the cerebellum detects specific spatio-temporal events in the mossy fiber input which correspond to specific events in the motor performance. In this scenario the parallel

---

[*] Present address: Washington University School of Medicine, Department of Anatomy and Neurobiology, Campus Box 8108, 660 S. Euclid Ave., St.Louis, Missouri 63110-1031.

fibers serve as delay lines and the Purkinje cells are sensitive coincidence detectors. Experimental evidence for the suggested role of parallel fibers and Purkinje cells will be discussed.

2. SOME GENERAL CONSIDERATIONS ON CEREBELLAR STRUCTURE AND FUNCTION

The principles of the cerebellar cortex wiring diagram have already been described by the Spanish neuroanatomist Ramon y Cajal in 1911. In contrast to the situation in the neo-cortex, where axons and dendrites do not follow a preferred direction but rather grow out in a seemingly random manner (Braitenberg, 1976; Braitenberg, 1978; Braitenberg & Schüz, 1991), neurons in the cerebellar cortex send their processes in one out of only three different directions: 1) dendrites all grow in a tangential direction towards the pial surface; 2) all excitatory fibers i.e., the parallel fibers, run in parallel to the pial surface in a latero-lateral direction; 3) inhibitory fibers also grow parallel to the pial surface but tangentially to the direction of the excitatory ones. An overview of the anatomy of the cerebellar cortex is given in Figure 1.

There are some more peculiarities about the cerebellar anatomy which should be mentioned since they are relevant for a functional interpretation: the dendritic trees of Purkinje cells and of two types of inhibitory interneurons named basket and stellate cells are flat and fan-shaped. They lie in a plane which is orthogonal to the direction of parallel fiber growth. The number of parallel fiber synapses on a single Purkinje cell is very high compared to the number of synapses on other neurons. (Estimations range up to 200.000 parallel fiber

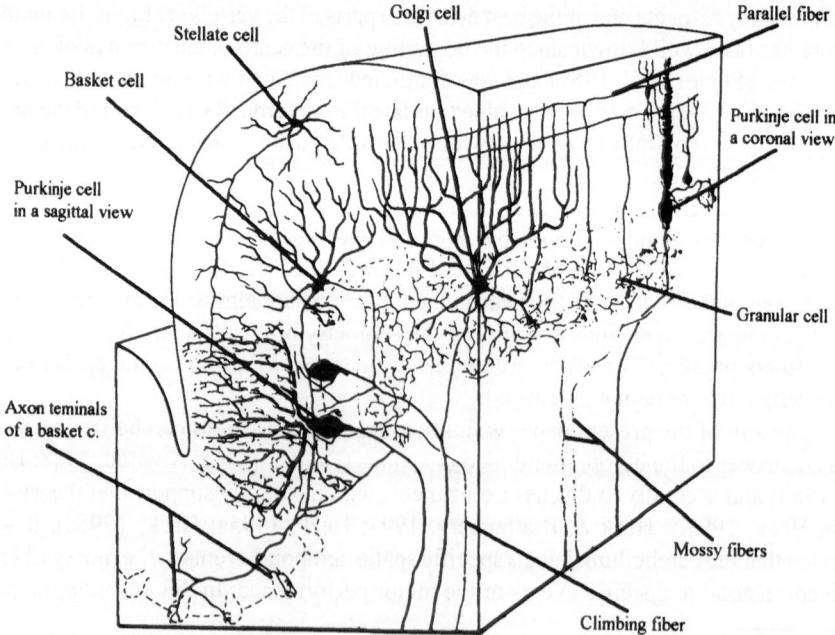

Figure 1. A three dimensional diagram of a part of the cerebellar cortex. The five major cell types (Purkinje, Golgi, stellate, basket and granular cells), the two types of input fibers (mossy and climbing fibers) are shown. The only output fiber is the Purkinje cell axon. The dotted lines in the lower left part indicate Purkinje cell somata. (Based on figure 1 in (Eccles *et al.* 1967); all changes made by the author).

synapses on a single Purkinje cell (Napper & Harvey, 1988) compared to about 10.000 to 20.000 synaptic inputs on neo-cortical or hippocampal pyramidal neurons (Braitenberg & Schüz, 1991; Trommald *et al.* 1995)). Besides the parallel fiber input Purkinje cells get a second kind of excitatory input which comes from the so called climbing fibers. Each Purkinje cell is contacted by only one single climbing fiber which, however, establishes several hundred synapses on the spineless dendritic shafts of the Purkinje cell (Llinás *et al.* 1969). Due to the high number of synapses each climbing fiber spike has a strong effect on the postsynaptic Purkinje cell (Granit & Phillips, 1956; Eccles *et al.* 1966). A single spike in a climbing fiber results in a strong and long-lasting depolarization of the Purkinje cell which has been termed a 'complex spike' (Thach, 1968; Thach, 1970) and which is mainly due to calcium influx into the dendritic tree (Ekerot & Oscarsson, 1981).

Interestingly, the cerebellum is also one of the oldest parts of the vertebrate brain and the basic principle of its wiring diagram has been preserved during the course of the more than 400 million years of vertebrate evolution.

With the presumption that there is information about function in structure, restrictions to a functional interpretation are most clearly defined in the cerebellar cortex. In order to give a functional interpretation of the structure it is necessary to define which structural details are considered as functionally relevant and consequently need to be explained in functional terms. The specific information processing going on in the cerebellar cortex must be related to its unique structural features. There are several such features distinguishing the cerebellar cortex form all other parts of the brain. The most obvious ones are the parallel fiber system and the fan-like geometry of the dendritic trees of Purkinje, stellate and basket cells. Ten features are listed below which I consider specific for the cerebellar cortex and which thus must find an explanation through theory (see also Braitenberg & Atwood, 1958; Braitenberg, 1967). Data which are still subject of current debate like, e.g., the somatotopic maps are disregarded. I have no doubt that the list is incomplete but also there should be no element in it which does not belong there.

1. The only excitatory axons in the cerebellar cortex, the so called parallel fibers, all run strictly in parallel with the latero-lateral direction.

2. The dendritic trees of Purkinje, stellate and basket cells are flat and fan-shaped. They spread in a plane orthogonal to the direction of parallel fiber outgrowth.

3. The number of granule cells, the axons of which are the parallel fibers, is very high. (Half of the nerve cells in the brain are granule cells.) With respect to the mossy fibers which are their only excitatory input there are 600 times more granule cells.

4. The cerebellar cortex is folded in a way such that the parallel fibers are never bent.

5. The axons of the inhibitory interneurons in the molecular layer grow out in a direction perpendicular to the direction of the parallel fibers.

6. There are anatomical and biochemical data showing that there is an overlying sagittal organization of the cerebellar cortex.

7. Every single Purkinje cell receives an unusually strong excitatory input from only one axon, the so-called climbing fiber.

8. The mossy fibers form a special type of synapse in the granular layer, the so-called glomeruli. In every glomerulus a mossy fiber contacts several dozens of granule cells. These dense synaptic structures additionally receive inhibitory axonal terminals from Golgi cells.

9. The only output of the cerebellar cortex is via the axons of the Purkinje cells, which are inhibitory.

10. Purkinje cells are spontaneously active at an average frequency of 50 Hz.

In one of the earliest theories trying to link structure and function of the cerebellar cortex it has been suggested that the parallel fibers might be interpreted as delay lines (Braitenberg & Atwood, 1958; Braitenberg, 1961). The basic idea of the authors was, that the cerebellum could produce time delays in the order of hundreds of milliseconds and that these delays would be used to control the precise timing required for the activation of muscles during movement. This 'timing theory' for the first time offered an explanation for the characteristic geometry of cerebellar cortical connectivity in that it explained why the parallel fibers had to run in parallel. In the course of the following years it was found, however, that the delays which could possibly be produced by the parallel fibers were much shorter than Braitenberg assumed in his 1961 paper. Parallel fibers have been shown to conduct spikes at a velocity of about 0.5 m/s (Eccles *et al.* 1967). Their length seems to vary between species but lies between 1 and 10 mm, so that each single branch can be maximally 5 mm long. Thus, the longest delay that might be produced by the parallel fibers is only about 10 ms.

Yet, although this idea was based on false assumptions about parameters unknown at that time, i.e., the length of the parallel fibers and their conductance velocity, the timing-hypothesis survived until today. A reason for that might be that there are experimental data which can be interpreted in support of the assumption that the cerebellum is involved in timing of muscle activity (e.g. Hore *et al.* 1991) or even in the explicit measurement of time intervals (e.g. Ivry, 1993). All these tasks would require the cerebellum to produce or measure time delays in the order of a hundred milliseconds or more. Apart from the fact that a correlation between timing disorders and cerebellar damage does not necessarily imply that the cerebellum is a time-measuring device, authors assigning a timing function to the cerebellum do not provide a solution to the above mentioned problem of too short delay lines.

3. THE TIDAL WAVE THEORY

A very elegant explanation of cerebellar anatomy is offered by the so-called tidal wave theory (Braitenberg, 1983; Braitenberg, 1987). It is a logical extension of the timing idea in that the parallel fibers are still interpreted as delay lines, but here they serve a different purpose and also produce much shorter delays. Basically, the parallel fibers transform sequential mossy fiber input into a synchronous synaptic input to Purkinje cells (Fig. 2). If the granule cells along a 'beam' of parallel fibers are sequentially excited by mossy fiber input, several parallel fiber spikes would be elicited. These spikes would all lie in a single plane parallel to the plane of Purkinje cell dendritic trees if the sequence of mossy fiber input was adequately ordered in space and time. This means that the mossy fibers would fire in a serial order such that the excitation seemingly 'moves' in a direction parallel to the parallel fibers and that the speed of the 'movement' would be the same as the speed of spike conductance in parallel fibers.

4. *IN VITRO* EXPERIMENTS

All experiments mentioned here were done in acute slices prepared from rat or guinea pig cerebellar cortices (Heck, 1993b; Heck, 1993a; Heck & Braitenberg, 1993; Heck, 1995a; Heck, 1995c). A detailed description of the experimental setup and the technical realization of the sequential stimulation is given in (Heck, 1995b).

The site of mossy fiber input, i.e., the granular layer in the slices, was stimulated with an array of 11 stimulating electrodes (Fig. 3). The responses of the parallel fibers electrodes

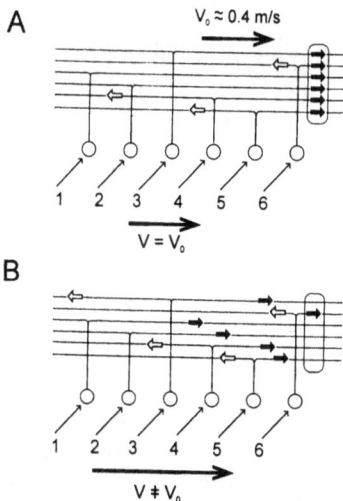

Figure 2. Schematic illustration of the mechanism by which a sequential mossy fiber input can lead to the build-up of a tidal wave of spikes in the parallel fiber system. Open circles indicate granule cells and their axons, the parallel fibers, which bifurcate in the molecular layer. Black and white arrows represent action potentials traveling to the right or left, respectively. The box on the right in A) and B) symbolizes the dendritic tree of a Purkinje cell. The arrow above the parallel fibers in A) indicates the conductance velocity in parallel fibers. In both figures A) and B) the effect of a 'moving' mossy fiber input is shown. In both cases the sequential stimulus 'moves' from position 1 to position 6, i.e. from left to right. In A) the stimulus moves exactly at the speed of conductance in parallel fibers. In this case, each new input excites a granular cell at exactly the time when the previously elicited spikes, which are already traveling in the molecular layer, reach the position of the newly excited granular cell. Only the spikes traveling in the direction of the stimulus 'movement', i.e. the black arrows, contribute to the tidal wave. As a result, when the stimulus finally reaches position 6 (i.e. excites the granule cell at position 6) all 6 spikes traveling in the direction of the stimulus 'movement' are gathered in the same plane and arrive synchronously at the Purkinje cell dendrites along their way. In the example shown in B) the velocity of the stimulus is higher than the velocity of spike conductance in parallel fibers. When the stimulus reaches the last granule cell at position 6, the spikes are in a position as shown. In this case the same number of spikes as in A) will arrive at the dendrite of the Purkinje cell but now asynchronously. The same is true for velocities slower than the velocity of spike conductance in parallel fibers.

to different input patterns were recorded with extracellular electrodes. The responses of Purkinje cells were recorded with sharp intracellular electrodes sticking in the soma or the proximal dendrite of the cell.

In a series of in vitro experiments predictions of the tidal wave theory have been confirmed: ordered sequential stimuli in the granular layer produce a population spike in the parallel fiber system, the amplitude of which is dependent on the 'direction' and on the velocity of the stimulus 'movement' (Fig. 4) (Heck, 1993b; Heck, 1993a).

As the distance covered by the movement is increased the amplitude of the parallel fiber population spike increases until it reaches a saturation for distances larger than about 1 mm (Heck & Braitenberg, 1993). Finally, and most important, the output neurons of the cerebellar cortex, the Purkinje cells, react specifically to sequential stimulation of the granular layer. Their responses reflect the mass activity of the parallel fibers and thus signal the occurrence of a sequential input to the granular layer (Fig. 5) (Heck, 1994; Heck, 1995b; Heck, 1995c). This clearly shows that Purkinje cells may be excited to supra-threshold voltage values by pure parallel fiber input, i.e., without synaptic input from the radial or rising parts of the granule cells axons, which other authors state to be the only source for supra-threshold excitation to Purkinje cells (Llinás, 1982; Bower & Woolston, 1983; Llinás & Sugimori, 1992).

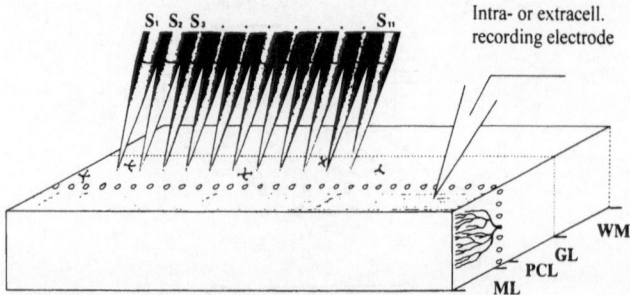

Figure 3. Schematic illustration of the *in vitro* experimental situation. The box symbolizes a part of a cerebellar cortex slice. A comb of 11 stimulating electrodes (S_1-S_{11}) sticks in the granular layer of the slice. The distance between two electrodes is 130-140μm. By sequentially switching the stimulus current from one electrode to the next, a 'movement' of the stimulus could be simulated. The direction of stimulus 'movement' can be inverted by inverting the electrode sequence. Different 'velocities' are generated by altering the time interval between consecutive stimuli. During both extra- and intracellular experiments the responses to stimulation with different 'velocities' and the two different directions were recorded. (S_1-S_{11} stimulating electrodes, ML molecular layer, PCL Purkinje cell layer, GL granular layer, WM white matter).

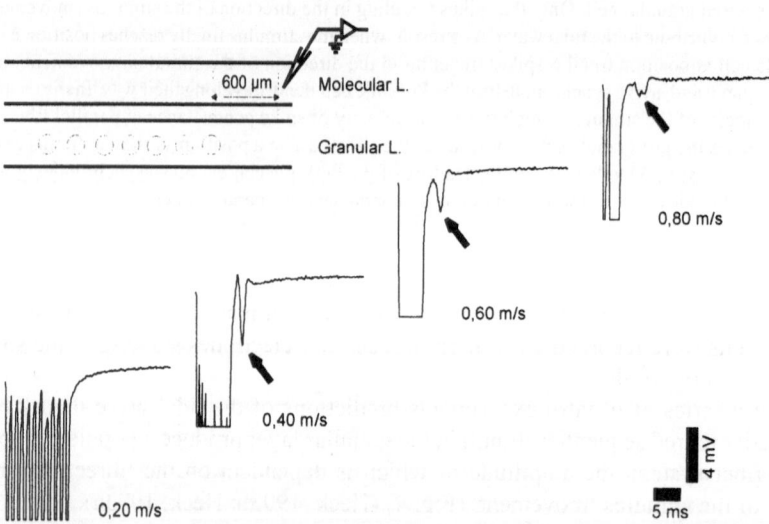

Figure 4. Extracellularly measured mass parallel fiber activity in response to stimulation of the granular layer with stimuli 'moving' at different velocities. The upper left inset schematically shows the experimental situation. The recording electrode was positioned in the molecular layer at a distance of about 600μm from the comb of stimulating electrodes. The open circles symbolize stimulating electrodes. The four traces shown are the responses measured at 'velocities' of 0.2, 0.4, 0.6 and 0.8 m/s. The initial downward deflections in every trace, which become narrower (i.e. shorter in time) with increasing velocity, are the 11 stimulus artifacts. The extracellular responses appear as negative deflections following the stimulus artifacts and are marked with the black arrows. There was now measurable response at all with a stimulus 'velocity' of 0.2 m/s. The response was maximal for 0.4 m/s, which is about the conductance velocity in parallel fibers. For higher velocities the response amplitude decreases again.

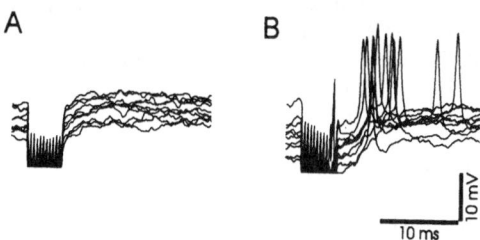

Figure 5. The responses of a guinea pig's Purkinje cell to sequential stimulation of the granular layer with stimuli 'moving' in two different directions. A superposition of 10 different trials is shown. In both cases the stimulus 'moved' at about the speed of conductance in parallel fibers (0.35 m/s). As in Fig. 4 the initial negative deflection is the stimulus artifact. A) The stimulus 'moved' away from the recorded Purkinje cell. The cell showed no measurable response to none of the stimuli. B) Response of the same cell to the same stimulus now only moving in the direction towards the Purkinje cell. In this case the cell responded with a spike to each of the 10 stimuli. In A) as well as in B) the same number of spikes should be elicited and arrive at the dendritic tree of the Purkinje cell. The only difference is that they arrive asynchonously at the Purkinje cell dendrite in A) and synchronously in B).

5. DISCUSSION

We have argued in this paper that the highly regular arrangement of neuronal elements in the cerebellar cortex strongly reduces the number of possible functions which may be performed by that structure. A key to cerebellar function is provided by those features which are unique for the cerebellar cortex. Outstanding among the cerebellar-specific features are the parallelism of the parallel fibers in combination with the flat and fan-shaped dendritic tree of the Purkinje cells. In the tidal wave theory these features find a convincing explanation: 1) parallel fibers have to be parallel in order to preserve the time relation between separate inputs; 2) the dendritic trees of Purkinje cells are flat because the interesting event that has to be detected is the formation of a tidal wave of parallel fiber spikes where many spikes group together in a single plane parallel to the plane of Purkinje cell dendrites. Experiments in slices of the cerebellum have shown that the cerebellar network responds to sequential stimuli exactly as predicted by the tidal wave theory. Still we do not understand what the cerebellum really does but the tidal wave theory together with the supporting experimental results provide a new perspective: the application of dynamic instead of static inputs has revealed properties of the cerebellar cortex hitherto unknown. We consider sequential inputs as a key to cerebellar function. Even if this view would prove to be false, it is very likely, however, that we will gain important new insights by further pursuing the investigation of dynamic processes in the cerebellum.

REFERENCES

BOWER, J. M. & WOOLSTON, D. C. (1983). Congruence of Spatial Organization of Tactile Projections to Granule Cell and Purkinje Cell Layers of Cerebellar Hemispheres of the Albino Rat: Vertical Organization of Cerebellar Cortex. *J.Neurophysiol.* **49**, 745-766.

BRAITENBERG, V. (1961). Functional Interpretation of Cerebellar Histology. *Nature* **190**, 539-540.

BRAITENBERG, V. (1967). Is the Cerebellar Cortex a Biological Clock in the Millisecond Range? In *Progress in Brain Research. Vol. 25. The Cerebellum*, eds. FOX, C. A. & SNIDER, R. S. pp. 334-346. Amsterdam: Elsevier.

BRAITENBERG, V. (1976). Real Neural Networks. In *Perspectives in Brain Research*, eds. CORNER, M. A. & SWAAB, D. F. pp. 197-205. Amsterdam: Elsevier.

BRAITENBERG, V. (1978). Cortical Architectonics: General and Areal. In *Architectonics of the Cerebral Cortex*, 443-465. New York: Raven Press.

BRAITENBERG, V. (1983). The cerebellum revisited. *Journal of Theoretical Neurobiology* 2, 237-241.

BRAITENBERG, V. (1987). The cerebellum and the physics of movement: some speculations. In *Cerebellum and neuronal plasticity*, eds. GLICKSTEIN, M., YEO, C. & STEIN, J. pp. 193-207. New York: Plenum.

BRAITENBERG, V. & ATWOOD, R. P. (1958). Morphological observations on the cerebellar cortex. *J.Comp.Neurol.* **109**, 1-33.

BRAITENBERG, V. & SCHüZ, A. (1991). *Anatomy of the Cortex. Statistics and Geometry*. Berlin,Heidelberg, New York: Springer Verlag.

ECCLES, J. C., LLINÁS, R. & SASAKI, K. (1966). The excitatory synaptic action of climbing fibres on the Purkinje cells of the cerebellum. *J.Physiol.(Lond.)* **182**, 268-296.

ECCLES, J. C., ITO, M. & SZENTÁGOTHAI, J. (1967). *The Cerebellum as a Neuronal Machine*. Berlin: Springer.

EKEROT, C.-F. & OSCARSSON, O. (1981). Prolonged depolarization elicited in Purkinje cell dendrites by climbing fiber impulses in the cat. *J.Physiol.(Lond.)* **318**, 207-221.

GILBERT, P. F. C. (1975). How the cerebellum could memorize movements. *Nature* **254**, 688-689.

GRANIT, R. & PHILLIPS, C. G. (1956). Excitatory and inhibitory processes acting upon individual Purkinje cells of the cerebellum in cats. *J.Physiol.(Lond.)* **133**, 520-547.

HECK, D. (1993a). Rat cerebellar cortex *in vitro* responds specifically to moving stimuli. *Neurosci.Lett.* **157**, 95-98.

HECK, D. (1993b). Specific responses of the cerebellar cortex to moving stimuli. In *Brain Theory: Spacio-Temporal Aspects of Brain Function*, ed. AERTSEN, A. M. H. J. pp. 127-130. Amsterdam: Elsevier.

HECK, D. (1994). Making sense of parallel fibers: further experimental evidence for the interpretation of the cerebellum as a sequence addressable memory. *Proceedings of the 22th Göttingen Neurobiology Conference* 565(Abstract)

HECK, D. (1995a). *Die Bedeutung raum-zeitlicher Dynamik für die Aktivität des Kleinhirnkortex und die Interpretation seiner Anatomie*. Hamburg: Verlag Dr. Kovac.

HECK, D. (1995b). Investigating dynamic aspects of brain function in slice preparations: Spatiotemporal stimulus patterns generated with an easy to build multi-electrode array. *J.Neurosci.Methods* **58**, 81-87.

HECK, D. (1995c). Sequential input to guinea pig cerebellar cortex *in vitro* strongly affects Purkinje cells via parallel fibers. *Naturwissenschaften* **82**, 201-203.

HECK, D. & BRAITENBERG, V. (1993). Specific Responses of Rat Cerebellar Cortex to Moving Input: Dependence of Distance Covered by the Movement. *Soci.Neurosci.Abstr.* **19**, 1589(Abstract)

HORE, J., WILD, B. & DIENER, H. C. (1991). Cerebellar dysmetria at the elbow, wrist and fingers. *J.Neurophysiol.* **65**, 563-571.

IVRY, R. (1993). Cerebellar involvement in the explicit representation of temporal information. *Ann.NY Acad.Sci.* **682**, 214-230.

LLINÁS, R., BLOEDEL, J. R. & HILLMAN, D. E. (1969). Functional characterization of neuronal circuitry of frog cerebellar cortex. *J.Neurophysiol.* **32**, 847-870.

LLINÁS, R. (1982). Radial Connectivity in the Cerebellar Cortex: A Novel View Regarding the Functional Organization of the Molecular Layer. In *The Cerebellum, New Vistas*, eds. PALAY, S. L. & CHAN-PALAY, V. pp. 189-192. New York: Springer.

LLINÁS, R. & SUGIMORI, M. (1992). The Electrophysiology of the Cerebellar Purkinje Cell Revisited. In *The Cerebellum Revisited*, eds. LLINÁS, R. & SOTELO, C. pp. 167-181. New York: Springer.

NAPPER, R. M. A. & HARVEY, R. J. (1988). Number of parallel fiber synapses on an individual Purkinje cell in the cerebellum of the rat. *J.Comp.Neurol.* **274**, 168-177.

RAMON Y CAJAL, S. (1911). *Histologie du système nerveux de l'homme et des vertébrés. Vol. II*. Paris: A. Maloine.

THACH, W. T. (1968). Discharge of cerebellar Purkinje and nuclear neurons during rapidly alternating arm movements in the monkey. *J.Neurophysiol.* **31**, 785-797.

THACH, W. T. (1970). Discharge of cerebellar neurons related to two maintained postures and two prompt movements. II.Purkinje cell output and input. *J.Neurophysiol.* **33**, 537-547.

TROMMALD, M., JENSEN, V. & ANDERSEN, P. (1995). Analysis of dendritic spines in rat CA1 pyramidal cells intracellularly filled with a fluorescent dye. *J.Comp.Neurol.* **353**, 260-274.

25

ARCHITECTURE FOR A REPLICATIVE MEMORY

Jacques Ninio*

Laboratoire de Physique Statitique
24 rue Lhomond
75231 Paris cedex 05, France

1. INTRODUCTION

Human memory works in a quite curious way. Formal experimental investigations of its properties, initiated by Ebbinghaus [2] have generated a substantial body of knowledge, well described in many textbooks (e.g., [15]). Although its basic principle is still unknown, it seems to differ from that of digital computer memories, and from that of the associative memories of formal neural networks.

For instance, while items are stored in a computer memory independently of their meaning, it is much easier for humans to store meaningful words, sentences or scenes than meaningless ones. A classical strategy for memorizing a string of digits (such as those composing a telephone number) or an odd foreign word, is to divide it in pieces, and establish a correspondence between each piece and a familiar item.

The associative memories of formal neural networks are sensitive to the content of the item to be memorised, but according to a peculiar "attractor logic." If a formal neural network has stored a stimulus A (i.e., it has generated an attractor corresponding to A), then a new stimulus B resembling A will fall into the attractor: it might be very difficult to store it as an independent attractor. For humans, on the other hand, similarity is rather helpful. For instance, a foreign language which is close to the native tongue is easy to learn, and similarity is not, in this case, a source of confusion.

Often, the debate on the nature of memory revolves around the opposition between localized and distributed memories. The two options are not easy to distinguish experimentally. As pointed out by Squire [11] evidence in favor of a distributed memory may be reinterpreted in terms of a localized form of memory, in which each item would be present in many replicas at various locations. I will in fact consider here a class of models in which memory works by replication. In the way they operate, these models have some analogy with a well-known search strategy in molecular biology, used for gene amplification and sequencing.

---

*Tel: (33) - 1 - 44 32 33 18; Fax: (33) - 1 - 44 32 34 33; E-mail: ninio@peterpan.ens.fr

While one often thinks of memory as a medium on which pieces of information are engraved, the emphasis here will be on the processes of storage and retrieval. Among the many clues that might betray the organisation of human memory, there are those concerned with "recall strategies." There is a body of knowledge on how to bring to consciousness bits of information that were thought to be lost, in amnesic patients and in people with normal memory as well (e.g., [12]). Many items seem to be linked in memory by unidirectional arrows: A might elicit B more efficiently than B elicits A. For instance, it is easier to associate the image of a face to a name, than to name the owner of a face. If you have difficulties in retrieving the name of a person A, then it is a good strategy to think about another person B having some connection to A, then think about some event linking the two persons [7].

In an attempt to depict to myself the mechanics of human memory as it seems to operate from the perspective of experimental psychology, I found that I could translate much of what I knew into a peculiar architectural model, requiring a precise connectivity between several layers of units [8]. My hope of course was that such a model would capture some essential aspects of our brain mechanisms, and this is why in presenting it I use words like "cell" or "neurone" to designate the units. I am fully aware that, as initially described, the model involves neurones that do not conform to neurophysiological orthodoxy. But the function which is assigned in the model to a neurone, may be, more realistically, embodied into larger assemblies of neurones. In particular I will suggest at the end that neurones of the model may be replaced with columns of neurones, provided the columns behave as coherent units.

2. THE MODEL

Let us consider the stage of processing of sensory stimuli where, according to Miyashita's elegant expression [6] "perception meets memory." I assume that the low-level analyses of sensory input have been carried out, and are expressed as the binary states of activity of a set of neurones belonging to a "representation" layer R. I assume that in this layer, each cell may be in one of two states, X or O. I consider now a single neurone S which sends dendritic processes making one-to-one contacts with the cells of layer R (Figure 1). I further assume that the dendrites can take one of two states, also labelled X and O, and that they interact with the cells they touch according to two modes:

- Reading mode: Each terminus of S takes state X or state O, according to whether the corresponding cell of layer R is in state X or state O.
- Writing mode: Each cell of layer R takes state X or state O according to whether the abutting terminus of cell S is in state X or state O.

Up until now, the function of the S neurone seems to be merely to capture a perception and restore it integrally. (A possible physiological role would be to maintain a stable representation of a scene while the perceptual processors are trying to analyse new inputs). Consider that there are several neurones of type S that are similarly connected to the layer R. Now, the model can work as a short-term memory. Ideally, we would like to have 7 plus or minus two neurones of type S [5]. The contents of the S neurones would be written on layer R during very short intervals of time, which would be separated by longer time intervals during which layer R would represent the inputs from the outside. One class of observations from experimental psychology related to this part of the model is that concerned with "mental imagery" (e.g., [3-4]) and, more generally with the equivalence between perception and memory. A casual example of this equivalence is when you receive on your ears a word from which a syllabe has been replaced with noise, and you do hear the complete word pronounced distinctly [14]. Auditory restoration has also been described for music [12].

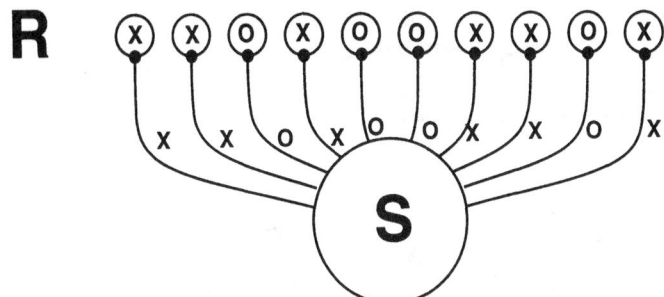

Figure 1. First stage of a memory model. It is assumed that a perception is represented as the binary (X or O) state of activity of a layer of cells "R." It is then assumed that a single unit "S" is able to reproduce, through its end-processes, the states of the cells in the whole layer R.

One way of connecting in good order several S neurones to a layer of R neurones would be to use an architectural principle found in the cerebellum. There, long axons protrude from a territory of granular cells and form bundles of long parallel fibers. A Purkinje cell sends dendrites in a plane that is orthogonal to a bundle of fibers, making one-to-one contacts with the parallel fibers. Therefore, the connectivity between Purkinje and granular cells in the cerebellum is just as requested for the interactions between S and R neurones in the first two layers of the memory model (Fig. 2). Note however that in the cerebellum, there is no output from the Purkinje cells onto the parallel fibers. But there is, mated to each Purkinje cell, a single climbing fiber having a ramified axon, whose branches embrace those of Purkinje's dendritic tree, making about 300 synaptic contacts with them. Whatever the

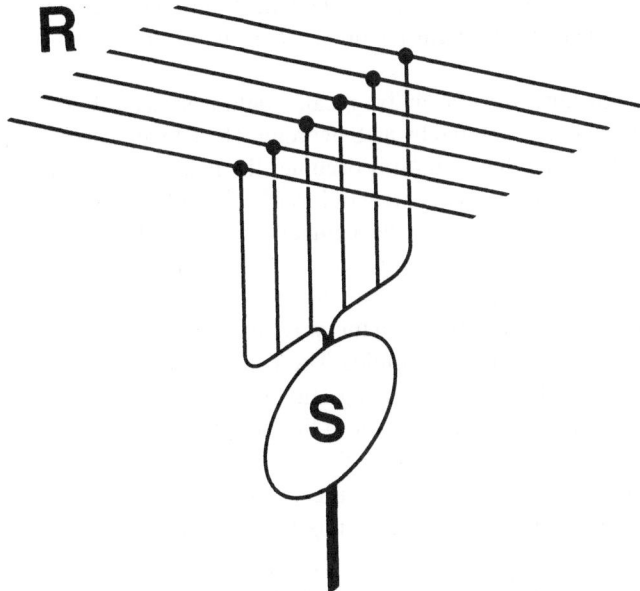

Figure 2. Spatial arrangement between a Purkinje cell (R) and parallel fibers (R) in the cerebellum. This 3D spatial arrangement is an elegant way of implementing the connectivity postulated in Fig. 1. It allows, in principle, several S neurones to be connected in good order to a same set of fibers R.

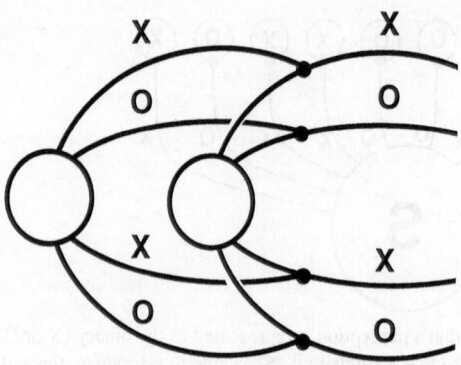

Figure 3. Intermediate stage of a memory model. It is postulated that a unit encoding a percept as the binary states of its end-processes can read the content of another similar unit, or write its own content onto the end-processes of the other unit.

function of this microcircuit (see e.g., [1]), I simply wish to point out here that circuits showing this kind of order do exist in Nature.

If a cell S writes its content on layer R, then another cell S' of the short-term memory reads the content of R, the content of S is transferred to S'. From the initial concept of a cell which stores the pattern of activity of a layer of cells, we have moved to the concept of a cell which replicates the information content of another cell. In Fig. 3, I show how, ideally, two such cells would be directly connected (instead of being indirectly connected, through layer R). The next stage of the model will be presented in terms of this simplified arrangement - keeping in mind that a different arrangement, as discussed in the end, might be more realistic neuroanatomically.

After the layer of short-term memory S neurones, I imagine several transit layers T1, T2, ... Tn. Each neurone in layer Ti has two or more parents in layer Ti+1. Each of these can reproduce the state of the neurone of Ti and, symmetrically, the neurone of Ti can reproduce the state of one of its parents. This kind of order also applies to the relationship between a neurone of S and the first T layer. At the other end, the layer Tn is in contact with an ultimate layer M, which is that of long-term memory, where items are stored for long durations (Fig. 4).

In this architecture, the associative recall would work as follows: A stimulus represented in layer R would be captured by an S neurone. Then it would propagate and multiply exponentially in the transit layers until it occupies the totality of the last transit layer Tn. There, it would be facing the entire content of layer M (Fig. 4). At this point, we need a third mode of interaction between neurones: the comparison mode. The contents of each neurone of layer M are compared -by direct interaction or through a third party-to the contents of the parent neurone in layer Tn. Above a certain amount of homology, the neurone of layer M writes its content on the neurone of Tn. In this way, many items of layer M may be collected by layer Tn in response to the initial stimulus. These items will now propagate towards layer R. Most of them will be lost during this migration since the number of available neurones decreases exponentially from layer to layer in the M to R direction. The surviving outputs would occupy layer R one at a time. I see the selection as based upon similarity, but in a probabilistic manner. Occasionally, a percept reaching Tn would be captured and stored for a long duration in a neurone of layer M. I see this also as an all-or-none probabilistic process.

The model is now explained in its bare principle. It can be refined in at least two ways. First, in order to make it work, one must add control devices that regulate the time-course of information propagation and in particular, partition the time between propagations from R to M and from M to R. Second, new functional properties can be implemented by supplementing the architecture with quite orthodox wiring.

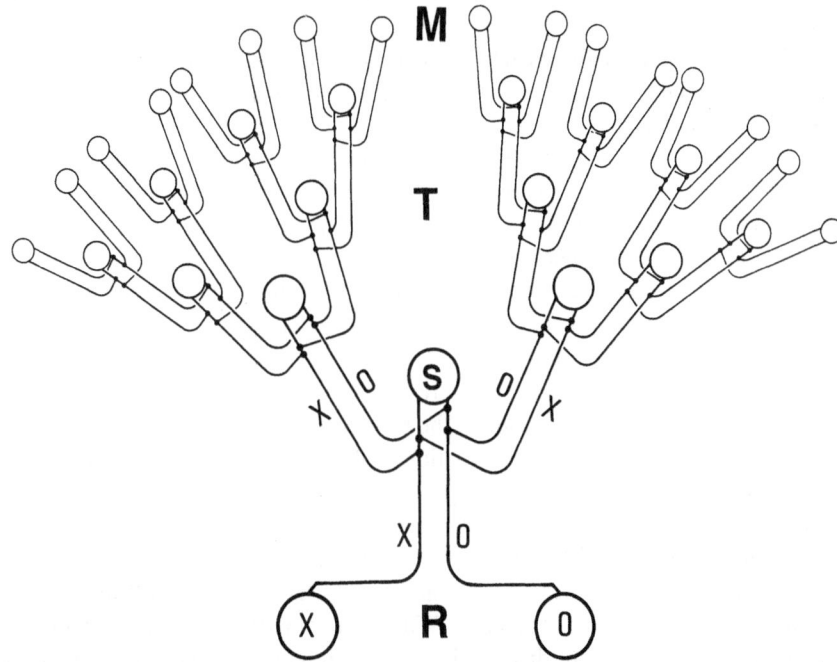

Figure 4. Architectural model of memory. Here, layer R has been reduced to a mere pair of cells, and the neurones in the other layers have been represented with just a couple of end processes. A perception represented in layer R is captured by neurone S, then it replicates in the intermediate layers T until it comes in contact with all the cells of the long-term memory layer M. Items having sufficient similarity with the entering percept replicate and migrate towards S and R.

One obvious possibility is implementing associative learning between items already in memory. This can be achieved by adding, after layer M, a standard neural network that would establish links between neurones of layer M. Another, more subtle possibility, is implementing some kind of "recursive encoding" or "relational encoding," as follows.

Consider that in layer R, there are cells which are directly connected to layer M, and provide a crude map of the regions of layer M that are affected by the recall. Then, these cells associate to the input, as analyzed by the perceptual processors, a kind of abstract code that describes what are the sectors of layer M with which it has the strongest relationships. This code may then work as part of an extended definition of the item, and be attached to it as it propagates again towards layer M, where it would be eventually stored. The possibility of encoding an item by its relations to other items may explain the "tip-of-the-tongue" phenomenon in which one cannot name a person, a place or a concept, and yet one has a clear idea of what this person, place or concept relates to.

3. DISCUSSION

How this architectural model may help us to understand the properties of human memory is discussed in detail elsewhere [9]. Here, I just wish to point out that this type of memory has subtle properties, and that it can work in both an erratic and an inspired manner. In robotics one usually thinks of memory as a device that must produce well-defined outputs in response to well-defined requests. Here memory appears as a general advisor, which, in any circumstance, brings to one's attention elements of information that may, occasionally, be relevant.

In Fig.4, we would ideally wish to use neurones with a hundred or a thousand, instead of two terminators. The major practical difficulty then is how to establish the needed wiring without generating inextricable entanglements. I see two solutions to the problem. One is to use the third dimension. Replace the couples of connectors in Fig. 4 by single neurone to neurone connections, and consider that Fig. 4 is just a cross section, the structure being repeated 100 or 1000 times in the third dimension. We would need here neurones with ramified end processes in thin sheets, like the Purkinje cells of the cerebellum. The neurone would be somewhat shaped like a comb. Alternatively, we might consider that the comb is made of piled up neurones that are anatomically distinct cells, but work functionally as single units, i.e., columns of neurones (Fig. 5).

Many variants of the model are conceivable. For instance, modules can be appended for the temporary storage of information at the time scale of the day or the week. Different units may be used for the transit of information in the R to M and M to R directions. The part of the memory requiring precise wiring may be split into several blocks, each one not larger than an insect's brain, etc. The model was elaborated to provide -at least- a mechanical metaphor for the psychological results. The neurophysiological results are, at present, less easy to accommodate. Are the neurones "responsive to faces" (e.g., [10]) to be identified with the long-term memory units of layer M? When a novel face is presented, their responses to the set of familiar faces is not disrupted, and their reponses to the novel face becomes "stable within a few presentations" [10]. If what is measured experimentally relates to what is encoded by the neurone, this result is an argument in favour of a distributed form of memory. On the other hand, if the experiments only provide some measure of how intensely the recorded neurone is coupled to some other neurone which is the main target of the stimulus, they are compatible with a localized form of memory.

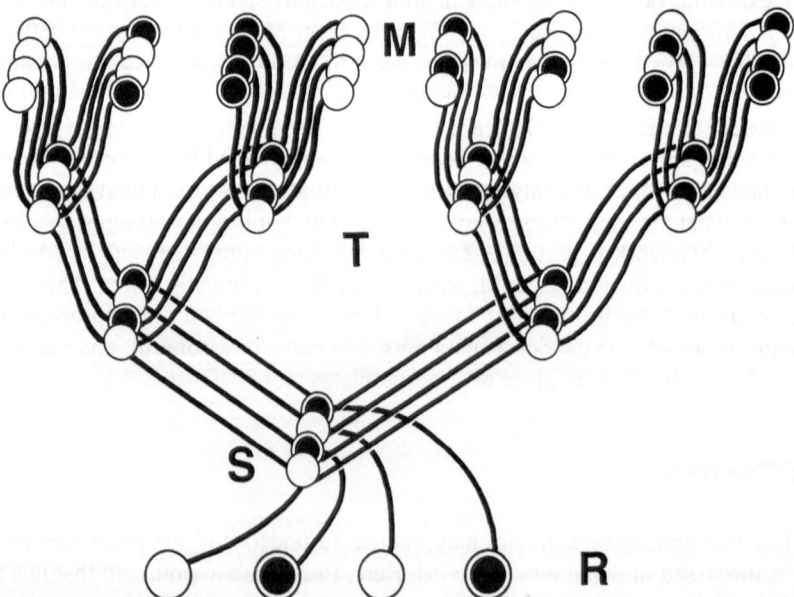

Figure 5. Architectural model of memory similar in principle to that of Figure 4, but in which each neurone of layers S, T, M has been replaced with a column of four neurones.

The retrieval strategy, in the proposed model, has some analogy with the current molecular biology gene sequencing techniques. When an investigator wishes to amplify and sequence a gene, he prepares on one side a crude extract of DNA containing the gene among many (and often a huge excess of) undesired genes This is the equivalent of layer M. On the other side, he prepares a "primer" which is a short section of RNA or DNA which, to the best of his knowledge, may have some resemblance with a section of the gene of interest. This corresponds to the entering percept of layer R. The molecules of primers are mixed with the DNA extract and the primers hybridize to various sub-sequences of the DNA, the probability of fixation being the largest when there is a nearly perfect match. This corresponds to the comparison between the contents of layers Tn and M. Then, a DNA replication enzyme is used to make a large number of copies of those regions in which the primer is hybridized, and which, hopefully, include the desired gene. This is the equivalent of the replicative propagation from layer M to layer R.

REFERENCES

1. Braitenberg, V (1993) The cerebellar network: attempt at a formalization of its structure. Network 4: 11-17
2. Ebbinghaus, H (1885) Über das Gedächtnis. Untersuchungen zur experimentellen Psychologie. Duncker & Humblot, Leipzig
3. Haber, RN (1969) Eidetic images. Scientific American 220 N°4: 36-44
4. Kosslyn, SM (1986) Ghosts in the mind's machine. W.W. Norton & Co. New-York, London
5. Miller, GA (1956) The magical number seven plus or minus two: some limits on our capacity for processing information. Psychol. Rev. 63: 81-97
6. Miyashita, Y (1993) Inferior temporal cortex: Where perception meets memory. Annu. Rev. Neurosciences 16: 245-263
7. Morton, J, Hammersley, RD, Bekerian, DA (1985) Headed records: A model for memory and its failures. Cognition 20:1-23
8. Ninio, J (1988) Modèle de mémoire iconique localisée, réplicable et associative, utilisant le temps partagé et trois modes de communications neuronales. Comptes-Rendus Acad. Sci. Paris, Série III, 306: 545-550
9. Ninio, J (1996) L'empreinte des sens. Perception, mémoire, langage (3rd edition) Odile Jacob, Paris
10. Rolls, ET (1994) Brain mechanisms for invariant visual recognition and learning. Behavioural Processes 33: 113-138
11. Squire, LR (1986) Mechanisms of memory. Science 232:1612-1619
12. Tulving, E, Shacter, DL (1990) Priming and human memory systems. Science 247: 301-306
13. Vicario, G (1982) Some observations in the auditory field. In: Beck, J (ed.) Organization and representation in perception. Lawrence Erlbaum, Hillsdale, New-Jersey, pp. 269-283
14. Warren, RM (1982) Auditory perception: a new synthesis. Pergamon Press, Elmsford.
15. Wingfield, A, Byrnes, DL (1981) The psychology of human memory. Academic Press, New-York

26

MEASURING INFORMATION FROM NEURONAL ACTIVITY

S. Panzeri,[1] A. Treves,[1] and D. Golomb[2]

[1]SISSA, Biophysics and Cognitive Neuroscience
Via Beirut 2-4, I-34013
Trieste, Italy
[2]Departments of Medicine and Electrical Engineering
Technion
Haifa 32000, Israel

1. INTRODUCTION

A quantification of the relation between neuronal responses and the events that have elicited them is important for understanding the brain. One way to do this in sensory systems is to treat a neuron as a communication channel[3] and to measure the mutual information between a set of stimuli presented to the animal and the neuronal response. In such experiments (e.g., Refs. 4 and 12) a set of S sensory (e.g., visual) stimuli is presented to the animal, each stimulus being presented for N_S trials. The neuronal response is quantified in various ways, such as the number of spikes in a certain time interval or a descriptor of the temporal course of the spike train, and the mutual information between the response and the set of stimuli is estimated.

In a typical neurophysiological experiment, due to practical reasons, only a limited number of stimulus repetitions can be carried out. However, a large number of trials is needed for accurate information calculations.[8,10,14] When the number of data samples is too small, there are systematic errors in estimating the mutual information by direct application of the mutual information definition. Larger numbers of samples are needed to make accurate information calculations as the dimensionality of the response increases. However, given that it is unlikely to gather very large data sets at any time in the near future, it is important to correct for the systematic error (bias) if we are going to be able to make effective use of mutual information measurements. Questions regarding the type of neuronal coding, especially the importance of temporal and collective effects, can be addressed with experimentally-measured information values only if the lack of accuracy due to the limited number of trials, resulting in both an average bias and fluctuation from the true values, is under control.

Different ad hoc correction terms[9] or regularization methods[6,7] have been proposed in the past, but on a heuristic basis. Recently, we have developed a different procedure to correct for such error,[14,10] consisting in the analytical calculation of the average error, its estimation from data, and its subtraction from raw information measures. The method can be adapted to the different regularization procedures used with neuronal data which are particularly useful when multidimensional codes or multi-cell recording are concerned.

In this paper, we will review the results obtained with our method when the simplest regularization of the response space, that is the discretization into a given number of intervals, is applied to raw data. Computer simulations are used to test the range of validity of the analytical results , and to suggest simple recipes to correctly calculate the information content of neuronal spikes measured in the laboratory. The comparison between the results with the procedure reported in this paper and the other correction methods present in the literature are fully discussed in Refs. 10 and 5.

2. ANALYTICAL EVALUATION OF THE BIAS

To be concrete, let us assume that we want to measure the average information carried by the response r of a neuron (or of several neurons) about a stimulus s presented to the animal. We assume that s is drawn at random from a discrete set S of S elements. Likewise, we initially require that the response space $\mathcal{R}$ be discretized, to include a total of R distinct responses. If the actual, raw responses are real numbers, (e.g., the weights of the firing train of one neuron on the principal components of the covariance matrix) we assume that they have been binned into R different intervals, by just assigning each response to the interval it falls in. This binning procedure satisfies an *independence* condition, i.e., the number of times a given bin is occupied depends only on the underlying probability of the given bin, and not on the occupancy of other bins. We stress that R is the total number of response bins, independently of what is the underlying dimensionality, if any, of the raw response space. If e.g., the raw responses are the firing rates of two cells, which are then discretized into R_1 and, respectively, R_2 bins, we set $R = R_1 \times R_2$. It is important to notice that, if the responses are continuous, in order to obtain an estimate of the mutual information from a finite set of N data a regularization of the raw data is always necessary; otherwise the finite number of responses will almost certainly be all different from each other, therefore each response will uniquely identify its stimulus, and, as a result one will obtain only a measure of the entropy of the stimulus set, and not of the mutual information. On the other hand, it should be also pointed out that the value of the information obtained *after* quantization, or regularization, is less than the value of information carried by the continuous responses, and in general information measures are dependent on the binning procedure adopted, and most importantly on the number of bins R. There is no way to estimate the difference between the unregularized and regularized values of the mutual information from first principles, but a good strategy to control these discrepancies can be to quantize the responses by successively increasing the value of R until the finite N measure, *after* the correction we are discussing, does not change very much. However, when the size of the data sample is small, a reasonable choice for R is a compromise between trying to keep the loss of information due to discretization as small as possible, which would require R large, and the need to control the finite-size distortion, which, as we shall see below, requires R small.

So, after the response discretization, the total amount of mutual information we aim at is:

$$I = \sum_{s \in S} \sum_{r \in \mathcal{R}} P(s,r) \log_2 \frac{P(s,r)}{P(s)P(r)} \tag{1}$$

where $P(s,r)$ is the underlying joint probability distribution of stimuli and responses, $P(s)$ and $P(r)$ the separate ones for stimuli and for responses, and obviously $P(r) = \sum_s P(s,r)$ while $P(s) = \sum_r P(s,r)$. In practice, we have N_s presentation of each stimulus, and thus a total of $N = S \times N_s$ experimental trials (i.e. stimulus-response pairs) available, so we get a raw estimate of the information

$$I_N = \sum_{s \in \mathcal{S}} \sum_{r \in \mathcal{R}} P_N(s,r) \log_2 \frac{P_N(s,r)}{P_N(s)P_N(r)} \tag{2}$$

where the P_N's are the experimental frequency-of-occupancy tables, e.g., $P_N(r) = n(r)/N$, or $P_N(r|s) = n(r|s)/N_s$, with $n(r|s)$ is the number of times response r occurred when the stimulus s is presented, and $n(r)$ the number of times response r occurred across all stimuli. The difference, or bias, between I_N and I of course fluctuates depending on the particular outcomes of the N trials performed. We can estimate the average of the difference, however, by averaging $(\langle \ldots \rangle)$ over all possible outcomes of the N trials, keeping the underlying probability distributions fixed. The procedure we use is first to rewrite $\langle I_N \rangle$ in a more convenient form:

$$\langle I_N \rangle = \left\langle \sum_s P_N(s) \sum_r \left\{ P_N(r|s) \log_2 P_N(r|s) - P_N(r) \log_2 P_N(r) \right\} \right\rangle \tag{3}$$

and then to use the replica trick to convert the logarithms into limits of a power:

$$\langle I_N \rangle = \left\langle \sum_s P_N(s) \frac{1}{\ln 2} \lim_{n \to 1} \frac{1}{n-1} \sum_r \left\{ P_N^n(r|s) - P_N^n(r) \right\} \right\rangle. \tag{4}$$

It is very important to note that in Eq. (4) one must exclude from the sum over response bins r, for each term of the sum over stimuli, the bins in which $P(r|s) = 0$ (in fact, in those bins the only permitted outcome is $P_N(r|s) = 0$ and they give a vanishing contribution to the average). Now the bias can be evaluated from Eq. (4) by decomposing $P_N^n(\cdot)$ as a binomial series, and then calculating term by term the averages of the integer powers of $P_N(\cdot)$. By finally performing the $n \to 1$ limit in Eq. (4) one finds that (see Ref. 14 for the details of the calculation) the average of the bias can be expressed as a formal expansion in $1/N$:

$$\langle I_N \rangle - I = \sum_{m=1}^{\infty} C_m \tag{5}$$

where C_m represents successive contributions to the asymptotic expansion of the bias (the term C_m is proportional to N^{-m}). Here we report just the leading term, which depends only on a few parameters characterizing the underlying probability distributions, and is expected to capture almost all the bias if N is large enough:

$$C_1 = \frac{1}{2N \log_2} \left\{ \left(\sum_s \tilde{R}_s \right) - \tilde{R} - S + 1 \right\}. \tag{6}$$

In Eq. ((6)) $\tilde{R}_s$ denotes the number of "relevant" response bins for the trials with stimulus s, which are the response bins with non-zero probability to be occupied during the presentation of the stimulus S. In the same way, $\tilde{R}$ is the number of response bins with non-zero occupancy probability across all stimuli.

Successive terms in the $1/N$ expansion, which depends instead on the full probability distributions, are of little use: either they are negligible with respect to the first term or, when N becomes too small, they can quickly explode, signalling that data are so scarce that the expansion becomes meaningless beyond the first term.[10,14]

3. ESTIMATION OF THE BIAS FROM THE DATA

At the end, to correct for the finite size problem we have to evaluate the correction term in Eq. (6), which depends on the underlying probabilities solely through the $\tilde{R}_s$ parameters, and thus in a much weaker way than the mutual information, which depends on the full distributions. Therefore, even though the parameters $\tilde{R}_s, \tilde{R}$ have to be, in turn, estimated from the data, this procedure is much more accurate than a direct estimate of the information.

To understand how one can estimate the number of "relevant" bins, we note that the number of relevant bins differs from the total number of bins allocated because some bins may never be occupied by responses to a particular stimulus. As a consequence, if $\tilde{R}_s$ is calculated using for each stimulus the total number of bins R, then the C_1 term, which is in this case equal to $(S - 1)(R - 1)/(2N \ln 2)$, turns out to overestimate the systematic error, whenever there are stimuli that do not span the full response set. On the other hand, the number of relevant bins differs also from the number of bins actually occupied, R_s, for each stimulus (with few trials), because more trials might have occupied additional bins. Again, it turns out that using the number of actually occupied bins R_s for calculating C_1 leads, when few trials are available, to an underestimate of the systematic error (the underestimation becoming negligible for $R/N_s \ll 1$ because R_s tends to coincide with $\tilde{R}_s$ for all stimuli).

It is clear that when N_S is small, more sophisticated procedures, such as Bayesian estimation, are needed to evaluate the quantities we are interested in.

Wolpert and Wolf[15] proposed the calculation, with the Bayes rule, of functions (in our case e.g., I) of the true probabilities *given* the experimental frequencies. This, which is in fact the original aim, (and which is obviously different from calculating the functions of the frequencies, e.g., our I_N) is feasible, however, only when a detailed knowledge of the *a priori* probability distributions of the probabilities is available.

If, for example, we want to measure a function $G(\{P(r)\})$ of the set of probabilities $\{P(r)\}$, and we know the prior probability distribution of the probabilities $\mathcal{P}(\{P(r)\})$, then the Bayesian estimate of the function $G(\{P(r)\})$ has the following expression as a function of the set of experimental data $\{n(r)\}$:

$$\hat{G}(\{n(r)\}) = \int \left(\prod_r dP(r) \right) \mathcal{P}(\{P(r)\}|\{n(r)\}) G(\{P(r)\}) \tag{7}$$

where $\mathcal{P}(\{P(r)\}|\{n(r)\})$ is the "posterior" conditional probability (calculated with the Bayes theorem) of the underlying probabilities *given* the experimental outcome (We refer to Ref. [15,10] for all the details of the procedure).

It is difficult, however, to see how to use this conceptually appealing approach to estimate the mutual information in cases, such as ours of stimulus-response pairs, when no reasonable assumption on the prior is self-evident. Nevertheless, here we show that a correction to the mutual information values depending only on a few parameters, such as $\tilde{R}_s, \tilde{R}$, can be well estimated also with a crude hypothesis about the prior probability functions. The idea is to use the Bayes theorem to reconstruct the true probabilities, supposing they are non-zero into $\tilde{R}_s$ intervals, and then choose a $\tilde{R}_s$ such that the expected number of occupied intervals (which can be calculated as a function of the Bayes estimate of the probabilities) matches the experimentally observed one.

The procedure we use here to evaluate $\tilde{R}_s$, for each stimulus s, is the following:

- We first pick for $\tilde{R}_s$ one of the allowed values, $R_s \leq \tilde{R}_s \leq R$.

- We construct the Bayes estimate $\widehat{P}(r|s)$ of the true probabilities given the experimental frequencies. The prior probability function $\mathcal{P}(\cdot)$ is chosen constant among the R_s non-empty bins, and for the other $\widetilde{R}_s - R_s$ empty bins is a different constant, fixed by requiring that the probability of that bin being empty is h_s times larger than the probability of its being occupied, where $h_s = \frac{N_s}{R_s}$. This last requirement simply reflects the fact that when the responses are concentrated into a few bins (i.e. high $\frac{N_s}{R_s}$), the probability in the empty bins should be less than the probability assigned by a prior function constant on all the $\widetilde{R}_s$ bins. We want to emphasize that we use the constant ansatz for the prior probability distribution only because this is the simplest one. Of course, if, in particular cases, some reasonable assumption on the prior probabilities is available, this more detailed assumption can be used, and the Bayes approach is expected to give better results.

- We pick other values for $\widetilde{R}_s$, and we finally choose as an estimate for $\widetilde{R}_s$ the value of $\widetilde{R}_s$ which gives the expectation value of the number of occupied bins:

$$\langle R_s \rangle = \sum_r \left[1 - (1 - \widehat{P}(r|s))^{N_s} \right] \tag{8}$$

closest to the experimental value of R_s.

- The procedure is the same for the evaluation of $\widetilde{R}$, the only difference being that the Bayesian estimate for $\widehat{P}(r)$ should be calculated from N, and not N_s, trials.

This estimation, although based on a very simple ansatz on the prior distributions, is sufficient to give, as we shall see below, good results even up to relatively small values of N. The reason of this good estimation, in our opinion, is in the fact that only the parameters $\widetilde{R}_s$ have to be estimated based on the arbitrary ansatz, and the information I depends on them only in the correction terms.

To support these analytical results, we performed explicit numerical simulations. We chose as "test" underlying probabilities Poisson distributions, which are fair simple models of the spontaneous activity of neurons under certain conditions.[1,11] We generated the distribution of mean firing rates $\bar{r}(s)$ corresponding to each stimulus s by selecting a random variable x from a flat distribution in the interval $[0, 1)$, and then setting

$$\bar{r} = -\log\left(1 - \frac{x}{2a}\right) \quad \text{if} \quad x < 2a, \quad \bar{r} = 0 \quad \text{if} \quad x > 2a, \tag{9}$$

The parameter a is, on average, the sparseness[13] of the firing rate distribution. The number of spikes n recorded on each trial over a period t ($t = 500$ msec. in the present simulations) followed the Poisson distribution

$$P(n|s) = \frac{[\bar{r}(s)t]^n \exp{-[\bar{r}(s)t]}}{n!}. \tag{10}$$

To measure, from N trials, the information carried by the firing rates generated in this way, we used the following regularization procedure: the range of responses was discretized into a preselected number R of bins, with the bin limits selected so that each bin contains the same number of trials within ± 1 (equipopulated bins). Figure 1 shows, for different values of N, the relative worth of using our correction by calculating the C_1 term with the different algorithms to evaluate the number of relevant bins explained before. The results for the raw information and for the values of the information after the subtraction of the differently evaluated correction terms are compared to the regularized values of the information, that

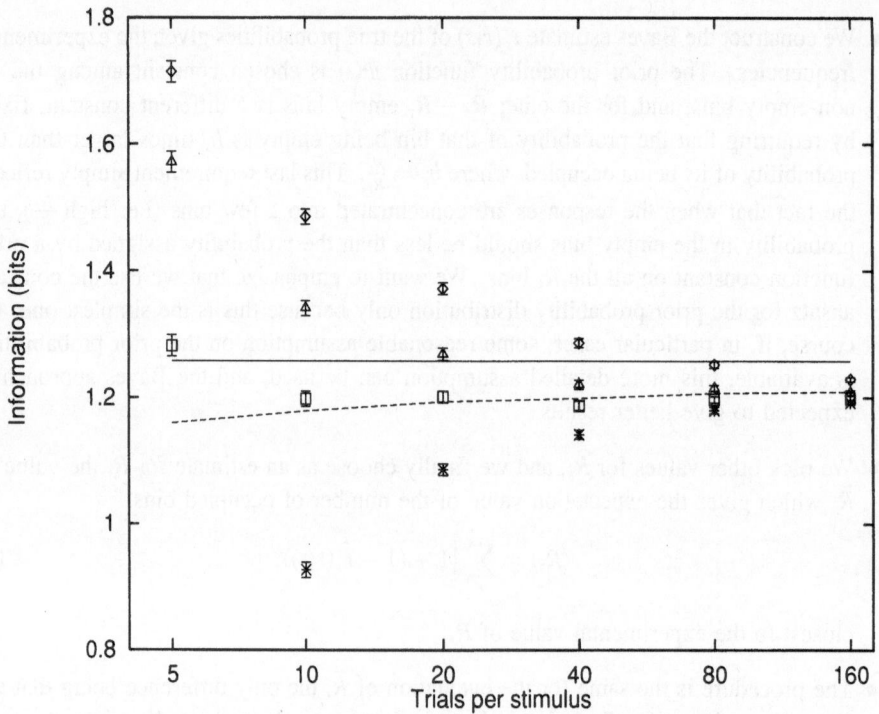

Figure 1. Mutual information values for the distribution of stimuli and Poisson responses described in the text (the sparseness of of mean firing rates is $a = 0.4$), with $S = 10$ and $R = 10$ and different values of N_s. The full line is the real value of the information in the distribution and the dashed line is the *regularized* value, that could be extracted from an infinite sample of data, after the prescribed regularization of the responses (this later value varies with N because the binning procedure is data dependent). Compared to these reference values are, for each N, the raw estimates ($\Diamond$), the estimates corrected by subtracting the C_1 term calculated by estimating the relevant bins by counting the number of actually occupied ones ($\triangle$), estimating the effective bins with the Bayesian procedure described in the text ($\square$), and taking all bins to be relevant ($\tilde{R}_s = R = 10$) ($\star$). Each value is plotted with the standard deviation of the mean of 100 measurements. Note that the N_s axis is on a logarithmic scale.

could be extracted from an infinite sample of data, after the quantization of the responses into R bins. As one can see, if one estimates $\tilde{R}_s$ and $\tilde{R}$ by simply counting the number of occupied bins, the correction procedure works well only if $\frac{N_s}{R} \gg 1$, whereas the use of a Bayesian estimation of the relevant bins significantly increases the range of validity of the method, which now gives reliable results even up to $\frac{N_s}{R} \sim 1$. The subtraction of the term $C_1 = \frac{1}{2N \ln 2}(S - 1)(R - 1)$ gives instead an overestimation of the bias, as expected from the theoretical analysis.

Once established the range of reliability of the method, the next question regards the choice of the "optimal" number of bins for an experiment with S stimuli and N_s trials per stimulus. A reasonable answer to this question can be to choose $R \sim N_s$, to be at the limit of the region where the correction procedure is expected to work, and thus still be able to control finite sampling, while minimizing the downward bias produced by binning into too few bins. This choice should effectively minimize the combined error due to regularization and finite sampling. In Figure 2 the information estimates obtained by choosing $R = N_s$ are compared, for different values of N_s, to the full, unregularized, value of the information carried by the Poisson distribution of responses. It can be noticed that, in this situation, results appear to be a reasonable estimate of the full value of the information already from $N_s = 10$, which can be thus considered, in this particular instance, the minimum number

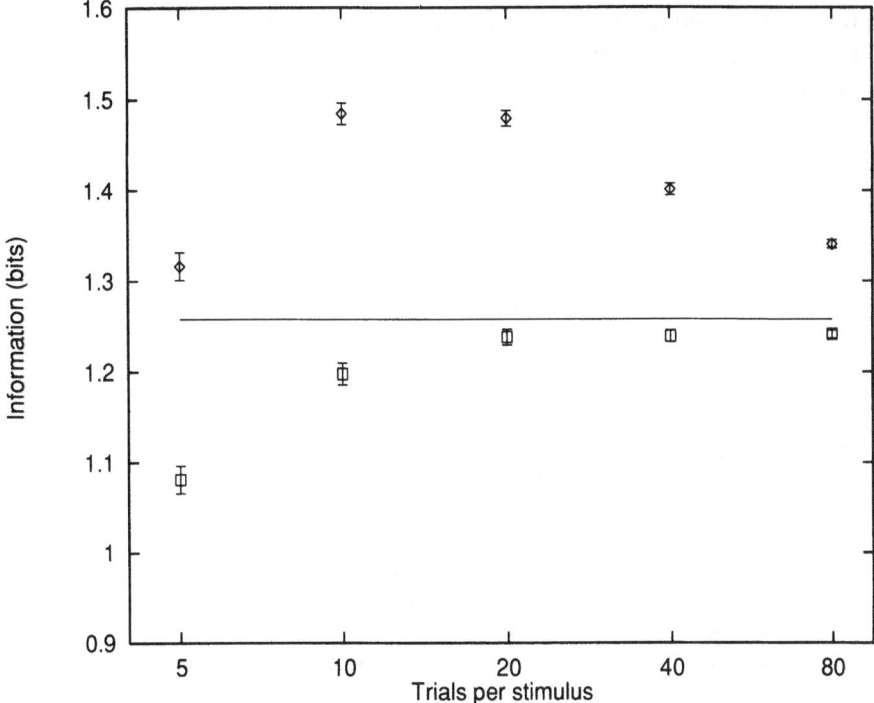

Figure 2. Mutual information values for the distribution of stimuli and Poisson responses described in the text. Here $a = 0.4, S = 10$ and the number of bins R is fixed equal to N_s. The symbols have the same meaning as in Fig. 1. Note that for $N_s = 10$, also $R = 10$ and the result is the same as shown in Fig. 1. For higher values of N_s, results approach the unregularized value of the information, whereas in Fig. 1 they approached the value regularized with $R = 10$ bins.

of trails per stimulus needed to measure the information. The correction procedure based on binning indicates how to work out this minimum number in other experiments. The correction functions reasonably up to $N_s \simeq R$, and the minimum number of response bins which may, if the appropriate code is used, not throw away information, is just the same as the number of stimuli, $R = S$. Therefore a minimum of $N_s = S$ trials per stimulus is a fair demand to be made on the design of experiments from which information estimates are going to be derived.

4. DISCUSSION

The work reported here is of practical importance, especially, although by no means solely, for the analysis of neuronal activity recorded in the mammalian nervous system *in vivo*. Measuring the information carried by neuronal activity has been avoided by many neurophysiologists because of the seemingly huge amount of data required to get reasonable statistics; and the outcomes of such measurements have been widely accepted only in a few instances, e.g., when performed in insects,[2] in which data collection is not a constraint, and the results appeared *hard* (just as hard-wired appear to be the nervous systems examined). Our work allows, instead, to obtain reliable estimate of information for a range of values of the data size, where a great deal of typical neurophysiological experiments lie. The data collected in such experiments is, then, available for information measurements, at the very limited cost of adding a routine or two to standard data analysis packages. The diffusion of the practice of measuring (accurately) the information content of neuronal activity is likely to greatly enhance our quantitative understanding of the processing of information in the nervous system.

REFERENCES

1. Abeles M, Vaadia E, Bergman H (1990) Firing patterns of single units in the prefrontal cortex and neural network models. Network 1:13-25
2. Bialek W, Rieke F, de Ruyter van Steveninck RR, Warland D (1991) Reading a neural code. Science 252:1854-1857
3. Cover, T.M., and Thomas, J.A. (1991) Elements of information theory. John Wiley, New-York.
4. Gawne T.J., and Richmond. B.J. (1993) How independent are the messages carried by adjacent inferior temporal cortical neurons? J. Neurosci. 13:2758-2771.
5. Golomb D, Hertz J, Panzeri S, Richmond B, Treves A (1995). How well can we estimate the information carried in neuronal responses from limited samples?, submitted
6. Hertz JA, Kjær TW, Eskandar EN, Richmond BJ (1992) Measuring natural neural processing with artificial neural networks. International Journal of Neural Systems 3, sup : 91-103
7. Kjær, TW, Hertz JA and Richmond BJ (1994) Decoding cortical neuronal signals: network models, information estimation and spatial tuning. Journal of Computational Neuroscience 1: 109-139
8. Miller, G.A. (1955) On the bias of information estimates. Information theory in psychology; problems and methods, II-B, 95-100.
9. Optican LM, Gawne TJ, Richmond BJ, Joseph PJ (1991) Unbiased measures of transmitted information and channel capacity from multivariate neuronal data. Biological Cybernetics 65: 305-310.
10. Panzeri S, Treves A (1995) Analytical estimates of limited sampling biases in different information measures. Network 7: in press
11. Softky WR, Koch C (1993) The highly irregular firing of cortical cells is inconsistent with temporal integration of random EPSPs. J. Neurosci. 13: 334-350
12. Tovee MJ, Rolls ET, Treves A, Bellis RP (1993) Information encoding and the responses of single neurons in the primate temporal visual cortex. Journal of Neurophysiology 70: 640-654.
13. Treves A (1990), Graded-response neurons and information encodings in autoassociative memories. Physical Review A 42: 2418-2430
14. Treves A, Panzeri S (1995) The upward bias in measures of information derived from limited data samples. Neural Computation 7: 399-407.
15. Wolpert D H and Wolf D R (1995) Estimating functions of probability distributions from a finite set of samples. preprint LA-UR 92-4369, SFI TR 93-07-046, submitted.

27

VISUAL PROCESSING IN THE TEMPORAL LOBE FOR INVARIANT OBJECT RECOGNITION

Edmund T. Rolls

Oxford University, Department of Experimental Psychology
South Parks Road, Oxford, OX1 3UD, England

1. INTRODUCTION

This paper draws together evidence on how information about visual stimuli is represented in the temporal cortical visual areas and the brain areas to which these are connected; on how these representations are formed; and on how learning about these representations occurs. The evidence comes from neurophysiological studies of single neuron activity in primates. It also comes from closely related theoretical studies which consider how the representations may be set up by learning, about the utility of the different representations found, and about how learning occurs in the brain regions which receive information from the temporal cortical visual areas. The recordings described are made mainly in non-human primates, firstly because the temporal lobe, in which this processing occurs, is much more developed than in non-primates, and secondly because the findings are relevant to understanding the effects of brain damage in patients, as will be shown. In this paper, particular attention will be paid to neural systems involved in processing information about faces, because with the large number of neurons devoted to this class of stimuli, this system has proved amenable to experimental analysis; because of the importance of face recognition and expression identification in the primate social behaviour; and because of the application of understanding this neural system to understanding the effects of damage to this system in humans on behaviour.

2. NEURONAL RESPONSES FOUND IN DIFFERENT TEMPORAL LOBE CORTEX VISUAL AREAS

Visual pathways project by a number of cortico-cortical stages from the primary visual cortex until they reach the temporal lobe visual cortical areas [2,33,71] in which some neurons which respond selectively to faces are found [15-18,22,37,45,46,53,55,56]. The inferior temporal visual cortex, area TE, is divided on the basis of cytoarchitecture, myeloarchitecture, and afferent input into areas TEa, TEm, TE3, TE2 and TE1. In addition there is

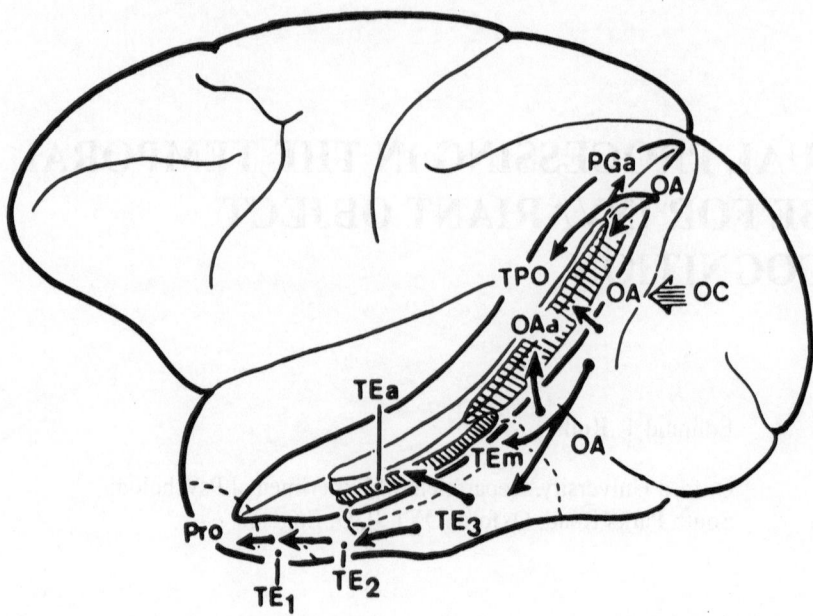

Figure 1. Lateral view of the macaque brain (left) and coronal section (right) showing the different architectonic areas (e.g. TEm, TPO) in and bordering the anterior part of the superior temporal sulcus (STS) of the macaque (see text). The coronal section is through the temporal lobe 133 mm P (posterior) to the sphenoid reference (shown on the lateral view). HIP - hippocampus; RS - rhinal sulcus.

a set of different areas in the cortex in the superior temporal sulcus [9,71] (see Fig. 1). Of these latter areas, TPO receives inputs from temporal, parietal and occipital cortex; PGa and IPa from parietal and temporal cortex; and TS and TAa primarily from auditory areas [71].

In order to investigate the information processing being performed by these parts of the temporal lobe cortex, the activity of single neurons was analysed in each of these areas in a sample of more than 2600 neurons in the rhesus macaque monkey during the presentation of simple and complex visual stimuli such as sine wave gratings, three-dimensional objects, and faces; and auditory and somatosensory stimuli [9]. Considerable specialization of function was found. For example, areas TPO, PGa and IPa are multimodal, with neurons which respond to visual, auditory and/or somatosensory inputs; the inferior temporal gyrus and adjacent areas (TE3,TE2,TE1,TEa and TEm) are primarily unimodal visual areas; areas in the cortex in the anterior and dorsal part of the superior temporal sulcus (e.g. TPO, IPa and IPg) have neurons specialized for the analysis of moving visual stimuli; and neurons responsive primarily to faces are found more frequently in areas TPO, TEa and TEm [9], where they comprise approximately 20% of the visual neurons responsive to stationary stimuli, in contrast to the other temporal cortical areas in which they comprise 4-10%. The stimuli which activate other cells in these TE regions include simple visual patterns such as gratings, and combinations of simple stimulus features [22,74]. Although face-selective neurons are thus found in the highest proportion in areas TPO within the superior temporal sulcus and TEa and TEm on the ventral lip of the sulcus, their extent is great in the anteroposterior direction (they are found in corresponding regions within the anterior half of the sulcus), and they are present in smaller proportions in many other temporal cortical areas (e.g. TE3, TE2 and TE1) [9]. Due to the fact that face-selective neurons have a wide distribution, it might be expected that only large lesions, or lesions that interrupt outputs of these visual areas, would produce readily apparent face-processing deficits. Further, as described below, neurons with responses related to facial expression, movement, and gesture are more likely to be found in the cortex in the superior temporal sulcus, whereas neurons with activity related to facial identity are more likely to be found in the TE areas (see also [24]). These neurophysiological findings suggest that the appropriate tests for the effects of STS lesions will include tests of facial expression, movement, and gesture; whereas facial identity is more likely to be affected by TE lesions.

3. THE SELECTIVITY OF ONE POPULATION OF NEURONS FOR FACES

The neurons described in our studies as having responses selective for faces are selective in that they respond 2-20 times more (and statistically significantly more) to faces than to a wide range of gratings, simple geometrical stimuli, or complex 3-D objects [8,9,46,55]. (In fact, the majority of the neurons in the cortex in the superior temporal sulcus classified as showing responses selective for faces responded much more specifically than this. For half of these neurons, their response to the most effective face was more than five times as large as to the most effective non-face stimulus, and for 25% of these neurons, the ratio was greater than 10:1. The degree of selectivity shown by different neurons studied is illustrated in Fig. 6 of Rolls [56], and by Baylis, Rolls and Leonard [8], and the criteria for classification as face-selective are elaborated further by Rolls [56].) The responses to faces are excitatory, sustained and are time-locked to the stimulus presentation with a latency of between 80 and 160 ms. The cells are typically unresponsive to auditory or tactile stimuli and to the sight of arousing or aversive stimuli. The magnitude of the responses of the cells is relatively constant despite transformations such as rotation so that the face is inverted or horizontal, and alterations of color, size, distance and contrast (see below). These findings indicate that explanations in terms of arousal, emotional or motor reactions, and simple visual feature sensitivity or receptive fields, are insufficient to account for the selective responses to faces and face features observed in this population of neurons [8,37,60]. Observations consistent with these findings have been published by Desimone et al. [18], who described a similar population of neurons located primarily in the cortex in the superior temporal sulcus which responded to faces but not to simpler stimuli such as edges and bars or to complex non-face stimuli (see also [22]).

In a recent study, further evidence has been obtained that these neurons are tuned to provide information about which face has been seen, but not about which non-face has been seen [68]. In this study a wide range of different faces (23) and non-face images (45) of real-world scenes was used. This enabled the function of this brain region to be analysed when it was processing natural scenes. The information available about which stimulus had been shown was measured quantitatively using information theory. This analysis showed that the responses of these neurons contained much more information about which (of 20) face stimuli had been seen (on average 0.4 bits) than about which (of 20) non-face stimuli had been seen (on average 0.07 bits). Multidimensional scaling to produce a stimulus space represented by this population of neurons showed that the different faces were well separated in the space created, whereas the different non-face stimuli were grouped together. The information analyses and multidimensional scaling thus provided evidence that what was made explicit in the responses of these neurons was information about which face had been seen. Information about which non-face stimulus had been seen was not made explicit in these neuronal responses. These procedures provide an objective and quantitative way to show what is "represented" by a particular population of neurons.

4. THE SELECTIVITY OF THESE NEURONS FOR WHOLE FACES OR FOR PARTS OF FACES

Masking out or presenting parts of the face (e.g. eyes, mouth, or hair) in isolation reveal that different cells respond to different features or subsets of features. For some cells, responses to the normal organization of cut-out or line-drawn facial features are significantly larger than to images in which the same facial features are jumbled [37]. These findings are consistent with the hypotheses developed below that by competitive self-organisation some neurons in these regions respond to parts of faces by responding to combinations of simpler visual properties received from earlier stages of visual processing, and that other neurons respond to combinations of parts of faces and thus respond only to whole faces. Moreover, the finding that for some of these latter neurons the parts must be in the correct spatial configuration show that the combinations formed can reflect not just the features present, but also their spatial arrangement.

5. ENSEMBLE ENCODING OF FACE IDENTITY

An important question for understanding brain function is whether a particular object (or face) is represented in the brain by the firing of one or a few gnostic (or "grandmother") cells [5], or whether instead the firing of a group or ensemble of cells each with somewhat different responsiveness provides the representation. We have investigated whether the face-selective neurons encode information which could be used to distinguish between faces and, if so, whether gnostic or ensemble encoding is used. First, it has been shown that the representation of which particular object (face) is present is rather distributed. Baylis, Rolls and Leonard [8] showed this with the responses of temporal cortical neurons that typically responded to several members of a set of 5 faces, with each neuron having a different profile of responses to each face. At the same time, the neurons discriminated between the faces reliably, as shown by the values of d', taken in the case of the neurons to be the number of standard deviations of the neuronal responses which separated the response to the best face in the set from that to the least effective face in the set. The values of d' were typically in the range 1-3. A measure of the breadth of tuning [73], which takes the value 0 for a local representation and 1 if all the neurons are equally active for every stimulus, had values that were for the majority of neurons in the range 0.7-0.95.

In the most recent study, the responses of another set of temporal cortical neurons to 23 faces and 45 non-face natural images was measured, and again a distributed representation was found [68]. The tuning was typically graded. The measure used of the tuning of the neurons was one useful in analyzing the quantitative implications of sparse representation in neuronal networks, namely

$$a = (\Sigma_{s=1,S} \, r_s/S)^2 \, / \, \Sigma_{s=1,S}(r_s^2/S)$$

where r_s is the mean firing rate to stimulus s in the set of S stimuli. If the neurons were binary (either firing or not to a given stimulus), then a would be 0.5 if the neuron responded to 50% of the stimuli, and 0.1 if a neuron responded to 10% of the stimuli. It was found that the sparseness of the representation of the 68 stimuli by each neuron had an average across all neurons of 0.65. This indicates a rather distributed representation. It is of interest to note that if neurons had a continuum of responses equally distributed between zero and maximum rate, a would be 0.75; while if the probability of each response decreased linearly, to reach zero at the maximum rate, a would be 0.67; and if the probability distribution had an exponentially decreasing probability of high rates, a would be 0.5. If the spontaneous firing rate was subtracted from the firing rate of the neuron to each stimulus, so that the changes of firing rate, i.e. the *active responses* of the neurons, were used in the sparseness calculation, then the 'response sparseness' had a lower value, with a mean of 0.33 for the population of neurons, or 0.60 if calculated over the set of faces rather than over all the face and non-face stimuli. Thus the representation was rather distributed. It is, of course, important to remember the relative nature of sparseness measures, which (like the information measures to be discussed below) depend strongly on the stimulus set used. Nevertheless, the results obtained are clearly not those expected for a local (i.e. grandmother cell) representation, in which each neuron codes for one object.

Complementary evidence comes from applying information theory to analyse how information is represented by a population of these neurons. The information required to identify which of S equiprobable events occurred (or stimuli were shown) is $\log_2 S$ bits. (Thus 1 bit is required to specify which of two stimuli was shown, 2 bits to specify which of 4 stimuli was shown, 3 bits to specify which of 8 stimuli was shown, etc. The important point for the present purposes is that if the encoding was local, the number of stimuli encoded by a population of neurons would be expected to rise approximately linearly with the number of neurons in the population. In contrast, with distributed encoding, provided that the neuronal responses are sufficiently independent, and are sufficiently reliable (not too noisy), the number of stimuli encodable by the population of neurons might be expected to rise exponentially as the number of neurons in the sample of the population was increased. The information available about which of 20 equiprobable faces had been shown that was available from the responses of different numbers of these neurons is shown in Fig. 2. First, it is clear that some information is available from the responses of just one neuron - on average approximately 0.34 bits. Thus knowing the activity of just one neuron in the population does provide some evidence about which stimulus was present. This evidence that information is available in the responses of individual neurons in this way, without having to know the state of all the other neurons in the population, indicates that information is made explicit in the firing of individual neurons in a way that will allow neurally plausible decoding, involving computing a sum of input activities each weighted by synaptic strength, to work (see below). Second, it is clear (Fig. 2) that the information rises approximately linearly, and the number of stimuli encoded thus rises approximately exponentially, as the number of cells in the sample increases [1,70].

This direct neurophysiological evidence thus demonstrates that the encoding is distributed, and the responses are sufficiently independent and reliable, that the representational capacity increases exponentially. The consequence of this is that large numbers of stimuli, and fine discriminations between them, can be represented without having to measure the activity of an enormous number of neurons. Although the information rises approximately linearly with the number of neurons when this number is small, gradually each additional neuron does not contribute as much as the first (see Fig. 2). In the sample analysed by Rolls, Treves and Tovee [70], the first neuron contributed 0.34 bits, on average, with 3.23 bits available from the 14 neurons analyzed. This reduction is however exactly what could be expected to derive from a simple ceiling effect, in which the ceiling is just the information in the stimulus set, or $\log_2 20 = 4.32$ bits, as shown in Figure 2. This indicates that, on the one hand, each neuron does not contribute independently to the sum, and there is some overlap or redundancy in what is contributed by each neuron; and that, on the other hand, the degree of redundancy is not a property of the neuronal representation, but just a contingent feature dependent on the particular set of stimuli used in probing that representation. The data available is consistent with the hypothesis, explored by Abbott, Rolls and Tovee [1] through simulations, that if the ceiling provided by the limited number of stimuli that could be presented were at much higher levels, each neuron would continue to contribute as much as the first few, up to much larger neuronal populations, so that the

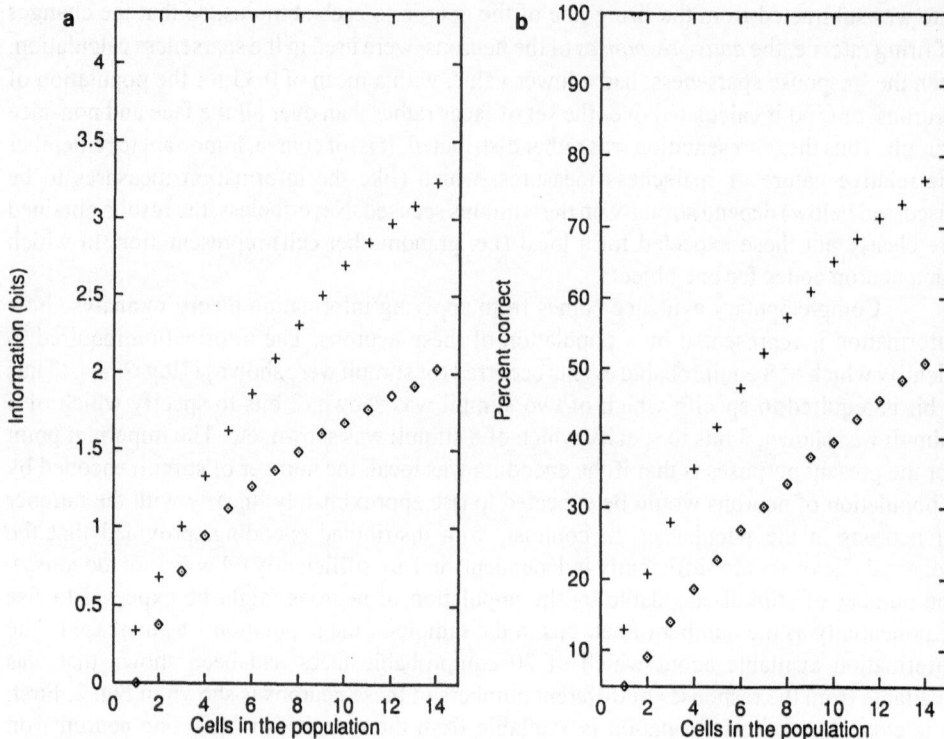

Figure 2. a. The values for the average information available in the responses of different numbers of these neurons on each trial, about which of a set of 20 face stimuli has been shown. The decoding method was Dot Product (DP, diamonds) or Probability Estimation (PE, crosses), and the effects obtained with cross validation procedures utilising 50% of the trials as test trials are shown. The remainder of the trials in the cross-validation procedure were used as training trials. The full line indicates the amount of information expected from populations of increasing size, when assuming random correlations within the constraint given by the ceiling (the information in the stimulus set, I=4.32 bits). **b**. The percent correct for the corresponding data to those shown in Fig. 4a. (From Rolls, Treves and Tovee, 1995).

number of stimuli that can be encoded still continues to increase exponentially even with larger numbers of neurons (Fig. 3; [1]). The redundancy observed could be characterised as flexible, in that it is the task that determines the degree to which large neuronal populations need to be sampled. If the task requires discriminations with very fine resolution between many different stimuli (i.e. in a high-dimensional space), then the responses of many neurons must be taken into account. If very simple discriminations are required (requiring little information), small subsets of neurons or even single neurons may be sufficient. The importance of this type of flexible redundancy in the representation is discussed below. The important point is that the information increases linearly with the number of cells used in the encoding, subject to a ceiling due to the fact that cells cannot add much more information as the ceiling imposed by the amount of information needed for the task is approached.

It may be noted that it is unlikely that there are further processing areas beyond those described where ensemble coding changes into grandmother cell (local) encoding. Anatomically, there does not appear to be a whole further set of visual processing areas present in the brain; and outputs from the temporal lobe visual areas such as those described, are taken to limbic and related regions such as the amygdala and via the entorhinal cortex the hippocampus. Indeed, tracing this pathway onwards, we have found a population of neurons with face-selective responses in the amygdala, and in the majority of these neurons, different

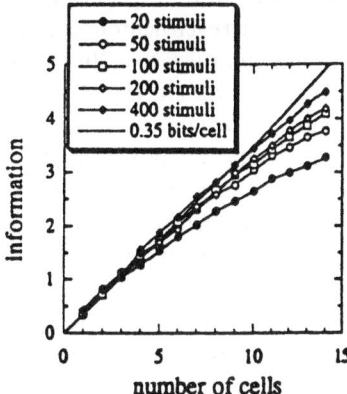

Figure 3. The information available about which stimulus was seen when the responses of many neurons to many stimuli are simulated (from Abbott, Rolls and Tovee, 1995).

responses occur to different faces, with ensemble (not local) coding still being present [29,54]. The amygdala in turn projects to another structure which may be important in other behavioural responses to faces, the ventral striatum, and comparable neurons have also been found in the ventral striatum [86].

6. ADVANTAGES OF THE DISTRIBUTED REPRESENTATION FOUND OF OBJECTS FOR BRAIN PROCESSING

Three key types of evidence that the visual representation provided by neurons in the temporal cortical areas, and the olfactory and taste representations in the orbitofrontal cortex, are distributed have been provided, and reviewed above. One is that the coding is not sparse [8,68]. The second is that different neurons have different response profiles to a set of stimuli, and thus have at least partly independent responses [8,68]. The third is that the capacity of the representations rises exponentially with the number of neurons [1,70]. The advantages of such distributed encoding are now considered, and apply to both fully distributed and to more sparse (but not to local) encoding schemes.

6.1. Exponentially High Coding Capacity

This property arises from a combination of the encoding being sufficiently close to independent by the different neurons (i.e. factorial), and sufficiently distributed. Part of the biological significance of such exponential encoding capacity is that a receiving neuron or neurons can obtain information about which one of a very large number of stimuli is present by receiving the activity of relatively small numbers of inputs from each of the neuronal populations from which it receives. For example, if neurons received in the order of 100 inputs from the population described here, they would have a great deal of information about which stimulus was in the environment. In particular, the characteristics of the actual visual cells described here indicate that the activity of 15 would be able to encode 192 face stimuli (at 50% accuracy), of 20 neurons 768 stimuli, of 25 neurons 3072 stimuli, of 30 neurons 12288 stimuli, and of 35 neurons 49152 stimuli ([1]; the values are for the optimal decoding case). Given that most neurons receive a limited number of synaptic contacts, in the order of several thousand, this type of encoding is ideal. It would enable for example neurons in the amygdala and orbitofrontal cortex to form pattern associations of visual stimuli with reinforcers such as the taste of food when each neuron received a reasonable number, perhaps in the order of hundreds, of inputs from the visually responsive neurons in the temporal cortical visual areas which specify which visual stimulus or object is being seen [52,54,55,58]. Such a representation would also be appropriate for interfacing to the hippocampus, to allow an episodic memory to be formed, that for example a particular visual object was seen in a particular place in the environment [49-51,81]. Here we should emphasize that although the sensory representation may have exponential encoding capacity, this does not mean that the associative networks that receive the information can store such large numbers of different patterns. Indeed, there are strict limitations on the number of memories that associative networks can store [63,80]. The particular value of the exponential encoding capacity of sensory representations is that very fine discriminations can be made as there is much information in the representation, and that the representation can be decoded if the activity of even a limited number of neurons in the representation is known.

One of the underlying themes here is the neural representation of objects. How would one know that one has found a neuronal representation of objects in the brain? The criterion we suggest that arises from this research [57] is that when one can identify the object or stimulus that is present (from a large set of stimuli, that might be thousands or more) with a realistic number of neurons, say in the order of 100, then one has a representation of the object. This criterion appears to imply exponential encoding, for only then could such a large number of stimuli be represented with a relatively small number of units, at least for units with the response characteristics of actual neurons. Equivalently, we can say that there is a representation of the object when the information required to specify which of many stimuli or objects is present can be decoded from the responses of a limited number of neurons.

The properties of the representation of faces, and of olfactory and taste stimuli, have been evident when the readout of the information was by measuring the firing rate of the neurons, typically over a 500 ms period. Thus, at least where objects are represented in the visual, olfactory, and taste systems (e.g. individual faces, odours, and tastes), information can be read out without taking into account any aspects of the possible temporal synchronization between neurons [19], or temporal encoding within a spike train [70,76]. Further, as shown in section 11, the information is available so rapidly in the responses of these neurons that temporal encoding is unlikely to be a fundamental aspect of neuronal spike trains in this part of the brain.

6.2. Ease with Which the Code Can Be Read by Receiving Neurons

For brain plausibility, it would also be a requirement that the decoding process should itself not demand more than neurons are likely to be able to perform. This is why when we have estimated the information from populations of neurons, we have used in addition to a probability estimating measure (PE, optimal, in the Bayesian sense), also a dot product measure, which is a way of specifying that all that is required of decoding neurons would be the property of adding up postsynaptic potentials produced through each synapse as a result of the activity of each incoming axon [1,70]. It was found that with such a neurally plausible algorithm (the Dot Product, DP, algorithm), which calculates which average response vector the neuronal response vector on a single test trial was closest to by performing a normalised dot product (equivalent to measuring the angle between the test and the average vector), the same generic results were obtained, with only a 40% reduction of information compared to the more efficient (PE) algorithm. This is an indication that the brain could utilise the exponentially increasing capacity for encoding stimuli as the number of neurons in the population increases. For example, by using the representation provided by the neurons described here as the input to an associative or autoassociative memory, which computes effectively the dot product on each neuron between the input vector and the synaptic weight vector, most of the information available would in fact be extracted [63,80].

6.3. Higher Resistance to Noise

This, like the next few properties, is in general an advantage of distributed over local representations, which applies to artificial systems as well, but is presumably of particular value in biological systems in which some of the elements have an intrinsic variability in their operation. Because the decoding of a distributed representation involves assessing the activity of a whole population of neurons, and computing a dot product or correlation, a distributed representation provides more resistance to variation in individual components than does a local encoding scheme.

6.4. Generalization

Generalization to similar stimuli is again a property that arises in neuronal networks if distributed but not if local encoding is used. The generalization arises as a result of the fact that a neuron can be thought of as computing the inner or dot product of the stimulus representation with its weight vector. If the weight vector leads to the neuron having a response to one visual stimulus, then the neuron will have a similar response to a similar visual stimulus. This computation of correlations between stimuli operates only with distributed representations. If an output is based on a single X,Y (input firing, synaptic weight) pair, then if the X or the Y is lost, the correlation drops to zero.

6.5. Completion

Completion occurs in associative memory networks by a similar process. Completion is the property of recall of the whole of a pattern in response to any part of the pattern. Completion arises because any part of the stimulus representation, or pattern, is effectively correlated with the whole pattern during memory storage. Completion is thus a property of distributed representations, and not of local representations. It arises for example in autoassociation (attractor) neuronal networks, which are characterised by recurrent connectivity. It is thought that such networks are important in the hippocampus in enabling incomplete recent episodic memories to be completed, and in the cerebral cortex, where the association fibres between nearby pyramidal cells may help the cells to retrieve a representation which depends on many neurons in the network [81].

6.6. Graceful Degradation or Fault Tolerance

This also arises only if the input patterns have distributed representations, and not if they are local. Local encoding suffers sudden deterioration once the few neurons or synapses carrying the information about a particular stimulus are destroyed.

6.7. Speed of Readout of the Information

The information available in a distributed representation can be decoded by an analyzer more quickly than can the information from a local representation, given comparable firing rates. Within a fraction of an interspike interval, with a distributed representation, much information can be extracted [70,72,79,82]. In effect, spikes from many different neurons can contribute to calculating the angle between a neuronal population and a synaptic weight vector within an interspike interval. With local encoding, the speed of information readout depends on the exact model considered, but if the rate of firing needs to be taken into account, this will necessarily take time, because of the time needed for several spikes to accumulate in order to estimate the firing rate. It is likely with local encoding that the firing rate of a neuron would need to be measured to some degree of accuracy, for it seems implausible to suppose that a single spike from a single neuron would be sufficient to provide a noise-free representation for the next stage of processing.

7. INVARIANCE IN THE NEURONAL REPRESENTATION OF STIMULI

One of the major problems which must be solved by a visual system is the building of a representation of visual information which allows recognition to occur relatively independently of size, contrast, spatial frequency, position on the retina, angle of view, etc. To investigate whether these neurons in the temporal lobe visual cortex are at a stage of processing where such invariance is being represented in the responses of neurons, the effect of such transforms of the visual image on the responses of the neurons was investigated.

To investigate whether the responses of these neurons show some of the perceptual properties of recognition including tolerance to isomorphic transforms (i.e. in which the shape is constant), the effects of alteration of the size and contrast of an effective face stimulus on the responses of these neurons were analysed quantitatively in macaque monkeys [60]. It was shown that the majority of these neurons had responses which were relatively invariant with respect to the size of the stimulus. The median size change tolerated with a response of greater than half the maximal response was 12 times. Also, the neurons typically responded to a face when the information in it had been reduced from 3D to a 2D representation in grey on a monitor, with a response which was on average 0.5 of that to a real face. (This reduction in amplitude does not by itself mean that the point in multidimensional space represented by the ensemble of neurons has moved. The point represented by a facial identity ensemble will move only to the extent that the responses of neurons in the facial identity ensemble are affected differently by this transform. The original data are shown in Rolls and Baylis [60].) Another transform over which recognition is relatively invariant is spatial frequency. For example, a face can be identified when it is blurred (when it contains only low spatial frequencies), and when it is high-pass spatial frequency filtered (when it looks like a line drawing). It has been shown that if the face images to which these neurons respond are low-pass filtered in the spatial frequency domain (so that they are blurred), then many of the neurons still respond when the images contain frequencies only up to 8 cycles per face. Similarly, the neurons still respond to high-pass filtered images (with only high spatial frequency edge information) when frequencies down to only 8 cycles per face are included [59]. Face recognition shows similar invariance with respect to spatial frequency [59]. Further analysis of these neurons with narrow (octave) bandpass spatial frequency filtered face stimuli shows that the responses of these neurons to an unfiltered face can not be predicted from a linear combination of their responses to the narrow band stimuli [61]. This lack of linearity of these neurons, and their responsiveness to a wide range of spatial frequencies, indicate that in at least this part of the primate visual system recognition does not occur using Fourier analysis of the spatial frequency components of images.

To investigate whether neurons in the inferior temporal visual cortex and cortex in the anterior part of the superior temporal sulcus operate with translation invariance in the awake behaving primate, their responses were measured during a visual fixation (blink) task in which stimuli could be placed in different parts of the receptive field [77]. It was found that in most cases the responses of the neurons were little affected by which part of the face was fixated, and that the neurons responded (with a greater than half-maximal response) even when the monkey fixated 2-5 degrees beyond the edge of a face which subtended 8 - 17 degrees at the retina. Moreover, the stimulus selectivity between faces was maintained this far eccentric within the receptive field. These results held even across the visual midline. It was also shown that these neurons code for identity and not fixation position, in that there was approximately six times more information in the responses of these neurons about which face had been seen than about where the monkey fixated on the face. It is concluded that at least some of these neurons in the temporal lobe visual areas do have considerable translation invariance so that this is a computation which must be performed in the visual system. Ways in which the translation and size invariant representations shown to be present in the brain by these studies could be built are considered below in section 13. It is clearly important that translation invariance in the visual system is made explicit in the neuronal responses, for this simplifies greatly the output of the visual system to memory systems such as the hippocampus and amygdala, which can then remember, or form associations about, objects. The function of these memory systems would be almost impossible if there were no consistent output from the visual system about objects (including faces), for then the memory systems would need to learn about all possible sizes, positions etc of each object, and there would be no easy generalization from one size or position of an object to that object when seen with another retinal size or position.

Until now, research on translation invariance has considered the case in which there is only one object in the visual field. The question then arises of how the visual system operates in a cluttered environment. Do all objects that can activate an inferior temporal neuron do so whenever they are anywhere within the large receptive fields of inferior temporal neurons? If so, the output of the visual system might be confusing for structures which receive inputs from the temporal cortical visual areas. To investigate this we measured the responses of inferior temporal cortical neurons with face-selective responses of rhesus macaques performing a visual fixation task. We found that the response of neurons to an effective face centred 8.5 degrees from the fovea was decreased to 71% if an ineffective face stimulus for that cell was present at the fovea. If an ineffective stimulus for a cell is introduced parafoveally when an effective stimulus is being fixated, then there was a similar reduction in the responses of neurons. More concretely, the mean firing rate across all cells to a fixated effective face with a non-effective face in the periphery was 34 spikes/s. On the other hand, the average response to a fixated non-effective face with an effective face in the periphery was 22 spikes/s. (These firing rates reflected the fact that in this population of neurons, the mean response for an effective face was 49 spikes/s with the face at the fovea, and 35 spikes/s with the face 8.5 degrees from the fovea.) Thus these cells gave a reliable output about which stimulus is actually present at the fovea, in that their response was larger to a fixated effective face than to a fixated non-effective face, even when there are other parafoveal stimuli ineffective or effective for the cell [69]. Thus the cell provides information biased towards what is present at the fovea, and not equally about what is present anywhere in the visual field. This makes the interface to action simpler, in that what is at the fovea can be interpreted (e.g. by an associative memory) partly independently of the surroundings, and choices and actions can be directed if appropriate to what is at the fovea (cf [4]). These findings are a first step towards understanding how the visual system functions in a normal environment.

8. A VIEW-INDEPENDENT REPRESENTATION OF VISUAL INFORMATION

For recognizing and learning about objects (including faces), it is important that an output of the visual system should be not only translation and size invariant, but also relatively view invariant. In an investigation of whether there are such neurons, we found that some temporal cortical neurons reliably responded differently to the faces of two different individuals independently of viewing angle [24], although in most cases (16/18 neurons) the response was not perfectly view-independent. Mixed together in the same cortical regions there are neurons with view-dependent responses (e.g. [23]). Such neurons might respond for example to a view of a profile of a monkey but not to a full-face view of the same monkey [39]. These findings, of view-dependent, partially view independent, and view independent representations in the same cortical regions are consistent with the hypothesis discussed below that view-independent representations are being built in these regions by associating together neurons that respond to different views of the same individual. These findings also provide evidence that the outputs of the visual system are likely to include representations of what is being seen, in a view independent way that would be useful for object recognition and for learning associations about objects; and in a view-based way that would be useful in social interactions to determine whether another individual is looking at one, and for selecting details of motor responses, for which the orientation of the object with respect to the viewer is required.

Further evidence that some neurons in the temporal cortical visual areas have object-based rather than view-based responses comes from a study of a population of neurons that responds to moving faces [24]. For example, four neurons responded vigorously to a head undergoing ventral flexion, irrespective of whether the view of the head was full face, of either profile, or even of the back of the head. These different views could only be specified as equivalent in object-based coordinates. Further, for all of the 10 neurons that were tested in this way, the movement specificity was maintained across inversion, responding for example to ventral flexion of the head irrespective of whether the head was upright or inverted. In this procedure, retinally encoded or viewer-centered movement vectors are reversed, but the object-based description remains the same. It was of interest that the neurons tested generalized across different heads performing the same movements.

Also consistent with object-based encoding is the finding of a small number of neurons which respond to images of faces of a given *absolute* size, irrespective of the retinal image size [60].

9. DIFFERENT NEURAL SYSTEMS ARE SPECIALIZED FOR RECOGNITION AND FOR FACE EXPRESSION DECODING

To investigate whether there are neurons in the cortex in the anterior part of the superior temporal sulcus of the macaque monkey which could provide information about facial expression [45-48,52], neurons were tested with facial stimuli which included examples of the same individual monkey with different facial expressions [24]. The responses of 45 neurons with responses selective for faces were measured to a set of 3 individual monkey faces with three expressions for each monkey, as well as to human expressions. Of these neurons, 15 showed response differences to different identities independently of expression, and 9 neurons showed responses which depended on expression but were independent of identity, as measured by a two-way ANOVA. Multidimensional scaling confirmed this result, by showing that for the first set of neurons the faces of different individuals but not expressions were well separated in the space, whereas for the second group of neurons, different expressions but not the faces of different individuals were well separated in the space. The neurons responsive to expression were found primarily in the cortex in the superior temporal sulcus, while the neurons responsive to identity were found in the inferior temporal gyrus. These results show that there are some neurons in this region the responses of which could be useful in providing information about facial expression, of potential use in social interactions [45-48,52]. Damage to this population may contribute to the deficits in social and emotional behavior which are part of the Kluver-Bucy syndrome produced by temporal lobe damage in monkeys [29,45-48,52].

A further way in which some of these neurons may be involved in social interactions is that some of them respond to gestures, e.g. to a face undergoing ventral flexion, as described above and by Perrett et al. [38]. The interpretation of these neurons as being useful for social interactions is that in some cases these neurons respond not only to ventral head flexion, but also to the eyes lowering and the eyelids closing [24]. Now these two movements (head lowering and eyelid lowering) often occur together when a monkey is breaking social contact with another, e.g. after a challenge, and the information being conveyed by such a neuron could thus reflect the presence of this social gesture. That the same neuron could respond to such different, but normally co-occurrent, visual inputs could be accounted for by the Hebbian competitive self-organization described below. It may also be noted that it is important when decoding facial expression not to move entirely into the object-based domain (in which the description would be in terms of the object itself, and would not contain information about the position and orientation of the object relative to the observer), but to retain some information about the head direction of the face stimulus being seen relative to the observer, for this is very important in determining whether a threat is being made in your direction. The presence of view-dependent representations in some of these cortical regions is consistent with this requirement. Indeed, it may be suggested that the cortex in the superior temporal sulcus, in which neurons are found with responses related to facial expression [23], head and face movement involved in for example gesture [24], and eye gaze [39], may be more related to face expression decoding; whereas the TE areas (more ventral, mainly in the macaque inferior temporal gyrus), in which neurons tuned to face identity [23] and with view-independent responses [24] are more likely to be found, may be more related to an object-based representation of identity. Of course, for appropriate social and emotional responses, both types of subsystem would be important, for it is necessary to know both the direction of a social gesture, and the identity of the individual, in order to make the correct social or emotional response.

Outputs from the temporal cortical visual areas reach the amygdala and the orbitofrontal cortex, and evidence is accumulating that these brain areas are involved in social and emotional responses to faces [52,54-57]. For example, lesions of the amygdala in monkeys disrupt social and emotional responses to faces, and we have identified a population of neurons with face-selective responses in the primate amygdala [29], some of which may respond to facial and body gesture [14]. We (observations of E.T.Rolls and H.D.Critchley), and Wilson et al. [87], have also found a small number of face-responsive neurons in the orbitofrontal cortex, and also in the ventral striatum, which receives projections from the amygdala and orbitofrontal cortex [86].

We have applied this research to the study of humans with frontal lobe damage, to try to develop a better understanding of the social and emotional changes which may occur in these patients. Impairments in the identification of facial and vocal emotional expression were demonstrated in a group of patients with ventral frontal lobe damage who had behavioural problems such as disinhibited or socially inappropriate behaviour [25]. A group of patients with lesions outside this brain region, without these behavioural problems, was unimpaired on the expression identification tests. The impairments shown by the frontal patients on these expression identification tests could occur independently of perceptual difficulties. Face expression identification was severely impaired in some patients whose recognition of the identity of faces was normal. Severe impairments on the vocal expression test (which consisted of non-verbal emotional sounds) were found in patients who produced excellent imitations of the sounds they could not identify, and whose identification of environmental sounds was also normal.

These findings suggest that some of the social and emotional problems associated with ventral frontal lobe or amygdala damage may be related to a difficulty in identifying correctly facial (and vocal) expression [25]. The question then arises of what functions are performed by the orbitofrontal cortex and amygdala with the face-related outputs they receive from the temporal cortical visual areas. The hypothesis has been developed that these regions are important in emotional and social behaviour because of their role in reward-related learning [47,48,52,58]. The amygdala is especially involved in learning associations between visual stimuli and primary (unlearned) rewards and punishments such as food taste and touch, and the orbitofrontal cortex is especially involved in the rapid reversal (i.e. adjustment or relearning) of such stimulus reinforcement associations. According to this hypothesis, the importance of projecting face-related information to the amygdala and orbitofrontal cortex is so that they can learn associations between faces, using information about both face identity and facial expression, and rewards and punishments. Now it is particularly in primate social behaviour that rapid relearning about individuals, identified by their face, and depending on their facial expression, must occur very rapidly and flexibly, to keep up with the continually changing social exchanges between different individuals and groups of individuals. It is crucial to be able to remember recent reinforcement associations of different individuals, and to be able to continually adjust these. It is suggested that these factors have led to the very major development of the orbitofrontal cortex in primates, to receive appropriate inputs (about identity from faces, and about facial expression), and to provide a very rapid and flexible learning mechanism for the current reinforcement associations of these inputs. Consistent with this, the same patients that are impaired in face expression identification are also impaired on stimulus-reinforcement relearning tasks such as visual discrimination reversal and extinction [67]. Moreover, this learning impairment is highly correlated with the social and behavioural changes found in these patients [67].

10. LEARNING OF NEW REPRESENTATIONS IN THE TEMPORAL CORTICAL VISUAL AREAS

Given the fundamental importance of a computation which results in relatively finely tuned neurons which across ensembles but not individually specify objects including individual faces in the environment, we have investigated whether experience plays a role in determining the selectivity of single neurons which respond to faces. The hypothesis being tested was that visual experience might guide the formation of the responsiveness of neurons so that they provide an economical and ensemble-encoded representation of items actually present in the environment. To test this, we investigated whether the responses of temporal cortex face-selective neurons were at all altered by the presentation of new faces which the monkey had never seen before. It might be for example that the population would make small adjustments in the responsiveness of its individual neurons, so that neurons would acquire response tuning that would enable the population as a whole to discriminate between the faces actually seen. We thus investigated whether when a set of totally novel faces was introduced, the responses of these neurons were fixed and stable from the first presentation, or instead whether there was some adjustment of responsiveness over repeated presentations of the new faces. First, it was shown for each neuron tested that its responses were stable over 5-15 repetitions of a set of familiar faces. Then a set of new faces was shown in random order (with 1 s for each presentation), and the set was repeated with a new random order over many iterations. Some of the neurons studied in this way altered the relative degree to which they responded to the different members of the set of novel faces over the first few (1-2) presentations of the set [62]. If in a different experiment a single novel face was introduced when the responses of a neuron to a set of familiar faces was being recorded, it was found that the responses to the set of familiar faces were not disrupted, while the responses to the novel face became stable within a few presentations. Thus there is now some evidence from these experiments that the response properties of neurons in the temporal lobe visual cortex are modified by experience, and that the modification is such that when novel faces are shown, the relative responses of individual neurons to the new faces alter. It is suggested that alteration of the tuning of individual neurons in this way results in a good discrimination over the population as a whole of the faces known to the monkey. This evidence is consistent with the categorisation being performed by self-organizing competitive neuronal networks, as described below and elsewhere [49-51].

Further evidence that these neurons can learn new representations very rapidly comes from an experiment in which binarized black and white images of faces which blended with the background were used. These did not activate face-selective neurons. Full grey-scale images of the same photographs were then shown for ten 0.5s presentations. It was found that in a number of cases, if the neuron happened to be responsive to that face, that when the binarized version of the same face was shown next, the neurons responded to it [64]. This is a direct parallel to the same phenomenon which is observed psychophysically, and provides dramatic evidence that these neurons are influenced by only a very few seconds (in this case 5 s) of experience with a visual stimulus.

Such rapid learning of representations of new objects appears to be a major type of learning in which the temporal cortical areas are involved. Ways in which this learning could occur are considered below. It is also the case that there is a much shorter term form of memory in which some of these neurons are involved, for whether a particular visual stimulus (such as a face) has been seen recently, for some of these neurons respond differently to recently seen stimuli in short term visual memory tasks [9,34], and neurons in a more ventral cortical area respond during the delay in a short term memory task [35].

11. THE SPEED OF PROCESSING IN THE TEMPORAL CORTICAL VISUAL AREAS

Given that there is a whole sequence of visual cortical processing stages including V1, V2, V4, and the posterior inferior temporal cortex to reach the anterior temporal cortical areas, and that the response latencies of neurons in V1 are about 40-50 ms, and in the anterior inferior temporal cortical areas approximately 80-100 ms, each stage may need to perform processing for only 15-30 ms before it has performed sufficient processing to start influencing the next stage. Consistent with this, response latencies between V1 and the inferior temporal cortex increase from stage to stage [75]. This seems to imply very fast computation by each cortical area, and therefore to place constraints on the type of processing performed in each area that is necessary for final object identification. We note that rapid identification of visual stimuli is important in social and many other situations, and that there must be strong selective pressure for rapid identification. For these reasons, we have investigated the speed of processing quantitatively, as follows.

In a first approach, we measured the information available in short temporal epochs of the responses of temporal cortical face-selective neurons about which face had been seen. We found that if a period of the firing rate of 50 ms was taken, then this contained 84.4% of the information available in a much longer period of 400 ms about which of four faces had been seen. If the epoch was as little as 20 ms, the information was 65% of that available from the firing rate in the 400 ms period [76]. These high information yields were obtained with the short epochs taken near the start of the neuronal response, for example in the post-stimulus period 100-120 ms. Moreover, we were able to show that the firing rate in short periods taken near the start of the neuronal response was highly correlated with the firing rate taken over the whole response period, so that the information available was stable over the whole response period of the neurons [76]. We were able to extend this finding to the case when a much larger stimulus set, of 20 faces, was used. Again, we found that the information available in short (e.g. 50 ms) epochs was a considerable proportion (e.g. 65%) of that available in a 400 ms long firing rate analysis period [78]. These investigations thus showed that there was considerable information about which stimulus had been seen in short time epochs near the start of the response of temporal cortex neurons.

The next approach was to address the issue of for how long a cortical area must be active to mediate object recognition. This approach used a visual backward masking paradigm. In this paradigm there is a brief presentation of a test stimulus which is rapidly followed (within 1-100 ms) by the presentation of a second stimulus (the mask), which impairs or masks the perception of the test stimulus. This paradigm used psychophysically leaves unanswered for how long visual neurons actually fire under the masking condition at which the subject can just identify an object. Although there has been a great deal of psychophysical investigation with the visual masking paradigm [13,26,83], there is very little direct evidence on the effects of visual masking on neuronal activity. For example, it is possible that if a neuron is well tuned to one class of stimulus, such as faces, that a pattern mask which does not activate the neuron, will leave the cell firing for some time after the onset of the pattern mask. In order to obtain direct neurophysiological evidence on the effects of backward masking of neuronal activity, we analysed the effects of backward masking with a pattern mask on the responses of single neurons to faces [65]. This was performed to clarify both what happens with visual backward masking, and to show how long neurons may respond in a cortical area when perception and identification are just possible. When there was no mask the cell responded to a 16 ms presentation of the test stimulus for 200-300 ms, far longer than the presentation time. It is suggested that this reflects the operation of a short term memory system implemented in cortical circuitry, the importance of which in learning invariant representations is considered below in section 13. If the mask was a stimulus which did not stimulate the cell (either a non-face pattern mask consisting of black and white letters N and O, or a face which was a non-effective stimulus for that cell), then as the interval between the onset of the test stimulus and the onset of the mask stimulus (the stimulus onset asynchrony, SOA) was reduced, the length of time for which the cell fired in response to the test stimulus was reduced. This reflected an abrupt interruption of neuronal activity produced by the effective face stimulus. When the SOA was 20 ms, face-selective neurons in the inferior temporal cortex of macaques responded for a period of 20-30 ms before their firing was interrupted by the mask [65]. We went on to show that under these conditions (a test-mask stimulus onset asynchrony of 20 ms), human observers looking at the same displays could just identify which of 6 faces was shown [66].

These results provide evidence that a cortical area can perform the computation necessary for the recognition of a visual stimulus in 20-30 ms. This provides a fundamental constraint which must be accounted for in any theory of cortical computation. The results emphasise just how rapidly cortical circuitry can operate. This rapidity of operation has obvious adaptive value, and allows the rapid behavioral responses to the faces and face expressions of different individuals which are a feature of primate social and emotional behaviour. Moreover, although this speed of operation does seem fast for a network with recurrent connections (mediated by e.g. recurrent collateral or inhibitory interneurons), recent analyses of networks with analog membranes which integrate inputs, and with spontaneously active neurons, shows that such networks can settle very rapidly [79,82].

12. VIEW-INVARIANT REPRESENTATIONS OF OBJECTS IN THE INFERIOR TEMPORAL VISUAL CORTEX

The majority of the neurophysiological experiments described above were performed with faces as stimuli, partly because cells which respond to faces can be found relatively reliably on tracks made into the temporal cortical visual areas, enabling systematic investigations to be performed. There may be many such cells, because many different individuals within the class face must be recognised. However, the principles of operation of the neural mechanisms which underlie the representation of objects may be quite similar, and these studies of the representation of faces were performed with the much broader objective of understanding visual representations of all types of stimuli in the higher parts of the visual system. In recent experiments, we (M.C.Booth and E.T.Rolls) have extended the types of analysis described above to objects. To investigate whether view-invariant representations of objects are encoded by some neurons in the inferior temporal cortex of the rhesus macaque, the activity of single neurons was recorded while monkeys were shown very different views of 10 objects. The stimuli were presented for 0.5 s on a colour video monitor while the monkey performed a visual fixation task. The stimuli were images of 10 real plastic objects which had been in the monkey's cage for several weeks, to enable him to build 3D representations of the objects. Control stimuli were views of objects which had never been seen as real objects. The neurons analyzed were in the TE cortex in and close to the ventral lip of the anterior part of the superior temporal sulcus.

In the experiments to date, many neurons have been found that respond to some views of some objects. In some cases, putative features that account for the responses of these neurons can be identified (e.g. overall shape, texture, or colour).

For a smaller number of neurons, the responses occur only to a subset of the objects, irrespective of the view of the objects. These neurons thus convey information about which object has been seen, independently of view, as confirmed by information theoretic analysis of the neuronal responses. Each neuron did not, in general, respond to only one object, but instead responded to a subset of the objects. They thus showed ensemble, sparse-distributed, encoding. The amount of information avaailable from an ensemble of these neurons about which object had been seen was shown to increase linearly with the number of neurons in the sample. This shows that, just as for faces, there is a view invariant representation of objects in the temporal cortical visual areas which can code information about objects by the firing rates of the different neurons, and that the capacity of the representation increases exponentially with the number of neurons in the sample. These neurons can be activated by the sight of the real 3D object, or by different 2D views of each object. The possibility that the representation of the objects is based on learning about the 2D views of each object is considered in the next two sections.

13. POSSIBLE COMPUTATIONAL MECHANISMS IN THE VISUAL CORTEX FOR OBJECT RECOGNITION

The neurophysiological findings described above, and wider considerations on the possible computational properties of the cerebral cortex [49,50,55,57], lead to the following outline working hypotheses on object recognition by visual cortical mechanisms. The principles underlying the processing of faces and other objects may be similar, but more neurons may become allocated to represent different aspects of faces because of the need to recognise the faces of many different individuals, that is to identify many individuals within the category faces.

Cortical visual processing for object recognition is considered to be organized as a set of hierarchically connected cortical regions consisting at least of V1, V2, V4, posterior inferior temporal cortex (TEO), inferior temporal cortex (e.g. TE3, TEa and TEm), and anterior temporal cortical areas (e.g. TE2 and TE1). (This stream of processing has many connections with a set of cortical areas in the anterior part of the superior temporal sulcus, including area TPO.) There is convergence from each small part of a region to the succeeding region (or layer in the hierarchy) in such a way that the receptive field sizes of neurons (e.g. 1 degree near the fovea in V1) become larger by a factor of approximately 2.5 with each succeeding stage (and the typical parafoveal receptive field sizes found would not be inconsistent with the calculated approximations of e.g. 8 degrees in V4, 20 degrees in TEO, and 50 degrees in inferior temporal cortex [12]) (see Fig. 4). Such zones of convergence would overlap continuously with each other (see Fig. 4). This connectivity would be part of the architecture by which translation invariant representations are computed. Each layer is considered to act partly as a set of local self-organising competitive neuronal networks with overlapping inputs. (The region within which competition would be implemented would depend on the spatial properties of inhibitory interneurons, and might operate over distances of 1-2 mm in the cortex.) These competitive nets operate by a single set of forward inputs leading to (typically non-linear, e.g. sigmoid) activation of output neurons; of competition between the output neurons mediated by a set of feedback inhibitory interneurons which receive from many of the principal (in the cortex, pyramidal) cells in the net and project back (via inhibitory interneurons) to many of the principal cells which serves to decrease the firing rates of the less active neurons relative to the rates of the more active neurons; and then of synaptic modification by a modified Hebb rule, such that synapses to strongly activated output neurons from active input axons strengthen, and from inactive input axons weaken (see [51]). (A biologically plausible form of this learning rule that operates well in such networks is

$$\delta w_{ij} = k \cdot f_i \cdot (r'_j - w_{ij})$$

where k is a learning rate constant, r'_j is the jth input to the neuron, f_i is the output of the ith neuron, w_{ij} is the jth weight on the ith neuron and f_i is a non-linear function of the output activation which mimics the operation of the NMDA receptors in learning; see [47-49]). Such competitive networks operate to detect correlations between the activity of the input neurons, and to allocate output neurons to respond to each cluster of such correlated inputs. These networks thus act as categorisers. In relation to visual information processing, they would remove redundancy from the input representation, and would develop low entropy representations of the information (cf. [6,7]). Such competitive nets are biologically plausible, in that they utilise Hebb-modifiable forward excitatory connections, with competitive inhibition mediated by cortical inhibitory neurons. The competitive scheme I suggest would not result in the formation of "winner-take-all" or "grandmother" cells, but would instead result in a small ensemble of active neurons representing each input [49-51]. The scheme has the advantages that the output neurons learn better to distribute themselves between the input patterns (cf. [11]), and that the sparse representations formed (which provide "coarse coding") have utility in maximising the number of memories that can be stored when, towards the end of the visual system, the visual representation of objects is interfaced to associative memory [49,50,63]. In that each neuron has graded responses centred about an optimal input, the proposal has some of the advantages with respect to hypersurface reconstruction described by Poggio and Girosi [44]. However, the system I propose learns differently, in that instead of using perhaps non biologically-plausible algorithms to optimally locate the centres of the receptive fields of the neurons, the neurons use graded competition to spread themselves throughout the input space, depending on the statistics of

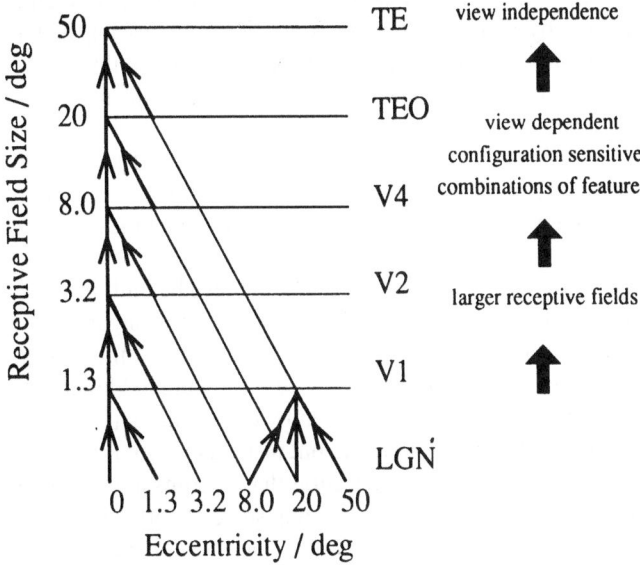

Figure 4. Schematic diagram showing convergence achieved by the forward projections in the visual system, and the types of representation that may be built by competitive networks operating at each stage of the system from the primary visual cortex (V1) to the inferior temporal visual cortex (area TE) (see text). LGN - lateral geniculate nucleus. Area TEO forms the posterior inferior temporal cortex. The receptive fields in the inferior temporal visual cortex (e.g. in the TE areas) cross the vertical midline (not shown).

the inputs received, and perhaps with some guidance from Backprojections (see below). The finite width of the response region of each neuron which tapers from a maximum at the centre is important for enabling the system to generalise smoothly from the examples with which it has learned (cf [43,44]), to help the system to respond for example with the correct invariances as described below.

Translation invariance would be computed in such a system by utilising competitive learning to detect regularities in inputs when real objects are translated in the physical world. The hypothesis is that because objects have continuous properties in space and time in the world, an object at one place on the retina might activate feature analyzers at the next stage of cortical processing, and when the object was translated to a nearby position, because this would occur in a short period (e.g. 0.5 s), the membrane of the postsynaptic neuron would still be in its "Hebb-modifiable" state (caused for example by calcium entry as a result of the voltage dependent activation of NMDA receptors), and the presynaptic afferents activated with the object in its new position would thus become strengthened on the still-activated postsynaptic neuron. It is suggested that the short temporal window (e.g. 0.5s) of Hebb-modifiability helps neurons to learn the statistics of objects moving in the physical world, and at the same time to form different representations of different feature combinations or objects, as these are physically discontinuous and present less regular correlations to the visual system. Foldiak [21] has proposed computing an average activation of the postsynaptic neuron to assist with the same problem. One idea here is that the temporal properties of the biologically implemented learning mechanism are such that it is well suited to detecting the relevant continuities in the world of real objects. Another suggestion is that a memory trace for what has been seen in the last 300 ms appears to be implemented by a mechanism as simple as continued firing of inferior temporal neurons after the stimulus has disappeared, as was found in the masking experiments described above (see also [65,66]). I also suggest that other invariances, for example size, spatial frequency, and rotation invariance, could be learned by a comparable process. (Early processing in V1 which enables different neurons to represent inputs at different spatial scales would allow combinations of the outputs of such neurons to be formed at later stages. Scale invariance would then result from detecting at a later stage which neurons are almost conjunctively active as the size of an object alters.) It is suggested that this process takes place at each stage of the multiple-layer cortical processing hierarchy, so that invariances are learned first over small regions of space, and then over successively larger regions. This limits the size of the connection space within which correlations must be sought.

Increasing complexity of representations could also be built in such a multiple layer hierarchy by similar mechanisms. At each stage or layer the self-organizing competitive nets would result in combinations of inputs becoming the effective stimuli for neurons. In order to avoid the combinatorial explosion, it is proposed, following Feldman [20], that low-order combinations of inputs would be what is learned by each neuron. (Each input would not be represented by activity in a single input axon, but instead by activity in a set of active input axons.) Evidence consistent with this suggestion that neurons are responding to combinations of a few variables represented at the preceding stage of cortical processing is that some neurons in V2 and V4 respond to end-stopped lines, to tongues flanked by inhibitory subregions, or to combinations of colours (see references cited by Rolls [53]); in posterior inferior temporal cortex to stimuli which may require two or more simple features to be present [74]; and in the temporal cortical face processing areas to images that require the presence of several features in a face (such as eyes, hair, and mouth) in order to respond (see above and [88]). (Precursor cells to face-responsive neurons might, it is suggested, respond to combinations of the outputs of the neurons in V1 that are activated by faces, and might be found in areas such as V4.) It is an important part of this suggestion that some local spatial information would be inherent in the features which were being combined. For example, cells might not respond to the combination of an edge and a small circle unless they were in the correct spatial relation to each other. (This is in fact consistent with the data of Tanaka et al. [74], and with our data on face neurons, in that some faces neurons require the face features to be in the correct spatial configuration, and not jumbled [66].) The local spatial information in the features being combined would ensure that the representation at the next level would contain some information about the (local) arrangement of features. Further low-order combinations of such neurons at the next stage would include sufficient local spatial information so that an arbitrary spatial arrangement of the same features would not activate the same neuron, and this is the proposed, and limited, solution which this mechanism would provide for the feature binding problem (cf. [31]). By this stage of processing a view-dependent representation of objects suitable for view-dependent processes such as behavioural responses to face expression and gesture would be available.

It is suggested that view-independent representations could be formed by the same type of computation, operating to combine a limited set of views of objects. The plausibility of providing view-independent recognition of objects by combining a set of different views of objects has been proposed by a number of investigators [27,30,42]. Consistent with the suggestion that the view-independent representations are formed by combining view-dependent representations in the primate visual system, is the fact that in the temporal cortical areas, neurons with view-independent representations of faces are present in the same cortical areas as neurons with view-dependent representations (from which the view-independent neurons could receive inputs) [24,40]. This solution to "object-based" representations is very different from that traditionally proposed for artificial vision systems, in which the coordinates in 3D-space of objects are stored in a database, and general-purpose algorithms operate on these to perform transforms such as translation, rotation, and scale change in 3D space (e.g. [32]). In the present, much more limited but more biologically plausible scheme, the representation would be suitable for recognition of an object, and for linking associative memories to objects, but would be less good for making actions in 3D-space to particular parts of, or inside, objects, as the 3D coordinates of each part of the object would not be explicitly available. It is therefore proposed that visual fixation is used to locate in foveal vision part of an object to which movements must be made, and that local disparity and other measurements of depth then provide sufficient information for the motor system to make actions relative to the small part of space in which a local, *view-dependent*, representation of depth would be provided (cf. [3]).

The computational processes proposed above operate by an unsupervised learning mechanism, which utilises regularities in the physical environment to enable representations with low entropy to be built. In some cases it may be advantageous to utilise some form of mild teaching input to the visual system, to enable it to learn for example that rather similar visual inputs have very different consequences in the world, so that different representations of them should be built. In other cases, it might be helpful to bring representations together, if they have identical consequences, in order to use storage capacity efficiently. It is proposed elsewhere [49,50] that the Backprojections from each adjacent cortical region in the hierarchy (and from the amygdala and hippocampus to higher regions of the visual system) play such a role by providing guidance to the competitive networks suggested above to be important in each cortical area. This guidance, and also the capability for recall, are it is suggested implemented by Hebb-modifiable connections from the backprojecting neurons to the principal (pyramidal) neurons of the competitive networks in the preceding stages [49,50].

The computational processes outlined above use coarse coding with relatively finely tuned neurons with a graded response region centred about an optimal response achieved when the input stimulus matches the synaptic weight vector on a neuron. The coarse coding and fine tuning would help to limit the combinatorial explosion, to keep the number of neurons within the biological range. The graded response region would be crucial in enabling the system to generalise correctly to solve for example the invariances. However, such a system would need many neurons, each with considerable learning capacity, to solve visual perception in this way. This is fully consistent with the large number of neurons in the visual system, and with the large number of, probably modifiable, synapses on each neuron (e.g. 5000). Further, the fact that many neurons are tuned in different ways to faces is consistent with the fact that in such a computational system, many neurons would need to be sensitive (in different ways) to faces, in order to allow recognition of many individual faces when all share a number of common properties.

14. A COMPUTATIONAL MODEL OF INVARIANT VISUAL OBJECT RECOGNITION

To test and clarify the hypotheses just described about how the visual system may operate to learn invariant object recognition, we have performed a simulation which implements many of the ideas just described, and is consistent and based on much of the neurophysiology summarized above. The network simulated can perform object, including face, recognition in a biologically plausible way, and after training shows for example translation and view invariance [84,85].

In the four layer network, the successive layers correspond approximately to V2, V4, the posterior temporal cortex, and the anterior temporal cortex. The forward connections to a cell in one layer are derived from a topologically corresponding region of the preceding layer, using a Gaussian distribution of connection probabilities to determine the exact

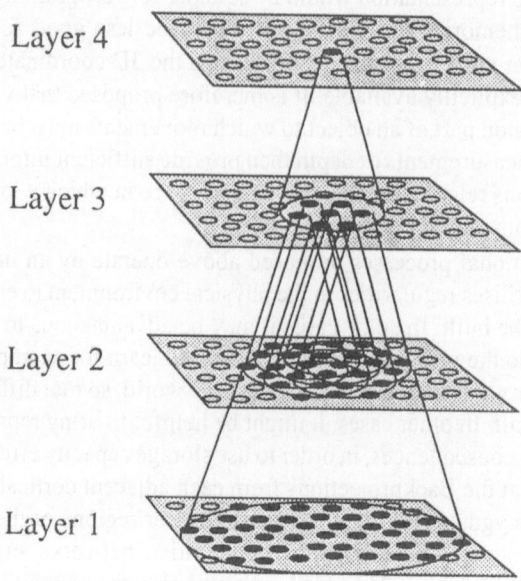

Layer 4

Layer 3

Layer 2

Layer 1

Figure 5. Hierarchical network structure of VisNet.

neurons in the preceding layer to which connections are made. This schema is constrained to preclude the repeated connection of any cells. Each cell receives 100 connections from the 32 x 32 cells of the preceding layer, with a 67% probability that a connection comes from within 4 cells of the distribution centre. Fig. 5 shows the general convergent network architecture used, and may be compared with Fig. 4. Within each layer, lateral inhibition between neurons has a radius of effect just greater than the radius of feedforward convergence just defined. The lateral inhibition is simulated via a linear local contrast enhancing filter active on each neuron. (Note that this differs from the global 'winner-take-all' paradigm implemented by Foldiak [21]). The cell activation is then passed through a non-linear cell output activation function, which also produces contrast enhancement of the firing rates.

In order that the results of the simulation might be made particularly relevant to understanding processing in higher cortical visual areas, the inputs to layer 1 come from a separate input layer which provides an approximation to the encoding found in visual area 1 (V1) of the primate visual system. These response characteristics of neurons in the input layer are provided by a series of spatially tuned filters with image contrast sensitivities chosen to accord with the general tuning profiles observed in the simple cells of V1. Currently, only even-symmetric (bar detecting) filter shapes are used. The precise filter shapes were computed by weighting the difference of two Gaussians by a third orthogonal Gaussian (see [84]). Four filter spatial frequencies (in the range 0.0625 to 0.25 pixels^{-1} over four octaves), each with one of four orientations (0° to 135°) were implemented. Cells of layer 1 receive a topologically consistent, localised, random selection of the filter responses in the input layer, under the constraint that each cell samples every filter spatial frequency and receives a constant number of inputs.

The synaptic learning rule used can be summarised as follows:

$$\delta w_{ij} = k \cdot m_i \cdot r'_j$$

and

$$m_i^t = (1 - \eta)f_i^{(t)} + \eta m_i^{(t-1)}$$

where r'_j is the jth input to the neuron, f_i is the output of the ith neuron, w_{ij} is the jth weight on the ith neuron, η governs the relative influence of the trace and the new input (typically 0.4 - 0.6), and $m_i^{(t)}$ represents the value of the ith cell's memory trace at time t. In the simulation the neuronal learning was bounded by normalisation of each cell's dendritic weight vector, as in standard competitive learning. An alternative, more biologically relevant implementation, using a local weight bounding operation, has in part been explored using a version of the Oja update rule [28,36]. To train the network to produce a translation invariant representation, one stimulus was placed successively in a sequence of 7 positions across the input, then the next stimulus was placed successively in the same sequence of 7 positions across the input, and so on through the set of stimuli. The idea was to enable the network to learn whatever was common at each stage of the network about a stimulus shown in different positions. To train on view invariance, different views of the same object were shown in succession, then different views of the next object were shown in succession, and so on.

One test of the network used a set of three non-orthogonal stimuli, based upon probable 3-D edge cues (such as 'T, L and +' shapes). During training these stimuli were chosen in random sequence to be swept across the 'retina' of the network, a total of 1000 times. In order to assess the characteristics of the cells within the net, a two-way analysis of variance was performed on the set of responses of each cell, with one factor being the stimulus type and the other the position of the stimulus on the 'retina'. A high F ratio for stimulus type (F_s), and low F ratio for stimulus position (F_p) would imply that a cell had learned a position invariant representation of the stimuli. The discrimination factor of a particular cell was then simply the ratio F_s / F_p (a factor useful for ranking at least the most invariant cells). To assess the utility of the trace learning rule, nets trained with the trace rule were compared with nets trained with standard Hebbian learning without a trace, and with untrained nets (with the initial random weights). The results of the simulations, illustrated in Fig. 6, show that networks trained with the trace learning rule do have neurons with much higher values of the discrimination factor. An example of the responses of one such cell are illustrated in Fig. 7. Similar position invariant encoding has been demonstrated for a stimulus set consisting of 8 faces. View invariant coding has also been demonstrated for a set of 5 faces each shown in 4 views [84,85].

In further investigations, we (E.T.Rolls, T.Milward and G.Wallis) have trained VisNet on much more difficult problems, to investigate whether its architecture is feasible for solving problems in which complex stimuli may fall in any position on the retina. We presented 7 face stimuli centered at every retinal location within a 33 x 33 training area of

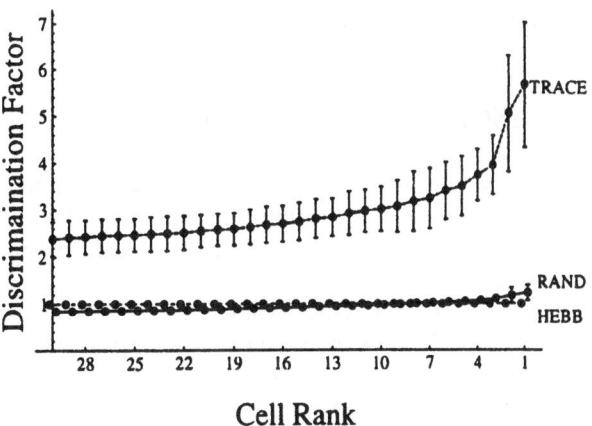

Figure 6. Comparison of network discrimination when trained with the trace learning rule, with a Hebb rule (No trace), and when not trained (Random). The stimuli were +, L and T shapes, presented in 9 different locations on the "retina." The discrimination factors (mean ± sem) (see text) for the 30 most invariant neurons for each type of learning rule are shown.

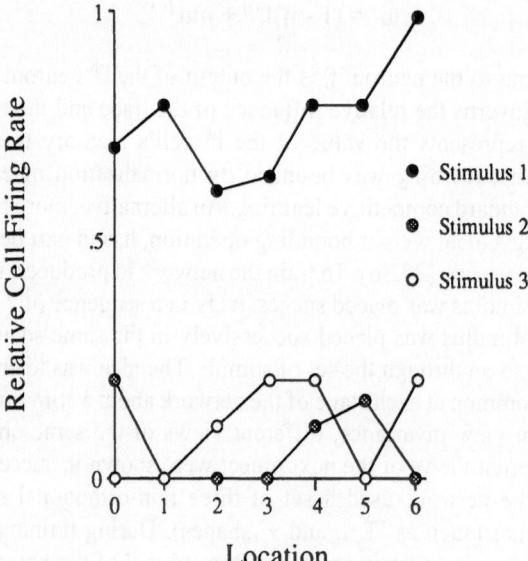

Figure 7. The responses of a layer 4 cell in the simulation illustrated in Fig. 6. The cell had a translation invariant response to stimulus 1.

the 128 by 128 "retina" of VisNet. The training locations were selected in sets of 49 for each stimulus before the next stimulus was selected. Within the 49 locations, the stimulus was moved in a way that might be produced by eye movements round a stimulus, and that would enable the trace rule to learn location invariance. In particular, the locations were selected from a random number generator in such a way that (e.g. 7) short movements were made, then a longer movement was made. The intention was to allow VisNet to learn invariance over both short and longer ranges of image translation.

Figure 8 shows the results for the 30 most selective cells when trained with the trace rule, when trained with a Hebb rule (a control not expected to lead to translation invariance), and when not trained (another control). The testing was with the stimuli centered at all 1089 locations within the 33 x 33 region of the retina in which the stimuli were centred. The results show that reasonable performance can be obtained, even with this very large number of training and test locations, provided that the trace rule is used.

These results show that the proposed learning mechanism and neural architecture can produce cells with responses selective for stimulus type with considerable position or view invariance. The ability of the network to be trained with natural scenes may also help to advance our understanding of encoding in the visual system.

15. DISCUSSION AND CONCLUSIONS

Mechanisms by which the brain could perform invariant recognition of objects including faces have been addressed neurophysiologically, and then a computational model of how this could occur was described. The investigations described here lead to the following conclusions.

Some neurons that respond primarily to faces are found in the macaque cortex in the anterior part of the superior temporal sulcus (in which region neurons are especially likely to be tuned to facial expression, and to face movement involved in gesture). They are also found more ventrally in the TE areas which form the inferior temporal gyrus. Here the

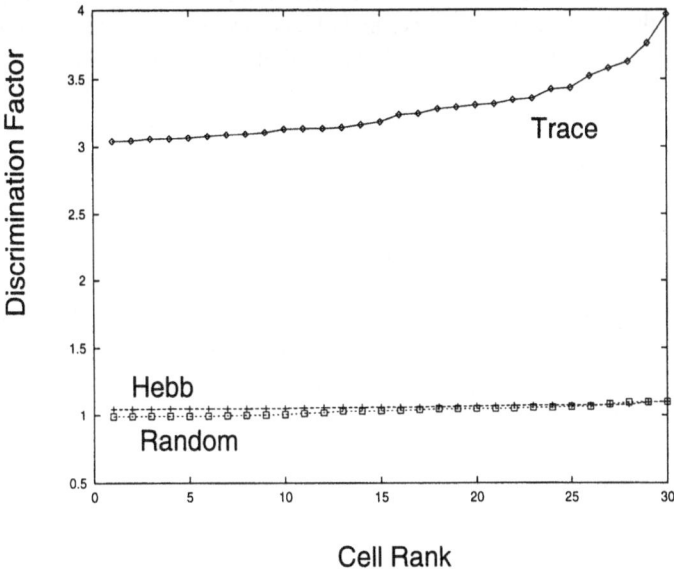

Cell Rank

Figure 8. Comparison of network discrimination when trained with the trace learning rule, with a Hebb rule (Hebb), and when not trained (Random). The stimuli were 7 faces presented in 1089 dfferent locations on the "retina." The discrimination factors (see text) for the 30 most invariant neurons for each type of learning rule are shown. The neurons operated with a sigmoid activation function.

neurons are more likely to have responses related to the identity of faces. These areas project on to the amygdala and orbitofrontal cortex, in which face-selective neurons are also found.

Quantitative studies of the responses of the neurons that respond differently to the faces of different individuals show that information about the identity of the individual is represented by the responses of a population of neurons, that is, ensemble encoding is used. The rather distributed encoding (within the class faces) about identity in these sensory cortical regions has the advantages of maximising the information in the representation useful for discrimination between stimuli, generalisation, and graceful degradation. In contrast, the more sparse representations in structures such as the hippocampus may be useful to maximise the number of different memories stored. There is evidence that the responses of some of these neurons are altered by experience so that new stimuli become incorporated in the network, in only a few seconds of experience with a new stimulus. It is shown that the representation that is built in temporal cortical areas shows considerable invariance for size, contrast, spatial frequency and translation. Thus the representation is in a form which is particularly useful for storage and as an output from the visual system. It is also shown that one of the representations which is built is view-invariant, which is suitable for recognition and as an input to associative memory. Another is viewer-centred, which is appropriate for conveying information about gesture. It is shown that these computational processes operate rapidly, in that in a backward masking paradigm, 20-40 ms of neuronal activity in a cortical area is sufficient to support face recognition.

Similar experiments have been started to analyse the representation of objects. Neurons have been found which provide a view-invariant representation of objects with which the monkey is familiar. The information in the representations shown by both the neurons that represent objects and those that represent faces increase approximately linearly with the number of neurons in the ensemble. Thus the number of stimuli that can be represented by a population of these neurons increases approximately linearly with the number of neurons in the sample. The remarkable point is that it is now possible to read off the code from the brain about how faces and objects are represented, by knowing the firing rates measured over a short period (50-500 ms) of the firing of the neurons.

In a clinical application of these findings, it is shown that humans with ventral frontal lobe damage have in some cases impairments in face and voice expression identification. These impairments are correlated with and may contribute to the problems some of these patients have in emotional and social behaviour.

To help provide an understanding of how the invariant recognition described could be performed by the brain, a neuronal network model of processing in the ventral visual system is described. The model uses a multistage feed-forward architecture, and is able to learn invariant representations of objects including faces by use of a Hebbian synaptic modification rule which incorporates a short memory trace (0.5 s) of preceding activity to enable the network to learn the properties of objects which are spatio-temporally invariant over this time scale.

ACKNOWLEDGMENTS

The author has worked on some of the investigations described here with P. Azzopardi, G.C. Baylis, M.C.Booth, P. Foldiak, M. Hasselmo, C.M. Leonard, G. Littlewort, D.I. Perrett, M.J. Tovee, A. Treves and G. Wallis, and their collaboration is sincerely acknowledged. This research was supported by the Medical Research Council, PG8513790.

REFERENCES

1. Abbott LA, Rolls ET, Tovee MJ (1996) Representational capacity of face coding in monkeys. Cerebral Cortex, in press.
2. Baizer JS, Ungerleider LG, Desimone R (1991) Organization of visual inputs to the inferior temporal and posterior parietal cortex in macaques. J Neurosci 11:168-190.
3. Ballard DH (1990) Animate vision uses object-centred reference frames. In: Eckmiller R (ed) Advanced Neural Computers. Amsterdam, North-Holland, pp 229-236.
4. Ballard DH (1993) Subsymbolic modelling of hand-eye co-ordination. In: Broadbent DE (ed) The Simulation of Human Intelligence. Blackwell, Oxford, ch 3, pp 71-102.
5. Barlow HB (1972) Single units and sensation: a neuron doctrine for perceptual psychology? Perception 1:371-394.
6. Barlow HB (1985) Cerebral cortex as model builder. In: Rose D, Dobson VG (eds) Models of the Visual Cortex. Chichester, Wiley, pp 37-46.
7. Barlow HB, Kaushal TP, Mitchison GJ (1989) Finding minimum entropy codes. Neural. Computat 1:412-423.
8. Baylis GC, Rolls ET, Leonard CM (1985) Selectivity between faces in the responses of a population of neurons in the cortex in the superior temporal sulcus of the monkey. Brain Res 342:91-102.
9. Baylis GC, Rolls ET, Leonard CM (1987) Functional subdivisions of temporal lobe neocortex. J Neurosci 7:330-342.
10. Baylis GC, Rolls ET (1987) Responses of neurons in the inferior temporal cortex in short term and serial recognition memory tasks. Exp Brain Res 65:614-622.
11. Bennett A (1990) Large competitive networks. Network 1:449-462.
12. Boussaoud D, Desimone R, Ungerleider LG (1991) Visual topography of area TEO in the macaque. J Comp Neurol 306:554-575.
13. Breitmeyer BG (1980) Unmasking visual masking: a look at the "why" behind the veil of the "how." Psychol Rev 87:52-69.
14. Brothers L, Ring B, Kling AS (1990) Response of neurons in the macaque amygdala to complex social stimuli. Behav Brain Res 41:199-213.
15. Bruce C, Desimone R, Gross CG (1981) Visual properties of neurons in a polysensory area in superior temporal sulcus of the macaque. J Neurophys 46:369-384.

16. Desimone R (1991) Face-selective cells in the temporal cortex of monkeys. J Cog Neurosci 3:1-8.
17. Desimone R, Gross CG (1979) Visual areas in the temporal lobe of the macaque. Brain Res 178:363-380.
18. Desimone R, Albright TD, Gross CG, Bruce C (1984) Stimulus-selective properties of inferior temporal neurons in the macaque. J Neurosci 4:2051-2062.
19. Engel AK, Konig P, Kreiter AK, Schillen TB, Singer W (1992) Temporal coding in the visual system: new vistas on integration in the nervous system. Trends in Neurosci 15:218-226.
20. Feldman JA (1985) Four frames suffice: a provisional model of vision and space. Behav Brain Sci 8:265-289 (see p. 279).
21. Foldiak P (1991) Learning invariance from transformation sequences. Neural Comp 3:193-199.
22. Gross CG, Desimone R, Albright TD, Schwartz EL (1985) Inferior temporal cortex and pattern recognition. Exp Brain Res Suppl 11:179-201.
23. Hasselmo ME, Rolls ET, Baylis GC (1989) The role of expression and identity in the face-selective responses of neurons in the temporal visual cortex of the monkey. Behav Brain Res 32:203-218.
24. Hasselmo ME, Rolls ET, Baylis GC, Nalwa V (1989) Object-centered encoding by face-selective neurons in the cortex in the superior temporal sulcus of the monkey. Exp Brain Res 75:417-429.
25. Hornak J, Rolls ET, Wade D (1996) Face and voice expression identification and their association with emotional and behavioural changes in patients with frontal lobe damage. Neuropsychologia 34: 247-261.
26. Humphreys GW, Bruce V (1989) Visual Cognition. Hove, Erlbaum.
27. Koenderink JJ, Van Doorn AJ (1979) Biol Cybern 32:211-217.
28. Kohonen T (1988) Self-organization and Associative Memory. New York, Springer-Verlag, 2nd Edition.
29. Leonard CM, Rolls ET, Wilson FAW, Baylis GC (1985) Neurons in the amygdala of the monkey with responses selective for faces. Behav Brain Res 15:159-176.
30. Logothetis NK, Pauls J, Bulthoff HH, Poggio T (1994) View-dependent object recognition by monkeys. Current Biol 4:401-414.
31. Malsburg, C. von der (1990) A neural architecture for the representation of scenes. In: McGaugh JL, Weinberger NM, Lynch G (eds) Brain Organization and Memory: Cells, Systems and Circuits. New York, Oxford University Press, ch 18, pp 356-372.
32. Marr D (1982) Vision. San Francisco: WH Freeman.
33. Maunsell JHR, Newsome WT (1987) Visual processing in monkey extrastriate cortex. Ann Rev Neurosci 10:363-401.
34. Miller EK, Desimone R (1994(Parallel neuronal mechanisms for short-term memory. Science 263:520-522.
35. Miyashita Y (1993) Inferior temporal cortex: where visual perception meets memory. Ann Rev Neurosci 16:245-263.
36. Oja E (1982) A simplified neuron model as a principal component analyzer. J Math Biol 15: 267-73.
37. Perrett DI, Rolls ET, Caan W (1982) Visual neurons responsive to faces in the monkey temporal cortex. Exp Brain Res 47:329-342.
38. Perrett DI, Smith PAJ, Mistlin AJ, Chitty AJ, Head AS, Potter DD, Broennimann R, Milner AD, Jeeves MA (1985a) Visual analysis of body movements by neurons in the temporal cortex of the macaque monkey: a preliminary report. Behav Brain Res 16:153-170.
39. Perrett DI, Smith PAJ, Potter DD, Mistlin AJ, Head AS, Milner D, Jeeves MA (1985b) Visual cells in temporal cortex sensitive to face view and gaze direction. Proc Roy Soc 223B:293-317.
40. Perrett DI, Mistlin AJ, Chitty AJ (1987) Visual neurons responsive to faces. Trends in Neurosc 10:358-364.
41. Poggio T (1990) A theory of how the brain might work. Cold Spring harbor Symposia in Quantitative Biology 55:899-910.
42. Poggio T, Edelman S (1990) A network that learns to recognize three-dimensional objects. Nature 343:263-266.
43. Poggio T, Girosi F (1990a) Regularization algorithms for learning that are equivalent to multilayer networks. Science 247:978-982.
44. Poggio T, Girosi F (1990b) Networks for approximation and learning. Proc IEEE, 78:1481-1497.
45. Rolls ET (1981) Responses of amygdaloid neurons in the primate. In: Ben-Ari Y (ed) The Amygdaloid Complex. Amsterdam, Elsevier, pp 383-393.
46. Rolls ET (1984) Neurons in the cortex of the temporal lobe and in the amygdala of the monkey with responses selective for faces. Human Neurobiol 3:209-222.
47. Rolls ET (1986a) A theory of emotion, and its application to understanding the neural basis of emotion. In: Oomura Y (ed) Emotions. Neural and Chemical Control. Tokyo and Karker, Basel, Japan Scientific Societies Press, pp 325-344.

48. Rolls ET (1986b) Neural systems involved in emotion in primates. In: Plutchik R, Kellerman H (eds) Emotion: Theory, Research, and Experience, Volume 3, Biological Foundations of Emotion. New York, Academic Press, ch 5, pp 125-143.

49. Rolls ET (1989a) Functions of neuronal networks in the hippocampus and neocortex in memory. In: Byrne JH, Berry WO (eds) Neural Models of Plasticity: Experimental and Theoretical Approaches. San Diego, Academic Press, ch 13, pp 240-265.

50. Rolls ET (1989b) The representation and storage of information in neuronal networks in the primate cerebral cortex and hippocampus. In: Durbin R, Miall C, Mitchison G (eds) The Computing Neuron. Wokingham, England, Addison-Wesley, ch 8, pp 125-159.

51. Rolls ET (1989c) Functions of neuronal networks in the hippocampus and cerebral cortex in memory. In: Coterill RMJ (ed) Models of Brain Function. Cambridge, Cambridge University Press, pp 15-33.

52. Rolls ET (1990) A theory of emotion, and its application to understanding the neural basis of emotion. Cog and Emot 4:161-190.

53. Rolls ET (1991) Neural organisation of higher visual functions. Curr Op Neurobiol 1:274-278.

54. Rolls ET (1992a) Neurophysiology and functions of the primate amygdala. In: Aggleton JP (ed) The Amygdala. New York, Wiley-Liss, ch 5, pp 143-165.

55. Rolls ET (1992b) Neurophysiological mechanisms underlying face processing within and beyond the temporal cortical visual areas. Phil Trans Roy Soc 335:11-21.

56. Rolls ET (1992c) The processing of face information in the primate temporal lobe. In: Bruce V, Burton M (eds) Processing Images of Faces. Ablex, Norwood, New Jersey, ch 3, pp 41-68.

57. Rolls ET (1994) Brain mechanisms for invariant visual recognition and learning. Behavioural Processes 33:113-138.

58. Rolls ET (1995) A theory of emotion and consciousness, and its application to understanding the neural basis of emotion. In: Gazzaniga MS (ed) The Cognitive Neurosciences. Cambridge, Mass, MIT Press, ch 72, pp 1091-1106.

59. Rolls ET, Baylis GC, Leonard CM (1985) Role of low and high spatial frequencies in the face-selective responses of neurons in the cortex in the superior temporal sulcus. Vis Res 25:1021-1035.

60. Rolls ET, Baylis GC (1986) Size and contrast have only small effects on the responses to faces of neurons in the cortex of the superior temporal sulcus of the monkey. Exp Brain Res 65:38-48.

61. Rolls ET, Baylis GC, Hasselmo ME (1987) The responses of neurons in the cortex in the superior temporal sulcus of the monkey to band-pass spatial frequency filtered faces. Vis Res 27:311-326.

62. Rolls ET, Baylis GC, Hasselmo ME, Nalwa V (1989) The effect of learning on the face-selective responses of neurons in the cortex in the superior temporal sulcus of the monkey. Exp Brain Res 76:153-164.

63. Rolls ET, Treves A (1990) The relative advantages of sparse versus distributed encoding for associative neuronal networks in the brain. Network 1:407-421.

64. Rolls ET, Tovee MJ, Ramachandran VS (1993) Visual learning reflected in the responses of neurons in the temporal visual cortex of the macaque. Soc Neurosci Abs 19:27.

65. Rolls ET, Tovee MJ (1994) Processing speed in the cerebral cortex, and the neurophysiology of visual backward masking. Proc Roy Soc B 257:9-15.

66. Rolls ET, Tovee MJ, Purcell DG, Stewart AL, Azzopardi P (1994) The responses of neurons in the temporal cortex of primates, and face identification and detection. Exp Brain Res 101:474-484.

67. Rolls ET, Hornak J, Wade D, McGrath J (1994) Emotion-related learning in patients with social and emotional changes associated with frontal lobe damage. J Neurol, Neurosurg Psychiat 57:1518-1524.

68. Rolls ET, Tovee MJ (1995a) Sparseness of the neuronal representation of stimuli in the primate temporal visual cortex. J Neurophys 73:713-726.

69. Rolls ET, Tovee MJ (1995b) The responses of single neurons in the temporal visual cortical areas of the macaque when more than one stimulus is present in the visual field. Exp Brain Res 103:409-420.

70. Rolls ET, Treves A, Tovee MJ (1996) The representational capacity of the distributed encoding of information provided by populations of neurons in the primate temporal visual cortex.

71. Seltzer B, Pandya DN (1978) Afferent cortical connections and architectonics of the superior temporal sulcus and surrounding cortex in the rhesus monkey. Brain Res 149:1-24.

72. Simmen MW, Rolls ET, Treves A (1996) On the dynamics of a network of spiking neurons. In Computations and Neuronal Systems: Proceedings of CNS95, eds. F.H.Eekman and J.M.Bower. Kluwer: Boston.

73. Smith DV, Travers JB (1979) A metric for the breadth of tuning of gustatory neurons. Chem Sens 4:215-229.

74. Tanaka K, Saito C, Fukada Y, Moriya M (1990) Integration of form, texture, and color information in the inferotemporal cortex of the macaque. In: Iwai E, Mishkin M (eds) Vision, Memory and the Temporal Lobe. New York, Elsevier, ch 10, pp 101-109.

75. Thorpe SJ, Imbert M (1989) Biological constraints on connectionist models. In: Pfeifer R, Schreter Z, Fogelman-Soulie F (eds) Connectionism in Perspective. Amsterdam, Elsevier, p 63-92.
76. Tovee MJ, Rolls ET, Treves A, Bellis RP (1993) Information encoding and the responses of single neurons in the primate temporal visual cortex. J Neurophysiol 70:640-654.
77. Tovee MJ, Rolls ET, Azzopardi P (1994) Translation invariance and the responses of neurons in the temporal visual cortical areas of primates. J Neurophysiol 72:1049-1060.
78. Tovee MJ, Rolls ET (1995) Information encoding in short firing rate epochs by single neurons in the primate temporal visual cortex. Visual Cognition 2:35-58.
79. Treves A (1993) Mean-field analysis of neuronal spike dynamics. Network 4: 259-284.
80. Treves A, Rolls ET (1991) What determines the capacity of autoassociative memories in the brain? Network 2:371-397.
81. Treves A, Rolls ET (1994) A computational analysis of the role of the hippocampus in memory. Hippocampus 4:374-391.
82. Treves A, Rolls ET, Tovee MJ (1996) On the time required for recurrent processing in the brain. In: Torre V, Conti F (eds) Neurobiology. Plenum: New York.
83. Turvey MT (1973) On the peripheral and central processes in vision: inferences from an information processing analysis of masking with patterned stimuli. Psych Rev 80:1-52.
84. Wallis G, Rolls ET, Foldiak P (1993) Learning invariant responses to the natural transformations of objects. Int Joint Conf on Neural Networks 2:1087-1090.
85. Wallis G, Rolls ET (1996) A model of invariant object recognition in the visual system.
86. Williams GV, Rolls ET, Leonard CM, Stern C (1993) Neuronal responses in the ventral striatum of the behaving macaque. Behav Brain Res 55:243-252.
87. Wilson FAW, O'Sclaidhe SP, Goldman-Rakic PS (1993) Dissociation of object and spatial processing domains in primate preforontal cortex. Science 260:1955-1958.
88. Yamane S, Kaji S, Kawano K (1988) What facial features activate face neurons in the inferotemporal cortex of the monkey? Exp Brain Res 73:209-214.

28

BIOPHYSICAL ASPECTS OF CORTICAL NETWORKS

Stefan Rotter

Max-Planck-Institut für Entwicklungsbiologie
Spemannstraße 35/IV, 72076 Tübingen, Germany

1. INTRODUCTION

Artificial neuronal networks provide attractive models for cortical function, in particular, if "cognitive" properties emerge from their structure. Unfortunately, it turns out difficult to set up classical models which are comparable to the biological system on the level of single neurons. We look at artificial neuronal networks from a fresh perspective, which has the potential to extend their merits to a detailed and quantitative description of physiological phenomena in nerve nets of spiking neurons. In fact, the framework of stochastic point processes provides the tools for the construction of mathematically consistent models, which allow for a direct comparison with electrophysiological recordings on the level of individual nerve cells, in particular, if these are part of a large network. Moreover, the estimation of model parameters from experiments becomes feasible, so that a quantitative theoretical treatment as well as computer simulations of large networks under realistic conditions can be undertaken.

The enormous richness of dynamical behavior of cortical nerve nets is not least due to the fact that different mechanisms operate at different scales in space and time—cooperative phenomena involving a large number of nerve cells are tightly intermingled with subthreshold intracellular processes. To achieve a systematic account for the dynamical repertory of such networks with the help of a model it is of particular importance to find the right level of abstraction.

Action potential firing is the most visible result of signal processing in neurons like the pyramidal cells of the mammalian neocortex. The tendency of nerve cells to generate, under certain conditions, stereotyped rapid depolarizations across their membrane has led to the metaphor of an elementary decision making unit, which can only vote "all-or-none". It is well established that different types of cortical nerve cells can generate temporal patterns of action potentials with very different characteristics. This diversity may come along with marked morphological or histochemical profiles. However, even more subtle variations in the dendritic geometry or only slight alterations in the ion channel equipment must be suspected to have impact on the dynamic repertory of a nerve cell (McCormick, Connors, Lighthall, Prince, 1985; Koch and Segev, 1989; Hille, 1992).

Neurobiology, edited by Torre and Conti,
Plenum Press, New York, 1996

Action potentials are fundamental for the communication of a cortical neuron with other neurons in its spatial vicinity and in distant cortical areas, and with neurons in the sensory periphery or subcortical structures as well. In the cortex, the degree of convergence and divergence is on the order of 10^3–10^5 depending on the species and the cell type (Braitenberg and Schüz, 1992). Action potentials may travel over very long distances by the use of active transport mechanisms in the axon. Upon their arrival at a presynaptic terminal, transmitter substance is released into the synaptic cleft. This causes changes in the kinetic parameters of specific ion channels in the postsynaptic neuron, thus altering the cell's properties with respect to action potential firing. In this way, action potential firing in one cortical nerve cell can modulate the patterns generated by a great number of other cells. The details of signal transmission may differ from neuron to neuron, and their realization may also be correlated with well-established characteristics like the transmitter system used, or with less accessible differences in some biophysical or biochemical parameter.

In contrast to the highly reproducible waveform of individual action potentials, the behavior of a nerve cell has a strong stochastic component as far as spike timing is concerned. Most cortical neurons show a considerable variability in their response to repetitive presentations of identical inputs. This is well known from electrophysiological recordings in both single cell preparations and intact brains of behaving animals. At the extreme, spike trains from individual neurons in the frontal cortex, which are generated in distinct trials of trained behavior, may have no systematic structure at all, leading to a completely flat PSTH (Vaadia, Bergman, Abeles, 1989). In contrast, it has been demonstrated recently that the precision of action potential timing can be very much improved, if the neuron is driven by appropriate current transients (Mainen and Sejnowski, 1995). Little is known, however, about the relative contributions of the various sources of stochasticity in spiking activity. Even less is known about the brain's strategy to control the fluctuations, and about the functional consequences of their presence. The model we are going to employ, based on combined mathematical and biophysical reasoning, is hoped to contribute some understanding to these issues.

2. METHODS

The source of our reasoning is threefold: mathematical theory, electrophysiology, and numerical simulation. An intertwined sequence of arguments is taken from the theory of stochastic processes, in particular point processes, from neurophysiology and neuroanatomy, both empirical and theoretical, and from numerical simulations of stochastic models for large neuronal networks. Probabilistic arguments imply a certain general class of network models, which come along with specific experimental paradigms and a set of methods for statistical inference. The results of these physiological experiments are in turn fed back to theory and constitute the basis for further analytical arguments. Large scale computer simulations are performed by again using the parameters found in the experiments, and these finally connect back to the physiology of networks and brains. We give a short outline for the scope of each class of arguments.

2.1. Stochastic Point Processes

The theory of stochastic processes is a branch of mathematical probability theory, which is concerned with non-deterministic phenomena in general spaces, and with stochastic dynamical systems in particular. The notion of a Markov process is of paramount

importance, since it comprises a class of stochastic dynamical systems which can be successfully treated with analytical methods. Stochastic point processes render a probabilistic
description for the distributed occurrence of point-like events—action potential firing in a
biological neuronal network is a textbook example for a physical point process in space
and time. The theory of such processes is very well developed and provides a concise
framework for both the quantitative description and the statistical analysis of spiking activity. We refer to Daley and Vere-Jones (1988) for a contemporary introduction into the
subject and for further guidance through the literature. An outline of our own contributions
to a dynamic theory of interacting point processes and a point process theory of neuronal
networks is deferred to the results section.

2.2. Slice Experiments in Rat Neocortex

The parameters of the stochastic model can be estimated from experiments in vitro,
where one has tight control over a neuron's input and output without too much affecting
its internal dynamics.

Experiments were made in acute coronal slices (400 μm) from visual cortex of 4–6
week old rats (Rotter, Heck, Aertsen, Vaadia, 1993). Using sharp glass micro-pipettes,
we performed intracellular recordings from layer II/III pyramidal cells. The pipettes were
filled with biocytin to allow for a morphological reconstruction of the recorded cells after
the experiment. The experimental paradigm required that all sources of external inputs to
the neuron be silent. To control for this, NMDA, non-NMDA, GABA_A, and GABA_B
mediated ionic currents were blocked by adding the appropriate receptor channel blockers
to the ACSF.

Substitute stimulation was presented to the neurons in terms of direct current input
through the electrode. Constant depolarizing current at different strengths (up to 500 pA)
was applied until the regular-spiking response of the neuron comprised a number of action
potentials which was sufficient for a statistical analysis of the spike train (a few hundred).
Several such spike trains, corresponding to different input currents, were the basis for the
parameter estimation procedure as outlined in the results section. For more details of the
preparation, the experimental paradigm, and the data analysis procedures we refer to Rotter
(1994).

2.3. Numerical Simulations of Large Networks

A variant of the general stochastic point process model is adapted to cortical networks and implemented on a computer. Effective simulations can be performed in discrete
time steps as well as in continuous time. Realizations of random variables like inter-spike
intervals are obtained by the use of common pseudo-random number generators. Some
technical details for the simulations are taken from the books by Ripley (1987) and Daley
and Vere-Jones (1988).

The basic "physiological" parameters including amplitudes and time constants of
after-hyperpolarization and postsynaptic currents are given random amplitudes and random
initial values from a reasonably large range, as marked off by the experiments. No attempt
is made to include specific topological structure into the simulated networks, the remaining
"anatomical" parameters like synaptic convergence and divergence, and synaptic delays are
also randomized in accordance with all available knowledge.

The simulations usually run over a few seconds or minutes physical time, where the
variables recorded depend on the purpose of the simulation. In addition to a full record of

action potentials from a number of selected cells it is also possible to access subthreshold phenomena by recording "intracellularly" from one or more neurons. Each particular experiment is usually run several times, each time fixing different random values for selected parameters. Thus, it is possible to explore the dynamic repertory of such networks empirically and compare the simulation results directly to the outcome of electrophysiological experiments as, for instance, simultaneous extracellular recordings from several nerve cells in an intact brain.

3. RESULTS

3.1. Cortical Networks and Point Processes

Spatio-temporal patterns of action potentials. The function of an individual cortical neuron is given by its strategy of transforming many incoming spike trains into one single outgoing spike train. Only information encoded in streams of spikes can be faithfully transmitted to other neurons. From this perspective, signalling in terms of spatio-temporal patterns of action potentials is indeed fundamental to nervous coding in the cortex. For the time being, we distinguish the "objects" handled by such a network from the "operators" used by the network to manipulate its objects. The objects are spatio-temporal patterns of action potentials—they are the words used by the brain to represent and share information. The operators are given by the response properties of the compound network—their physical basis lies in the anatomy and physiology of its neurons, of the synaptic connections between them, and of the supporting structures. We will give more specific biophysical arguments and, at the same time, make this distinction mathematically explicit.

By all what is known from physiology, it is safe to idealize a cortical action potential as a point on the time axis, which has no temporal extension in itself, and which carries no amplitude information. Stochastic point processes then appear to be a suitable mathematical abstraction for a quantitative approach to spatio-temporal spike patterns in neuronal networks (Perkel, Gerstein, Moore, 1967). We claim that stochastic point processes are indeed fundamental for a theory of cortical function which accounts for the specific properties of a code relying on identical point-like events distributed in space and time, and which can in particular address the problem of controlling the degree of its stochasticity.

The occurrence of a single action potential is fully specified by the time epoch of its observation and the identity of the generating neuron. The combination of both data represents a point in space–time. A full realization of a point process is given by the complete record of all points generated during a specific experiment. It will be termed a "spatio-temporal pattern". For a mathematical point process it is required that any finite time interval carries only a finite number of points. This, of course, is an obvious property of neuronal signals in physical networks. The process itself is fully specified by a probability law on the space of all admissible spatio-temporal patterns, with the additional requirement that counting points in any given time interval be a "measurable" procedure, in a mathematically well-defined sense. Thereby, the outcome of an observation of the process, which is a spatio-temporal pattern, turns into a random variable. Again, we refer to Daley and Vere-Jones (1988) for the mathematical details.

Temporal evolution of patterns. A point process in space and time can alternatively be considered as a stochastic dynamical system where point patterns gradually evolve in time. This is achieved mathematically by successive conditioning on the internal history of the process, rather than looking at the probabilities for complete realizations. In fact, the full spatio-temporal pattern generated before a given time epoch can be viewed as the state of the system in that very moment. These states will be qualified as "external" to distinguish them from "internal" states, which will be introduced later. One obtains a Markov process, whose dynamics is given in terms of probabilities for transitions among states. The infinitesimal parameters of the process—then called point process intensities— are given by the probability for a point in one particular node normalized for an infinitesimal time interval ("response"), conditional on the pattern previously generated by all nodes ("input"). If these parameters exist, the resulting master equation is readily solved, see Appendix for more details.

The very nature of a Markov system is that its future evolution depends only on the present state, independent of the past. Incidentally, for a point process, each state includes a complete account of the internal history. The causal structure which is implicit in this description completely parallels the above statement that the function of the cortical network is defined by its strategy of transforming incoming spike trains into outgoing spike trains. In other words, the cortical hardware implements the set of transition probabilities, which have been termed operators above. From this point of view, the Markov approach to point processes is functional and therefore best suited for an interpretation in a biophysical context. This general framework sets the stage for a quantitative assessment of many otherwise untractable network phenomena.

Physical representation of states. Models for the function of biological neuronal networks are usually constructed bottom-up, on the basis of known biophysical properties of isolated nerve cells. They rely on the argument that the determinants of spike generation must also have some faithful representation in terms of physical properties of matter. The list of possibilities includes electrical properties of the cell membrane, metabolic states of its organelles, space-dependent concentrations of ions and neurotransmitters, conformational states of ion channels, and the like. It is hard to tell from the beginning which of the above should be considered for an adequate caricature of neuronal action and interaction. It is even less obvious which input features are actually represented in terms of physical states of single neurons. As a result, the potential of such models to grasp the essentials of neuronal network function can hardly be predicted and must be obtained from expensive numerical simulations.

The situation is different for the type of model we would like to put forward. The temporal evolution of any system of interacting point process has the spatio-temporal pattern of all previous points as a natural state variable. For each individual node, the point process intensity is a function of the external state, specifying the probability rate that it generates a point in response to the input. Thereby, the intensity functions provide a complete characterization for the system's stochastic dynamics, and any particular point process model amounts to making them explicit. The use of this concept for applications is only limited by the fact that the full space of patterns is somewhat unwieldy, it cannot be described by a finite number of parameters. For important examples of concrete point processes, including neuronal processes, it also seems to be unduly general and complicated as a physical system's state space, so that one may well ask for more specific, and simpler, models.

Mathematically, the spatio-temporal pattern of previously generated action potentials determines the instantaneous behavior of all neurons in the network. One cannot expect, however, that each individual neuron in a biological network with limited resources keeps

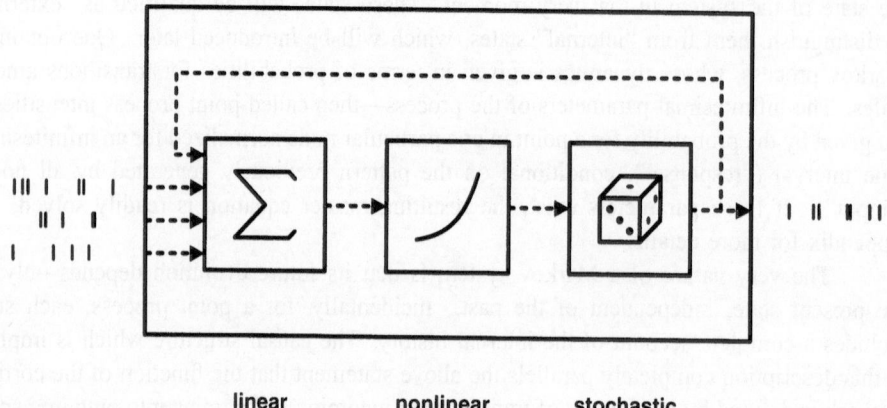

linear   **nonlinear**   **stochastic**

Figure 1. Schematic representation of single neuron function in the stochastic point process model. Inputs from other neurons in the network are first linearly integrated. The resultant internal state of the neuron is then non-linearly transformed into the point process intensity, which governs the stochastic generation of action potentials. The universal applicability of this scheme follows from pure mathematical reasoning without invoking physical or biological arguments. A biophysical interpretation of the parameters, however, leads to quantitative models of neuronal network function.

a full record of the previous history. For example, gross anatomy poses clear restrictions which forbid that the complete pattern is accessible to all neurons. On the other hand, some extent of information reduction is highly desirable for any device devoted to the processing of information. Elementary feature extraction, for instance, takes place in the processing of sensory information even on the level of single cortical cells, as is well known. In any case, the set of different physical states a single neuron can take at a given time instant may be expected to be small as compared to the set of all admissible external states. We will introduce additional features into the single neuron model to account for this.

"Integrate-and-fire" is universal. We discuss a formal way to achieve a model of reduced complexity by systematically accounting for redundancies in the space of external states, without appealing to biological mechanisms or relations among physical variables in the first place. In fact, purely algebraic manipulations on the space of patterns allow yet another interpretation of the probabilistic parameters of the system. For each neuron separately, we call two external states congruent, if they yield the same intensity in that neuron, as subpatterns of any larger pattern, and at all times. We collect all congruent external states into a single "internal state". Without going into the details of this construction, suffice it to say that one is led to a notion of internal states which can be guaranteed to be of reduced complexity, as compared to the external states. Nevertheless, internal states carry all information to fully determine the point process intensities of their nodes and, therefore, the stochastic dynamics of the system. The map which puts an internal state to its corresponding point process intensity will be termed "characteristic function". Finally, the projection from external states to internal states respects the "superpositionality" of patterns, which is important for the construction of models.

Each node now operates according to an "integrate-and-fire" scheme where the input pattern is first "linearly" projected onto some internal state and is then "nonlinearly" transformed into the point process intensity. Figure 1 illustrates the idea. A small internal state space obviously narrows the dynamical repertory of the process. The initial problem is reduced to the question how the internal states are represented in terms of physical properties of the material nodes. Despite the fact that these states must remain unidentified in biophysical terms for the time being, one can safely pinpoint a number of correspondences between the biological network and its abstract point process model.

We assert that any internal single neuron state should correspond to some "minimal set of physical properties" of this particular neuron, which is sufficient to completely char- acterize the cell's instantaneous readiness to generate an action potential. The projection of the external state onto the internal state corresponds to "dendritic integration", where the personality of the neuron leads to a possibly very specific evaluation of the inputs obtained from other neurons. Each neuron's dynamics also interferes with its own previous history. The superpositionality of internal states means, among other things, that any such state can be built from a less compound one by "adding" the effects of some extra action poten- tials. It may turn out that the temporal range of these "elementary" effects is finite or at least rapidly decreasing with time, as is typical for postsynaptic or after-hyperpolarization effects. In other cases, the induced changes in the neuronal behavior may be long-lasting or even permanent, an important example of this is long term potentiation and depres- sion of synaptic gains upon appropriate stimuli. The resting state of the neuron may be identified with the "null-element" in the internal state space, because its effect on action potential firing equals the effect of the empty pattern, where definitely any input is absent. Finally, spike initiation, which is commonly thought to take place in the soma or at the axon hillock (Stuart and Sakmann, 1994), is governed by the point process intensity. The latter is obtained by evaluating the actual internal state using the characteristic function.

The class of models we have obtained as a mathematical consequence of very general assumptions represent only a slight generalization of some ad hoc models, which were invented to describe networks of synaptically coupled neurons and other biological systems (McCulloch and Pitts, 1943; Little, 1974; Hopfield, 1982; Mirollo and Strogatz, 1990). In these models, the input to a neuron is first evaluated by linearly adding all elementary contributions in terms of some internal state variable. Common views aim at the membrane potential or membrane currents, but it is not at all obvious whether these two are necessary or sufficient coordinates for a reasonable description of neuronal action potential firing. The neuron's behavior is then determined by a non-linear transform of the internal variable, which specifies its instantaneous firing rate. Sometimes, this notion of rate coincides with our notion of intensity. In most cases, however, it has been interpreted as the expected number of spikes for a non-infinitesimal time interval. Then, no explicit action potential generation is assumed, and one clearly has no point process. In the classical case, all determinants of such an integrate-and-fire model must be constructed on the basis of indirect (and vague) external reasoning. In addition, common choices for the mapping of these operations on physical properties and mechanisms are biased by the methods of physiology.

3.2. A Stochastic Model for Cortical Pyramidal Cells

The situation for point process models is again different. Namely, in some cases, the physical nature of the abstract internal single neuron states can be identified with the help of experiments. As pointed out previously, any specific model amounts to an interpretation of the infinitesimal parameters in terms of biophysical properties of nerve cells. Each cell has its own state space with a linear structure and a nonlinearity transforming states into rates. The rates—in contrast to the situation in some classical models—do not specify averages over some extended period of time. Rather, they are considered as instantaneous measures of neuronal excitation, very much like any other physical property of the neuron as the membrane potential or the number of open sodium channels. Braitenberg (1974) discusses an interpretation of conditional probabilities, and how they could be used to quantify neuronal action and interaction in the brain.

One is left with the task to estimate the time course of the point process intensity from experimental observations. Using the explicit knowledge of the solutions for the general master equation (see Appendix) one can obtain numerical fits for the parameters of integrate-and-fire models to experimental observations (Rotter, 1994). In fact, having available an estimate for the point process intensity as a function of time during an experiment opens the possibility to check the consistency of the implied hypotheses with the requirements of the model (see Figure 2).

By injecting direct current of different strengths and evaluating the systematic changes of the neuron's response, it is possible to decide on the physical nature of the internal state space of a regular-spiking neuron, and to give quantitative estimates of the parameters as well. Indeed, one can identify membrane current, rather than membrane potential, as the physical state variable containing more complete information on the neuron's state of excitation. Once this is established, it is possible to extract the explicit time course of (hypothetical) post-spike hyperpolarizing currents from the spike train statistics. Moreover, one obtains the approximate shape of the characteristic function translating the total membrane current into point process intensity, at least as far as it is covered by the data. Together with the substitute input current injected through the electrode, these data provide a complete stochastic characterization for a single neuron, according to the scheme represented by Figure 1. This simple parametric model is able to account for many details of spike generation upon direct current injection into regular-spiking pyramidal neurons, which constitute the major cell type in the neocortex (McCormick et al., 1985). The range of parameters covered by different subtypes of cells in different parts of the cortex, however, remains to be established. Finally, there is strong experimental evidence that this current-based description extends to synaptic interaction on the basis of spike-induced postsynaptic currents (Reyes and Fetz, 1993), which were pharmacologically blocked in our experiments. Generally, results of this type give new relevance to models of the integrate-and-fire type.

3.3. Toward a Network Model of the Cortex

Further analysis of the model equipped with parameters as extracted from experiments in real neurons should be expected to also yield information on the role of subthreshold processes for the emergence of collective behavior. For the present discussion, we pick three examples in support of this assertion.

Stability in cortical networks. A mathematical analysis of the "integrate-current-and-fire-spike" model and the corresponding stochastic point process contributes to the as yet unsettled problem of stability in recurrent neuronal nets. It turns out that a model network of regular-spiking nerve cells maintains a stable low level of activity, if all characteristic functions are non-decreasing, and if the matrix of synaptic couplings has a "dominant" diagonal. This is the case, for instance, whenever in each neuron the electric charge transported by after-hyperpolarization exceeds the total charge transported by excitatory and inhibitory synaptic currents. Both this assertion and the underlying hypothesis are in

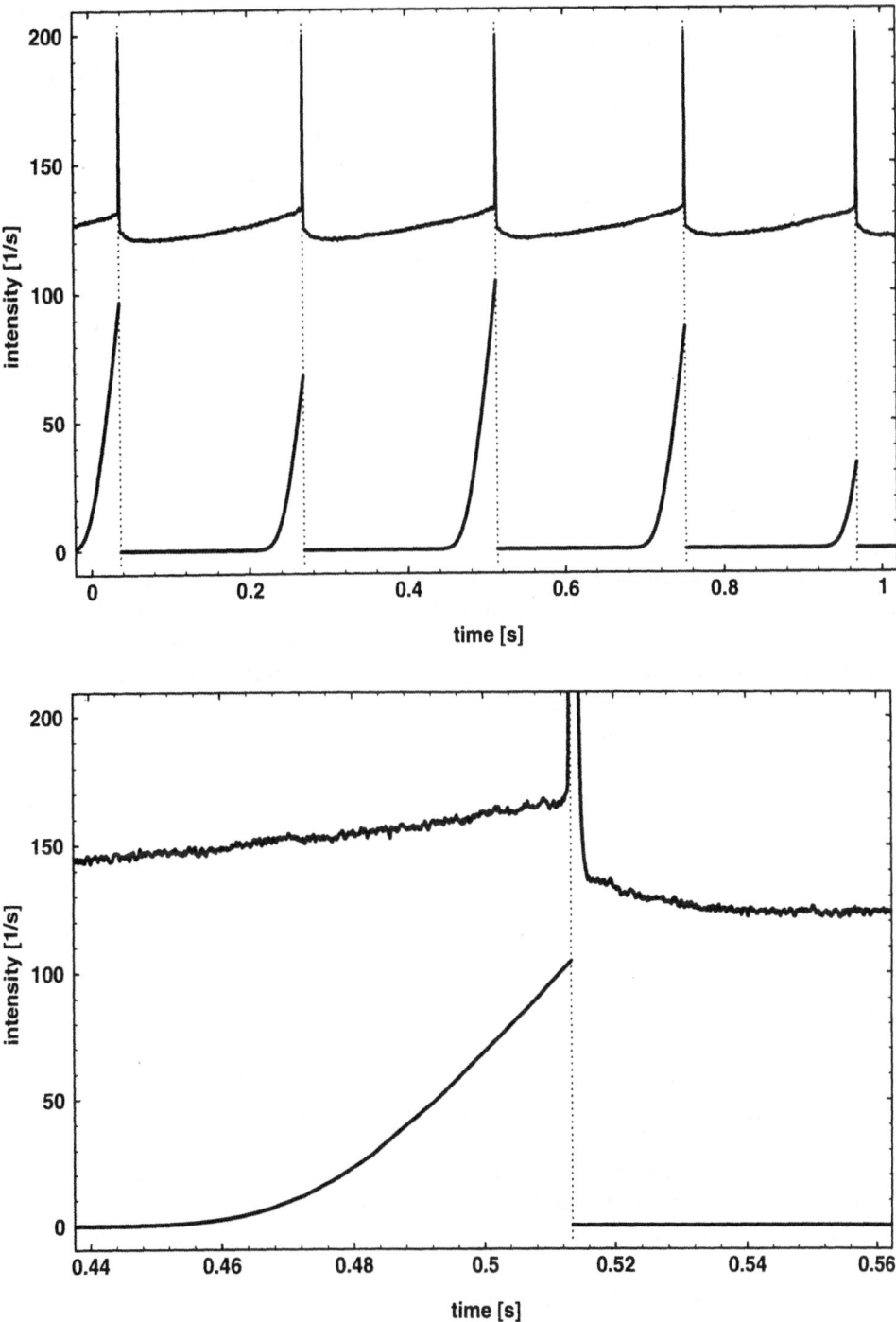

Figure 2. Empirical intensity of a neuronal point process from a current injection experiment. The intracellular recording (upper trace in both figures) shows the regular-spiking response to direct current stimulation, which is typical for cortical pyramidal cells. The corresponding point process intensity (lower trace in both figures) gives the probability for a spike normalized for an infinitesimal time interval. It is zero as long as the neuron is refractory or deeply hyperpolarized right after a spike, and it increases continuously until the action potential occurs. For the analysis, it is assumed that no spike but the most recent one influences the neuron's instantaneous excitability. The second figure shows the same data at a smaller time scale.

accordance with our knowledge of cortical physiology that synapses are sparse and post-synaptic currents are weak and fast, whereas after-hyperpolarization is strong and slow. Surprisingly though, inhibitory neurons are not required for stability, all activity control is taken care of by "auto-inhibition". Synaptic inhibition does contribute to achieve the low rates typically found in the cortex. An estimate of the maximal number of synapses which can be compensated for by after-hyperpolarization (10^2–10^5) is also in very good accordance with the anatomical figures (Braitenberg and Schüz, 1992).

Another interesting aspect of this parameter constellation is that the coefficient of variation for the inter-spike interval distribution of single neuron spike trains increases with the total amount of recurrent excitation in the network. Again, no inhibitory neurons are necessary to achieve coefficients as obtained from an analysis of in vivo data. This result is in contrast with some recent claims in the literature (Softky and Koch, 1993; Shadlen and Newsome, 1994).

Two time scales of synchronization. To organize and understand cooperative phenomena in larger networks, it turns out useful to characterize a single cortical pyramidal cell as a stochastic oscillator. The reason for this is not only its regular firing behavior upon direct current stimulation, but also its response to transient inputs. In fact, the susceptibility of the model neuron to synaptic inputs shows a distinct phase preference, leading to pronounced resonance phenomena. In addition, the characteristic function translating current into neuronal excitation turns out convex, which causes the neuron to prefer synchronous inputs over asynchronous ones. From this, one correctly predicts that transient synchronization of groups of neurons should be part of the dynamic repertory for recurrent networks of regular-spiking neurons.

Computer simulations of sparsely connected nets with random topology confirm that one can indeed distinguish two time scales of synchronization phenomena. At the macroscopic time scale extending over tens or hundreds of milliseconds, subgroups of neurons exhibit transient states of loose spike synchronization, with no appreciable effect on the average rates. Such "assemblies" desynchronize and reorganize themselves periodically upon persistent stimulation. Within the periods of enhanced group activity, one observes complex patterns of action-potentials extending in space and time, which recur with a precision in the millisecond range. This is all the more surprising as cell-cell interactions of the simulated networks were randomized in all their parameters including delay, amplitude, and time constant. Similar findings, both at the macroscopic time scale (Vaadia et al., 1995) and at the level of millisecond spike patterns (Abeles et al., 1993), have been reported from multi-neuron recordings in the prefrontal cortex of the behaving monkey.

On the plasticity of time structure. The ability of the cortex to keep its representations plastic are most valuable for the adaptive control of an agent's behavior in a highly variable environment. We began to investigate the possibilities of plasticity in time structure of neuronal signals by introducing physiologically inspired Hebb-like synapses into our stochastic networks. In the context of point process models with physiological parameters, this leads to a learning rule of the covariance type (see Appendix). From numerical simulations we conclude that the apparent weakening of synapses connecting out-of-phase neurons can serve to maintain a stable total amount of excitation within such networks. In contrast to most contemporary artificial neuronal systems in use, no artificial normalization of synaptic strengths is necessary. By employing a measure for the distance between spatio-temporal patterns, which naturally emerges from point process theory, we can demonstrate that the mere presence of plastic synapses has already useful consequences. One observes

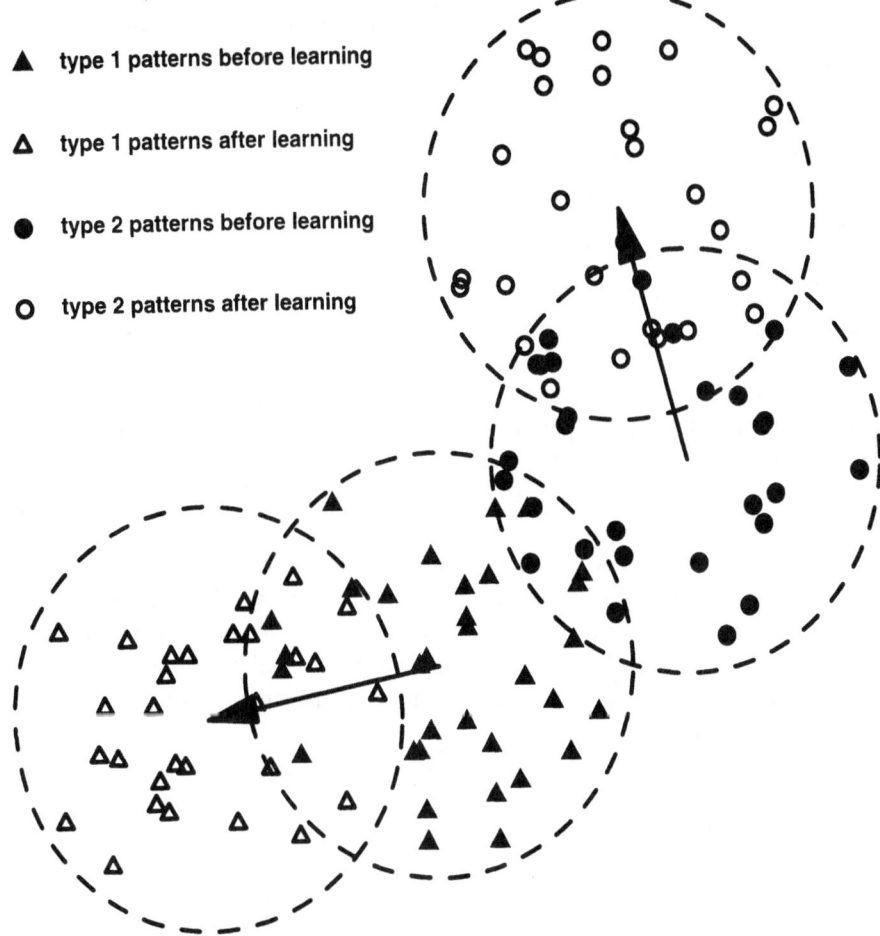

Figure 3. Representation of stimuli as spatio-temporal patterns. We show a schematic diagram which illustrates the effect of learning in a stochastic network. A small subnetwork responds to two different stimuli from surrounding neurons in two different ways ("type 1" and "type 2" responses) with a certain jitter and some overlap. This is indicated by the overlapping dashed circles comprising filled symbols. The mere presence of plastic synapses leads to a slow change of the representations for the two stimuli, which is finally non-overlapping while avoiding an increase in the jitter of the representation. This is indicated by the non-overlapping dashed circles comprising open symbols. The movement of the circles in space, as indicated by the arrows, stands for an increase in distance, or, equivalently, a decrease in similarity of the spatio-temporal patterns used for the representations.

the imprinting of spatio-temporal representations for temporally extended stimuli, as well as the automatic generation of discriminating spatio-temporal representations for several distinct stimuli (see Figure 3).

4. DISCUSSION

4.1. Summary

The spatio-temporal dynamics of cortical spike activity may be studied by using an abstract network model compatible with the notion of interacting stochastic point processes.

Any such system can be viewed as a Markov process, whose state at a given time instant is the spatio-temporal pattern of all previously generated spikes. The transition probabilities specify how a pattern gradually evolves in time. Neuronal network models which use a slightly generalized integrate-and-fire dynamics are a direct mathematical consequence of the assumption that neurons communicate by the use of action potentials. Assuming the existence of infinitesimal parameters, the corresponding dynamic equations are completely solved, and the solutions are used to identify important model parameters from electrophysiological recordings of real neurons. A simple parametric characterization of single neuron function is in fact achieved by fitting the model to the regular-spiking behavior of cortical pyramidal cells. A number of fundamental properties of recurrent cortex-like networks assembled from such neurons can be predicted, most notably their ability to maintain stable low rates of activity without the help of inhibitory neurons.

Computer simulations indicate that high precision spatio-temporal patterns, embedded in periods of enhanced cooperative group activity, may play a role for coding and computation in such networks. This is true, even if neither the anatomy of the network nor the physiology of its neurons are in any sense specially designed for that purpose. Plasticity of the temporal structure of such patterns is achieved by introducing Hebb-like synapses into the network. The resulting properties bring the point process model close to what more abstract neuronal networks are known to be capable of. By further exploiting our model system's known mathematical structure we expect to derive quantitative predictions which can also be applied to more complex experimental paradigms, involving, for instance, neural structures with specialized topology and plastic properties.

4.2. The Dynamic Repertory of Neuronal Processes

The notion of the "dynamic repertory" of a point process must remain vague until a mathematically well-defined formulation can be given. This is not possible so far, however, what we mean intuitively is something like the "true dimensionality of the system's phase space" or the "number of substantially different trajectories the process can take in a controlled way". Generally, the two determinants for the dynamic repertory of a system of interacting point processes are its dictionary, that is, the set of reachable spatio-temporal patterns, and its syntax, that is, the collection of rules governing the evolution of patterns. Clearly, both a more voluminous dictionary and a more flexible syntax increase the number of words at the system's disposal.

Among the many properties of the cortical network which contribute to its rich dynamic repertory we emphasize the sheer number of neurons, and the complexity of the network's organization as far as the processing of signals is concerned. The latter is reflected, among other things, by the degree of synaptic divergence and convergence, and by the degree of plasticity of the anatomical substrate during development. Another obvious source of dynamic diversity are cell-cell interactions, where extra complexity is added by postsynaptic currents with different time characteristics and possibly different reversal potentials, as well as by spike-induced modulatory currents. Yet another aspect, which is really at the heart of the stochastic model, is the precision in the control of timing as determined by the relationship between the strength of inputs and the stochastic properties of their integration. It is not unlikely that certain aspects of dendritic geometry and of the distribution of ion channels on the cell's soma and dendrite come into play at this point (Clay and DeFelice, 1983; Bernander, Douglas, Martin, Koch, 1991). Finally, we mention the non-stationary aspects of interactions among cells, the local or non-local characteristics of synaptic plasticity, and the depth and the time constants of memory. No doubt, all these parameters will greatly affect the capacities of a neuronal network as an abstract language analyzer and generator.

In principle, all factors which contribute to the dynamic repertory of the neuronal process play a potential role for the evolution of cortical structure and function. One general evolutionary strategy could be to increase the number of distinguishable words in the dictionary to choose from, simply to provide for more powerful representations. This can be achieved by an increased number of independently operating mechanisms to elicit action potentials in single cells and mechanisms to synchronize spikes into spatial patterns which involve many cells. A related strategy could be to increase the flexibility and the adaptedness of the syntax to represent relations among objects. This might involve, for instance, plasticity of the interactions between cells by means of synaptic learning rules which evaluate correlations and similarities among the inputs.

5. APPENDIX

5.1. Point Process Dynamics

A spatio-temporal pattern is a list of points from space–time $\{1, \ldots, n\} \times \mathbb{R}$

$$x = \{(i_1, t_1), (i_2, t_2), \ldots\}.$$

such that any finite time interval carries only a finite number of points. A pattern at some time t has no points with time coordinates later than t. We consider the case where the probability for a transition of a pattern x at time s into another pattern y at time t is given by a density function

$$p_{s,t}(x, y)$$

which is then usually called point process likelihood. The point process intensity

$$\lambda_t^i(x)$$

represents the probability for a point at (i, t) normalized for infinitesimally small time intervals ("node i generates a spike in a small time interval around t"). The master equation of the process of patterns

$$\frac{\partial}{\partial t} p_{s,t}(x, y) = p_{s,t}(x, y) \left(-\sum_{i=1}^{n} \lambda_t^i(y) \right)$$

has the general solution

$$p_{s,t}(x, y) = \prod_{(u,j) \in y \setminus x} \lambda_u^j(y_u) \prod_{i=1}^{n} \exp\left(-\int_s^t \lambda_v^i(y_v) \, dv \right)$$

where y_u means the restriction of pattern y to times earlier than u. The explicit knowledge of these functions can be used for further theoretical analysis as well as for parameter estimation from experimental data.

5.2. A Synaptic Learning Rule

The following rule is used in simulations of stochastic networks

$$\Delta W_{ij}(t) = S_j(t)A_{ij}\left[W_{ij}^*\left(U_i(t)\right) - W_{ij}(t)\right]$$

where the parameters have the following meaning:

i	postsynaptic neuron
j	presynaptic neuron
W_{ij}	synaptic strength
ΔW_{ij}	change in synaptic strength
S_j	assumes 1 upon the arrival of a presynaptic spike, otherwise 0
A_{ij}	amplitude for the change in synaptic strength
W_{ij}^*	pivot for synaptic strength
U_i	postsynaptic depolarization

Depending on the depolarization of the postsynaptic neuron, this rule allows for both increase and decrease of synaptic efficacies, very much as found in hippocampal and cortical nerve cells (Artola, Bröcher, Singer, 1990). In the case of a non-decreasing transfer function it amounts to an iterative on-line measurement of the covariance between presynaptic and postsynaptic activity.

REFERENCES

1. Abeles M, Bergman H, Margalit E, Vaadia E (1993) Spatiotemporal firing patterns in the frontal cortex of behaving monkeys. J Neurophysiol 70:1629–1638
2. Abeles M, Prut Y, Bergman H, Vaadia E, Aertsen A (1993) Integration, synchronicity and periodicity. In: Aertsen A (ed.) Brain Theory: Spatio-Temporal Aspects of Brain Function. 149–181. Amsterdam, New York, London, Tokyo: Elsevier
3. Artola A, Bröcher S, Singer W (1990) Different volatage-dependent thresholds for inducing long-term depression and long-term potentiation in slices of rat visual cortex. Nature 347:69–72
4. Bernander Ö, Douglas RJ, Martin KAC, Koch C (1991) Synaptic background activity influences spatiotemporal integration in single pyramidal cells. Proc Natl Acad Sci USA 88:11569–11573
5. Braitenberg V (1974) On the representation of objects and their relations in the brain. In: Conrad M, Güttinger W, Dallin M (eds.) Physics and Mathematics of the Nervous System. 290–298. Berlin, Heidelberg, New York: Springer.
6. Braitenberg V, Schüz A (1991) Anatomy of the Cortex—Statistics and Geometry. Studies of Brain Function, vol. 18. Berlin, Heidelberg, New York: Springer
7. Clay JR, DeFelice LJ (1983) Relationship between membrane excitability and single channel open-close kinetics. Biophys J 42:151–157
8. Daley DJ, Vere-Jones D (1988) An Introduction to the Theory of Point Processes. Springer Series in Statistics. New York: Springer
9. Hille B (1992) Ionic Channels of Excitable Membranes (2nd ed.) Sunderland (MA): Sinauer
10. Hopfield JJ (1982) Neural networks and physical systems with emergent collective computational abilities. Proc Natl Acad Sci USA 79:2554–2558
11. Koch C, Segev I (1989) Methods in Neuronal Modeling—From Synapses to Networks. Cambridge: MIT-Press
12. Little WA (1974) The existence of persistent states in the brain. Math Biosci 19:101–120
13. Mainen ZF, Sejnowski TJ (1995) Reliability of spike timing in neocortical neurons. Science 268:1503–1506
14. McCormick DA, Connors BW, Lighthall JW, Prince DA (1985) Comparative electrophysiology of pyramidal and sparsely spiny stellate neurons of the neocortex. J Neurophysiol 54:782–806

15. McCulloch WS, Pitts W (1943) A logical calculus of the ideas immanent in nervous activity. Bull Math Biophys 5:115–133

16. Mirollo RE, Strogatz SH (1990) Synchronization of pulse-coupled biological oscillators. SIAM J Appl Math 50:1645–1662

17. Perkel DH, Gerstein GL, Moore GP (1967) Neuronal spike trains and stochastic point processes. I. The single spike train. Biophys J 7:391–418

18. Perkel DH, Gerstein GL, Moore GP (1967) Neuronal spike trains and stochastic point processes. II. Simultaneous spike trains. Biophys J 7:419–440

19. Reyes AD, Fetz EE (1993) Effects of transient depolarizing potentials on the firing rate of cat neocortical neurons. J Neurophysiol 69:1673–1683

20. Reyes AD, Fetz EE (1993) Two modes of interspike interval shortening by brief transient depolarizations in cat neocortical neurons. J Neurophysiol 69:1661–1672

21. Ripley BD (1987) Stochastic Simulation. Wiley Series in Probability and Mathematical Statistics. New York: Wiley

22. Rotter S, Heck D, Aertsen A, Vaadia E (1993) A stochastic model for networks of spiking cortical neurons: Time-dependent description on the basis of membrane currents. In: Elsner N, Heisenberg M (eds.) Gene—Brain—Behavior. No. 491. Stuttgart: Thieme

23. Rotter S (1994) Wechselwirkende stochastische Punktprozesse als Modell für neuronale Aktivität im Neocortex der Säugetiere. Reihe Physik, vol. 21. Frankfurt: Harri Deutsch

24. Shadlen MN, Newsome WT (1994) Noise, neural codes and cortical organization. Curr Opin Neurobiol 4:569–579

25. Softky WR, Koch C (1993) The highly irregular firing of cortical cells is inconsistent with temporal integration of random epsps. J Neurosci 13:334–350

26. Stuart GJ, Sakmann B (1994) Active propagation of somatic action potentials into neocortical pyramidal cell dendrites. Nature 367:69–72

27. Vaadia E, Haalman I, Abeles M, Bergman H, Prut Y, Slovin H, Aertsen A (1995) Dynamics of neuronal interactions in the monkey cortex in relation to behavioral events. Nature 373:515 518

28. Vaadia E, Bergman H, Abeles M (1989) Neuronal activities related to higher brain functions—theoretical and experimental implications. IEEE Transact Biomed Eng BME-36:25–35

Acknowledgments

I would like to thank Dr. Detlef Heck (Washington University Medical School, St. Louis) for carrying out the slice experiments in vitro, which were the basis for parameter estimation in the stochastic model. I also thank Profs. Moshe Abeles and Eilon Vaadia (Hadassah Medical School, Hebrew University, Jerusalem) for providing the data from an in vivo experiment in the monkey VIGA, which entered the paper implicitly.

29

ON THE TIME REQUIRED FOR RECURRENT PROCESSING IN THE BRAIN

Alessandro Treves, Edmund T. Rolls and Martin J. Tovee

S.I.S.S.A. — Cognitive Neuroscience
via Beirut 2-4, 34013 Trieste, Italy
U. of Oxford — Experimental Psychology
South Parks Road, Oxford OX1 3UD, UK
U. of Newcastle — Psychology
Ridley Building, Newcastle upon Tyne NE11 7RU, UK

1. INTRODUCTION

Can one tell which connections are involved in processing information of a given sort in the cortex, by looking at the temporal course of the responses of individual neurons? Thorpe and Imbert[32] have argued that the speed with which neurons in the early visual system appear to be able to produce selective responses, e.g. when tested with different orientations of a stimulus, indicates that the processing is of a feedforward nature. They have recorded from orientation selective neurons in the primary visual cortex of the monkey, and orientation tuning curves appear very similar if computed from the firing rate over a 300 ms period, or only over a 50 ms period after response onset. These results have been interpreted as ruling out the involvement of *feedback loops*, i.e. that a given neuron may alter its response over time, due to the firing of other cells which it itself influences (whether directly or polysynaptically). Examples of some classes of network in which feedback effects may be important include networks with recurrent excitatory connections, such as are found between cortical pyramidal cells, and networks with recurrent lateral inhibition (see Fig. 1). Very similar interpretations to those of Thorpe and Imbert have been drawn by Oram and Perrett,[24] from an analysis of the responses of "face" cells in the temporal lobe cortex.

A series of experiments we have carried out[34] has addressed the question of processing speed in higher visual areas, in the temporal cortex of macaques, using information-theoretical quantitative methods.[25,23,22] We have estimated the average information, in bits, contained in the response of each of several cells about which one of a set of stimuli (faces) had been presented to the awake animal. By measuring the firing rate over several different poststimulus time intervals, we were able to contrast the amounts of information available in those intervals. The main finding is that a large proportion of the information is present if the firing rate is measured in a period as short as a few tens of ms taken near the onset

Neurobiology, edited by Torre and Conti,
Plenum Press, New York, 1996

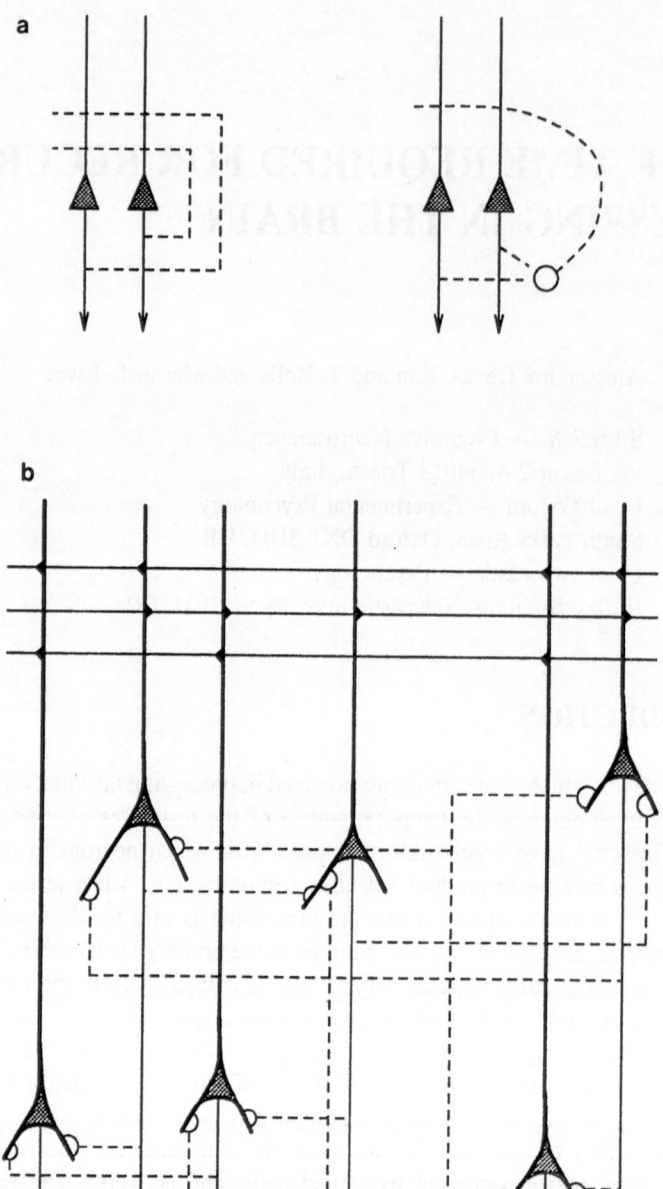

Figure 1. Recurrent networks. a) Schematic diagrams of a network with recurrent collaterals between pyramidal cells (left) and with lateral inhibition (right, the empty circle denotes an inhibitory interneuron). Recurrent connections are drawn as dashed lines. b) Slightly more detailed representation a patch of neocortex that may implement, locally, an autoassociative memory. Only the "skeleton"[5] of pyramidal cells is represented. Dendrites are drawn as thick lines, axons as thin ones, and axon collaterals as dashed lines. Black triangles denote afferent synaptic connections, white semicircles synapses on recurrent collaterals, and arrows indicate efferents leaving the cortex into the white matter. It is estimated[3,6] that the average number of synapses made on a pyramidal cell by recurrent collaterals is broadly of the same order as the number of synapses made by extrinsic (mainly cortico-cortical) afferents. The former terminate primarily on basal dendrites, the latter on apical dendrites.

of the neuronal response. For example, computing the rate over a 20 ms interval beginning 100 ms after stimulus onset (i.e. starting after the response latency of virtually all recorded cells) yielded on average 59% of the information obtained by computing the rate over a 500 ms interval beginning at the same time, while using a 50 ms interval yielded on average 78%. Further, the firing rate itself (a unidimensional measure of the response), when calculated over a long interval, contained nearly all the information (73%) that could be extracted from the same neurons by analyzing the full temporal course of their responses (a highly multidimensional and potentially much richer measure). In addition, experiments on the neurophysiology of masking show that neurons in the inferior temporal visual cortex fire for as little as 20ms when a stimulus can just be recognized.[29,28] Taken together, these findings suggest that in higher visual areas, in which cells are selective to complex patterns of visual stimulation, the processing involved e.g. in object recognition is very rapid. The initial output, produced by each neuron shortly after receiving afferent inputs, is already information-rich, and is quickly refined within for example 50ms of the initial response.[34]

Do such experimental results really rule out, or severely limit, the role of feedback in networks in the brain during rapid object recognition? A feedback loop, it is to be noted, can be precisely defined only when referring to a specific cell A. It presupposes the emission of a sequence of spikes (σ) by a chain of synaptically connected neurons: $\sigma(A) \rightarrow \sigma(B) \rightarrow \cdots \rightarrow \sigma(C) \rightarrow \sigma(A)$. The extent to which feedback loops involving different chains determine the response of a whole population remains, however, an intrinsically ill-defined quantity. It is possible to resolve the ambiguity by using the more comprehensive notion of recurrent processing: the response of a local population of neurons is dependent on recurrent processing if it would be altered by removing lateral connections (those for which both the pre- and the post-synaptic cell belong to the population, as in Fig. 1). The dependence can be quantified by quantifying the difference between the responses produced when such connections are present and when they are removed. What we ask, then, is whether recurrent processing may contribute, already within e.g. 20 ms, to generate the experimentally observed selective responses.

Previous consideration of this issue relied, explicitly or implicitly, on considerably simplified dynamical models, in which either time was discretized into artificial time steps ("cycles"),[30] or else the spiking, discontinuous nature of the dynamics, and its dependence on conductance kinetics, were neglected altogether in favor of a description in terms of continuously varying firing rates.[40] Models of that kind reinforced the intuitive conclusion that cycling information through lateral connections (see Fig. 1) necessarily involves using up precious time "cycles"; and that, hence, the observed speed indicates a streamlined processing mode, in which the response of each cell is determined by afferent inputs weighted by the strengths of synapses on afferent connections, and is quickly fed forward to the next cortical station.

We have developed a new analytical method for treating, in continuous time, the dynamics of a network of spike-emitting cells. The method considers a circumscribed population of neurons in which there are three sets of synaptic connections: afferents, efferents, and *recurrent collateral* ones. Here we present the results of detailed analyses,[36,2] and we apply them to address the question of recurrent processing in the visual system. This leads to new conclusions about the time constraints that the experimental data place on the type of processing.

2. METHODS

The analytical method we use is articulated as follows. First, we choose a suitable formal representation for individual units (neurons and synapses), that, though simplified, includes the salient features of their dynamical behavior. We note that this model leads to behavior similar to that of real neurons recorded in vitro. Second, the network is defined in terms of its connectivity and synaptic strengths. Third, an analytical treatment of the behavior of the network is described. Although a full dynamical treatment is complicated, we are able to analyze the approach to a type of asymptotic (long-time) behavior. We then describe this asymptotic behavior itself, in which the firing rate of each cell is stationary (provided that external inputs are kept, after an initial transient, constant), and also the transient modes, each with its own time constant, through which this behavior is approached. A simple analytical expression yields the full spectrum of time constants for a given network, and, by allowing calculation of these time constants, explicitly relates them to the biophysical parameters introduced in defining the model.

2.1. Integrate-and-Fire Neurons

A simple electrotonically compact model neuron is considered, whose membrane potential, in between spikes, follows the usual RC equation[21]

$$C_i \frac{dV_i(t)}{dt} = g_i^0(V_i^0 - V_i(t)) + g_i^K(t)(V_i^K - V_i(t)) + \sum_\alpha g^\alpha(t)(V^\alpha - V_i(t)) \tag{1}$$

with the symbols defined in Table 1.

The process of spike emission is faster than the time scales of interest, and hence it is not modelled in detail; rather, as the potential reaches a threshold value, a spike is added to the spike count, and the potential itself is reset to a hyperpolarization potential, from which the evolution resumes as in Eq. (1)*. In Eq. (1) we have introduced an intrinsic potassium conductance in order to produce a simplified version of the phenomenon of rate adaptation. Its dynamics is simply described by assuming it to decay exponentially in between spikes and to be incremented by a fixed amount during each spike emission,

$$\frac{dg_i^K(t)}{dt} = -\frac{g_i^K(t)}{\tau^K} + \Delta g_i^K \sum_k \delta(t - t_{k,i}). \tag{2}$$

2.2. Synapses

Synaptic conductances, both from recurrent and afferent connections, are modelled by a conventional α-function behavior.[8] A spike emitted at time $t_{k j_\alpha}$ by a presynaptic cell j_α activates, after a short interval (Δt) summarizing axonal and synaptic delays, a conductance on the postsynaptic membrane which then follows the equation

$$\tau^\alpha \frac{d^2 g^\alpha(t)}{dt^2} + 2\frac{dg^\alpha(t)}{dt} = -\frac{g^\alpha(t)}{\tau^\alpha} + \Delta g^\alpha \sum_k \delta(t - \Delta t - t_{k j_\alpha}) \tag{3}$$

where τ^α is the corresponding synaptic time constant, characterizing the exponential return of the conductance to its closed state (see Fig. 2).

---

*While one may consider also an absolute refractory period (lasting eg. 1–2 ms), we neglect that, for simplicity. In theory such a simplification makes possible infinite spiking rates, but in practice actual rates will be regulated, and at much lower levels, by other mechanisms such as inhibition,[1] and the absolute refractory period will not therefore be influential.

Table 1. Notation Used in the Formal Analysis

variables	
$V_i(t)$	membrane potential of cell i
$g^\alpha(t)$	conductance associated with synapse α
$g_i^K(t)$	time-dependent potassium conductance of cell i
$t_{k,i}$	time of emission of the kth spike by cell i
cell parameters (chosen according to each cell's class)	
V^0	resting potential
V^{thr}	threshold potential
V^{ahp}	reset potential
V^K	potassium conductance equilibrium potential
C	membrane capacitance
τ^K	potassium conductance time constant
g^0	leak conductance
Δg^K	extra potassium conductance following a spike
Δt	conduction and synaptic delay
synaptic parameters (chosen according to the classes of pre- and post-synaptic cells)	
V^α	synapse equilibrium potential
τ^α	synaptic conductance time constant
Δg^α	conductance contributed by one presynaptic spike

2.3. Network Architectures

Cells of the same class (i.e. with similar properties) are placed in a homogeneous group, and the network consists of a number N_C of such groups of cells. No specific spatial structure is associated with such architecture. The parameters characterizing each class, and those characterizing synapses between cells from any two classes, are specified. If the connectivity is chosen such that the existence of a synapse and the magnitude and time constant of its associated conductance depend only on the type of the pre- and post-synaptic cells (ie., for example, all layer 2/3 pyramidal cells make synapses, of identical characteristics, with all inhibitory stellate cells), then the partition into classes may correspond to a physiological classification into cell types.[20,18] Alternatively, when for example considering an associative memory model in which information is stored in the detailed synaptic structure, one has to partition a physiologically homogeneous population of cells of one type into as many classes as required so that synaptic strengths (eg. the magnitude of the relative conductances) are fully determined by the classes of the pre- and postsynaptic cells. This may result in many classes consisting of only few cells, or even one. Examples of both kinds may be treated with the present method.[36]

2.4. Mean-Field Analysis

A mean-field description is obtained by summing up the equations describing the dynamics of individual units to get (fewer) equations that describe collective behavior.[11] Thus grouping Eqs. (1) results in N_C functional equations describing the evolution in time of the fraction of cells of a particular class that at a given instant have a given membrane

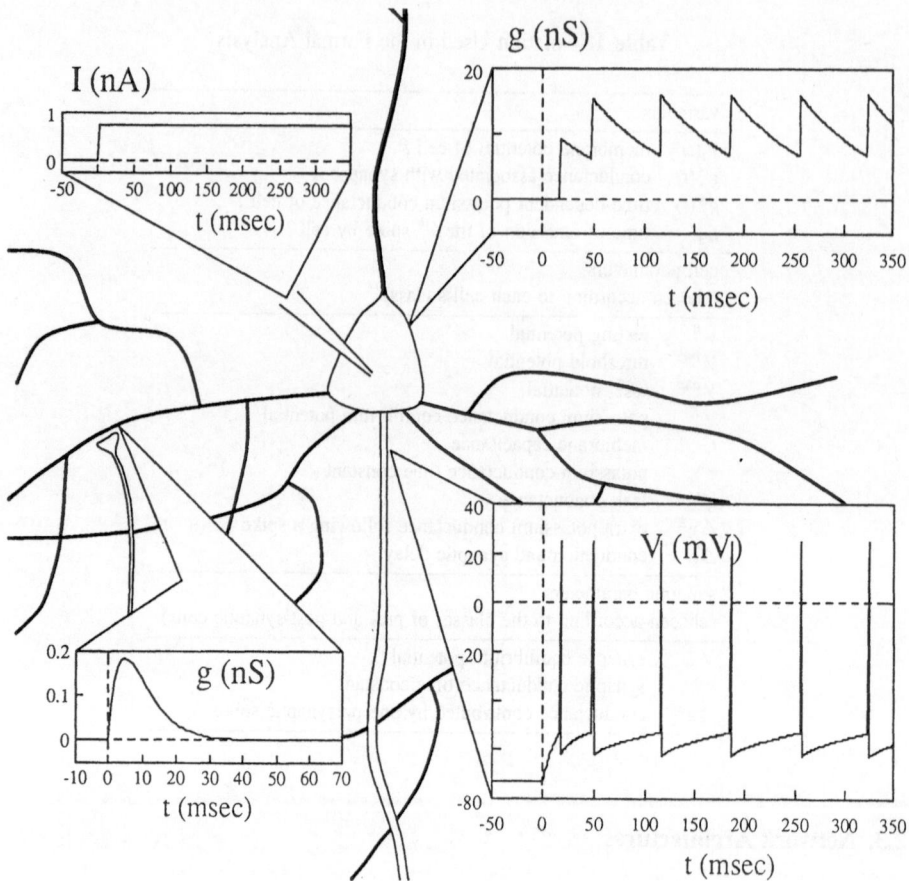

Figure 2. The integrate-and-fire model. The evolution of the membrane potential (bottom right, and Eq. (1)) and of the intrinsic potassium conductance (top right, and Eq. (2)) model the behavior of e.g., a pyramidal cell in response to *in vitro* current injections in the cell body (top left). The model includes adaptation effects in the firing rate. The conductance opened in response to a synaptic input (bottom left, and Eq. (3)) follows an α-function dynamics. The time constant of the α-function in the model has to be longer than measured locally, to take into account the effects of dendritic propagation to the soma, neglected in the point-like model.[15,16,31,39,4]

potential, while grouping Eqs. (3) results in $N_C \times N_C$ equations describing the dynamics of the summed conductance opened on the membrane of a cell of a particular class by all the cells of another given class. The system of mean-field equations has stationary solutions characterized by a constant firing rate for each cell class. As the neuronal current-to-frequency transfer function under stationary conditions is rather similar to a threshold-linear function, while each synaptic conductance is constant in time, the stationary solutions are essentially the same as can be obtained using much simpler, non-dynamical, methods, with threshold-linear units.[14,13,35,38] The advantage of the dynamical treatment, however, is that one can describe the approach to the asymptotic solutions, characterized by transient modes that decay with an exponential time dependence $e^{\lambda t}$. By linearizing the dynamical equations near a stable stationary solution and manipulating them algebraically, as explained in detail in Ref. 36, one obtains an equation, Eq. (4) below, yielding all the possible values of λ (they must all have a negative real part for the solution to be stable).

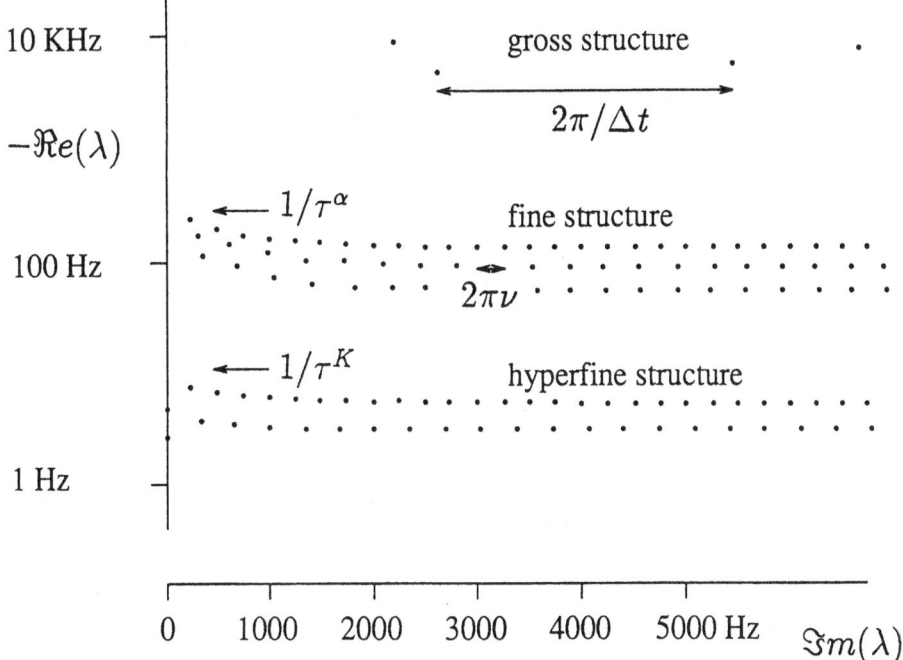

Figure 3. The spectrum of time constants. Schematic representation of the spectrum of time constants characterizing collective dynamics, representing the dependence on the main model parameters, as found by numerically solving Eq. (4). See text.

3. RESULTS

For any possible stationary asymptotic firing behavior of the network of formal neurons described under "Methods", there is an infinite set of transient modes, each associated with a complex time constant λ. The full λ spectrum for a given stationary state is obtained as the set of solutions of the complex equation

$$\Delta = \left|\mathbf{Q}(\lambda) - e^{\lambda \Delta t}\mathbf{1}\right| = 0 \qquad (4)$$

where Δ denotes the determinant of a matrix of rank N_C (the number of classes), composed of the matrix $\mathbf{Q}(\lambda)$ and the identity matrix $\mathbf{1}$. The form of $\mathbf{Q}(\lambda)$ (see the Appendix for details) is determined by the description chosen to model single units, i.e. neurons and synapses. Once that description, and all the parameters that accompany it, are specified, one has to solve Eq. (4) to find the time constants that govern, asymptotically, the collective behavior of the network. This last step may involve a straightforward numerical procedure, if N_C is small, or further analytical manipulation followed by numerical evaluation, if N_C is large or, in particular, tends to infinity.[36] In either case, the set of λ values obtained reflects the parameters describing cells and synapses, and thus explicitly links the single unit dynamics (the level at which the model attempts to capture relevant biophysical features) with network dynamics (where contact with neurophysiological experiment can be made).

The dependence of the spectrum on the underlying parameters is in general very complicated, but may be characterized as follows. The spectrum, plotted on the complex plane $\Re e(\lambda)$, $\Im m(\lambda)$, where both axes are measured in Hz, presents a gross, a fine and a hyperfine structure (see Fig. 3). The gross structure consists of those λ values which satisfy Eq. (4), whose real part is in the kHz range and beyond. These values correspond thus to fast time scales, and are determined by fast time parameters (eg. 1 ms) such as the delay

Δt. They are also very sensitive to global conditions like the detailed balance of excitation and inhibition, and in fact an instability associated with an imbalance of this sort may show up as one or more of the λ's of the gross structure acquiring a positive real part. There are two reasons, however, for not focusing one's attention on the gross structure of the spectrum. One is that, inasmuch as fast phenomena characterizing the real neural system have either been omitted altogether from the model (such as the Hodgkin-Huxley details of spike emission) or crudely simplified (such as unifying all axonal and synaptic delays into a unique delay parameter Δt), the model itself is not likely to be very informative about the fast dynamics of the real system. Second, in the presence of a stable stationary solution and of transients (those associated with the fine and hyperfine structures) lasting longer than, say, 5 ms, the faster transients are not very significant anyway.

The fine and hyperfine structures both consist of orderly series of λ values satisfying Eq. (4), whose real parts are similar (within a series), while the imaginary parts take discrete values ranging from zero all the way to infinity, with approximate periodicity $\Delta \Im m(\lambda) \simeq 2\pi\nu$, where ν is the firing rate of a particular group of cells. The (negative) real parts are complicated functions of all the parameters, but in broad terms they are determined by (and similar in value to) the inverse of the conductance time constants.[36] The fine and hyperfine structures differ in the magnitude of their real parts, i.e. in their time constants, in that relatively fast inactivating conductances (eg. those associated with excitatory transmission via AMPA receptors) produce the fine structure, while slower conductances (e.g., the intrinsic ones underlying firing rate adaptation,[7] or $GABA_B$ inhibitory ones[9]) produce the hyperfine structure. The fine structure, which therefore corresponds to time scales intermediate between those of the gross and hyperfine structures, is perhaps the most interesting, as it covers the range in which the formal model best represents dynamical features of real neuronal networks.

In principle, having assigned a definite starting configuration for the model network, i.e. the distributions of membrane potentials of the cells and the degree of synaptic activation, at a given time, one should be able to follow in detail the dynamical evolution, thereby quantifying the relative importance of different time scales. In practice however, the explicit characterization of the modes, with their spectrum of time constants, is only exact asymptotically close to a stationary state, and thus applies only to starting configurations infinitesimally close to the stationary state itself. For arbitrary initial conditions, a behavior increasingly similar to that described by the modes and their time constants will progressively set in, provided the network approaches a stable stationary state of the type considered here.

What does the analysis of the formal model imply for the behavior of real networks of neurons, e.g. in visual cortex? With the proviso mentioned in the last paragraph, the essential conclusions, for the model, are that i) the approach to stable firing configurations is governed by a variety of time scales of widely different orders of magnitude; and ii) an important subset of those time scales is determined by the inactivation times of synaptic conductances. These conclusions are expected to be valid, as well, for the dynamics of local cortical networks. With excitatory synapses between pyramidal cells,[19] the time course of inactivation, even taking into account the spread due to the finite electrical size of the postsynaptic cell (neglected by our treatment based on point-like neurons), may be as short as 5–10 ms. Therefore a substantial component in the "settling down" of the distribution of firing activity into a stable distribution, under the influence of interactions mediated by lateral connections, may be very rapid, that is it may occur in little more than 5–10 ms.

4. DISCUSSION

What about the intuitive notion we referred to earlier, suggesting that feedback in such neuronal networks would be associated with much slower time scales? Such reasoning might proceed like this: a) the stabilization of a firing configuration mediated by lateral connections is an inherently feedback effect; b) for feedback to take place, the minimal requirement is that a cell contributes to the firing of another cell, then receives back a spike from that cell and fires again, using up in the order of two interspike intervals (one each at the current firing rate of each of the two cells); c) the process is iterated a few times in order to lead to a steady activity distribution, taking up in the order of several hundreds of ms.

Such reasoning, which incidentally neglects both the spread in the latencies of activation of the cells in the network and, more importantly, their ongoing spontaneous activity, is a sequel of three incorrect steps. In fact, a) lateral connections may contribute much of their effect well before genuine feedback occurs, in fact even before most cells have had the time to fire a single action potential in response to a new stimulus. For example, a spike emitted by cell A may immediately influence the firing of a cell B that happened to be close to threshold, while more time may be needed for the firing of B to have a feedback effect on the emission of a subsequent spike by A. b) The influence, reciprocated or not, of a given cell onto another is only minor for the collective behavior of the cell population, which is determined by the concerted action of the thousands of different inputs to each cell. c) Reasoning in terms of time steps, or "iteration cycles", is meaningless anyway, as the dynamics of the real system is in continuous time and, unless a special synchronization mechanism sets in[*], is asynchronous.

A formal model based on firmer assumptions leads, as we have seen, to different conclusions[†]. These may be understood intuitively, if one considers that within a large population of cells, firing at low but non-zero rate, as occurs in cerebral cortex,[27] incoming afferent inputs will find some cells that are already close to threshold. The slight modification in the firing times of those cells (brought forward or postponed depending on whether each cell is excited or inhibited by the afferents) will be rapidly transmitted through lateral connections, and (for example in associative networks) contribute to initiate a self-amplifying process of adjusting the firing pattern that would have been determined by external inputs alone. The process will proceed at a pace set by the time course of synaptic action, which in practice, in a situation in which transmission delays are negligible, means by the time constants of synaptic conductances. The fact that the speed of this "collateral effect"[17] is essentially independent of prevailing firing rates is somewhat analogous to, and no more puzzling than, the propagation speed of elastic waves being distinct from the speed of motion of the individual particles carrying the wave.

Information-rich neuronal responses emerging at relatively early stages may therefore reflect not only feedforward but also, to a considerable degree, lateral interactions. As discussed above, lateral interactions may or may not involve, for any given cell, the propagation of spikes around feedback loops. The sequential activation of most such loops

---

[*]As postulated by theoretical notions motivated by the recent wave of experiments on synchronization,[12,10] but cf. Ref. 33.

[†]Approaches based on simulations rather than analytical techniques are computationally expensive and less transparent, but they can also be brought to bear on this issue. Moreover they can easily incorporate additional elements of realism, such as fluctuations in the parameters of, and in the inputs to, each individual cell. Preliminary results from one such project suggest that under certain conditions recurrent collaterals do in fact exert the rapid effect indicated by our analysis (L Cangiano, M Mehta and A Treves, unpublished).

will require more than 20 ms, but the population response will already within that time be dependent on the activation of recurrent connections. The ability of recurrent connections to contribute usefully to information processing is due to the fact that part of their effects occurs within a time determined primarily by the time constants of the synaptic conductances. Slower time scales, partly intrinsic and partly also associated with lateral interactions, such as those involving slow inhibition and those underlying adaptation processes, may produce, later on, further adjustments in the pattern of firing activity. Those adjustments may possibly, but not necessarily, increase the information content of neuronal responses. In addition, in the longer time scale of 50 or more ms, long range feedback loops, such as those which may be implemented by backprojections within and to the cerebral neocortex,[26] may be important in a number of types of top-down influence on information processing.

An experimental check of the implications of our analysis will clearly require techniques for the selective inactivation of recurrent collaterals, that leave intact afferent and efferent transmission. In parallel, in order to be a quantitative check in the sense of information theory, it requires the recently developed (see Ref. 37, see also Panzeri *et al*, this volume) refinements in the statistical techniques for extracting information measures from very short samples of neuronal firing .

It is interesting to note that our conclusions depend on the use of a formal network model in which we have introduced, in simplified form, important aspects of real neuronal and synaptic behavior, such as conductance dynamics. This suggests that the evolution in time of neuronal activities in the brain may follow very different patterns, even qualitatively, from those that could be inferred from the behavior of artificial dynamical systems in which those aspects are absent.

REFERENCES

1. Abbott, L. F. (1991) Realistic synaptic inputs for model neural networks, *Network* 2: 245-258.
2. Abbott, L. F. and van Vreeswijk, C. (1993) Asynchronous states in networks of pulse-coupled oscillators, *Phys. Rev. E* 48: 1483-1490.
3. Abeles, M. (1991) *Corticonics* (Cambridge Univ Press, NY).
4. Agmon-Snir, H. and Segev, I. (1993) Signal delay and input synchronization in passive dendritic structures, *J. Neurophysiol.* 70: 2066-2085.
5. Braitenberg, V. (1978) Cortical architectonics: general and areal, in *Architectonics of the Cerebral Cortex*, eds. Brazier, M. A. B. and Petsche, H. (Raven Press, New York) pp. 443-465.
6. Braitenberg, V. and Schütz, A. (1991) *Anatomy of the Cortex: Statistics and Geometry* (Springer, Berlin).
7. Brown, D. A., Gähwiler, B. H., Griffith, W. H. and Halliwell, J. V. (1990) Membrane currents in hippocampal neurons, *Progr. Brain Res.* 83: 141-160.
8. Brown, T. H. and Johnston, D. (1983) Voltage-clamp analysis of mossy fiber input to hippocampal neurons, *J. Neurophysiol.* 50: 487-507.
9. Connors, B. W., Malenka, R. C. and Silva, L. R. (1988) Two inhibitory post-synaptic potentials, and $GABA_A$ and $GABA_B$ receptor-mediated responses in neocortex of rat and cat, *J. Physiol.* 406: 443.
10. Eckhorn, R., Bauer, R., Jordan, W., Brosch, M., Kruse, W., Munk, M. and Reitbaeck, H. J. (1988) Coherent oscillations - a mechanism of feature linking in the visual cortex; multiple electrode and correlation analysis in the cat, *Biol. Cybern.* 60: 121-130.
11. Frolov, A. A. and Medvedev, A. V. (1986) Substantiation of the "point approximation" for describing the total electrical activity of the brain with the use of a simulation model, *Biophysics* 31: 332-337.
12. Gray, C. M., König, P., Engel, A. K. and Singer, W. (1989) Oscillatory responses in cat visual cortex exhibit intercolumnar synchronization which reflects global stimulus properties, *Nature* 338: 334-337.
13. Hadeler, K. P. and Kuhn, D. (1987) Stationary states of the Hartline-Ratliff model, *Biol. Cybern.* 56: 411-417.
14. Hartline, H. K. and Ratliff, F. (1974) *Studies of Excitation and Inhibition in the Retina* (Chapman and Hall, London).

15. Hestrin, S., Nicoll, R. A., Perkel, D. J. and Sah, P. (1990) Analysis of excitatory synaptic action in pyramidal cells using whole-cell recoprding from rat hippocampal slices, *J. Physiol.* 422: 203-225.

16. Jonas, P. and Sakmann, B. (1992) Glutamate receptor channels in isolated patches from CA1 and CA3 pyramidal cells of rat hippocampal slices, *J. Physiol.* 455: 143-171.

17. Marr, D. (1971) Simple memory: a theory for archicortex, *Phil. Trans. R. Soc. Lond.* B 262: 23-81.

18. Mason, A. and Larkman, A. (1990) Correlations between morphology and electrophysiology of pyramidal neurons in slices of rat visual cortex, *J. Neurosci.* 10: 1415-1428.

19. Mason, A., Nicoll, A. and Stratford, K. (1991) Synaptic transmission between individual pyramidal neurons of rat visual cortex *in vitro*, *J. Neurosci.* 11: 72-84.

20. McCormick, D. A., Connors, B. W., Lighthall, J. W. and Prince, D. A. (1985) Comparative electrophysiology of pyramidal and sparsely spiny stellate neurons of the neocortex, *J. Neurophysiol.* 54: 782-806.

21. McGregor, R. J. (1987) *Neural and Brain Modelling* (Academic Press, San Diego).

22. Optican, L., Gawne, T. J., Richmond, B. J. and Joseph, P. J. (1991) Unbiased measures of transmitted information and channel capacity from multivariate neuronal data, *Biol. Cybern.* 65: 305-310.

23. Optican, L. and Richmond, B. J. (1987) Temporal encoding of two-dimensional patterns bysingle units in primate inferior temporal cortex. III: information theoretic analysis, *J. Neurophysiol.*, 57: 162-178.

24. Oram, M. W. and Perrett, D. I. (1992) Time course of neural response discriminating different views of the face and head, *J. Neurophysiol.* 68: 70-84.

25. Richmond, B. J. and Optican, L. (1987) Temporal encoding of two-dimensional patterns bysingle units in primate inferior temporal cortex. II: Quantification of response waveform, *J. Neurophysiol.* 57: 147-161.

26. Rolls, E. T. (1989) Functions of neuronal networks in the hippocampus and neocortex in memory, in *Neural Models of Plasticity*, eds. Byrne, J. H. and Berry, W. O. (Academic press, San Diego), pp. 240-265.

27. Rolls, E. T. (1992) Neurophysiological mechanisms underlying face processing within and beyond the temporal cortical visual areas, *Phil. Trans. R. Soc. Lond.* B 335: 11-21.

28. Rolls, E. T. and Tovee, M. J. (1994) Processing speed in the cerebral cortex and the neurophysiology of visual masking, *Proc. Roy. Soc.* B 257: 9-15.

29. Rolls, E. T., Tovee, M. J., Purcell, D. G., Stewart, A. L. and Azzopardi, P. (1994) The responses of neurons in the temporal cortex of primates, and face identification and detection, *Expl. Brain Res.* 101: 474-484.

30. Rumelhart, D. E. and McClelland, J. L. (1986) *Parallel Distributed Processing* (MIT Press, Cambridge, Mass.).

31. Stern, P., Edwards F. A. and Sakmann, B. (1992) Fast and slow components of unitary EPSCs on stellate cells elicited by focal stimulation in slices of rat visual cortex, *J. Physiol.* 449: 247-278.

32. Thorpe, S. J. and Imbert, M. (1989) Biological constraints on connectionists models, in *Connectionism in Perspective*, eds. Pfeifer, R., Schreter, Z. and Fogelman-Soulie, F., pp. 63-92.

33. Tovee, M. J. and Rolls, E. T. (1992) Oscillatory activity is not evident in the primate temporal visual cortex with static stimuli, *Neuroreport* 3: 369-372.

34. Tovee, M. J., Rolls, E. T., Treves, A. and Bellis, R. P. (1993) Information encoding and the responses of single neurons in the primate temporal visual cortex, *J. Neurophysiol.* 70: 640-654.

35. Treves, A. (1990) Graded-response neurons and information encodings in autoassociative memories, *Phys. Rev. A* 42: 2418-2430.

36. Treves, A. (1993) Mean-field analysis of neuronal spike dynamics, *Network* 4: 259-284.

37. Treves, A. and Panzeri, S. (1995) The upward bias in measures of information derived from limited data samples, *Neural Comp.* 7: 399-407.

38. Treves, A. and Rolls, E. T. (1991) What determines the capacity of autoassociative memories in the brain?, *Network* 2: 371-397.

39. Williams, S. H. and Johnston, D. (1991) Kinetic properties of two anatomically distinct excitatory synapses in hippocampal CA3 pyramidal neurons, *J. Neurophysiol.* 66: 1010-1020.

40. Wilson, H. R. and Cowan, J. D. (1974) Excitatory and inhibitory interactions in localized populations of model neurons, *Biophys. J.* 12: 1.

Acknowledgments

We benefited from collaborations and discussions with L.F. Abbott, M. Imbert, L. Cangiano, G. Major, M. Mehta, S. Panzeri and S. Thorpe. Financial support was in part from an E.E.C. B.R.A.I.N. Initiative grant, from the M.R.C. of the U.K. and from the C.N.R. of Italy.

Appendix

The form of the matrix $\mathbf{Q}$ in a simple case.

Consider the relatively simple case in which no firing adaptation effects are included, i.e. the intrinsic potassium conductance is omitted from the single cell model. It simplifies the analysis to express all characteristic potentials in terms of adimensional variables, such as

$$x(t) = \frac{V(t) - V^{ahp}}{V^{thr} - V^{ahp}} \tag{5}$$

which measure the excursion of the membrane potential between spikes. $x(t)$, for cells of a given class F, follows the equations

$$\dot{x}(t) = A_F(t) - x(t)B_F(t), \tag{6}$$

where $A_F(t)$ and $B_F(t)$ represent the appropriate combinations of intrinsic and synaptically driven conductances.

At steady state each cell of class F fires at a constant frequency

$$\nu_F^0 = B_F^0 \left[\ln \frac{A_F^0}{A_F^0 - B_F^0} \right]^{-1}, \tag{7}$$

which results in a system of self-consistent equations, given that A_F^0 and B_F^0 depend in turn on the distribution of frequencies $\{\nu_G^0\}$.

The time scales of the transients, through which the steady state is approached, can be shown[36] to be given by the solutions to Eq. (4), with a matrix $\mathbf{Q}$ of elements

$$Q_F^G(\lambda) = \frac{\nu_F^0 \omega_F^G}{\tau_F^G(\lambda + 1/\tau_F^G)^2(B_F^0 + \lambda)}$$
$$\times \left\{ 1 + \lambda \frac{1 + x_F^G[e^{-(\lambda + B_F^0)/\nu_F^0} - 1]}{[A_F^0 - B_F^0][e^{-\lambda/\nu_F^0} - 1]} \right\}. \tag{8}$$

In the synaptic parameters x (adimensional equilibrium potentials), τ (time constants) and ω (which measure efficacy, and have the dimensions of a frequency), the superscript denotes the class of the presynaptic cell, the subscript that of the postsynaptic one.

30

PERCEPTION OF LUMINANCE AND COLOR

Comparing Functional Properties of Detection and Induction in Human Vision

Thomas Wachtler[*] and Christian Wehrhahn[†]

Max-Planck-Institut für biologische Kybernetik
Tübingen, Germany

1. INTRODUCTION

Luminance and color of surfaces provide fast and reliable means for the visual recognition and identification of objects (Dobkins & Albright, 1994; Sun & Perona, 1996). The perceptions of both color and luminance seem to share many properties. They are established essentially on the basis of local contrasts but subject to variations due to more global variations of illumination (Daw, 1984; Shapley & Reid, 1985; Reid & Shapley, 1988). Long range phenomena like constancy and induction appear to modify the perception of both luminance and color in a comparable way.

In this article we review psychophysical experiments establishing a possibility to quantitatively compare the perception of luminance and color. We describe two sets of experiments in which the relative strengths of these quantities in identical tasks were determined: In the first experiment we investigate the ability of human observers to detect luminance and color of small surfaces. In a second experiment we examine the differences in the induction strength of either luminance or color between two test areas separated by a transition area. This transition consists of a ramp, succeeded by a step and then another ramp. The two ramps are of equal slope and amplitude and the amplitude of the step inbetween them is equal to the total amount bridged by the ramps, but of opposite sign. Depending on the nature of the transitory area for most human observers the two test fields differ in their luminance or color. This latter phenomenon is a simplified version of the well-known Craik-Cornsweet illusion (review: Daw, 1984).

The means to quantitatively compare color and luminance is brought about by determining the contrast of the stimuli on the basis of the excitation of the cones sensitive to light of medium and long wavelengths.

---

[*] Present address: The Salk Institute for Biological Studies, La Jolla, California 92186.

[†] Send correspondence to: Christian Wehrhahn, Max-Planck-Institut für biologische Kybernetik, Spemannstrasse 38, D-72076 Tübingen, Germany.

Luminance stimuli activate both long- and middle-wavelength cones in the same way (as well as short-wavelength cones, which are neglected here (Eisner & MacLeod, 1980)),whereas isoluminant chromatic stimuli are achieved by changing the relative excitation of long- and middle-wavelength cones in opposite directions. The magnitude of the change in luminance and color is expressed in cone contrast, calculated as the square root of the sum of squares of relative changes of long- and middle-wavelength cone excitations. All stimuli were created on a calibrated color monitor. Details are described elsewhere (Wachtler & Wehrhahn, 1996).

Luminance stimuli activate both long- and middle-wavelength cones (as well as short-wavelength cones, which are neglected here (Eisner & MacLeod, 1980)) and are quantified by the sum of their respective excitations. Isoluminant chromatic stimuli are achieved by changing the relative excitation of long- and middle-wavelength cones in opposite directions and are therefore measured by determining the difference of their respective excitations. All stimuli were created on a calibrated color monitor, the details of which are described elsewhere (Wachtler & Wehrhahn, 1996).

2. PROCEDURES AND RESULTS

2.1. Detection Experiment

Both experiments were carried out by testing the performance of human subjects in psychophysical experiments. Subjects had normal or corrected to normal vision and were color normal as established by a suitable test (Farnsworth Hue 100). Subjects were asked to observe patterns displayed on a monitor by means of a computer (486) and a graphics interface (ELSA). With the exception of the stimuli the surface of the monitor was a light grey of 22 cd m^{-2} (CIE x = 0.28, y = 0.36). In the first experiment stimulation started with the display of a dark grey fixation point in the middle of the monitor screen. A stimulus (i.e. a rectangle differing from the surround by luminance or color) appeared at random either to the left or to the right side of the fixation point. The edge of the stimuli next to the fixation point was 10 arcmin away from the fixation point in all cases. Subjects were asked to signal the side at which the rectangle had appeared by pressing a corresponding mouse button. Stimuli were generated by an increase in luminance or an isoluminant increase in redness. Stimulus strength varied in different steps of cone contrast with respect to the background for each condition. Color changes were isoluminant. Values for isoluminant color contrasts were determined individually by a flicker test following a procedure described earlier (Livingstone & Hubel, 1984; Hubel & Livingstone, 1990). Psychometric thresholds were achieved by fitting a cumulative Gaussian function to the response probabilities for different contrasts. Thresholds were defined as that contrast value at which subjects correctly identified the position of the stimulus in 75% of the cases. With this method we established a quantitative relation for the detection of color and luminance stimuli.

Fig. 1 shows averaged results for four subjects tested with isoluminant red stimuli, as well as with luminance stimuli. Thresholds for color and luminance stimuli are qualitatively similar to each other, but color thresholds are up to one order of magnitude lower than luminance thresholds. Moreover, closer inspection of the two curves indicates that that for increasing presentation times color thresholds decrease faster than luminance thresholds. When the same data are plotted as a function of stimulus size little, if any, variation can be detected. This effect is seen more clearly in Fig. 2. Here the ratio of luminance and color thresholds from Fig. 1 is shown. It is plotted a) as a function of presentation time and b) as a function of stimulus size, respectively. Equal increases in presentation time improve color

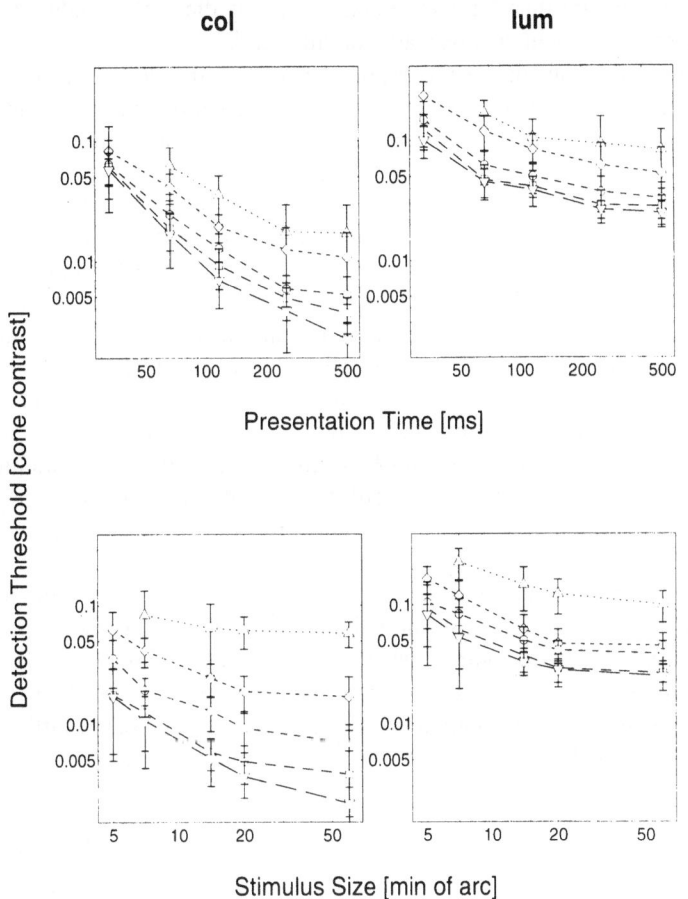

Figure 1. Detection thresholds for color (left) and luminance stimuli (right) averaged for four subjects plotted as a function of presentation time (upper panel) and stimulus size (lower panel). Dash length indicates stimulus size or presentation time, respectively. Bars indicate 95% confidence intervals. Color thresholds are up to an order of magnitude lower than luminance thresholds. This means that human subjects are much more sensitive to color contrasts than to luminance contrasts.

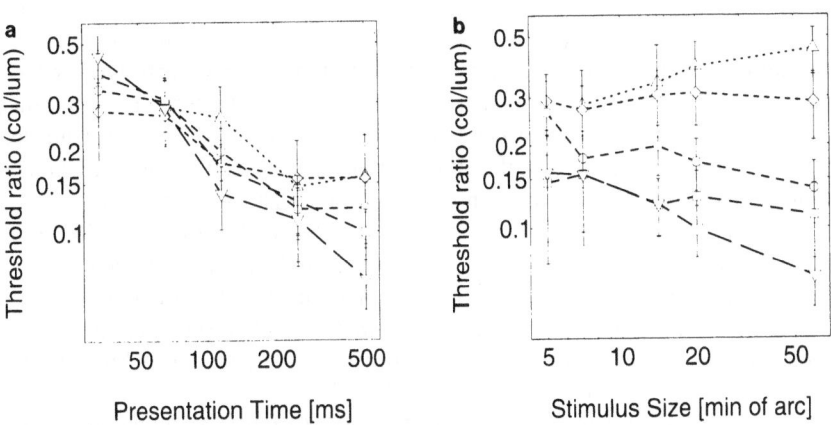

Figure 2. Ratio of averaged detection thresholds for luminance and color plotted as a function of either presentation time (a) or stimulus size (b). Other details as in Figure 1.

detection much stronger than luminance detection, regardless of stimulus size. Increasing stimulus size seems to yield no advantage for either task.

We conclude from the results reported here that the detection of an isoluminant change in color requires much less cone contrast than an achromatic increment in luminance (Chaparro, Stromeyer, Kronauer & Eskew, 1994). This phenomenon is independent of the direction of color change (Wachtler & Wehrhahn, 1996).

The geniculo-cortical part of the visual pathway in primates is thought to be the input to conscious visual perception (review: Hubel, 1988). A subdivision of the ganglion cell population connecting eye and brain into a "magnocellular" (MC)-pathway and a "parvo-cellular" (PC)-pathway has gained much attention for more than a decade (review: Kaplan, Lee & Shapley, 1993). MC-pathway cells respond transiently to achromatic stimulus changes and do so with high contrast sensitivity especially at low contrasts (Kaplan & Shapley, 1987). PC-pathway cells exhibit a transient and a tonic response component. The tonic component lasts up to several hundred Milliseconds and is strong to chromatic and rather weak to achromatic stimuli (Lee, Pokorny, Smith & Kremers, 1994). We used isoluminant stimula-tion in the experiments described here in order to separate MC- and PC-pathway responses and hence isolate presumptive downstream functions of the respective pathways (for other attempts see: Livingstone & Hubel, 1984; Dobkins & Albright, 1994). Parvocellular gan-glion cells transmit the signals evaluated for isoluminant color tasks while magnocellular ganglion cells do not. This is consistent with our observation of a relative increase in gain for the sensitivity to color contrast when presentation time increases. Perhaps not all components of the psychophysical responses seen in the results of the first experiment can be understood on the basis of ganglion cell function. This is the goal of further experiments involving physiological responses from MC- and PC-pathway cells in the retina of macaque monkeys.

2.2 Induction Experiment

In these experiments the experimental set up was identical to that in the previous section (Wachtler & Wehrhahn, 1993). The overall size s of the stimulus was between 80 and 320 arcmin and was displayed to the center of gaze of the observers. Two homogeneous rectangles appeared on either side of a transition area situated inbetween, whose size T varied between 20 and 160 arcmin. The transition area consisted of two parts in which luminance and color changed with a saw tooth like function shaped like two identical ramps interrupted by a step. The amplitude of the step inbetween the ramps was equal to the total amount bridged by the ramps but of opposite sign (see inset of Fig. 3 for an illustration). As in the discrimination experiment described in the previous section color changes were isoluminant and both luminance and color changes were expressed in cone contrasts. On appearance of a stimulus subjects were asked to signal the side on which the test field adjacent to the transitory area T appeared either darker (for luminance tests) or redder (for color tests) by pressing the corresponding button of a mouse. Response strengths were determined by a nulling method as a function of cone contrast amplitudes. This means that a resulting value indicates the amount of green contrast which had to be added to the test field on the right hand side in order to compensate for the amount of red induced through the transition area. In other words, the results of this section are a measure for the strength of the induced luminance or color. Several different stimulus geometries and presentation times were used. Three color normal subjects as assessed by testing with the Farnsworth Hue 100 test, and with normal or corrected to normal vision were tested.

Fig. 3 shows the result of an experiment where different stimulus geometries were presented. When plotted as a fuction of T/s thresholds collapse onto one line for different values of s. The larger the ratio T/s, the stronger is the induced luminance or color. This

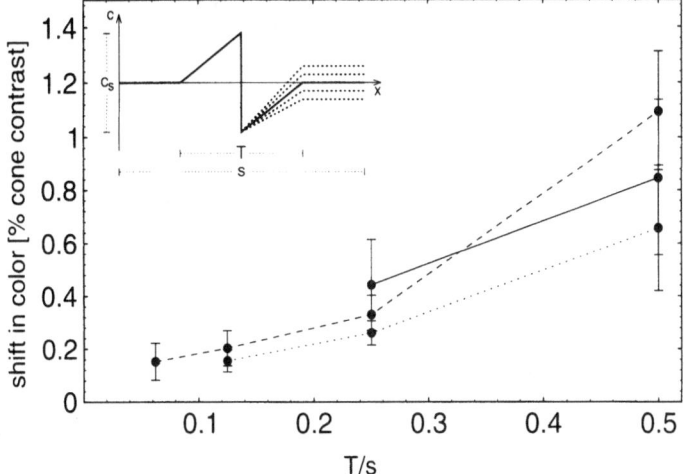

Figure 3. Strength of the induced effect for color and luminance conditions with various stimulus geometries (see inset) and a contrast amplitude of .15 for the transitory area. Stimulus size s was 80 (solid), 160 (dotted), and 320 arcmin (dashed). Averaged data from 3 subjects.

means that induction increases with increasing width of the transition area T and with decreasing width of the homogeous test fields adjacent to it. Moreover, for all subjects tested a much stronger induction strength is observed for luminance stimuli than for color stimuli as can be seen by inspecting Fig. 4. Different strengths for induction phenomena in luminance and color have been mentioned earlier but rarely quantitatative data are reported. There is no straight forward candidate which can be pointed out as the neural basis of this difference. Neural signals observed in higher cortical areas are processed in complex ways (Zeki, 1984;

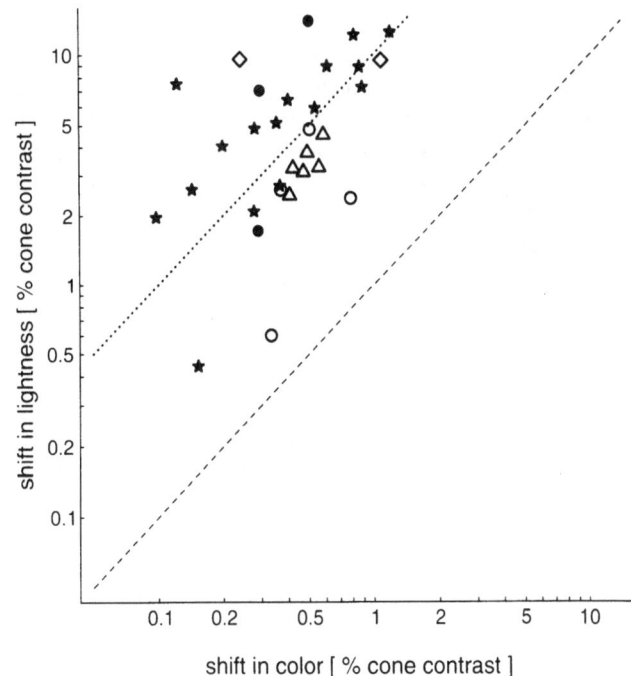

Figure 4. Strength of the induced effect for luminance conditions plotted against color conditions for different stimulus geometries and presentation times. The strength of the induced effect is larger by up to a factor of 10.

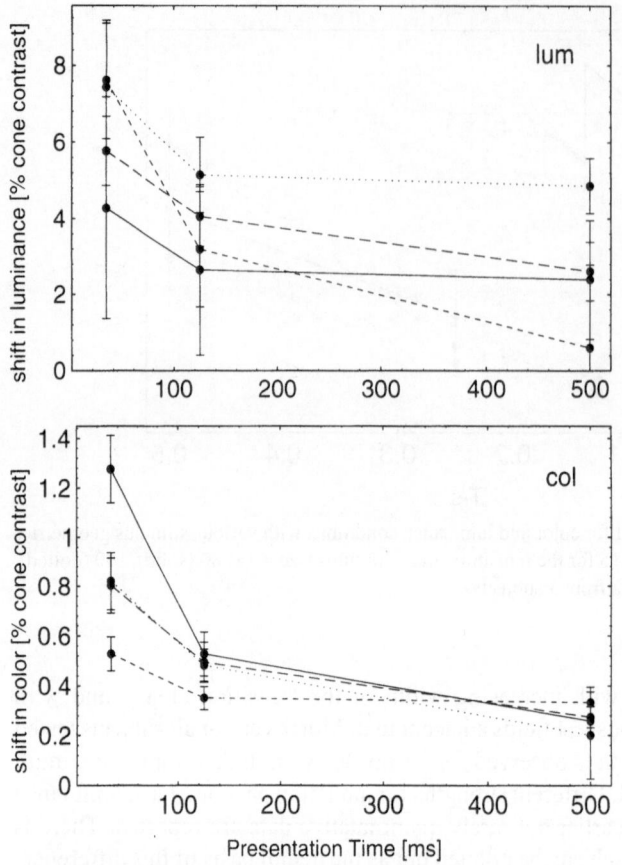

Figure 5. Strength of the induced effect for luminance (upper panel) and color (lower panel) conditions plotted as a function of presentation time. Strength decreases for increasing presentation time, both for the luminance and for the color condition. Data from 4 subjects. Note different scales on y-axes.

Ts'o & Gilbert, 1988, Roe & Ts'o, 1995). Thereby signals similar to those measured in perceptual experiments are generated from retinal ganglion cell signals.

Fig. 5 shows the dependence of the induced effect as a function of presentation time. In all subjects we found stronger induction effects for shorter presentation times. We would like to stress two aspects in these results. First, as in the previous task subjects showed much stronger induction strength to luminance changes as opposed to color changes. Second, for increasing presentation time a decrease in induction strength is observed. This second observation has been reported earlier for the tilt illusion (Westheimer & Wehrhahn, 1994) and may be common to many induced phenomena following the logic: the more time subjects are given to judge the strength of an illusion the less they are inclined to take it for real.

3. DISCUSSION

The results reviewed in this article indicate that the neural mechanisms underlying the two tasks are very likely to be different. While functional properties of the responses observed with the first task can be attributed to functional properties of MC- and PC-pathway ganglion cells, those observed with the second task cannot. Thus higher processes involved in the perception of color differ markedly from more elementary ones. However, these results also show that isolated consideration of the results of these and other experiments may lead to estimations about the strengths of neural processes far from the physiological reality. Comparison of the strengths of the psychophysical responses determined in the first experiment clearly show that contrast sensitivity for color surfaces is up to an order of magnitude higher than for corresponding luminance stimuli. Considering this fact, the — apparently weak — strength of the induced response in the color task analyzed in the second experiment becomes much stronger or equal to that observed in the luminance task. Thus the reliability of the color signals is as high as that of the luminance signals. This means humans can perceive the weak contrast changes induced in the illusion used in the second experiment as clearly and reliably as they perceive the much higher luminance contrast values.

4. ACKNOWLEDGMENTS

Thomas Wachtler is supported by a grant from the Deutsche Forschungs Gemeinschaft. We thank Harald Teufel for carrying out part of the experiments and Dietmar Rapf and Reinhard Feiler for early contributions to the software used in the second experiment.

5. REFERENCES

Chaparro, A., Stromeyer III, C. F., Kronauer, R. E. & Eskew jr., R. T. (1994) Separable red-green and luminance detectors for small flashes. *Vision Research* 34, 751-762.

Daw, N. (1984). The psychology and physiology of color vision. *TINS September 1984*, 330-335.

Dobkins, K. R. & Albright, T. D. (1994) What happens if it changes color when it moves? The nature of chromatic input to macaque visual area MT. *Journal of Neuroscience* 14, 4854-4870.

Eisner, A. & MacLeod, D. I. A. (1980) Blue-sensitive cones do not contribute to luminance. *Journal of the Optical Society of America* A 70, 121-123.

Hubel, D. H. (1988) *Eye, Brain, and Vision*. Freeman & Co, New York.

Hubel, D. H. & Livingstone, M. S. (1990) Color and contrast sensitivity in the lateral geniculate body and primary visual cortex of the macaque monkey. *Journal of Neuroscience* 10, 2223-2237.

Kaplan, E. & Shapley, R. M. (1986) The primate retina contains two types of ganglion cells, with high and low contrast sensitivity. *Proceedings National Academy of Sciences USA* 83, 2755-2757

Kaplan, E., Lee, B. B. & Shapley, R. M. (1993) New views of primate retinal function. Ch. 7. In *Progress in retinal research (9th edition)*, ed. Osborne, N. & Chader, J. pp. 273-335. Pegamon Press, Oxford and New York.

Kremers, J., Lee, B. B., Pokorny, J. & Smith, V. C. (1993) Responses of Macaque Ganglion Cells and Human Observers to Compound Periodic Waveforms. *Vision Research* 33, 1997-2011.

Livingstone, M. S. & Hubel, D. H. (1984) Anatomy and physiology of a color system in the primate visual cortex. *Journal of Neuroscience* 4, 309-356.

Reid, C. & Shapley, R. (1988) Brightness induction by local contrast and the spatial dependence of assimilation. *Vision Research* 28, 115-132.

Roe, A. W. & Ts'o, D. Y. (1995) Visual topography in primate V2: Multiple representation across functional stripes. *Journal of Neuroscience* 15, 3689-3715.

Shapley, R. & Reid, C. (1985) Contrast and assimilation induce perception of brightness. *Proceedings National Academy of Sciences USA* 82, 5983-5986.

Sun, J. & Perona, P. (1996) Early computation of shape and reflectance in the visual system. *Nature* 379, 165-168.

Ts'o, D. Y. & Gilbert, C. D. (1988) The organization of spatial and chromatic interactions in the primate striate cortex. *Journal of Neuroscience* 8, 1712-1727.

Wachtler, T. & Wehrhahn, C. (1993) Long-range interactions in perception of lightness and color induced by local changes of luminance and chrominance. *Perception* 22, 60.

Wachtler, T. & Wehrhahn, C. (1996) Human foveal contrast sensitivity to color and luminance stimuli. Submitted.

Westheimer, G. & Wehrhahn, C. (1994) Discrimination of direction of motion in human vision. *Journal of Neurophysiology* 71, 33-37.

Zeki, S. M. (1984) The construction of colors by the cerebral cortex. *Transactions of the Royal institution of Great Britain*. 56, 231-257.

INDEX

Adenosine, 64
Adrenaline, 99, 175
Amino acid
  alanine, 1
  asparagine, 1
  glutamate, 1
  serine, 1
  tyrosine, 97
ATP, 12, 18, 64, 132, 190

Brain
  regions
    amygdala, 335
    brainstem, 233, 295
    cerebellum, 97, 303, 311
    cortex, 95, 162, 167, 330, 355, 371
    forebrain, 97
    hindbrain, 97
    hippocampus, 36, 63, 335, 277
    hypothalamus, 97, 170
    lateral reticular nucleus, 176
    midbrain, 95
    olfactory bulb, 156, 166
    temporal lobe, 325
    thalamus, 64, 97
  slices, 63, 286, 304, 357

Caffeine, 109
Ca^{2+}hysteresis, 109
Carbachol, 68
CCD camera, 265, 297
Cell
  cultured dissociated neurons, 251
  dialysis, 34
  glial cell, 97
  in the auditory system
    hair cell, 129, 185, 193
    stellate cell, 140
  in the cerebellum
    Golgi cell, 303
    granular cell, 311
    Purkinjie cell, 311
  in the cortex
    face selective neurons, 326

Cell (*cont.*)
    pyramidal cell, 355, 372
    stellate cell, 375
  in the hippocampus
    CA1 pyramidal cell, 63, 289
    CA3 pyramidal cell, 36
    dentate granule cell, 286
  in the insect antennae
    receptor neurons, 217
  in the leech
    anterior pagoda, 270
    Retzius cell, 270
    sensory cell, 270
  in the midbrain
    DA neurons, 95
  in the olfactory system
    olfactory receptor neurons, 75, 155
  in the retina
    cone, 76, 201, 383
    ganglion cell, 262, 386
    rod, 2, 75, 201
  in the spinal cord
    nociceptive neurons, 176
  in the vomeronasal organ
    olfactory neurons, 166

Channels
  CNG, 1, 75, 156
  ion selective
    Ca^{2+}, 23, 193
    ion, 41, 298
    K^{+}, 183, 193
    Na^{+}, 42, 134, 156
  Mitochondrial, 11
  pore, 18, 41, 75
  properties
    conductance, 2, 34, 45, 63
    desensitization, 7
    inactivation, 186, 196
    ionic selectivity, 2
    kinetic, 81, 196
    modulation, 34
    permeability, 34